PHYSICAL THOUGHT

from the Presocratics to the Quantum Physicists

AN ANTHOLOGY

Physical Thought

FROM THE PRESOCRATICS TO THE QUANTUM PHYSICISTS

AN ANTHOLOGY

SELECTED, INTRODUCED, AND EDITED BY

SHMUEL SAMBURSKY

PICA PRESS
New York

Published in the United States of America in 1975 by
PICA PRESS
Distributed by Universe Books
381 Park Avenue South, New York, N.Y. 10016

© Shmuel Sambursky 1974

Library of Congress Catalog Card Number: 74-12946

ISBN 0-87663-712-8

Printed in the United States of America

CONTENTS

CONTENT ANALYSIS

SECTION IV THE ROYAL SOCIETY TO LAPLACE

SECTION V DALTON TO MACH

PREFACE

This anthology is intended for all those interested in the history of scientific ideas and in the interaction between science and philosophy. The anthology may be useful to the historian or philosopher, both in his own studies and for reference and source material in his teaching. To the general reader who wishes to acquaint himself with the evolution of the physical sciences the anthology may serve as a guide.

The selection of original texts in a work of this kind is necessarily determined by the subjective approach and the particular scientific attitude of the editor. This arbitrariness is no less than that involved, for instance, in an anthology of poetry; every anthologer must have recourse to some principles of selection, to set sensible and practical limits to the size of the work. To reduce the enormous amount of material available to manageable single volume scope, and to prevent the anthology from growing into an encyclopedia, the editor has selected texts dealing with only certain aspects of the many branches of physics, astronomy, cosmology and the philosophy of science (though a few passages are concerned with chemistry and mathematics), which serve to illustrate the evolution of physical thought. Even so, the material contained here barely covers the essential topics.

A different but important consideration imposed a stringent restriction on the selection of material: technical detail, such as mathematical derivations and experimental methods, had to be kept to a minimum in the texts selected. This limitation applies particularly to those chapters dealing with more recent work, where scientific publications are increasingly complicated and unintelligible to the layman. However, this particular restriction is balanced by the development of a non-technical scientific literature during the past hundred years or so. From the early nineteenth century we find summaries and retrospective surveys from some of the greatest scientists. Distinguished men of science have described their discoveries or the philosophical consequences of their research in essays and books deliberately intended for the non-specialist. Some examples of this category of scientific literature have been included in the last two sections of this anthology.

The purpose of my General Introduction is to try to provide some of the background necessary for those readers who have not made science and the history of science the main focus of their interest. Limitations of space again imposed some restrictions. Some aspects of the subject are treated at length, some only superficially and others are omitted completely. In the special introductions to each of the six sections of the anthology, the discussion emphasizes the main problems of each particular period, with special reference to the illustrative texts which follow.

If one bears in mind that this anthology ranges across the continuous activity of an important aspect of human thought during a span of some 2,500 years (from the mid-sixth century B.C. to the first third of the twentieth century A.D.), one sees clearly that not even half a dozen anthologies similarly devoted to physical thought would by any means exhaust the subject.

SHMUEL SAMBURSKY, 1974

GENERAL INTRODUCTION

The purpose of this Anthology

This anthology is intended to serve several purposes, the first of which is its use as an illustration, however incomplete and fragmentary, of the development of physical thought as a chapter in the history of ideas. The passages selected here are thus concerned mainly with concepts and theories, and to a certain extent with questions of method; but they deal only rarely with problems of experimental technique, and mathematical derivations are omitted almost entirely. In brief: the procedural aspect of science is kept in the background, and the emphasis is on its conceptual aspect. The reader will thus receive a much more one-sided picture of the development of physics than that found in, for example, the usual kind of source-book provided for the physicist; on the other hand there are quite a number of texts which reflect the interaction of physical ideas with those of such other spheres of thought as philosophy and religion, thus giving the general reader some indication of the role of physical thinking in the evolution of human culture during the two and a half thousand years from the time of the Presocratics to the first half of the twentieth century.

The selection of material for any anthology—one concerned with science not less than one of poetry—will be subject to a considerable degree of arbitrariness. The very necessity of reducing the bulk of available material to an amount suitable for one volume forces the editor to apply stringent rules of selection, according to certain guiding principles—but the scale of priorities must remain a matter of personal judgement.

Looking at the history of science as part of the history of ideas means seeing it primarily as a continuous attempt by men to arrive at a rational comprehension of natural phenomena, and to construct a logically consistent picture of nature. The main purpose of this Anthology is to reflect some aspects of this struggle to conceive of nature as the sum total of phenomena given to rationalization. The story of the struggle is on the whole the story of a continuous progress—this will, I hope, be revealed to some extent by the passages quoted in the Anthology, even though it is restricted to only a few of the many cross-sections through the widely ramified structure of the whole body of physical thought. This progress is not merely one from the kind of beginnings exemplified by the aphoristic writings of Greeks of the sixth century B.C. to the crystallization of a systematic discipline, the various branches of which have gradually intertwined to form the framework of a world-view in which it is already possible to discern, in many respects, the outlines of a unified picture; it is also progress in the sense that the extension and deepening of our world-picture can be thought of as the development of a series of representations of an ideal ultimate reality, each of which has

I

more nearly corresponded with the ideal than did its predecessor, though perhaps none will ever do so exactly. This gradual approach towards an ideal, in the process of unification of the world-picture, can be seen in the way that every successful representation contains the earlier one within its conceptual framework as a still usable theory whose predictions approximate closely to those of the newer one under certain limited conditions. Obvious examples are such fundamental theories as the theory of relativity—with the predictions of which those of classical mechanics agree well in the case of velocities small compared with that of light—and quantum mechanical theory, where classical mechanics gives results which approximate closely to those predicted by the newer theory in the case of actions compared with which the quantum of action is small enough to be neglected.

It is important to emphasize that the use of the term 'approximation' must not be taken to imply that the successive representations of reality have grown from one another in a smooth progression, each of them giving a slight correction of the results of its predecessors. On the contrary, the relativity and quantum theories provide good examples of one of the most characteristic features in the development of scientific ideas—namely the fact that every major advance, resulting in a new representation which *post factum* can be seen to have reduced the earlier picture to one whose results approximate closely to those of the newer one in special cases, has been connected with a revolutionary change in outlook, and with a radical revision of the epistemological and metaphysical foundations of the earlier picture. It is at such turning-points that scientific thought is most clearly revealed as creative speculation, kept within certain boundaries, and corrected, by facts and experimental evidence. In these supreme and often unexpected moments the working of the human mind in its confrontation with the physical world presents itself most strikingly as akin to that sphere of inspiration which brings about the great creations of art: both constitute sudden and unpredictable insights into reality which no artificial and mechanical devices, such as computers, could ever achieve. Even in those cases where conceptual clarification and the emergence of a victorious theory came only after a long struggle between opposing hypotheses—the history of the theories of light is a good illustration—science in the making can be seen to be as much an experiment with ideas as a search after significant experimental data.

Even though—in contradistinction to the history of, for instance, religious or purely philosophical ideas—the history of scientific ideas is on the whole the story of a steady advance, it is necessary to bear in mind that the different phases of this advance, from the time of the Ionian philosophers of nature up to the present, have neither been equally successful nor progressed at the same pace.

The birth of science in Greek antiquity and its renaissance in modern times
I shall not attempt to give a rational explanation—even if such an explanation were really possible—of the explosive event that took place in the middle of the

sixth century B.C. on the west coast of Asia Minor and which is usually associated with the birth of science as we know it today. But the suddenness of the transition from *mythos* to *logos*, measured against the time-span of human history, should not be allowed to obscure the fact that the transformation of mythological concepts into their scientific counterparts was still a gradual process. This is one of the many instances of a general phenomenon inherent in the evolutionary process of concept formation in the history of science. The development of new concepts, made necessary by the emergence of a novel situation created by new discoveries, depends largely on the conceptual framework and terminology of the preceding period. This dependence has dialectical implications; on the one hand it facilitates further development by presenting scientists with a ready-made set of terms and concepts from which they can draw according to their new needs, but on the other it can become a source of ambiguity and lack of clarity apt to hamper the progress of thought. The history of modern science teems with such examples, but it may be worth while to draw special attention to an earlier instance—the role of the mythological precursor of the conceptual framework of Greek science. An essential feature of mythology in its confrontation with nature was the personification of things, forces, qualities and states; heaven and earth, winds, the ocean, day and night were all personified, and gods were identified with the elements in which they dwelt. Thus the intuitive phase of *mythos* provided a set of terms for use in the conceptual phase of early science. Later on, however, as the need arose for a greater differentiation of concepts, the old terminology hampered further clarification: an example of this is the idea of the four elements, which persevered for centuries by the sheer force of mental inertia. It is interesting, also, to note how the genealogical stories of Greek theogony intuitively anticipated certain features of the scientific phase which began once cosmogony had been separated from theogony and the emphasis was laid on the problem of matter. This can be illustrated by interpreting Hesiod's story of Gaia, the goddess of Earth, and Uranus, the god of Heaven, begetting Oceanus, as an allegorical description of the interplay of the elements which later figured so prominently in the scientific writings of the Presocratic philosophers. Similarly, a tale of Day and Light being born from Night and Darkness can be seen as a pre-scientific precursor of the problem of the opposites and their interaction which played such an important part in Greek science from the time of the natural philosophy of the Pythagoreans onwards. It is very likely that such phenomena as the unceasing succession of day and night and rotation of the starry heavens have impressed themselves from time immemorial on the unconscious mind of early man and have led to the creation of archetypal ideas like that of the opposites in general or of the circle and circular motion as the perfect shape and motion.

After science in Greece emerged from mythology, a process that occurred in an incredibly short time, its development from the sixth to the end of the third centuries B.C. showed an astonishing depth of imagination and intuitive grasp

3

of the great problems. The discovery and evolution of the scientific approach, simultaneously with the great achievements of the Greeks in philosophy, literature and the arts, will always remain a historical phenomenon bordering on the miraculous. In those few centuries—to mention only the exact sciences—were laid the foundations of mathematics, in particular of geometry, and of mathematical astronomy, on which rested the whole building of these branches of science when the great renaissance began in the sixteenth century A.D. The elements of geometrical optics and basic physical facts such as the corporeality of air were discovered and stated, and fundamental ideas, such as that of an element, were conceived. Moreover, theory formation began on a high level and in a very elaborate manner, and two physical systems of prime significance—atomism and continuum theory—were constructed which have recurred again and again, at different levels of comprehension, until our own time. Both are concerned with the concept of matter as well as with the mode of physical action; but whereas the atomic theory of Leucippus, Democritus and later on of Epicurus centred on a particle model and was associated with impact, grouping and shapes, the continuum theory of the Stoics centred on the model of an all-pervading *pneuma*—the scientific counterpart of the idea of an omnipresent God—and was associated with the notion of tension and the superposition of tensional motions. These two rival systems foreshadowed the centuries-old antithesis between the corpuscular and field concepts which has played a prominent part in physics and first reached a synthesis in the modern quantum mechanical interpretation forty years ago.

Towards the beginning of the second century B.C. the progress of scientific thought and the rate of discovery began to slow down—though some of the great scientists of Greek antiquity, such as Hipparchus, Galen and Ptolemy, did belong to the later period covering the last two centuries B.C. and the first two centuries A.D. The slackening of scientific tension during this period heralded the decline of the classical spirit of science and the advent of the Middle Ages, in which the investigation of nature was no longer one of the central issues occupying the human mind, its place having been taken by theology and inquiry about the place of man in the universe and his relationship to his fellow men and to God.

The course of Neoplatonic philosophy from the third to the sixth centuries A.D. is a lucid illustration of the gradual suppression of the rational outlook by the intrusion of the irrational, in which process the influence of oriental religions served as a strong catalyst. Nevertheless, the writings of Proclus on mathematics and astronomy show that the scientific tradition was still alive, and these centuries also produced such amazing men as Philoponus, whose ideas anticipated by a thousand years the advent of a new age.

For the early Christians and Moslems, nature and the problems of science were no more than subsidiary subjects which were dealt with only insofar as they could be of interest to the main concerns of religion. The fact that science survived at all during these centuries can be attributed to two main factors: to a certain continuity of the Greek philosophical and scientific tradition in Byzantium and in the Islamic renaissance which began in the ninth century, and to the subsequent incorporation of the Aristotelian philosophy—including its scientific aspects—into the world-views of the three monotheistic religions. Islamic science, at first following in the footsteps of Greek science but later developing some original research of its own, leant heavily on the Aristotelian world-picture. The western Scholastics, taking their initiative from Islam, but later making direct use of the ancient sources, kept the flame of science fitfully alight within the framework of the Aristotelian doctrine.

For more than a thousand years the rationalization of nature was perverted and distorted by being preserved—on the whole—as a bookish body of learning which was treated and taught as a given system of doctrines and concepts about nature and the physical world, and this body of learning came to be regarded to a large extent as an integral part of the world-view of Christendom. When, therefore, in the sixteenth and early seventeenth centuries scientific inquiry came once more to be based on direct contact with nature itself, this rediscovery of nature as something which had to be questioned directly in a rational way if a sensible answer was to be obtained was just as revolutionary a change as that from *mythos* to *logos* in the sixth century B.C. But whereas in ancient Greece the transition had involved a shift to a completely different set of categories while both mythology and science were concerned with the same objects—the same wind and water and fire, the same stars and their movements—at the beginning of the new era in the sixteenth century the same set of rational concepts held and the same frame of logical argument was applied—*prima facie* at least—while the change involved, and was made difficult by, a shift in the objects of interest.

When once again a direct approach was made to nature and its phenomena and contact with experience was re-established, the rate of progress increased with the same bewildering speed that the fifth and fourth centuries B.C. had witnessed. It is appropriate to note here the central role played by astronomy in the development of the physical sciences in the sixteenth century as well as in the fifth century B.C., and also during a considerable part of the Middle Ages. An obvious reason for this was the periodicity of the daily rotation of the heavens and of the movements of the sun, moon and planets, in which nature supplied men with simple laboratory conditions which enabled them to repeat the same observations a practically unlimited number of times. The same patterns of events occurred again and again and gave generations of observers the opportunity to check and to improve their findings. Although observation with the naked eye had already reached its limit of precision in antiquity the fascination of this endless periodicity remained one of the main driving forces in astronomical observation, another being the need to improve the calendar—a practical reason which kept astronomy going through the many centuries of the Middle Ages. But the overriding importance of astronomy for the beginning of the new era must be seen not only in this and in the fact that the heliocentric system of Copernicus opened up a new approach to mechanics and dynamics, but also in that, for the first time in history, science was confronted with two radically different comprehensive theories or, rather, world systems—the old Ptolemaic and the new Copernican—both based on the same set of known facts. This meant that the respective merits and disadvantages of the two rival explanations could be appraised only on philosophical grounds involving an epistemological evaluation and raising the deeper metaphysical question as to the truth content of the theories; that is, the problem of how far the new theory offered a better approximation to physical reality than did the old one—in connection with these questions a comparison of Osiander's Introduction to Copernicus' book *De Revolutionibus Orbium Coelestium* with Copernicus' own Dedication is most illuminating. This novel situation gave rise to a conception of physical science in which theory, experiment and observation interacted in a quite new way to contribute to the rapid growth of knowledge.

The expansion which, from the seventeenth century onwards, established science as an autonomous field of human thought, penetrating all departments of life and deeply affecting and transforming them, did not occur at an equal rate in all branches of physical science. Mechanics, together with astronomy, made the most spectacular progress in one grand sweep in the period between the end of the sixteenth and that of the eighteenth century—even though, curiously enough, the first modern scientific work which combined experimental study with theoretical insight was not one in either of these fields but on the magnet (W. Gilbert, 1600), a subject that, because of the striking phenomenon of attraction, had already occupied the minds of many ancient scientists.

There are several reasons why mechanics developed rapidly, outpacing the growth of all other branches of physics. One is that the starting-point of mechanics is kinematics, the study of motion, and the motions of celestial bodies had been studied by astronomers since antiquity. After Galileo broke down the barrier between celestial and terrestrial phenomena, kinematics became the science of motion in general, and the conceptual and mathematical tools needed for its development were worked out in great detail. Later, the concept of force as the cause of acceleration having been introduced and successfully established by Newton, dynamics emerged as the central subject of mechanics in the seventeenth and eighteenth centuries and provided the foundation for the theory of universal gravitation and for the explanation of a whole complex of other phenomena which could be described mechanically. Another very important reason is the fact that the observation of 'push and pull', impact, and generally every direct contact between bodies, suggests the most intuitively obvious means of explanation for all kinds of physical occurrences—in other words, the fact that mechanical models lend themselves more readily than any others as analogies for the demonstration and clarification of 'how things tick'.

The mechanistic mode of thought has been prominent in physical explanations since antiquity, and from the time of Descartes the trend towards such a way of thinking has been intensified and expanded into a mechanical view of the world. There are, no doubt, deep psychological roots to this, going back to the anthropomorphic world of early childhood; the most immediate and concrete contact of the child with its environment, earlier than visual and other contacts, is the mechanical, tactile one which follows from the movement of its limbs. From the time of Anaximander in the sixth century B.C., mechanical models have been used for the explanation of celestial—and, in particular, planetary—movements, in various hypotheses about the nature of gravitation, and later on in theories of magnetism and electricity as well as in atomic and molecular theories. So deeply rooted has been the mechanical approach to physical phenomena that even up to the end of the nineteenth century eminent physicists held the view that a scientific theory had not been really proved if it had not been demonstrated in terms of mechanical analogies.

Those branches of physics which, in the early nineteenth century, were lagging behind mechanics in their development, benefited greatly from the advanced mathematical techniques used in it. This was particularly striking in the case of magnetism and electricity, but in many respects it was true also for the theories of heat and light. Conceptually, on the other hand, the mechanical approach constituted a considerable obstacle to advance in these other fields, and a turning-point was reached in the history of physical thought when, in 1864, Maxwell dissociated himself from all mechanical pictures in the final version of his electromagnetic theory. This step was the harbinger of an even more fundamental breakaway from mechanical ways of thought, which became unavoidable when Planck's quantum hypothesis turned out to provide the only satisfactory explanation of the distribution of heat radiation from a black body, and afterwards of the hydrogen spectrum.

Together with the quantum theory, Einstein's special theory of relativity inaugurated a new phase of intensive interaction between science and philosophy, in that both theories grew from an epistemological analysis of the foundations of physics which led to a thorough revision of its fundamental assumptions and gave rise to the renaissance of a philosophical approach to science comparable only with that of the seventeenth century. All these developments considerably accelerated the pace of progress in all fields of physical inquiry during the first quarter of the present century. It should be noted, however, that towards the middle of this century—that is, after the terminal point set for this Anthology—the pace has slackened again as far as the theoretical explanation of the immensely rich results of experimental research in atomic and nuclear physics is concerned; the same applies to cosmological theories based on the general theory of relativity.

Continuity and discontinuity in scientific thought

One major purpose of this preliminary and very general survey of the progress of physical thought is to throw into relief the continuity of science as an intellectual activity throughout the ages. Once the scientific way of thinking had been initiated by the Ionian philosophers, it went on deepening and widening and opening new vistas, and even surviving the vicissitudes of the great social, political and spiritual changes of late antiquity and the Middle Ages. Continuity is one of the most conspicuous aspects of the development of science, and the passages in this Anthology have been selected with a view to reflecting this aspect in several ways. It shows itself in the recurrence of the same basic ideas, which appear again and again at different levels of understanding and accumulated knowledge. The opposing concepts of particles and the continuum are of course the most striking examples, but there are others—such as the continuous discussion about, and elaboration of, the principles of scientific inquiry in the light of past experience and as a pointer to the direction of possible future developments. Continuity can also be seen in the manner in which scientists, and often the greatest among them, recapitulate the achievements of their predecessors, or summarize the progress of knowledge in an earlier period on which they built their own theories and which formed the basis for their own researches. Such passages, in which scientists of a former generation look back in perspective at the scientific history of a still earlier generation, are most revealing as specimens of the process of gradual clarification of concepts and crystallization of theories in the unbroken chain of scientific development.

Superimposed on the continuous development, and sometimes not immediately discernible in its full significance, is the other principal aspect of scientific thought—its discontinuous character, exhibited in the sudden emergence of new ideas, in the 'quantum jumps' of insight into physical reality, of which the Copernican revolution is an outstanding instance. The most obvious examples of such flashes of insight which created new historical situations are the discoveries of new phenomena—that is, the discovery, and realization of the significance of, observed or experimentally induced physical facts. Among these are Newton's discovery of the separation of a ray of sunlight into its components of different refrangibility—in other words, into its different colours; or Ørsted's discovery of the deflection of a magnetic needle by an electric current; or Rutherford's discovery of the sharp deviation of some of the α-particles when a beam of them is passed through a metal foil. Rømer's measurement, by an astronomical method, of the finite velocity of light was a discovery of fundamental importance, but the idea that the velocity of light might be finite had been anticipated by Galileo forty years earlier—I shall return presently to the problem of anticipation and priority in science.

Among the hypotheses or theories which inaugurated a new era of scientific thought must be counted the purely speculative doctrine of the Greek Atomists because of their ingenious postulate of the existence of indivisible and immutable elements, Newton's theory of universal gravitation because of its foundation on a general principle of force, and Carnot's thought-experiment with his cyclic

8

machine because it opened up a completely new field whose significance, subsequently formulated in the second law of thermodynamics, was as much philosophical as physical. I hesitate to include in this list the law of conservation of energy, for here the ground had been prepared gradually and thus, in spite of Robert Mayer's ingenious paper and of Helmholtz' great and authoritative analysis, I am inclined rather to put the formulation of this law in the category of 'continuity of science'. But certainly Planck's hypothesis of a quantum of action must be reckoned among the great 'quantum jumps' in the history of scientific thought, as must Einstein's interpretation of the proportionality of inertial and heavy mass and de Broglie's association of wavelengths with particles. The development of these last three ideas led to a whole world of new conceptions and inventions, comparable with that which followed when Maxwell likened the changing electric field between the plates of a condenser in an oscillating electric circuit to an electric current—the 'displacement current'. That idea had triggered off the development of electromagnetic and wireless technology.

There is only a small number of analogies as brilliant as Maxwell's which can be regarded as turning-points in the history of scientific ideas. One of comparable greatness, although it had no technical implications whatsoever, was made nearly two and a half thousand years ago, when Anaxagoras compared the size of the sun with that of the Peloponnese. The implication that celestial events could be understood by analogy with those on earth was by far the most important breach made in the fortress of mythology in an age when the sun reigned supreme as one of the great divinities. Seen in a historical context, it freed man from long-standing inhibitions and led him eventually to see celestial and terrestrial phenomena as a physical unity. These inhibitions were removed once and for all only in the time of Galileo, but Anaxagoras' attitude had no doubt facilitated—consciously or unconsciously—the attempts of men such as Aristarchus, Eratosthenes and Poseidonius to calculate celestial distances and dimensions with the help of geometrical methods. Another important analogy was the graphical representation of time as a linear extension, used by Galileo in 1638. Graphical representation as such cannot claim a single inventor; it emerged gradually in the later Middle Ages. But the conception of time as a coordinate made possible a new way of thinking, that of seeing changing quantities as variables depending on time, now so familiar to all of us that it seems almost trivial. It also led to another fundamental invention, that of the calculus, an indispensable tool of the age of dynamics; this invention will always be associated with the names of Newton and Leibniz, despite its foreshadowing in the work which led up to it.

A good example of the difficulties that arise when a discovery is attributed to one particular person, or when it is regarded as a unique event in the history of ideas, is provided by the statement of the law of inertia. If this law were thought of in the context of the basic laws of mechanics formulated by Newton, its discovery would have to be dated in 1687, the year of publication of the *Principia*, in which case Descartes', and to some extent Galileo's, contributions to the problem would have to be regarded as just anticipations of Newton's. But to do so would certainly be unfair to Galileo, who had laid much of the groundwork for the Newtonian statement of the law and on whose shoulders Newton rested at least in respect of this problem. To give priority for discovery of the law to Galileo, on the other hand, would be to ignore its slightly later formulation by Descartes which, although put forward on grounds which would today be regarded as less scientific than those on which Galileo argued, was more general than the latter's. Further: if the problem of inertia is considered more generally, in terms of the comparison of the state of rest with a given kind of motion, provided that it is persistent and geometrically uniform, then Philoponus—who lived in the sixth century A.D.—should perhaps be regarded as the father of the idea, comparing as he did the endless and uniform circular motion of the daily rotation of the heavens with rest: 'Persevering in the same motion *is* rest.' It would be better to think of all these pre-Newtonian statements not so much as anticipations of the law of inertia, but rather as different versions of the idea of inertia, stated at different times and in different contexts, but all of them correct on the point of principle.

There have, however, been several occasions during the long period of growth of science when later developments have genuinely been anticipated. Such anticipations were the work of men who understood the essential features of an important problem and formulated them in a way which could have provided a point of departure for developments which might then have occurred earlier than they did and could have changed the course of the history of ideas. What was missing in such cases was the congenial mental climate, the resonance of contemporary thought, the ready response which could have led to an expansion of the whole complex of ideas associated with the problem at issue. In brief: something must be 'in the air' for a novel suggestion to become generally accepted and fit into the main stream of intellectual development. Where an idea has been accepted in this sense, it is often found to have been expressed independently by several people, whereas premature ideas, anticipations, have usually been suggested in isolated, single utterances. This Anthology contains a selection of these, such as Aristotle's statement about things disintegrating by themselves with time but not improving by themselves with time—the first recognition of the 'arrow of time' and implicitly of the law of entropy. Then there was Philoponus' insistence on the unity of heaven and earth, echoing the view of some of the Presocratics and anticipating by eleven hundred years the Galilean position. Daniel Bernoulli's explanation of gas pressure, illustrated by a diagram, anticipated the kinetic theory of gases by a century. Prout's paper suggesting that the atomic weights of the elements were multiples of the weight of the hydrogen atom appeared a century before the advent of the proton in atomic theory. Finally, Mendeleev's suggestion that the failure of Prout's hypothesis might be due to a mass-energy equivalence, though expressed in terms which could not be so concise as those used fifty years later, was a remarkable anticipation of Einstein in 1905 and of the discovery of the packing effect in atomic nuclei in 1920.

The development of scientific language and style is one of the aspects of scientific development which are brought out in this Anthology. Clarity of expression in explaining scientific findings and theories must necessarily evolve gradually, in step with the slow maturing and crystallization of the ideas themselves. A later generation, familiar with the accepted achievements, can often express them more lucidly and with greater elegance than their originators. But just this wrestling with newly formed concepts, this difficulty of making a new idea understood or of talking clearly about a newly discovered fact, is a good illustration of science in the making. It goes without saying that, apart from this, lucidity of style is a quality depending very much on the personality of the author. It is well known that there is no correlation between the capacity for research and for teaching, and that some of the most brilliant scientists have been among the worst of lecturers; but when both these gifts happen to be united in one and the same person the resulting combination is a source of supreme intellectual pleasure. An anthology selected with a view to presenting scientific prose at its best would look quite different from the present one, but even then the writings of a Galileo or a Laplace would be exceptional for their quality. In contrast to these, the scientific publications of some distinguished authors in the second half of the nineteenth century are often disappointing because of their pedestrian style, and only during the twentieth century can an improvement in this respect be observed. However, many popular expositions of these periods, not included in the Anthology, excel in their brilliance. Among them are some of Maxwell's and Boltzmann's articles and the writings of Einstein and Eddington.

Guiding principles in the physical sciences—structural and dynamical approaches

I want now to survey the history of physical thought from the point of view of the different basic ways of understanding nature which have been developed and which present a picture of contrary views, sometimes conflicting with and sometimes complementing each other. The most striking contrast is that between the idea of nature—inorganic as well as organic—acting according to design and purpose and the concept of a purely mechanical necessity underlying the happenings of the physical world. The first, the teleological approach, was predominant in ancient Greece, especially in the work of Aristotle and his school; the other, the mechanical, or causal, approach, played an important role for shorter periods in antiquity, in the Presocratic and later in the Hellenistic periods, but it could not hold its place against the teleological view. In modern times, however, it has steadily gained ground—at least in the physical sciences—

and, with the short-lived exceptions found in the work of Leibniz and Maupertuis, has reigned supreme since the eighteenth century.

The idea of purpose in nature was foreshadowed in the work of the Presocratic philosopher, Diogenes of Apollonia, when he said that things as they were, for example the succession of seasons, were arranged in the best possible way. Similarly, Plato's allegorical description of the creation of the universe by the demiurge, which was the core of the *Timaeus*, his dialogue on natural science, lent itself quite naturally to the idea that the world was the result of a planning and designing mind. But the concept of the universe as the sum total of events that tend towards a goal and in which an earlier event happens for the sake of a later—the later being seen as the result of purpose and design—was the creation of Aristotle. He built from it a system of the physical world which became the dominating world-view for nearly two thousand years, until it was finally overthrown in about A.D. 1600 as the result of the work of, especially, Galileo and Descartes. The clearest and most pointed expression of the teleological position was given by Aristotle himself when he compared nature to a supreme creative artist, to an architect who builds a house according to a well-designed plan.

This analogy of Aristotle's is illuminating in that it shows the relation of the teleological view to another fundamental approach to nature, which can be described as the structural approach. This approach—in contradistinction to the dynamical, which puts the emphasis on forces and the resulting movements—sees nature primarily as a well-ordered whole, as a structure whose parts are related to the whole and to each other in some definite pattern. The structural conception of the universe is not necessarily linked with teleology—indeed, it existed before the emergence of the teleological view—and during the last hundred years, with teleology long since banned from the physical sciences, it has become increasingly dominant in them. But it is not hard to imagine how an emphasis on the structural aspects of nature could lead to the suggestion of a master plan, a design, the following of which would gradually bring about the realization and preservation of the structure, in analogy with the relationship of the architect and his plan to the house.

The structural conception of the universe underlay the whole Pythagorean philosophy of nature. It showed itself in the basic tenet of that school that number was the essence of physical reality, and that numbers and numerical relations, rather than a material substrate, revealed to men the very foundations of that reality. The great discovery of the Pythagoreans that musical harmonies could be expressed in terms of simple ratios—such as the octave by $1 : 2$, the quint by $2 : 3$, and the quart by $3 : 4$—was generalized by a quite natural analogical argument to form the basis of the explanation of events in the universe in terms of harmony and number.

But it was another early Pythagorean notion that, even more than that of number, lent itself to the development of a structural conception of nature; this was the idea of the universe as exhibiting pairs of opposites, such as odd-even, light-dark, hot-cold and rest-motion. Seeing the universe as composed of such polarities was the most elementary form of seeing it as an entity exhibiting structural differentiations. The significance of this conception is thrown into relief by some of the early cosmogonies, for example that of Anaximander in which the creation of the universe began with the separation of the opposites hot and cold out of the unlimited and undefined substrate.

The introduction of polarities and, more generally, of structural features, was fundamental to the Pythagorean explanation not only of how things could be, but also of how events could happen in a regular, ordered way. The existence of a structure, either merely as a sum of polarities or as a differentiated whole, involved the existence of movement whose tendency was to remove those polarities or to abolish the tensions created by the differentiations. Structure and dynamics thus came to be interdependent, and the main difference between the structural and the dynamical conceptions of the universe has been in the emphasis put on the one or the other. In the Newtonian age, in which the dynamical conception prevailed, differences of potential or gradients of fields of force were regarded as causing changes in states of motion, but the theory of electricity and magnetism again introduced polarities in the guise of opposite electric charges or opposite magnetic poles.

The structural conception of the Pythagorean School was taken over by Plato in his philosophy of nature, in which harmony was identified with the world-soul, and by the Neoplatonists, for whom harmony became the main pillar of their idea of nature. As the Greek approach to nature was basically a biological one, an attitude strengthened by the Platonic tradition of seeing the universe as a living creature, the cosmos was for the Greeks a highly developed organic structure whose harmony consisted in the continuous interaction of all its parts with the whole. This view was extended and generalized in the notion of sympathy, which expressed the belief that each part of the universe was affected by all the others, even the most remote ones. In later antiquity the idea of universal sympathy was tainted by the mystical tendencies of Neoplatonism, and the religious belief in the essential unity and harmony of the universe had its ramifications in the mystical and cosmological doctrines of the Middle Ages. At the beginning of the modern era, Kepler was profoundly influenced by this Neoplatonic view of the universe as a perfectly harmonious whole, and although his astronomical observations and his three laws were an essential contribution to the foundations of modern science, he remained a Pythagorean and Platonist in his structural approach to nature.

Kepler's world-view was the last important manifestation of the structural conception of the universe for nearly three centuries. His contemporary, Galileo, initiated the age of the dynamical conception, which might also be called the Newtonian age, whose decline coincided with a renewal of the structural approach on a much higher level, associated with quantum physics.

But before reviewing these recent developments, I want to refer back to antiquity and discuss briefly the physical doctrine of Aristotle, whose approach to nature was a structural as well as a teleological one. Although Aristotle rejected the Pythagorean and Platonic philosophy of harmony, his physical system had a definite structural character and was especially important because it determined the world-view of the Middle Ages until the advent of Galileo. An important feature of its structural character was the dichotomy of heaven and earth, i.e. the division of the closed universe into celestial and sublunar regions. The celestial region contained the aetherial spheres and the stars attached to them, whose movements were endless and circular. In the sublunar region were bodies composed of the four elements fire, air, earth and water; they might be either stationary or moving in straight lines or in an irregular way. The kinetic consequences of the particular structure of the sublunar region are given in Aristotle's theory of natural motion, which asserts that any body, whether 'heavy' or 'light', will move towards its natural place and remain there at rest provided that it is not prevented from doing so, or removed from it, by a 'forced' movement. Aristotle's views on dynamics must be seen as part of his teleology: a heavy body, for example, which is not in its natural place strives to pass from its state of potential heaviness into actual heaviness, and it does this by falling downwards. Natural movement, conceived as movement towards the actualization of a natural state, is thus analogous to the development of a seed into the tree of which it was the seed, seen as the movement from being potentially a tree to being actually a tree; and both are instances of Aristotle's teleological philosophy of nature.

The first dynamical conception of the universe was elaborated in the physical doctrine of the Stoics, in the Hellenistic period. Like the Platonists, they thought of the cosmos as having an organic structure similar to that of a living being, but they regarded this structure as a continuum pervaded and governed by inherent forces which caused the coherence of the whole structure as well as the movements of its parts. These forces were thought to be of a continuous nature, deriving from the all-pervading *pneuma*, a mixture of air and fire whose tensions determined all physical happenings.

Historically, the idea of a *pneuma* was an anticipation of the modern concept of a field of force, the central concept of Faraday's and Maxwell's physics. Since the Stoics made no use whatsoever of mathematics, their doctrine remained a purely qualitative anticipation, but it exerted a deep influence on the thinking of the seventeenth and eighteenth centuries, especially on ideas concerning the physical continuum and on the aether theories of that period.

The dynamical approach of the modern era was inaugurated by Newton. However, Galileo's study of a particular kinematic phenomenon, namely the accelerated motion of freely falling bodies, paved the way for this approach. Galileo's work was a challenge to Kepler's structural conception and to the all-embracing cosmological theories of ancient physical thought. His research was remarkable in many respects, bearing all the marks of a modern experiment. Instead of measuring the acceleration due to gravity by direct observation of freely falling bodies, which with the means at his disposal would not have given sufficiently accurate results, he calculated it from measurements of the rate of fall of bodies rolling down inclined planes. He increased the accuracy of his results by systematic repetition of his measurements and, finally, gave a detailed description of the measuring instruments and the arrangements involved in the experiment. His graphical representation of time as a linear coordinate gave rise to the modern conception of time as an independent variable, which has become indispensable for every description of physical change as a function of time. His clear definition of velocity and acceleration, and their description in mathematical terms, were the first instances of the formulation of exact physico-mathematical notions. It could well be said that Galileo's work in mechanics was a turning-point in the history of physics in three fundamental respects: with it there began modern experimental science, the advance of scientific knowledge through concentration on limited problems, and the Newtonian age of dynamics.

Another turning-point, nearly coinciding with that brought about by Galileo's work, was Descartes' mechanistic conception of the universe, which had even more far-reaching consequences for the world-picture of the subsequent centuries than did his great achievements in mathematics. Descartes was a lesser physicist than Galileo, but his philosophy of science was of supreme importance in demolishing the Aristotelian finalistic view of physical events. Descartes' theory of vortices, by which he explained the planetary motions, his mechanical interpretation of the propagation of light, his mechanistic approach to biological phenomena, although all were superseded by other theories, profoundly influenced modern scientific thought by restricting its modes of explanation to purely causal considerations and mechanical analogies, taking 'mechanical' here to mean 'non-teleological'. From then on teleological tendencies were banned from the physical world-picture, with the above-mentioned exception of a brief but important episode in the eighteenth century. Descartes attempted to apply to the physical picture of the world his conception of matter as pure extension, but he failed to carry out his programme of a geometrization of physics, and the first man to achieve the successful mathematization of physical concepts and extend the introduction of mathematics into the procedures of physical science was Newton.

Newton and the era of classical physics

Newton was by far the most prominent of the astonishingly large number of great men of science produced in the seventeenth century; indeed, he stands out as one of the greatest scientists of all times, and his *Philosophiae Naturalis Principia Mathematica* (1687) is one of the major milestones in the history of science. This work, as well as his *Opticks* (1704), is remarkable in several respects, some of which I shall discuss further on, but in the present context I just want to point out that the *Principia* furnished most of the basic conceptual tools for the elaboration of the dynamical approach to the physical world. The concepts of acceleration, force and mass, together with the law of inertia, the law of force and the law of action and reaction contained implicitly all the essentials for the subsequent development of physics up to the nineteenth century; and the infinitesimal calculus, invented independently by Newton and Leibniz, provided the mathematical algorithm for the whole range of phenomena which could be treated in terms of dynamics. It could well be said, adapting a well-known statement of Whitehead's, that all the mechanics of the eighteenth century was but a marginal note to the *Principia* of Newton. The main progress in that century consisted in technical rather than conceptual innovations—in the introduction of differential equations to supplant cumbersome geometrical methods, in the elaboration and clear definition of such fundamental concepts as momentum, angular momentum, work, potential and kinetic energy, and in the formulation of general principles from which the laws of Newton could be derived in a more fruitful and elegant way.

However, the historical significance of the *Principia* is much greater than this. In this work the first serious attempt was made to construct mechanics, and dynamics in particular, as a systematic scientific discipline, starting from definitions and axioms, followed by theorems and conclusions of a general nature, making use of the whole body of theoretical and experimental knowledge at the disposal of seventeenth-century physics. Implicit in this attempt was an analysis of the metaphysical presuppositions on which Newtonian physics was founded—such as the concepts of absolute space and absolute time, which were overthrown only by Einstein's theory of relativity. The *Principia* also contained Newton's theory of gravitation, which was the most important application of his basic laws, and his explanation, by this theory, of the planetary movements, the orbits of comets and the phenomena of the tides.

The fact that the mathematical theory of gravitation made accurate predictions possible in the field of astronomy had an immense influence on the world-picture of the eighteenth and nineteenth centuries, especially in establishing a deterministic view of nature. By a bold extrapolation, scientists like Laplace held the opinion that men, by following Newton's method, would be able to predict every event in the whole range of natural phenomena. This determinism was closely linked to the picture of the universe as a clock-work or a machine, and the description of a hypothesis in terms of a mechanical model was seen by many scientists as strong evidence for its truth. However, Newton's theory of gravitation was accepted even though Newton did not provide for any mechanical explanation of gravitational attraction, as Descartes had attempted to do in his doctrine of aetherial vortices, which was discarded by Newton. Still, it was significant that Newton himself during his long life looked for such an explanation, and tried to construct a model of an aether whose density gradients could explain gravitation as well as certain optical phenomena. From the time of Descartes' mechanical hypothesis of the aether until the second half of the nineteenth century, physicists continued to look for mechanical models to explain physical phenomena such as gravitation, magnetism and electricity. Mechanical analogies still serve physical science, but since Maxwell moved away from any mechanical explanation of electromagnetism they have come to be regarded as mere makeshifts and temporary aids rather than as pictorial descriptions of reality.

Newton's *Opticks* also had a great impact on scientific and philosophical thought. His discovery of the optical spectrum resulting from the refraction of light through a glass prism, and his definition of a given colour by a number expressing the refrangibility of the associated ray of light, provided the objective counterpart to the subjective sensation of colour as propounded in Locke's theory of secondary qualities.

It took about a hundred and fifty years, from the publication of the *Principia* until the first half of the nineteenth century, for Newtonian mechanics to be elaborated and equipped with the mathematical tools necessary for it to ripen and crystallize into the first mathematically and conceptually self-consistent discipline of the physical sciences, which later would serve as a model for the working out of other branches of physics which were still in an embryonic state at the end of the eighteenth century.

After Newton's contemporary, the great Dutch physicist Huygens, had made his important contributions to the science of mechanics, many of the foremost scientists of the eighteenth century shared in the clarification and systematic representation of the principles of mechanics. Among them were Leibniz, d'Alembert, the Bernoullis, Euler, Laplace and Lagrange, and finally, at the beginning of the nineteenth century, Hamilton.

The discovery and interpretation of one of these principles, the principle of least action, led to the renewal of the discussion which seemed to have come to an end since the rise of Descartes' mechanistic approach to the phenomena of nature. Already Fermat, in the middle of the seventeenth century, returning to the old Greek principle that 'nature does nothing in vain', had stated that a ray of light moving between two points would always take the shortest possible time. Leibniz in 1712, and a few decades after him Maupertuis, saw in the principle of least action the very essence of nature's working, deriving not from mechanical necessity and efficient causes, but rather from purpose and final causes —in short, teleology as the antithesis of mechanical causality was revived again during the crucial period of the scientific foundation of mechanics. From the point of view of the history of ideas this revival is of no little interest, because in its formulation by Hamilton the principle of least action has proved to be the most flexible and fruitful of all general principles of physics and has also been of very great value in its application to non-mechanical phenomena such as those of thermodynamics, electromagnetism and quantum theory. Leibniz went so far as to state that the laws of motion could not be fully explained by considering the efficient causes alone—that is, by using only a principle of mechanical necessity—but only by also having recourse to a principle of purpose or design.

The teleological interpretation of the principle of least action is based on the fact that it presupposes a given goal towards which a system moves from a given point; it states that among all the possible paths connecting these two points the actual path has the property of making the action (the sum of the products of mass, length and velocity) of the system a minimum. However, the physicists of the nineteenth century declined to associate the special role played by the principle of least action with the metaphysical idea of a purpose prevailing in nature and stuck to Descartes' belief that science can make do with a mechanical or causal explanation. Nevertheless, more than two hundred years after Leibniz, Planck, while adhering to the view that the principle of least action did not introduce an essentially new notion into statements about the laws of nature, remarked that 'it is worth while noting that theoretical physics in its historical development has led to a formulation of physical causality which displays a decidedly teleological character'.

The revival of the controversy concerning purpose *versus* mechanical necessity was but a brief interlude in the history of modern science. The conflict between a structural and a dynamical approach to nature, however, went on into the twentieth century. But before I continue this story I want at this point to discuss another more fundamental philosophical conflict which scientific thought has gradually succeeded in merging into a synthesis by a long process of trial and error—the conflict between empiricism and rationalism. Concerning the beginnings of both these rival philosophical systems one could perhaps comment that the successful introduction of a new point of view often involves a certain degree of exaggeration. While the empiricists maintained that the human mind, before being exposed to experience, resembled a blank sheet of paper which sense perception only gradually covered with script, the rationalists declared that even the first experience was acquired through the medium of innate ideas pre-existing in the mind, and that mind alone could offer the only safe guarantee of truth as far as assertions about the external world were concerned.

Francis Bacon, who, though no scientist himself, influenced many eminent scientists in the seventeenth century by his philosophy, strongly rejected the speculative approach and purely deductive reasoning of medieval scholasticism and was the first to stress the necessity for observational and experimental evidence in every scientific inquiry. He also had the great merit of warning scientists against certain kinds of prejudices which could easily be pitfalls on the road to scientific understanding. But his reliance on induction as the principal feature of scientific method was exaggerated. Reflection on the centuries-long discussion of the problem of induction would suggest that a satisfactory solution cannot be reached without having recourse to some sort of assumption of a non-empirical nature, as for example that of a regularity underlying all natural phenomena—which is an assumption of a definitely metaphysical character.

Descartes was the first modern protagonist of rationalism—that is, of the assertion that mind, and not sense perception, is the only certain foundation on which knowledge of the external world can rest. His view was that every genuine perception of an object was more than a purely sensory one, and that only after the integration of all sense data by the critical faculties of our mind could we arrive at the conception of an object as an entity, uniting the haphazard and changing information transmitted by our senses. His firm belief in the superiority of mind over the senses led Descartes, one of the greatest geometers of all times, to the conclusion that deduction was the backbone of science, and that inquiry into nature must, like any other science, be done *more geometrico*. In other words: we have to take as premisses some truths which our minds have proved to be absolutely certain and to proceed by means of the absolutely certain principles of logical deduction to the demonstration of further truths.

The conflicting philosophical tenets of empiricism and rationalism were brought to a practical synthesis in the course of the development of the exact sciences during the last three hundred years. A survey of the history of physics since the age of Galileo and Newton shows in particular that the formation of scientific theories begins with the putting forward of hypotheses, after a minimum of empirical data has been accumulated. Then the consequences of the hypotheses are deduced, and put forward as factual assertions to be compared with, and checked against, experimental and observational results. Thus theory and experiment constantly interact with, and mutually support, each other, and the question of which came first is of no relevance. It should, however, be emphasized that this synthesis cannot be described as a fusion of two 'pure' components, namely of 'pure facts' and 'pure concepts'. What actually happened in the course of the long process of merging 'factual' and 'conceptual' elements into a higher unity was that the former gradually came to be seen less as 'pure facts', immediately dependent on sense perception, and more as 'higher order facts', the understanding of which tacitly presupposed the knowledge of simpler facts as well as an increasing theoretical element. The conceptual components, on the other hand, became more and more remote from the elementary abstractions of the world of commonly accepted concepts and changed into scientific constructs which combined the results of purely theoretical considerations with the knowledge of facts of a higher order. In brief: the synthesis of the factual and conceptual components of scientific knowledge emerged as the result of a long and gradual evolution, during which each of these components itself turned out to be the product of a synthesis of factual and theoretical elements.

The gradual emergence of a physical picture of the world, including the scientific constructs which are regarded as elements of physical reality in no less a degree than observed facts, became more obvious from the early nineteenth century onwards, when scientific observations and theoretical deductions began to support and confirm one another in fields other than pure mechanics. The physicists succeeded in putting the phenomena of light, electricity, magnetism and heat on a consistent and systematic basis. Further, it gradually became clear that several of these partial domains of reality were interconnected, as in the linking of heat with mechanics, the fusion of electricity and magnetism into electromagnetism, and the explanation of the essential features of light within the framework of electromagnetism. This successive fusion of previously partial pictures into one picture of increasingly universal validity appeared in itself as a confirmation of the synthetic method of the physical sciences and as a kind of verification of the principles on which these sciences were founded. Thus it became increasingly unlikely that one of the major laws of physics could be falsified without affecting the rest. Indeed, modern science is like an edifice which could easily collapse if one of its essential building-stones were removed.

In this context it must be remembered that this process of unification and consolidation was greatly helped by the increasing mathematization of physics which made it possible to discover and present the inner logical connections of the exact sciences in a way hitherto unknown. Whereas in Greek antiquity and later, until the end of the seventeenth century, geometry and the theory of proportions were the main mathematical tools of science, the discovery of calculus and all its mathematical algorithms and applications enormously enriched mathematics as the language best suited to the deductive reasoning of physics. Differential equations, the calculus of variations, and other methods of higher mathematics developed in the eighteenth and nineteenth centuries, became the backbone of mathematical physics. Mathematical derivations and mathematical analogies increasingly supplemented and even replaced the pictorial analogies of mechanical models. Important as had been Faraday's intuitive conception of the field of force, which again brought to life the dynamical continuum concept of the Stoics and the aether hypothesis of the seventeenth century, although with more precision and greater depth than its precursors, it was only through the mathematical language of Maxwell's equations that the idea of the electromagnetic field with all its implications and consequences received its definitive shape, and the impact of the modern continuum concept was felt in many other departments of physical thought. Thus Einstein, in his special theory of relativity, found immediate confirmation for the connection of energy and momentum in his new mechanics by the fact that the same connection held also for Maxwell's electromagnetic field. The equivalence of mass and energy, perhaps the most important result of the special theory of relativity, was solidly established within the mathematical framework evolved so lucidly and so elegantly by H. Minkowski.

The nineteenth century—towards a unitary world-picture

The importance of the contribution of nineteenth-century physics and chemistry to the unification of the physical world-picture can hardly be overestimated. The growing accuracy and sophistication of experimentation with its keen observation of relevant detail together with the formation of concepts which lent themselves to an increasing degree of quantification, and the subsequent establishment of hypotheses linking these concepts together in a wider framework of theories—all this rapidly transformed many branches of science from the backward state in which they existed in the second half of the eighteenth century into coherent bodies of knowledge—a goal reached by classical mechanics in the eighteenth century. Chemical research was put on a strictly scientific basis by Lavoisier, whose skilful and exact experiments and careful use of the analytical balance inaugurated the modern era of chemistry and in particular demolished the notion of phlogiston—that mysterious 'stuff' of negative weight so surprisingly similar to the Aristotelian element of 'absolute lightness'. This development paved the way for Dalton's scientific revival of the purely speculative atomic doctrine of Democritus and Epicurus. A quantitatively definable concept of a chemical element now emerged, and scientists began to distinguish between an atom—that is, the smallest unit of an element not divisible by chemical means—and a molecule—that is, the smallest unit of a non-elementary substance, consisting of a definite combination of several atoms, chemically separable. When the subsequent development of organic chemistry revealed that different molecules composed of the same combination of atoms could exhibit quite different properties, there followed the discovery of differences in molecular structure, and this brought about the revival of a structural approach to nature. At first it was confined to chemical research, but gradually extended later on to physics in the discoveries concerning atomic structure, the structure of the nucleus, and the physics of elementary particles. This later development, however, belongs to the twentieth century, and one might say that physical science in this century is characterized by a synthesis of the dynamical and the structural concepts of physical phenomena, since the structural explanation is based on, and supported by, investigations of the dynamical manifestations of the various structures.

A novel feature of the modern structural conception of matter, one not anticipated in earlier approaches to nature, is the idea of periodicity. When chemical elements were discovered in increasing numbers, the characteristic property of an element which lent itself most readily to classification as well as to quantification was that of the atomic weight, the weight of an atom relative to that of the lightest element, hydrogen. Even Lavoisier knew twenty-three different elements, and in the first third of the nineteenth century the number known increased to more than fifty, in the second third to more than sixty, reaching the number ninety-two at the beginning of the twentieth century. This threatening multiplication of entities, to which science has been averse at all times, was overcome by a development dating back to 1869, when Mendeleev discovered that the elements, when arranged according to their atomic weights, showed periodicities in their physical and chemical properties. For instance, when arranged in this manner in rows of suitable lengths, elements with similar properties were found to be placed in the same column. The discovery of the Periodic Table of the elements was the harbinger of a series of further discoveries which revealed the structure of the atom and its periodic features, such as the shell structure of its electronic envelope, and later the details of the nuclear structure. The discovery of the constituents of these structures enabled the physicists to explain the existence of the multitude of elements in terms of a small number of more basic entities. This successful reduction gives rise to the hope that the structural conception will again enable scientists to overcome the problem of multiplication of entities which is at present threatening the theory of elementary particles.

A similar development almost contemporaneous with the quantification of chemistry was that of electricity. The hazy concept of 'electrical fluid' of the eighteenth century was replaced by the concept of 'electric charge', and the formal analogy of the law of electrical attraction and repulsion with that of gravitational attraction facilitated the development of a mathematical theory of electrostatics which proved to be the first consistent chapter of mathematical physics after classical mechanics.

In developments of the nineteenth century two different trends, which ran side by side almost independently, can be distinguished. The one is the growth of the modern physical continuum concept, from Faraday's field theory and Maxwell's electromagnetic equations to Einstein's theory of gravitation. The other is the particle concept, with which further discoveries of structural features in matter were associated after it became apparent that properties such as chemical affinity were closely related to electrical phenomena. Faraday's discovery of the laws of electrolysis is a conspicuous example of the growing awareness of the fact that matter, electrically neutral as a whole, is built up of electrically charged particles. Subsequently, greatly improved vacuum techniques and the application of electric and magnetic forces to moving charges finally established the reality of the electron and of positive ions, and of the simplest of these, the nucleus of the hydrogen atom, or proton. The kinetic theory of gases and statistical mechanics completed the picture of the particle structure of matter based on classical mechanics. Yet it took decades before physicists acquiesced in the reality of atoms; the majority regarded them as a mere construct, a convenient fiction or a helpful product of scientific imagination. This reluctance to accept atoms as entities just as real as visible ones is of great historical interest, since it is a significant aspect of the spell cast by positivist thought over science in the nineteenth century. The positivist trend had been an inherent feature of science since its emancipation from the tutelage of institutional religion and its establishment as an autonomous activity of the human mind. The need to draw a sharp demarcation line between belief and knowledge was an important factor in the purging of 'metaphysical elements' from scientific thought. German *Naturphilosophie*, an offspring of romanticism at the end of the eighteenth century, gave an added impulse to this trend. The writings of Schelling and others revived neo-Platonic ideas, including cosmological speculations and comprehensive doctrines about the structure of nature as a whole, thus antagonizing scientists at a time when the systematic and detailed application of theory and experiment was beginning to bear fruit in all branches of scientific research. What made matters worse was that in the wake of these doctrines Newton's physics, and in particular his optics, was sharply attacked by men like Goethe and Schopenhauer. Hegel's *Encyclopedia of the Philosophical Sciences*, in which he treated physics and biology as an integral part of dialectical philosophy, increased the rift between science and philosophy still further, despite some brilliant ideas which it contained.

The reaction of the community of scientists to the various utterances of the protagonists of *Naturphilosophie* made itself felt in many ways throughout the nineteenth century. In this context it will suffice to give two conspicuous examples. One is the reluctant and almost hostile attitude of many leading men of science to the first publications dealing with the law of conservation of energy. After the observations of Rumford and the quantitative experiments of Joule had demolished the conception of heat as another kind of 'tenuous matter', physicists returned to the idea, already suggested by Daniel Bernoulli one hundred years earlier, that heat consisted in the kinetic energy of the ultimate particles of matter. This paved the way for the first general formulations of the law of conservation of energy or, as it is also called, the first law of thermodynamics. However, scientists were deeply suspicious of the kind of universal law of nature exemplified by this and regarded it as a relapse into the comprehensive conception of nature of ancient Greece, akin to *Naturphilosophie*, and as an attempt to discover a 'world formula' of a clearly metaphysical kind. The opposition to Robert Mayer's paper can easily be explained by his philosophical arguments in favour of the principle, but even Helmholtz' publication *On the Conservation of Force* (1847), with its careful exposition based on purely scientific considerations, aroused misgivings among many of his colleagues.

The other example of the strong influence of positivism is the protracted argument, in the last decades of the nineteenth century, between Ostwald and Mach on the one hand and Boltzmann on the other, as to the reality of atoms. The former regarded the concept of atoms as a useful device invented by scientists to help them to explain certain observed results in macrophysics, or as a mere convention by the help of which theoretical derivations could be carried out with greater ease. Boltzmann was convinced that atoms were real objects, no less than chairs and tables, and that they were physical facts whose reality was established by sound inference, even if they were not directly accessible to sense perception. Boltzmann's point of view has been fully vindicated by the development of atomic and nuclear physics. There is no longer any doubt as to the reality of atoms and elementary particles. One may perhaps call them 'facts of a higher order', presupposing a considerable number of other experiences and involving a greater number of tacit theoretical assumptions than the so-called 'simple facts' confirmed by the evidence of the senses, but still undisputably facts.

Positivism in the nineteenth century was found among philosophers as well as among scientists, and a notable instance of this attitude was that of J. S. Mill. The gist of his logic of science, followed later in the more sophisticated works of other philosophers, was that the final goal of the scientist was only a methodological one, namely the establishment of laws asserting the co-existence and sequence of phenomena. In this, he believed, philosophy could guide science by setting up a general canon of experimental method on the basis of induction, a canon valid for every scientific theory, beginning from the first statement of fact and the first formation of a scientific concept and ending with its last consequences. The basic tenet of positivism, namely that science is concerned exclusively with methodology, was still accepted by many scientists even after the emergence of the two great systems of modern physics, relativity and quantum theory, had shown theory formation to be a creative act which far transcended methodological rules and essentially depended on intuitive anticipation of further conceptual and factual developments. Einstein as well as Planck emphasized repeatedly that the belief in an ultimate reality had always been the main stimulus to, perhaps the prime mover of, scientific research. It is a non-rational belief which can neither be proved nor disproved by experience, but without which the great theoretical systems of science could never have been conceived. But over and above this, the whole process of scientific understanding is ample proof of the fact that in the establishment of scientific laws as well as in the formation of more comprehensive theories, belief and knowledge are intimately connected and intertwined. No scientist, experimental or theoretical, can make a start without working hypotheses and tacit assumptions which, though subject to corroboration by experiment and observation, have in the first instance the character of beliefs. It goes without saying that the boundaries between believing and knowing will always be shifting in the development of understanding.

In spite of the efforts of positivism to divorce science from philosophy—or, rather, from metaphysics—the philosophical implications of scientific discoveries inexorably impressed themselves upon scientists as well as philosophers. A striking example is the second law of thermodynamics which was formulated mathematically shortly after the law of conservation of energy. It centres round the concept of entropy, an abstract mathematical expression concerning the quantities of heat and temperature which characterize the state of a given physical system. Clausius' statement that the sum of entropies in a closed system always increased, or at any rate did not decrease, immediately brought up the problem of time, in contrast to purely mechanical phenomena whose mathematical description remained unchanged if the positive and negative directions of time were interchanged. Did this indicate a general running-down of the physical universe, and was the end, characterized by the state of maximum entropy, unavoidable; and what were the implications for questions about the age of the universe and its beginning? Time and its flux were pre-eminently metaphysical concepts, yet answers, albeit tentative ones, to questions like these had to be found. After the introduction of the atomic hypothesis and the application of statistical mechanics to such phenomena as heat diffusion, the second law was transformed from a mere description of dynamical phenomena into a structural statement. Its central concept became that of order, and the assertion that in every isolated thermodynamical process the total disorder of the systems involved in the process always increased bore a definitely structural character. Positivists tried to dismiss any association of the second law with questions about the state of the universe as a whole. However, the history of astrophysics and cosmology in the twentieth century has clearly refuted this narrow point of view, and physicists are seriously asking, on the ground of observational evidence as well as theoretical considerations, whether there may not be parts of the universe in which the entropy is decreasing or, to put it in more sensational terms, where time is running backwards. Metaphysicians, psychologists, and biologists are joining physicists in their search for suitable answers.

The influence of twentieth-century physics on philosophy

The second law of thermodynamics turned out to be only the stormy petrel of an era marked by a much closer relationship between science and philosophy. The most prominent feature of this new venture was the fact that science itself took the lead from philosophy when physicists, on the advent of relativity and quantum physics, had a new look at the foundations of their subject and were forced to revise their epistemology of science. In the light of the scientific world-picture which has emerged from these efforts, nobody would deny any longer the metaphysical nature of those foundations. There are so many facets to this revolution in physical thought that within the limits of an introduction some of them can be dealt with only summarily while others can hardly be mentioned at all.

Einstein's special theory of relativity constitutes a revision of the fundamental concepts of absolute space and absolute time postulated in Newton's *Principia*. These postulates were essential presuppositions of classical mechanics, which explained forces as causes of accelerations—that is, of deviations from either a state of rest or of rectilinear motion at a constant speed. Since, in Newtonian mechanics, laws of force were regarded as a most important class of the basic laws of nature, it was vital to ascertain that a given observed phenomenon was in fact explicable in terms of such a law and was not merely generated by, and dependent on, some special state of the observer. Only an observer who was himself in a state of rest or of inertial motion in absolute space could decide whether an observed movement was genuinely accelerated, and only he would be in a position to discover a real force. Moreover, by its very statement of the law of inertia, Newtonian mechanics led to the more general conclusion that all observers who were in a state of inertial motion relative to each other were in that privileged position in which 'real' and 'fake' forces could be distinguished from each other. This was the classical principle of relativity, which partially abolished the old Aristotelian contrast between rest and movement and replaced it by the contrast between inertial motion and motion under the influence of a force. The classical principle of relativity could also be formulated in another way: laws of nature were invariant with respect to all observers who were in a state of inertial motion relative to each other. The linking of the concept of a law of nature to the concept of invariance, which lent itself to mathematical presentation, gave a new lease of life to the old Pythagorean belief that the main ideas of nature could be represented mathematically.

However, the Newtonian idea of absolute space remained as a very dubious foundation for these otherwise sound considerations which led to the classical principle of relativity. Not only was it impossible to discover any point of reference which would make absolute space somehow 'observable' in the positivist sense; even the state of motion of the aether, which was supposed to fill the vast interstellar space as well as the interstices between the particles composing matter, could not be determined unambiguously. This gave a deadly blow to the long list of hypotheses which had attributed so many conflicting properties to the last remaining of the tenuous 'stuffs' of physics. Nineteenth-century physicists clung to it tenaciously because the wave theory of light and of electromagnetism strongly suggested the aether as the medium of propagation for these waves, in analogy with elastic bodies and sound waves. Einstein's interpretation of the failure to detect the aether was revolutionary in many respects; by discarding the aether he implied that there was a limit to the value of mechanical models and analogies—a conclusion already reached by Maxwell as a result of his attempts to describe the electromagnetic field in mechanical terms—and that the propagation of electromagnetic waves had to be considered without having recourse to the idea of a medium. His conclusion that the velocity of light was a universal constant valid for any observer, irrespective of his state of motion, amounted to the discovery of a basic law of nature as well as to the acceptance of this velocity, though mathematically finite, as an unattainable physical magnitude, an upper limit for the velocity of propagation of physical signals. The velocity of light was thus established as a universal constant which entered into explanations in all branches of physics, mechanical as well as electrical, and became a concrete indication of a decisive unification of the physical picture of reality. Indeed, the great significance of Einstein's conception was the combination of the classical principle of relativity with the principle of the constancy of the velocity of light in his principle of relativistic mechanics. It was one of the great triumphs of deductive physical reasoning, completely vindicated by its success; it put classical mechanics on a conceptually firmer basis, reconciled the laws of mechanics and of electromagnetism, and led to the discovery of such a fundamental law as that of the equivalence of mass and energy.

Einstein's principle of relativity deeply affected philosophical and, in particular, epistemological thought in that it abolished the concepts of absolute space and time, which were superseded by that of the absolute magnitude of the velocity of light. Intervals of length and time became relative magnitudes, depending on the state of relative motion of observer and observed body. The relativity of the flux of time—in particular the so-called retardation of moving clocks, later experimentally confirmed through observation of the decay of mesons in the atmosphere—led to a new physical and philosophical evaluation of the concept of time. The relativization of the concept of synchronous events had already in the early days of the theory of relativity caused a shock even greater than had the discovery, one hundred and fifty years earlier, that the velocity of light was finite. Space by itself and time by itself were, in Minkowski's happy phrase, reduced to mere shadows, and only the space-time interval between two events in the four-dimensional world was held as an invariant. This again was a result of supreme philosophical importance, showing, as it did, that only man's anthropocentric environment, that of observers moving with velocities small compared with that of light, deceived him into splitting the real four-dimensional world-continuum into the apparently independent three spatial dimensions and the one time-like dimension. The obvious conclusion was that physical reality was of a more abstract nature than the world of 'common sense' made familiar through daily experience.

The abstract nature of the physical world-picture was made greater by the general theory of relativity. Einstein's general principle of relativity, postulating that laws of nature must be invariant with regard to all observers, including those who were in a state of accelerated motion, led immediately to a new theory of gravitation. Its starting-point was a reappraisal of the fact, known since the time of Galileo, that all bodies, irrespective of their weight, fell freely with the same acceleration under the influence of gravity. Here was another conspicuous example of how a new interpretation of a well-established fact could give scientists a new look at the physical world. The great classical example was of course that of Copernican astronomy, because his heliocentric system was the starting-point for the development of the whole of modern mechanics and physical science. In the case of the general theory of relativity the philosophical implications were perhaps of even greater importance than the scientific ones. Since the gravitational field has the property of giving all bodies in it the same acceleration, an observer accelerated appropriately in the absence of such a field will observe the same dynamical phenomena as another one who is in a state of rest in the field. Einstein's conclusions led him to discard the antithesis between inertia and force in the case of the force of gravity. This was another step in the direction of the conceptual synthesis of phenomena, involving the abstract concept of a four-dimensional universe of non-Euclidean metric in which masses perform exclusively inertial movements along geodetic lines. The special character of the metric in a given part of the universe depends on the distribution

27

of masses in that part, and only at a sufficiently great distance from such masses does the metric become 'plane'—that is, does space-time exhibit a Euclidean geometry.

The geometrization of a physical concept like that of the force of gravitation gave rise to many philosophical discussions. Non-Euclidean geometries have been known since the early nineteenth century, when mathematicians constructed geometrical systems in which Euclid's parallel postulate was replaced by other axioms, and the subject attracted further attention through the papers of Helmholtz and Riemann. Earlier, Gauss had attempted to measure the angles of a triangle determined by the peaks of three mountains in order to find out whether space was Euclidean. But in the general theory of relativity the empirical nature of the geometry of space became for the first time an integral part of the foundations of a physical theory. It was supported by observational evidence; in addition to the facts explained by Newton's law of gravitation, it predicted three small but important effects which could be observed and measured.

Philosophers as well as scientists were reluctant to accept the idea of the non-Euclidean nature of empirical space. Many Kantians, for instance, had maintained that the concept of space as an *a priori* intuition necessarily included its Euclidean nature. However, at a time when the abstract nature of physical reality was making itself felt in many branches of science, the argument that the existence of space curvature was to be rejected because it could not be perceived directly, failed to carry scientific weight. On the other hand, the main tenet of general relativity, that space and matter were interdependent, was philosophically much more satisfactory than the Newtonian concept of space as an inert receptacle of matter. Furthermore, the unification of inertia and gravitation led to the plausible conclusion that the inertial mass of a body was determined by the presence of all the other masses in the universe. Apart from the philosophical aspect of this result there followed implications of great scientific significance. General relativity led to new thinking in cosmology—that is, to new theories about the structure and evolution of the universe. Einstein himself made the first recent attempt at such a cosmology, a hundred and sixty years after Kant's paper, based on Newtonian physics, on the evolution of the solar system. Einstein's model of the universe is now of only historical interest, but it included the first reasonable estimate of the total amount of matter contained in a space of given extension, and it was the starting-point in the present history of cosmologies which after fifty years is still in its infancy. The wealth of new astrophysical facts and the tentative nature of their interpretation does not yet allow for the definite choice of a cosmological model completely satisfactory from the point of view and state of knowledge of present-day physics. However, the plausible interpretation of the nebular red shift as the result of an expansion of the universe has revived interest in the idea of a finite age of the world as we know it. The spatial counterpart to this temporal finiteness of the universe is the restriction of observational space to a depth where the apparent speed of recession of galaxies approaches the limit set by the velocity of light. Thus a picture is emerging of a finite universe in a state of evolution, and the problem of the arrow of time has come up again on a cosmic scale.

In the wake of the geometrization of the force of gravity and its explanation as a result of the particular metric structure of the space-time continuum, attempts have been made to explain away the electromagnetic forces by interpreting them as additional structural features of the metric field. In particular, Einstein himself devoted the last twenty-five years of his life to attempts to construct, in mathematical terms, a 'unified field theory', which have proved unsuccessful. Finding the common roots of such universal forces as those of electricity and gravitation by means of a suitable mathematical construction was certainly a goal worth the work of a lifetime. Einstein has occasionally given expression to the philosophy underlying his task: 'Our experience hitherto justifies us in believing that nature is the realization of the simplest conceivable mathematical ideas.' It was the age-old vision of the Pythagoreans and Plato, of Descartes and Kepler, sometimes expressed as the *credo* of the rationalist, and at other times as the belief of the mystic. On a grander scale, it was the revival of Newton's secret and reluctantly revealed hopes that the aether might be the common cause of the phenomena of light and gravitation. Faraday cherished similar hopes when he devised experiments to test whether electricity could be induced by the acceleration due to gravity. But Einstein's undertaking to derive the ultimate synthesis of physical reality from mathematical symbols and equations surpassed in its breadth of conception all the former attempts. It was once again the search for the 'cosmic formula', and it failed, not because of the structural conception associated with it, but doubtless because it carried the idea of continuity to a level of perfection which defied realities. Seen from the point of view of the history of ideas, the theory of general relativity was the realization of two conceptions, that of the structural approach to nature and that of the idea of a physical continuum. The first, in which structures and invariants were taken to be the absolute permanent elements underlying the changing phenomena in space and time, was most probably an important step in the right direction; the second, on the other hand, was doomed to failure, because the notion of a metric continuum is necessarily linked with that of a continuity in physical events—that is with determinism.

Since the advent of quantum theory, however, the evidence for the discontinuous nature of physical events has been accumulating to an extent where it amounts to certainty. The quantum of action, another universal constant of physics, was introduced by Planck in 1900 as an *ad hoc* hypothesis, but it soon proved to be of universal significance. During the first phase of quantum physics, the experimental techniques of atomic physics developed rapidly and led to important discoveries which left no doubt as to the atomic structure of energy as well as of matter, and confirmed in this way the idea of the equivalence of matter and energy. The discontinuous character of all micro-events manifested itself in the interactions in which energy quanta and particles of every kind were involved, and in the fact that the energy states of complex structures like atoms and molecules consisted of a set of discrete values, suggesting jump-like changes from one state to another. The first phase of quantum physics is an illuminating example of the development of scientific understanding. Physicists attempted to apply to atomic events and structures the principles of classical mechanics, which had served macro-physics so well over such a long period, while they took into account at the same time the newly discovered empirical quantum rules. Niels Bohr's famous model of the hydrogen atom (1913) was a conspicuous instance of this state of compromise, which did not last long because the methods applied could not be generalized for more complex cases and led to inconsistencies and to results clashing with experience. Einstein, in 1915, had found in tensor analysis the mathematical tool for the successful development of his general theory of relativity, and in 1926 mathematics was again to be the guide which led science out of the impasse. In that year Schrödinger adapted a mathematical algorithm which had previously been used in another branch of theoretical physics to the description of atomic phenomena. The success of this adaptation, known by the name of wave mechanics and shortly afterwards improved by Dirac, initiated the second phase of quantum physics, but only after a far-reaching discussion involving the interpretation of certain mathematical expressions. The interpretation of Born and Jordan, adopted by the majority of leading physicists, though not by Einstein or Schrödinger, was of epoch-making importance. It amounted to the interpretation of a certain mathematical expression, the square of a wave-function, as a measure of the probability of a given physical event or state. The quantum physicist, as a consequence of this interpretation, accepts the primary laws of physics as probability laws allowing only for statistical predictions, which amounts to an admission that a single observation cannot be subject to a causal law. He regards the deterministic laws of classical physics as merely secondary, since they apply only to macroscopic events consisting of the superposition of a large number of single events. Classical determinism, so lucidly stated by Laplace and for a long stretch of time one of the main pillars of the scientific world-view, had to be abandoned—a result of very great philosophical significance. However, this does not entail the renunciation of the principle of causality in a general sense, that in which an empirically verifiable mathematical deduction of a whole class of phenomena can be understood as a causal description, because quantum mechanics has shown that quantum phenomena, too, although being subject only to statistical predictions, can be presented mathematically in the framework of a consistent scheme of deductive reasoning.

The new conceptual situation can again, as in the case of relativity theory, be analysed from an operational angle, which for quantum mechanics involves the relationship between the observer of a micro-phenomenon and the phenomenon observed. The theories of pre-quantum physics, including relativity theory, were deterministic in the sense that from the state of a system at a given moment they derived mathematically its state at any other—earlier or later—moment. In classical mechanics the initial state of a system was determined if two independent sets of data, the positions and velocities of the bodies constituting that system, were known. It was tacitly assumed that there were no limits to the exactness of this knowledge, because the experimental procedure of macrophysics could be so arranged that the inevitable interference with the object of any measurement caused by the measuring observer could be practically neglected. The main reason for this is that for macroscopic events the means of observation can always be made small enough compared with the observed system, and the uncertainties associated with the measuring process can be reduced so as to remain below the usual limits of errors of observation. Things are different, however, in the micro-region where the means of observation and the observed object are of the same order of magnitude, since atoms and elementary particles such as protons, electrons and nucleons are involved in both. The disturbing effect of observation can never be eliminated in these cases by definite corrections, and it cannot be reduced below a size comparable with Planck's quantum of action. Any measurement must therefore be regarded as a small but finite interaction between the observing subject and the observed object, and the very act of observation will interrupt the causal nexus of the events preceding it and those following it because this interaction will, in a way which cannot be determined beforehand, affect both the state of the means of observation and that of the observed system.

There is an interesting parallel between the operational aspects of the conceptual revolutions caused by relativity and by quantum mechanics. If the speed of transmission of signals is assumed to be infinite, not finite, the classical picture of absolute space and time, and the discarded concept of an absolute simultaneity of events, is restored; if Planck's quantum of action is assumed to be zero, but not small and finite, the classical picture of determinism and a continuous sequence of physical events is likewise restored.

The most important philosophical consequence of quantum mechanics is the revision of the concept of a phenomenon. If the definition of a given datum includes both observed object and observing subject, an observed phenomenon cannot be completely divorced from the experimental arrangement leading to that observation. The necessity of describing the state of the events connected with it will always force the scientist to make an arbitrary division between observer and observed object, but this division does not correspond to a genuine separation between an 'objective' phenomenon, independent of the observer, and the observer himself. The well-known duality of corpuscle and wave, or rather that of the discontinuous-particle aspect and the continuous-field aspect of light and matter, represents the experimental confirmation of the idea that the observer is part and parcel of the phenomenon. The centuries-old argument whether light consists 'in reality' of waves or of particles has thus become meaningless. There have been other instances in the history of physical thought where conceptual progress has deprived long-standing arguments of any significance—heated arguments between the followers of d'Alembert and Leibniz about the primacy of momentum or *vis viva* (kinetic energy) in mechanics were resolved in this way—and the history of the debate about particles and waves will probably not be the last one.

The new concept of a phenomenon, denying as it does the existence of an 'objective' reality independent of the observer, has shed new light on the relation between physical thought and other fields of human knowledge. Niels Bohr, by coining the notion of complementarity, has drawn attention to the universal significance of the new attitude to the problem of man *versus* the outside world, and in general to the philosophical problem of the relation between subject and object. His interpretation of Heisenberg's uncertainty relations, the mathematical expressions of the indeterminacy of a physical state, centred round the notion of a pair of complementary opposites, the two sets of data characterizing that state, which were the position and velocity (or alternatively time and energy) values of the components of a system. The members of these pairs are not independent, but are coupled by a reciprocal relation such that each necessarily complements the other in their determination of the state; at the same time, they oppose each other in that a greater exactness in determining the one entails a lesser exactness in determining the other. Bohr's contention is that physical complementarity represents a logical extension of the classical notion of causality, and that it has to be regarded as a special case of a universal phenomenon, pertaining to the world of human experience. He pointed at several pairs of complementary opposite concepts which are of a non-physical nature, such as thinking and feeling, which are both indispensable for the description of human situations, and in this connection he referred to examples taken from ethics, aesthetics, epistemology and sociology—such as the problem of freedom of the will, or that of the relationship between justice and love.

Bohr's extremely illuminating remarks prove the extent to which quantum physics and its conceptual analysis of physical phenomena have moved closer to the domain of the humanities and to their mode of describing some aspects of human experience. This closer relationship between the worlds of matter and of mind should cause no surprise if it is borne in mind that the indivisibility of the quantum of action introduced into physics the concept of wholeness, thus exhibiting one of the essential features of mental and psychical phenomena. Human consciousness, being a whole, cannot subdivide a phenomenon into partial phenomena without essentially influencing that phenomenon in its totality. The impossibility of drawing a clear-cut line between the subject and object of a physical phenomenon thus reflects a much more general state of affairs, namely the fact that the objects of human experience and the consciousness of the experiencing person are continually merging into one another, and in consequence of this are apt at any moment to change a given situation in a way which cannot be determined beforehand.

In the forty years since the last great theoretical breakthrough, an enormous wealth of new experimental data has accumulated in elementary particle physics, astrophysics and cosmology. Theory has been lamentably lagging behind experiment in recent decades. Many eminent physicists of the last generation, notably Bohr and Pauli, have expressed the opinion that quantum mechanics may well be the first of many more steps which will lead physics away from familiar classical concepts. A major theoretical advance could possibly be achieved only through ideas involving further radical renunciations of some notions to which physicists had become accustomed in the age of determinism, which was also the age of mechanical conceptions and which saw the zenith of the dynamical approach to nature.

The structural approach which began with modern chemistry and is characterizing the present theory of matter is steadily making headway. Although physicists engaged in the study of elementary particles observe dynamical phenomena, such as the paths of moving particles and their collisions, and make use of dynamical concepts, such as the classification of different types of interacting forces, considerations of symmetry are gradually becoming a guiding principle in the physics of matter. All these symmetries and the symbols associated with them clearly reflect the structural approach predominating today. In this connection it is interesting to note that in this latest phase of physical research certain contrarieties play a central role, calling to mind the Pythagorean opposites of the ancient structural approach—such as the contrarieties of positive and negative electric charges, of odd and even, of left and right, and of future and past. However, behind this recurrence of formal patterns of thought in the course of history, a steady progress to higher levels of understanding is unmistakably discernible. No doubt, with every step forward new questions will arise which will have to be answered and new riddles will appear which will have to be solved. One is reminded of Pascal's magnificent aphorism comparing human knowledge with an ever-expanding sphere: the more its circumference grows, the greater the number of its points of contact with the unknown.

SECTION I

Antiquity

Section I: Antiquity

INTRODUCTION

SCIENTIFIC thought, in what may be called Greek antiquity, spans a period of perhaps eleven hundred years, nearly half of the lifetime of what we habitually call science. On the shores of Asia Minor, in the sixth century B.C., the beginnings of a scientific language developed in the place of a mythological one to describe natural phenomena. Mythological notions for the explanation of the physical world were replaced by scientific and rational categories (such as those of force and of elementary substance), by philosophical principles and by mechanical models. A wealth of factual material, accumulated in former generations, was brought into a new conceptual framework; it became the raw material for a rational world-picture, emerging from a process of questions and answers which are essentially the same as those that occupy us to this day.

Only a small number of original fragments from the so-called Presocratic period (600–450 B.C.) have been preserved; the greatest part of the documentary material is to be found in the writings of later authors. Most of the quotations in this Anthology are original passages; even so, it is difficult to prevent anachronisms in the presentation of early Greek thought. The main reason is that the fragments are too short to give us an understanding of their full meaning, without taking recourse to some kind of interpretation.

There was one question of an eminently epistemological (or rather metaphysical) character that occupied the minds of those early philosophers. 'What is the immutable principle that underlies changing phenomena—what is the single root of the multiplicity of things in the physical world?' This question received two kinds of answer. They came from those who presupposed a basic principle that was material in essence, and, on the other side, from those who regarded measure, or number, as the foundation of all that exists.

The first thinkers to maintain a material first principle were extreme monists. They were divided, however, in the question of the specific quality of this primordial matter. Thales of Miletus, of whose writings not one original fragment remains, took this to be water, whereas Anaximenes believed it to be air, and Heracleitus fire. The solution of Anaximander, a contemporary of Thales, is of special interest. He called primordial matter 'the unlimited', an ambiguous expression that can be used qualitatively as well as quantitatively. In this conception Anaximander displayed his profound understanding of the problem— the analysis of a defined quality cannot result in another defined quality for this would confront us again with the question of basic principles. We must instead descend to that level at which matter loses all its qualities and properties, and where the primordial element appears wholly undefined and unlimited. The

37

original fragment of Anaximander considers the generation and decay of things in a poetical manner. There is a regularity in the balance between creation and decay which establishes order in the sequence of these phenomena in the course of time.

The additional facts of continuous creation and unceasing change in the whole fabric of material structure were strongly emphasized by Heracleitus. The phenomenon of change constitutes a continuum the signs of which can be perceived in all the events of the visible world. Heracleitus saw a hidden harmony in the phenomena, and this is *his* true reality behind all visible changes. Harmony appears too in the polarities that exist in the structure of the physical world, and these are considered the source of all that happens (as the tension of the strings in the lyre or the bow). More explicitly, Empedocles introduced the idea of opposing forces acting in the world—such as attraction and repulsion, which, in his poetical language, he called 'love' and 'hate'. Their dynamic equilibrium maintains the world by preserving it from the destruction which would result from the predominance of one single force.

Empedocles, in common with many of the other Presocratic philosophers, presents a mechanistic and materialistic conception. He thought that physical actions are propagated by effluences emanating from the pores of bodies, and moving from one to another. The theory of effluences is also the basis of his scientific explanation of magnetism; similarly he regarded light as a fluid substance emanating from a luminous body, such as the sun. Renouncing the extreme monism of his predecessors, Empedocles introduced the theory of the four elements earth, water, air, and fire, whose combinations constitute all the different forms of matter in nature. The principle of the four elements prevailed until the end of the Middle Ages.

Anaxagoras' conception of matter was pluralistic in the extreme. He assumed that of each and every part of a substance, even the smallest particle contained a mixture of all the different kinds of matter. This conception, which Aristotle called the 'theory of the equal parts', had its origin in the attempt to explain metabolism, the process by which in the body food is transformed into sinews, muscle, bones and blood; the assumption was that all food contains in itself the whole of organic matter ('Everything is contained in everything'). According to Anaxagoras metabolism is nothing else but the splitting up of the elements of food into their different constituents. Anaxagoras' materialistic approach is most striking when he applies it to subjects hallowed by age-old beliefs, such as astrology. He held the sun and the stars to be glowing stones or rocks, set alight by air-resistance due to the velocity of their rotation around the earth. His cosmogony, too, is based on a physical principle—small hot elements were split up from closely packed cold ones by the impetus of rotation. However, in his physical theory there is a non-materialistic principle, introduced to explain the cause of motion, in particular of circular motion. This principle he called 'Mind' and described it as something distinctly different from matter.

The highest achievement in the Presocratic theories of matter is the atomic theory of Leucippus and Democritus. Of this, unfortunately, very little has come down to us in the original. Greek atomism was speculative, but its basic assumptions have in fact been proven by modern physics. Its scientific simplicity is admirable—the atoms, differentiated by their shapes, move in an infinite vacuum, and everything that happens is subject to the law of necessity. The theory attempts to explain by logical reasoning that macroscopic phenomena are the results of combinations of invisible factors. The atomistic theory of sense perception was in part influenced by Empedocles' ideas.

A different answer regarding the nature of the basic principle was given by the Pythagorean School. These philosophers held that physical events were not determined by matter, for the basis of reality was number. Philolaus and Archytas formulated with great clarity the idea of the structural significance of the integers. Musical harmonies, which the Pythagoreans investigated in particular, are based on the simple numerical ratios of the numbers one to four. Arithmetic and music appear as closely linked scientific disciplines, while the third basic science is astronomy—the regularity of the periods of the planets also display harmonies, and the spheres of the heavens as a whole can be compared to the strings of the lyre and their harmonic accords. The numerical harmony that pervades all parts of the universe also determines the shapes of all things; this is shown by the five perfect bodies which were discovered towards the end of the Pythagorean period. Harmony, however, does not deny the existence of the opposites, conceived in Heracleitus' theory. On the contrary, the 'opposite' is an essential part of harmony; the Pythagoreans listed ten contrarieties, such as even-odd, movement-rest, light-dark, etc., governing all cosmic phenomena. The Pythagorean physician Alcmaeon declared this balance ot opposites as the necessary condition for human health. From an historical point of view, the Pythagorean conception is of the utmost importance, because it provided the main stimulus for the development of mathematics and also assisted in that of astronomy. These are the two scientific disciplines in which Greek contribution to science was the greatest. However, we have to bear in mind that in the philosophy of number there is a kernel of mysticism for it is inclined to discover a symbolical significance in any numerical ratio. A typical example of this is the introduction of the holy 10 as the number of principal heavenly bodies in Pythagorean cosmology, contrived by the addition of the central fire and of the invisible star, the counter-earth, to the sun, the moon, the earth and the five planets. The Pythagorean School underwent a dangerous crisis when it was discovered that the integers were not able to explain all geometrical forms and their ratios. The discovery that the ratio of the diagonal of a square and the length of its side cannot be expressed in integers (in other words, the discovery of irrational numbers) confronted them with the problem of the continuum. The difficulty was exemplified most clearly and instructively by Zeno of Elea in his paradoxes, which also show the difficulties inherent in the concept of motion as a continuous phenomenon. Some of the questions raised by Zeno remained subjects of discussion until the end of the nineteenth century at least.

In the fourth century B.C., a new chapter of Greek scientific thought was initiated through the physical theories of Plato and Aristotle. In his writings Plato emphasizes the importance of mathematics and its central rôle in the exploration of nature. He was strongly influenced by the Pythagorean School. In Plato's *Timaeus*, which contains his cosmology, he also expounds his scientific approach in general. Plato recognizes only one science as unquestionably true and dependable: his theory of ideas. All the other sciences that describe the physical world with its perpetual changes can produce only probable explanations—a plausible picture of the universe and all it contains. Plato's belief in the possibility of a rational explanation of the universe is to be found in his allegory of the struggle of reason and the stark necessity of matter, where reason can overcome necessity by persuasion, but will never achieve a complete victory. An instructive illustration of this 'plausible story' is the geometrical theory of matter in the *Timaeus* where Plato develops his ideas on mathematics as the language of science. This theory is the beginning of mathematical physics, an approach which uses mathematical symbols for the explanation of physical reality. Most aspects of Plato's importance in the history of astronomy are to be found in the *Timaeus*, the *Republic* and the *Laws*. His challenge to 'save the phenomena' had an immense influence on mathematical astronomy (on thinkers from his pupil Eudoxus to Ptolemy), with its presentation of the movements of the planets as combinations of regular circular motions, those perfect motions which alone are proper for heavenly bodies. In the eternal rhythm of a periodically recurring motion Plato saw a sensible expression of the World Soul. In his approach to the motions of the heavenly bodies there is the beginning of the dichotomy of heaven and earth, the division of the universe into the terrestrial sphere beneath the moon and the celestial region. This dichotomy is one of the principal aspects of Aristotle's physics. It prevailed for many generations until the modern era, having a decisively retarding influence on the development of the natural sciences; only Galileo broke its dominance.

Another aspect of Aristotle's physics is anticipated in Plato's cosmology. This is the teleological approach to natural phenomena, clearly expressed in the allegory of the demiurge who creates the universe purposely. In Aristotle's conceptual framework this approach is manifest in the principle of purposeful reason whereby all phenomena in nature, animate as well as inanimate, can be compared to the works of an artist who creates according to a pre-conceived plan. A hint of this concept had appeared already in the Presocratic period, in a fragment by Diogenes of Apollonia which considers the 'reasonable' division of the seasons of the year. We must bear in mind always that Aristotle's original research was in the field of biology; his whole conception of nature was biological and influenced by a way of thought which today does not seem applicable to inanimate objects.

Aristotle's universe has infinite extension in time, but is finite in space. It is a globe whose outer region is the sphere of the planets and whose centre is the centre of the earth. This cosmos is a physical continuum consisting of the four elements earth, water, air and fire in the sublunar region, with the fifth element, the aether, in the heavens. The existence of a vacuum Aristotle negates absolutely for a number of reasons, for instance because everything that happens is defined only by the medium where the event takes place. This is the source of his definition of space as the place of a body in relation to its physical surroundings, in other words to the bodies around it. In his profound analysis of the essence of time we can see Aristotle's efforts to define and isolate the pointlike *Now* in the perpetual stream moving from the past to the future. This *Now* cannot be detached completely from the concept of motion, which contains a change of place as well as quantitative and qualitative changes. Qualitative changes are in the main transitions from one element to another. As the four elements are combinations of the four sensible qualities (the two pairs of opposites warm–cold and wet–dry), every transition from one element to another can thus be explained as the exchange of one component by another one. Every change, in other words all creation and decay, is in the last resort nothing but the mixing or separation of qualities alone or in combination. A body created by the combination or mixture of other bodies will contain their components in the form of potential bodies, and, when it disintegrates, these will reappear as actual bodies. However, Aristotle was not consistent in this respect—he assumed that if the quantity of one component was much greater than that of the other, the weaker would be, so to speak, assimilated by the stronger. Thus a drop of wine in a large body of water would 'lose its identity' and, in fact, turn into water. As far as change of place is concerned, we have to distinguish between natural motions, in other words motions of the bodies towards their natural places impelled by their innate tendencies, and forced motions—those caused by an external force. The natural place of heavy bodies, of earth and water in other words, is in or near the centre of the earth, while that of light bodies (fire and air) is on or near the border between the sublunar and the heavenly regions. Thus natural motions are rectilinear and finite because of the limited nature of the sublunar region; the natural motion of the fifth element, however, is circular and eternal, reflecting in geometrical form the eternal character of the aether which is devoid of qualities.

A body moving in forced motion, such as a thrown missile, was thought to move as long as there is direct contact with the body causing the motion. In other words, there is no forced motion without a constantly acting force. Aristotle had great difficulty in trying to explain this false law by the assumption of a finite chain of movers and moved. Yet it is worth noticing that in his deliberations there is a first hint of that notion of impetus which began to crystallize only in the Hellenistic period. Of interest also is the fact that when Aristotle linked the time of the motion, the mass of the body moved, and the distance covered, he gave the first quantitative formula of any law of motion, and this law held good until it was disproved by Newton.

Aristotle's theory of the spheres is characteristic of his realistic approach and his attempt to construct a unified system. Whereas the concentric spheres of Eudoxus and Callipus were a geometrical contrivance for 'saving the phenomena' of the planets and for the reduction of the irregular movements of each planet to a combination of uniform and circular motions, Aristotle turned these spheres into physical entities consisting of the fifth element, the aether. Moreover, he described all the planets (including sun and moon) as one single system, from Saturn, whose sphere is nearest to that of the fixed stars, to the moon whose sphere is closest to the terrestrial region. The assumption of such a unity forced Aristotle to introduce a number of 'unrolling spheres' between each sphere and those nearest to it. The task of these 'unrolling spheres' was the unification of all the circular movements into the movement of one single system. The resulting number of spheres for the seven planets totalled fifty-five—a high price for Aristotle to pay for this unity. Finally he explained the heat of the sun as a result of the friction of the sun's sphere arising from its rapid rotation. This was necessary as the aether of which the sun consists is devoid of all qualities and thus cannot be hot.

Several of the passages from Aristotle provided here shed some light on his scientific epistemology. For example his profound ideas on the scientific approach towards an understanding of nature are demonstrated, as is his analysis of the sentence '*Tomorrow there will be a naval battle*', where we can see that at least as far as human events were concerned, Aristotle's attitude was not deterministic. The problem of freedom of will versus fate appeared in all its importance only in Hellenistic times. It arose first with Epicurus, whose physical explanation for freedom of will is based on the indeterminate deviations of the atoms from their paths. Epicurus' philosophy is described in Lucretius' didactic poem *On the Nature of Things* (*De Rerum Natura*). This work, written 250 years after Epicurus, unrolls before our eyes the whole essence of the theory of atoms, and its poetical form does not in the least detract from the clarity and accuracy of the representation of the problems. Lucretius' verses supplement the sayings of Epicurus that have been preserved in his letters to his pupils. His notions on the scientific explanation of celestial phenomena bring out the weak point of Epicurean philosophy which preferred peace of mind to a clear-cut physical approach. On the other hand, we find in Epicurus' theory of atoms a conceptual development over and above that of Democritus, the limitation of the number of shapes of atoms to a finite though very large value. This restriction arises from Epicurus' demand that the size of atoms be always below the limit of visibility, because an atom devoid of qualities but visible would be a contradiction in itself.

The rival of the Epicurean physical theory—the continuum theory of the Stoics—survived for a long period, as did the philosophy of the Stoa which prevailed for five hundred years. Stoic physics, characterized by its scientific logic and its unity of thought, is in the main the work of Zeno, Chrysippus and Poseidonius. They regarded the physical world as a dynamic continuum, made coherent by the *pneuma* that totally pervades the universe and all matter contained in it. The *pneuma* is an exceedingly tenuous substance, a mixture of fire and air, which imparts to matter its structure and all physical qualities. Through its tension the *pneuma* also constitutes a kind of elastic medium for the propagation of physical action. The extreme concept of the continuity of the universe and its happenings made the Stoics consistent determinists. They tried to reconcile freedom of will and determinism by saying that man's will and actions were indeed part of the causal nexus of all the events in the world, but, being determined by the personal parameters of each one and arising from each one's particular nature, they impart a feeling of inner freedom. In the spirit of this Stoic approach, Spinoza said that, if a falling stone were conscious of itself, it would see itself as falling of its own free will. The concept of the dynamic continuum brought the Stoics to the threshold of the calculus, although for their speculations they did not use mathematical terms.

The golden age of Greek science is connected with the names of Euclid, Archimedes, Eratosthenes, Apollonius of Perga and Hipparchus. In their investigations ancient science reached its peak. From the writings of Archimedes we have included here two passages. One is from his treatise *The Sandreckoner*, where he develops a method for expressing very large numbers. He applied this method as an example to the calculation of the number of grains of sand in a sphere, the radius of which is the distance of the earth from the sun as it was then known. Here he relates the heliocentric hypothesis of his contemporary Aristarchus, which was not accepted by the majority of ancient astronomers, but which is of interest as an anticipation of Copernican ideas. In his treatise *The Method*, Archimedes introduced the use of mechanical considerations for the discovery of geometrical propositions. In his words we also find a hint of the idea of scientific progress, expressed more specifically in the writings of the Stoic Seneca. While speaking about comets he says that scientific research is a task for many generations, because no one generation can solve all the problems of science. Different scientific ideas of the several philosophical schools existing in the Hellenistic period are reflected in the writings of Plutarch (first century A.D.). His treatise *On the Face of the Moon* is a conversation on scientific and religious subjects between friends who are protagonists of different philosophical systems. The main topic is the nature of the moon. We find here astrophysical conceptions of an amazing insight and an approach close to modern ideas, such as hypothesis on the universal nature of gravity.

Ptolemy's work *The Mathematical Syntaxis* (*Almagest* in the Arabic version), written in the middle of the second century A.D., was *the* astronomical textbook for 1,400 years, until the time of Copernicus. The passages published in this Anthology, dealing with the theoretical and experimental foundations of the geocentric hypothesis, do not give a full picture of the immense importance of the book in the history of astronomy through its discussion of the mathematical problems connected with the regular movement of the moon, sun and five planets. Ptolemy, who was a great practical astronomer as well as a great mathematician, summarizes in the book his own researches and the achievements of his predecessors, in particular those of Hipparchus (second century B.C.). In the beginning of the Hellenistic period it had already become evident from observation that Eudoxus' hypothesis of concentric spheres was untenable. This was because the changes in the brightness of the planets indicated that their distances from the earth are not constant. Apollonius of Perga, Hipparchus, and Ptolemy still held on to the axiom that the movements of the stars had to be explained as combinations of uniform circular motions. The devices they used to conform to this demand involved the method of eccentric circles, that of epicycles, or a combination of the two. Ptolemy generally resorted to the combination, and assumed, for each planet, a main circle around the earth the centre of which did not coincide with that of the earth, and a secondary circle (or epicycle) along which the planet moved. The centre of this epicycle passed along the circumference of the main circle. The process of adapting the details of this system to observations which became more and more exact in the course of time caused the Ptolemaic system to become more and more complicated. In principle, however, it succeeded to 'save the phenomena' to such a degree that it could hardly be improved before the invention of the telescope.

In spite of the important results achieved by Ptolemy's theory, he was not really satisfied with this system and the complicated details which could not be reduced to the simplicity and unity of the principle of concentric spheres. At the end of his book he defends his model, so clumsy compared to the Pythagorean–Platonic hypothesis of the movement of the stars. Even more weighty doubts concerning the Ptolemaic system were expounded in the middle of the fifth century in Proclus' book on astronomy. In the introduction and also at the end of the work we see the exertions of an orthodox Platonist confronted with the original Platonic concept of the simplicity and regularity of the heavenly motions on the one hand, and the calculations and hypotheses of the mathematical astronomers on the other. Proclus does not see in the Ptolemaic system a true picture of reality. He accepts it, however, as a rational device to assist astronomers towards the achievement of practical results.

The heritage of scientific thought from antiquity and Hellenism survived mainly in the philosophical literature of later times, as for instance in the writings of the Neoplatonic school. The Christian philosopher Johannes Philoponus was raised in the atmosphere of this philosophy. He taught in Alexandria in the sixth century and was one of the last great commentators of Aristotle. Philoponus, with an extensive knowledge of the natural sciences, was a philosopher of great independence of mind. In his commentaries to Aristotle's physical writings he often differs from him, here and there enlarging and complementing his assumptions conceptually. Philoponus sometimes used most instructive examples which show the force of his scientific imagination. In his attempts to reconcile phenomena that are not concordant with the idea of universal harmony there is discernible the influence of the Stoics as well as the Neoplatonists. Philoponus extended the central Aristotelian concept of the transition from the potential to the actual by using the notion of 'fitness'. Such a transition is impossible, if a body does not possess the physical fitness for such a change. Of interest also is Philoponus' mechanistic approach; in this context he describes the human body as a machine that cannot be reactivated by the soul once it has stopped functioning. Of historical importance is Philoponus' conception of the laws of motion, opposed to that of Aristotle, and in particular his adoption of the idea of impetus many hundreds of years after Hipparchus had first expressed it. The Arabic philosophers took this idea over from Philoponus, and some of the medieval European Scholastic Schools preserved the tradition; in modern science it appears as the vectorial concept of momentum or the scalar concept of kinetic energy. Philoponus also made use of the notion of impetus in his theory of light, when explaining the rays emanating from a luminous body as a special case of it. However, of greatest importance is Philoponus' cosmology, based on his monotheism. Believing that heaven and earth were both created by God *ex nihilo* he vehemently attacked Aristotle's assumptions with regard to the eternity of the universe and its dichotomy into a heavenly and a sublunar region. In particular he tried to disprove by physical considerations Aristotle's belief that the sun and the stars consisted of aether, and claimed that they were sources of fire of the same kind as terrestrial fires, being like those subject to creation and decay. Moreover, he declared that all matter everywhere is nothing but tri-dimensional extension and in this respect, too, there is no difference between heaven and earth. Philoponus' philosophy found no echo in his time, and twelve hundred years had to pass until the impact of Galileo's ideas brought about a complete change in scientific thought.

I: ANTIQUITY

THE PRESOCRATICS

ANAXIMANDER

1

Among those who posit one thing, unlimited and in motion, the Milesian Anaximander, son of Praxiades, and the successor and pupil of Thales, said that the first principle and element of things in existence is the unlimited, and it was he who introduced this term 'first principle'. He maintains that it is not water, nor any other of the things called elements, but its nature is something different from these, being unlimited; and out of it come all the heavens, and the world orders in them. Into these things from which all that exists is generated comes destruction also, according to necessity; for they pay the penalty and make retribution one to another for their injustice, in accordance with the ordering of time.

ANAXIMENES

2

The Milesian Anaximenes, son of Eurystratus, who associated with Anaximander, also says, as he does, that there is one underlying original something which is unlimited, but he does not, like Anaximander, think that it is indeterminate, but of a definite character, identifying it with air; rarefaction and condensation account for the differences in things. Air in its rarefied state becomes fire, but when it condenses it becomes wind, then cloud, and after further condensation water, then earth and then stones; everything else comes from these. He also posits eternal motion, which causes the series of changes.

3

As our soul, which is air, holds us together, so breath and air surround the whole universe.

HERACLEITUS

4

We step and do not step into the same rivers, we exist and do not exist.

5

They do not understand how what is in conflict with itself is in accord; harmony consists of opposing tension, as in the bow and the lyre.

6

The unseen harmony is stronger than the seen.

7

All things are an exchange for fire, and fire for all things, as are goods for gold, and gold for goods.

8

Nature likes to hide.

THE PYTHAGOREAN SCHOOL

9

Pythagoras of Samos, son of Mnesarchus, who introduced the term 'philosophy', gives as first principles numbers and the proportions they contain, which he also calls harmonies; combinations of both are elements, termed geometrical. Moreover, among the first principles he puts the one and the indefinite dyad. Of these principles one tends to the efficient and formal cause, which is mind and god, and the other to the passive and material, which is the visible world. Ten is the basis of number, for all Greeks and foreigners count as far as ten, and when they reach it they go back again to one. And he says further that the power of ten rests in the four, the tetrad. And this is why: if anyone, going from one to four, adds together these numbers, he will have made up the number ten.

10

It is evident that these thinkers suppose number to be the first principle, both as the material of existing things, and as explaining their temporary and permanent characters. They assert that the elements of number are the even, which is unlimited, and the odd, which is limited; the one comes from both of these (for it is both even and odd), and number from the one; and the whole heaven, as has been said, is numbers.

Another group of Pythagoreans says that there are ten principles, and they arrange them in a column of pairs: limit and unlimited, odd and even, one and many, right and left, male and female, rest and motion, straight and curved, light and dark, good and bad, square and oblong.

11

It is hard to say whether time repeats itself, as some maintain, or not [. . .] If we were to believe the Pythagoreans, that the same events recur in numerical order, and that once more I shall be lecturing to you with this pointer in my hand, and you sitting there, and everything else will be as it is now, then it is reasonable to suppose that time repeats itself. For earlier and later of one repeated motion, and likewise of many things repeated, are one and the same, and also the number of them. With all things repeated then, time is as well.

12

It is clear from this that the theory of a harmony resulting from the movement of the stars, that is that the sounds they produce are concordant, although set out with remarkable skill by its supporters, is nonetheless not true. Some people think that the movement of such large bodies must give out a sound, because this is what happens in our experience with bodies smaller in size and moving at a slower speed. Since the sun and the moon, as well as the vast number of stars of great size, are moving with so swift a motion, it is impossible for there not to be an extraordinarily loud noise. Starting from this premiss, taken together with the observation that the speeds, judged from the distances, show the same ratios as musical concordances, they conclude that the sound made by the circular movement of the stars is a harmony. The reason they give for the apparently surprising fact that we do not hear this music is that the sound is with us right from our birth, and therefore cannot be distinguished from its contrary silence. Sound is only discerned by contrast with silence, and vice versa, so that in this respect men are like coppersmiths, who do not notice the noise of the forge because they are used to it.

13

Although most people say that the earth lies at the centre [. . .] the Italians called Pythagoreans assert the opposite; they say that fire is at the centre, and that the earth is one of the stars moving in a circle round the centre, and so making night and day. In addition they set up another earth, opposite to ours, which they call 'counter-earth', and in doing so they are not looking for explanations and reasons to account for observed facts, but forcibly trying to make what they see fit their own theories and opinions. But there are many others who seek confirmation in theory rather than in observation, and they too would say that the earth should not be given the central position. They think that what is most honourable should have the most honourable place, and that fire is more honourable than earth, and the limit, that is the circumference and the centre, more so than the intermediate position.

ALCMAEON

14

The preservation of health is the 'equal distribution of rights' among the powers of wet and dry, cold and hot, bitter and sweet, and the rest, and the cause of disease is 'supreme rule' among them; for the supreme rule of either one in any pair brings destruction. And illness arises sometimes directly from an excess of heat or cold, sometimes indirectly because of too much or too little food, and sometimes it is centred in the blood or marrow or brain. Occasionally it arises in these areas from external causes as well, such as certain kinds of moisture, environment, fatigue, distress and the like. But health is the mixture of the qualities in proportion.

PHILOLAUS

15

All the things that can be known have number, for otherwise nothing could be understood or known.

16

This is how nature and harmony are to be explained. The substance of things is eternal, and divine not human knowledge is needed to understand its nature. Moreover no existing thing known to us could be generated if there were not the bases of things, out of which the universe was composed, the limiting and the unlimited. But since these first principles are not alike or related, it would have been impossible for a world to have been made from them if there had not been the addition of harmony, whatever way it came into existence. Things alike and akin had no need of harmony, but those which were unlike and unrelated and unequally arranged had to be fastened together by harmony of a kind which would enable them to prevail in the universe.

17

Harmony is a unity of what is very mixed, and an agreement among the disagreeing.

18

The sphere has five physical elements: fire, water, earth and air in the sphere, and the fifth is the hull of the sphere.

ARCHYTAS

19

I think that mathematicians make some excellent distinctions, and it is not surprising that they come to correct conclusions about the character of individual things; for having been highly perceptive about the nature of the whole they were bound to see clearly what the parts were like. And they actually have handed on to us a clear explanation about the speed of the stars and their rising and setting, as well as about geometry, arithmetic, spherics and in particular music. These branches of mathematics appear to be sister sciences, for they are concerned with the two primary forms of existence, which are related.

ZENO

20

Zeno does away with motion when he says: 'A moving body does not move in the place in which it is, nor in that in which it is not.'

21

There are four arguments of Zeno about motion which give trouble to those who try to solve the problems that they present. First is the argument that there can be no motion because an object must reach the halfway point of its journey before it reaches the end.

22

Second is the 'Achilles', as it is called. It is this: the slowest runner will never be overtaken by the fastest because the pursuer must first reach the point where the pursued started off—consequently the slower is inevitably always some distance ahead.

23

Third is that the arrow in flight is stationary. This arises from assuming that time is a sequence of present instants.

24

Fourth is that concerning the solid objects moving past solid objects equal in size to themselves at equal speed on a race-course from the opposite direction, that is, some moving away from the end of the course and the others from the turning point of the course. The result here is that half an amount of time becomes equal to its double.

EMPEDOCLES

25

A twofold tale I shall tell: at one time it grew to be only one from many, at another it divided again to be many from one. There is a double generation of what is mortal, and a double passing away; for the uniting of all things brings one generation to birth and destroys it, and the other grows up and is scattered as they are again being divided. And these things never cease continually changing position, at one time all coming together into one through Love, and again at another being borne apart from each other by Strife's repulsion. In this way, insofar as one is accustomed to come from many and many spring again from one as it is being divided, to this extent they come to be and have no abiding life; but insofar as they never cease their continual change of position, so far they are always unaltered in the cycle. [. . .] Fire and water and earth and the immense height of air, with dread Strife apart from these, matched to them in every direction, and Love among them, their equal in length and breadth. [. . .] All these are equal and of like age, but each has a different prerogative and its particular character, and they prevail in turn as the time comes round.

26

As when painters, men well-taught by wisdom in the practice of their craft, decorate temple offerings: when they take pigments of various colours in their hands, having mixed them in harmony, more of some and less of others, they produce from them shapes resembling all things, creating trees, and men and women, animals and birds and the water-nourished fish, and the long-lived gods too, highest in honour; so let not error convince your mind that there is any other source for the countless perishables that are seen.

27

They prevail in turn as the cycle moves round, and decline into each other and grow great in appointed succession. For these are the only things in existence, and as they run through each other they become men and the kinds of other animals, at one time coming into one order through Love, at another again being borne apart from each other by Strife's repulsion, until they grow into one and become completely subdued.

Why the earth is at rest

28

Others like Empedocles say that the motion of the heavens holds back the motion of the earth by its circular rotation and its greater speed, just like the water in a cup. For when the cup is revolved in a circle the water, even when it often comes beneath the bronze, is prevented, for the same reason, from moving downwards despite its natural tendency to do so.

The doctrine of 'effluences'

29

Some think that each case of being affected is due to the last agent, the agent in the strictest sense, entering into certain pores; in this way, they say, we see and hear and perceive through all the other senses. Moreover we can see through air and water and other transparent media because they have pores which, although invisible on account of their minute size, are close set in rows, and the more so, the more transparent the body. Some, like Empedocles, put forward this explanation in certain contexts, but not exclusively in connection with action and suffering action; he says that mixing also takes place whenever the pores are symmetrically aligned.

30

Why the magnet attracts iron. Empedocles' theory is that they both give off effluences, and, since the pores of the magnet are symmetrical with the effluences from iron, iron is drawn towards the magnet. The effluences from the magnet push and move away the air around the pores of the iron which had covered them like a lid; once this gap has been made, the iron follows the continual flow of the effluence. The effluences from it move towards the pores of the magnet, and, since they are symmetrical with them and fit exactly, the iron also, along with the effluences, follows and moves towards the magnet.

31

It was Empedocles who said that light, being body, is an effluent substance emitted from the luminous body which first enters the region intermediate between the earth and the Heavens and then reaches us, but that this movement of light is such that we fail to notice it because of its speed.

ANAXAGORAS

The doctrine of like parts

32

He said that the first principles are homeomeries; for as gold is made up from gold-dust, so all the world is an aggregation of minute bodies, the parts of which are like the whole. The first principle of movement is Mind.

33

There is not a smallest part of what is small, but there is always a smaller (for it is impossible for what exists not to exist)—and also there is always something larger than what is large, equal in sum to the small. But each thing, with respect to itself, is both large and small.

34

Since this is the case we must suppose that there are many things of every sort, and seeds of all things with every kind of shape and colour and taste, in all that is coming together.

35

Because the portions of the large and of the small are equal in number, so too would all things be in everything. Separate existence is impossible, but all things have a share of everything. Since there cannot be a smallest part, nothing could be separated or come to exist of itself, but as they were in the beginning so now are all things all together. And in all things there are many ingredients, in equal number in the larger and the smaller of what is being separated off.

'Mind' controls the motions of matter

36

Other things have a portion of everything, but Mind is unlimited and self-controlled, and is mixed with nothing, but is by itself all alone. If it were not by itself, but mixed with some other thing, it would have a share of all things if mixed with any one, for in everything there is a portion of everything, as I have stated before. If things were mingled with Mind they would hinder it, so that it would not have control over anything in the same way as it has, being by itself, alone. It is the finest and purest of all things, and has complete knowledge of everything and the greatest power, and Mind has control over all the things that have life, the greater and the smaller. Mind controlled the whole rotation, so that there was rotation in the beginning. First it started to rotate from a small area, and now rotates over a wider area, and will rotate over one still wider. The things that are being mingled and separated and divided off—Mind knows them all, and such things as were going to be and were, but are not now, and such as are now and will be—Mind arranged them all, as well as this rotation in which now rotate the stars, the sun and the moon, and the air and aether which are being separated off; and this rotation caused the separating off. The dense is being separated off from the rare, the hot from the cold, the bright from the dark, and the dry from the moist. But there are many portions of many things, and nothing is completely separated off, nor is one thing divided from another except Mind.

37

When Mind initiated motion, separation began from everything which was being moved, and all that Mind moved was divided off; and as things were moved and divided the rotation greatly increased the process of division.

38

From these as they were separated off earth is made solid, for water is separated off from the clouds, and earth from water; from earth stones are solidified by the cold, and these, more so than water, tend to move outwards.

The nature of celestial bodies

39

A huge stone fell down, according to popular belief, from heaven at Aegos-potamoi. Even now it is displayed as an object of religious awe among the inhabitants of the Chersonese. Anaxagoras is reported publicly to have declared that when there is some landslide or quake in one of the celestial bodies, a piece knocked off will be hurled, and fall, for not one of the stars is in its original natural position. Being of heavy stony substance, they shine by resisting the pressure of the air upon them and, held in the whirl and tension of their circular movement, they are drawn along by force

40

The sun and moon and all the stars are fiery stones carried round with the rotation of the aether. [. . .] The sun is bigger than the Peloponnese. The light of the moon is not its own, but comes from the sun.

41

The moon is inhabited and has ridges and ravines.

Sense perception and reality

42

Because of the feebleness of our senses we cannot discern the truth.

43

Things seen give a glimpse of the unknown.

DIOGENES OF APOLLONIA

44

Such a distribution would not have been possible without Intelligence, that all things should have their measure: winter and summer and night and day and rains and winds and periods of fine weather; other things also, if one will study them closely, will be found to have the best possible arrangement.

LEUCIPPUS

The atomistic explanation of matter

45

Leucippus and his associate Democritus say that the full and the empty are the elements, calling the one being and the other non-being—the full and solid being being, the empty non-being (whence they say being no more is than non-being, because the solid no more is than the empty); and they make these the material causes of things. And as those who make the underlying substance one generate all other things by its modifications, supposing the rare and the dense to be the sources of the modifications, in the same way these philosophers say the differences in the elements are the causes of all other qualities. These differences, they say, are three—shape and order and position. For they say the real is differentiated only by 'rhythm' and 'inter-contact' and 'turning'; and of these rhythm is shape, inter-contact is order, and turning is position; for A differs from N in shape, AN from NA in order, ⊐ from H in position.

46

Leucippus says in his work *On Mind*: 'Nothing happens haphazardly, but everything from reason and by necessity.'

DEMOCRITUS

Atomism

47

Democritus thinks that the nature of the perpetual things consist of small particles infinite in number. For these he postulates another space of infinite size. He designates space by the terms 'the Void', 'Nothingness' and 'the Infinite', and each of the particles by the terms 'the Something', 'the Solid' and 'the

Being'. He thinks the particles are so small as to be imperceptible to us, and take all kinds of shapes and all kinds of forms and differences of size. Out of them, like out of elements, he now lets combine and originate the visible and perceptible bodies. They move in confusion in the void because of their dissimilarity and the other specified differences, and while in motion they collide and become interlocked in entanglement of a kind which causes them to be juxtaposed and in proximity to one another without actually forming any real unity whatsoever. The reason he gives is that it would be quite absurd that two or more entities should ever become one. The cause of the continuance of aggregations of particles for some period of time, he says, is their fitting into one another in bondage—for some of them are uneven, others barbed, some concave and some convex and others possessed of innumerable other distinctive qualities. He thinks they hold together and continue to do so until the time when some stronger force coming from the environment disrupts and disperses them in different directions.

48

In the theory of Democritus, sight is by image, and this he explains in his own way. The image, he says, does not appear directly in the pupil, but the air between the eye and the object of sight is contracted, and receives an impression from both the object of sight and the seer; for a kind of effluence is continually being given off from everything. [. . .]

Democritus distinguished heavy and light by size; for if each object were broken down into units, these, even if they differed in shape, would naturally have weight according to their size. Among compounds, however, those with more void are lighter, and those with less heavier. This was his explanation in some places, but in others he maintains simply that fine is light. He speaks of hard and soft in roughly the same way. For what is compact, he says, is hard, and what is loose is soft, and the intermediate stages proportionately so. But there is a difference between hard and soft, and heavy and light, in the position and arrangement of the empty spaces. That is why iron is harder than lead, but lead is the heavier. For iron is an uneven substance, with many large empty spaces; it is compact in some parts, but altogether it has more empty spaces than lead. Lead on the other hand has less empty space but is of a uniform texture, so that lead is heavier, but softer, than iron. [. . .]

With reference to tastes he says that 'sharp' has a shape that is 'angular' and 'much-twisted', as well as small and thin. Because of its piercing character it swiftly slips in everywhere, and being rough and 'angular' it draws in and holds the rest. In this way it heats the body by making empty spaces in it, for more heat is generated where there is more empty space. 'Sweet' is made up from round atoms which are quite large. It therefore relaxes the body completely as it passes through it gently and unhurriedly. [. . .]

He states that there are four primary colours. What is smooth is white, for everything which is not rough or hard to penetrate, and which does not cast a shadow, is bright. [. . .]

Black is composed of atoms of the opposite type, rough, crooked and uneven, for these cast shadows, and the pores are not straight or easy to penetrate. Their effluences too are sluggish and confused, for the intervention of air changes the character of the effluence, and makes some difference to the presentation. Red consists of the same atoms as heat, but larger; and when the aggregations are larger but with the same shapes, red predominates. Evidence for red being made of such atoms is the fact that we redden when we become hot, as do objects put in the fire until they have a fiery appearance. Things made of large atoms have more red, for example the flame and embers of green wood are redder than those of dry. Iron and the like redden when put in fire; for the objects with most fire, and the finest, are brightest, but the redder objects have coarser and less fire. The redder therefore are less hot, for the fire is hot.

The limitations of knowledge

49

This argument shows that in fact we know nothing about anything, but opinion, for each one is an influx.

50

Moreover it will be apparent that truly it is impossible to know what each thing is.

51

'There are two kinds of knowledge, one genuine and the other bastard. Sight, hearing, smell, taste and touch are all part of bastard knowledge, but genuine knowledge is separate from this.' Then, after rating genuine knowledge above the bastard, he continues: 'Whenever the bastard knowledge is no longer able to see something quite small, or hear or smell or taste it, or perceive it by touch, but a finer instrument is needed, then genuine knowledge comes into its own, since it has the finer instrument for knowing.'

52

After Democritus had attacked sensation by saying that colour exists by convention, sweet by convention, bitter by convention, atoms and void exist in reality, he lets the senses say the following words against the mind: 'Miserable mind, you get your evidence from us and do you try to overthrow us? The overthrow will be your downfall.'

CLASSICAL PERIOD
PLATO

The rational explanation of the universe

53

For the generation of this universe was a mixed result of the combination of Necessity and Reason. Reason overruled Necessity by persuading her to guide the greatest part of the things that become towards what is best; in that way and on that principle this universe was fashioned in the beginning by the victory of reasonable persuasion over Necessity.

54

SOCRATES: Now we may, I think, divide the knowledge involved in our studies into technical knowledge, and that concerned with education and culture, may we not?

PROTARCHUS: Yes.

SOCRATES: Then taking the technical knowledge employed in handicraft, let us first consider whether one division is more closely concerned with knowledge and the other less so, so that we are justified in regarding the first kind as the purest, and the second as relatively impure.

PROTARCHUS: Yes, we ought to so regard them.

SOCRATES: Should we then mark off the superior types of knowledge in the several crafts?

PROTARCHUS: How so? Which do you mean?

SOCRATES: If, for instance, from any craft you subtract the element of numbering, measuring, and weighing, the remainder will be almost negligible.

PROTARCHUS: Negligible indeed.

SOCRATES: For after doing so, what you would have left would be guesswork and the exercise of your senses on a basis of experience and rule of thumb, involving the use of that ability to make lucky shots which is commonly accorded the title of art or craft, when it has consolidated its position by dint of industrious practice.

The doctrine of geometrical atomism

55

In the first place, then, it is of course obvious to anyone that fire, earth, water, and air are bodies; and all body has depth. Depth, moreover, must be bounded by surface; and every surface that is rectilinear is composed of triangles. Now all triangles are derived from two, each having one right angle and the other angles acute. Of these triangles, one has on either side the half of a right angle, the division of which is determined by equal sides (the right-angled isosceles); the other has unequal parts of a right angle allotted to unequal sides (the right-angled scalene). This we assume as the first beginning of fire and of the other bodies, following the account which combines likelihood with necessity; the principles yet more remote than these are known to Heaven and to such men as Heaven favours. Now, the question to be determined is this: What are the most perfect bodies that can be constructed, four in number, unlike one another, but such that some can be generated out of one another by resolution? If we can hit upon the answer to this, we have the truth concerning the generation of earth and fire and of the bodies which stand as proportionals between them. For we shall concede to no one that there are visible bodies more perfect than these, each corresponding to a single type. We must do our best, then, to construct the four types of body that are most perfect and declare that we have grasped the constitution of these things sufficiently for our purpose.

Now, of the two triangles, the isosceles is of one type only; the scalene, of an endless number. Of this unlimited multitude we must choose the best, if we are to make a beginning on our own principles. Accordingly, if anyone can tell us of a better kind that he has chosen for the construction of these bodies, his will be the victory, not of an enemy, but of a friend. For ourselves, however, we postulate as the best of these many triangles one kind, passing over all the rest; that, namely, a pair of which compose the equilateral triangle. The reason is too long a story; but if anyone should put the matter to the test and discover that it is not so, the prize is his with all good will. So much, then, for the choice of the two triangles, of which the bodies of fire and of the rest have been wrought: the one isosceles (the half-square), the other having the greater side triple in square of the lesser (the half-equilateral). We must now be more precise upon a point that was not clearly enough stated earlier. It appeared as though all the four kinds could pass through one another into one another; but this appearance is delusive; for the triangles we selected give rise to four types, and whereas three are constructed out of the triangle with unequal sides, the fourth alone is constructed out of the isosceles. Hence it is not possible for all of them to pass into one another by resolution, many of the small forming a few of the greater and vice versa. But three of them can do this; for these are all composed of one triangle, and when the larger bodies are broken up, several small ones will be formed of the same triangles, taking on their proper figures; and again when several of the smaller bodies are dispersed into their triangles, the total number made up by them will produce a single new figure of larger size, belonging to a single body. So much for their passing into one another. The next thing to explain is, what sort of figure each body has, and the numbers that combine to compose it.

First will come the construction of the simplest and smallest figure (the pyramid). Its element is the triangle whose hypotenuse is double of the shorter side in length. If a pair of such triangles are put together by the diagonal, and this is done three times, the diagonals and the shorter sides resting on the same point as a centre, in this way a single equilateral triangle is formed of triangles six in number. If four equilateral triangles are put together, their plane angles meeting in groups of three make a single solid angle, namely the one (180°) that comes next after the most obtuse of plane angles. When four such angles are produced, the simplest solid figure is formed, whose property is to divide the whole circumference into equal and similar parts.

A second body (the octahedron) is composed of the same (elementary) triangles when they are combined in a set of eight equilateral triangles, and yield a solid angle formed by four plane angles. With the production of six such solid angles the second body is complete.

The third body (the icosahedron) is composed of one hundred and twenty of the elementary triangles fitted together, and of twelve solid angles, each contained by five equilateral triangular planes; and it has twenty faces which are equilateral triangles. Here one of the two elements, having generated these bodies, has done its part. But the isosceles triangle went on to generate the fourth body, being put together in sets of four, with their right angles meeting at the centre, thus forming a single equilateral quadrangle.

Six such quadrangles, joined together, produced eight solid angles, each composed by a set of three plane right angles. The shape of the resulting body was cubical, having six quadrangular equilateral planes as its faces. [. . .] Let us next distribute the figures whose formation we have now described, among fire, earth, water and air.

To earth let us assign the cubical figure; for of the four kinds earth is the most immobile and the most plastic of bodies. The figure whose bases are the most stable must best answer that description; and as a base, if we take the triangles we assumed at the outset, the face of the triangle with equal sides is by nature more stable than that of the triangle whose sides are unequal; and further, of the two equilateral surfaces respectively composed of the two triangles, the square is necessarily a more stable base than the triangle, both in its parts and as a whole. Accordingly we shall preserve the probability of our account, if we assign this figure to earth; and of the remainder the least mobile to water, the most mobile to fire, and the intermediate figure to air. Again, we shall assign the smallest body to fire, the largest to water, and the intermediate to air; and again the body with the sharpest angles to fire, the next to air, the third to water.

Now, taking all these figures, the one with the fewest faces (pyramid) must be the most mobile, since it has the sharpest cutting edges and the sharpest points in every direction, and moreover the lightest, as being composed of the smallest number of similar parts; the second (octahedron) must stand second in these respects, the third (icosahedron), third. Hence, in accordance with genuine reasoning as well as probability, among the solid figures we have constructed, we may take the pyramid as the element or seed of fire; the second in order of generation (octahedron) as that of air; the third (icosahedron) as that of water.

Now we must think of all these bodies as so small that a single body of any one of these kinds is invisible to us because of its smallness; though when a number are aggregated the masses of them can be seen. And with regard to their numbers, their motions, and their powers in general, we must suppose that the god adjusted them in due proportion, when he had brought them in every detail to the most exact perfection permitted by Necessity willingly complying with persuasion. Now, from all that we have said in the foregoing account concerning the kinds, the following would be the most probable description of the facts.

Earth, when it meets with fire and is dissolved by its sharpness, would drift about—whether, when dissolved, it be enveloped in fire itself or in a mass of air or of water—until its own parts somewhere encounter one another, are fitted together and again become earth; for they can never pass into any other kind.

But (1) when water is divided into parts by fire, or again by air, it is possible for one particle of fire and two of air to arise by combination; and (2) the fragments of air, from a single particle that is dissolved, can become two particles of fire. And conversely, (3) when a little fire, enveloped in a large quantity of air or water or (it may be) earth, is kept in motion within these masses which are moving in place, and makes a fight, and then is overcome and shattered into fragments, two particles of fire combine to make a single figure of air. And (4) when air is overpowered and broken small, from two and a half complete figures, a single complete figure of water will be compacted. Let us reconsider this account once more as follows. (a) When one of the other kinds is enveloped in fire and cut up by the sharpness of its angles and edges, then (α), if it is recombined into the shape of fire, there is an end to the cutting up; for no kind which is homogeneous and identical can effect any change in, or suffer any change from, that which is in the same condition as itself. But (β) so long as, passing into some other kind, a weaker body is contending with a stronger, the resolution does not come to an end. And, on the other hand, (b) when a few smaller particles are enveloped in a large number of bigger ones and are being shattered and quenched, then, (α) if they consent to combine into the figure of the prevailing kind, the quenching process comes to an end: from fire comes air, from air, water. But (β) if they (the smaller particles) are on their way to these (air or water), and one of the other kinds meets them and comes into conflict, the process of their resolution does not stop until either they are wholly dissolved by the thrusting and escape to their kindred, or they are overcome and a number of them form a single body uniform with the victorious body and take up their abode with it. [...] In this way, then, the formation of all the uncompounded and primary bodies is accounted for. The reason why there are several varieties within their kinds lies in the construction of each of the two elements: the construction in each case originally produced its triangle not of one size only, but some smaller, some larger, the number of these differences being the same as that of the varieties in the kinds. Hence, when they are mixed with themselves or with one another, there is an endless diversity, which must be studied by one who is to put forward a probable account of Nature.

Reason versus observation in science

56

And now, Socrates, I will praise astronomy on your own principles, instead of commending its usefulness in the vulgar spirit for which you upbraided me. Anyone can see that this subject forces the mind to look upwards, away from this world of ours to higher things.

Anyone except me, perhaps, I replied. I do not agree.

Why not?

As it is now handled by those who are trying to lead us up to philosophy, I think it simply turns the mind's eye downwards.

What do you mean?

You put a too generous construction on the study of 'higher things'. Apparently you would think a man who threw his head back to contemplate the decorations on a ceiling was using his reason, not his eyes, to gain knowledge. Perhaps you are right and my notion is foolish; but I cannot think of any study as making the mind look upwards, except one which has to do with unseen reality. No one, I should say, can ever gain knowledge of any sensible object by gaping upwards any more than by shutting his eyes and searching for it on the ground, because there can be no knowledge of sensible things. His mind will be looking downwards, though he may pursue his studies lying on his back or floating on the sea.

I deserve to be rebuked, he answered. But how did you mean the study of astronomy to be reformed, so as to serve our purposes?

In this way. These intricate traceries in the sky are, no doubt, the loveliest and most perfect of material things, but still part of the visible world, and therefore they fall far short of the true realities—the real relative velocities, in the world of pure number and all perfect geometrical figures, of the movements which carry round the bodies involved in them. These, you will agree, can be conceived by reason and thought, not seen by the eye.

Exactly.

Accordingly, we must use the embroidered heaven as a model to illustrate our study of these realities, just as one might use diagrams exquisitely drawn by some consummate artist like Daedalus. An expert in geometry, meeting with such designs, would admire their finished workmanship, but he would think it absurd to study them in all earnest with the expectation of finding in their proportions the exact ratio of any one number to another.

Of course it would be absurd.

The genuine astronomer, then, will look at the motions of the stars with the same feelings. He will admit that the sky with all that it contains has been framed by its artificer with the highest perfection of which such works are capable. But when it comes to the proportions of day to night, of day and night to month, of month to year, and of the periods of other stars to Sun and Moon and to one another, he will think it absurd to believe that these visible material things go on for ever without change or the slightest deviation, and to spend all his pains on trying to find exact truth in them.

Now you say so, I agree.

If we mean, then, to turn the soul's native intelligence to its proper use by a genuine study of astronomy, we shall proceed, as we do in geometry, by means of problems, and leave the starry heavens alone.

That will make the astronomer's labour many times greater than it is now. [. . .]

Only we must constantly hold by our own principle, not to let our pupils take up any study in an imperfect form, stopping short of that higher region to which all studies should attain, as we said just now in speaking of astronomy. As you will know, the students of harmony make the same sort of mistake as the astronomers: they waste their time in measuring audible concords and sounds one against another.

Yes, said Glaucon, they are absurd enough, with their talk of 'groups of quarter-tones' and all the rest of it. They lay their ears to the instrument as if they were trying to overhear the conversation from next door. One says he can still detect a note in between, giving the smallest possible interval, which ought to be taken as the unit of measurement, while another insists that there is now no difference between the two notes. Both prefer their ears to their intelligence.

You are thinking of those worthy musicians who tease and torture the strings racking them on the pegs. I will not push the metaphor so far as to picture the musician beating them with the plectrum and charging them with faults which the strings deny or brazen out. I will drop the comparison and tell you that I am thinking rather of those Pythagoreans whom we were going to consult about harmony. They are just like the astronomers—intent upon the numerical properties embodied in these audible consonances: they do not rise to the level of formulating problems and inquiring which numbers are inherently consonant and which are not, and for what reasons.

CLASSICAL PERIOD
ARISTOTLE

Principles for the cognition of nature

57

When the objects of an inquiry, in any department, have principles, conditions, or elements, it is through acquaintance with these that knowledge, that is to say scientific knowledge, is attained. For we do not think that we know a thing until we are acquainted with its primary conditions or first principles, and have carried our analysis as far as its simplest elements. Plainly therefore in the science of Nature, as in other branches of study, our first task will be to try to determine what relates to its principles.

The natural way of doing this is to start from the things which are more knowable and obvious to us and proceed towards those which are clearer and more knowable by nature; for the same things are not 'knowable relatively to us' and 'knowable' without qualification. So in the present inquiry we must follow this method and advance from what is more obscure by nature, but clearer to us, towards what is more clear and more knowable by nature.

Now what is to us plain and obvious at first is rather confused masses, the elements and principles of which become known to us later by analysis. Thus we must advance from generalities to particulars; for it is a whole that is best known to sense-perception, and a generality is a kind of whole, comprehending many things within it, like parts.

58

It seems that perceptible things require perceptible principles, eternal things eternal principles, corruptible things corruptible principles; and, in general, every subject matter principles homogeneous with itself.

59

Of things that exist, some exist by nature, some from other causes.

'By nature' the animals and their parts exist, and the plants and the simple bodies (earth, fire, air, water)—for we say that these and the like exist 'by nature'.

All the things mentioned present a feature in which they differ from things which are *not* constituted by nature. Each of them has *within itself* a principle of motion and of stationariness (in respect of place, or of growth and decrease, or by way of alteration). On the other hand, a bed and a coat and anything else of that sort, *qua* receiving these designations—i.e. insofar as they are products of art—have no innate impulse to change. But insofar as they happen to be composed of stone or of earth or of a mixture of the two, they *do* have such an impulse, and just to that extent—which seems to indicate that *nature is a source or cause of being moved and of being at rest in that to which it belongs primarily*, in virtue of itself and not in virtue of a concomitant attribute.

60

Now that we have established these distinctions, we must proceed to consider causes, their character and number. Knowledge is the object of our inquiry, and men do not think they know a thing till they have grasped the 'why' of it (which is to grasp its primary cause). So clearly we too must do this as regards both coming to be and passing away and every kind of physical change, in order that, knowing their principles, we may try to refer to these principles each of our problems.

In one sense, then, (1) that out of which a thing comes to be and which persists, is called 'cause', e.g. the bronze of the statue, the silver of the bowl, and the genera of which the bronze and the silver are species.

In another sense (2) the form or the archetype, i.e. the statement of the essence, and its genera, are called 'causes' (e.g. of the octave the relation of $2 : 1$, and generally number), and the parts in the definition.

Again (3) the primary source of the change or coming to rest; e.g. the man who gave advice is a cause, the father is cause of the child, and generally what makes of what is made and what causes change of what is changed.

Again (4) in the sense of end or 'that for the sake of which' is a thing done, e.g. health is the cause of walking about. ('Why is he walking about?' we say. 'To be healthy', and, having said that, we think we have assigned the cause.) The same is true also of all the intermediate steps which are brought about through the action of something else as means towards the end, e.g. reduction of flesh, purging, drugs, or surgical instruments are means towards health. All these things are 'for the sake of' the end, though they differ from one another in that some are activities, others instruments.

This then perhaps exhausts the number of ways in which the term 'cause' is used.

61

Now that which is must needs be when it is, and that which is not must needs not be when it is not. Yet it cannot be said without qualification that all existence and non-existence is the outcome of necessity. For there is a difference between saying that that which is, when it is, must needs be, and simply saying that all that is must needs be, and similarly in the case of that which is not. In the case, also, of two contradictory propositions this holds good. Everything must either be or not be, whether in the present or future, but it is not always possible to distinguish and state determinately which of these alternatives must necessarily come about.

Let me illustrate. A sea-fight must either take place tomorrow or not, but it is not necessary that it should take place tomorrow, neither is it necessary that it should not take place, yet it is necessary that it either should or should not take place tomorrow. Since propositions correspond with facts, it is evident that when in future events there is a real alternative, and a potentiality in contrary directions, the corresponding affirmation and denial have the same character.

This is the case with regard to that which is not always existent or not always non-existent. One of the two propositions in such instances must be true and the other false, but we cannot say determinately that this or that is false, but must leave the alternative undecided.

All things are designed to fulfil a purpose

62

Moreover, it is impossible that any abstraction can form a subject of natural science, seeing that everything that Nature makes is means to an end. For just as human creations are the products of art, so living objects are manifestly the products of an analogous cause or principle, not external but internal, derived like the hot and cold from the environing universe. And that the heaven, if it had an origin, was evolved and is maintained by such a cause, there is therefore even more reason to believe, than that mortal animals so originated. For order and definiteness are much more plainly manifest in the celestial bodies than in our own frame; while change and chance are characteristic of the perishable things of earth. Yet there are some who, while they allow that every animal exists and was generated by nature, nevertheless hold that the heaven was constructed to be what it is by chance and spontaneity; the heaven, in which not the faintest sign of haphazard or of disorder is discernible! Again, whenever there is plainly some final end, to which a motion tends should nothing stand in the way, we always say that such final end is the aim or purpose of the motion; and from this it is evident that there must be a something or other really existing, corresponding to what we call by the name of Nature.

63

A difficulty presents itself: why should not nature work, not for the sake of something, nor because it is better so, but just as the sky rains, not in order to make the corn grow, but of necessity? What is drawn up must cool, and what has been cooled must become water and descend, the result of this being that the corn grows. Similarly if a man's crop is spoiled on the threshing-floor, the rain did not fall for the sake of this—in order that the crop might be spoiled—but that result just followed. Why then should it not be the same with the parts in nature, e.g. that our teeth should come up of *necessity*—the front teeth sharp, fitted for tearing, the molars broad and useful for grinding down the food—since they did not arise for this end, but it was merely a coincident result; and so with all other parts in which we suppose that there is a purpose? Wherever then all the parts came about just what they would have been if they had come to be for an end, such things survived, being organized spontaneously in a fitting way; whereas those which grew otherwise perished and continue to perish, as Empedocles says his 'man-faced ox-progeny' did.

Such are the arguments (and others of the kind) which may cause difficulty on this point. Yet it is impossible that this should be the true view. For teeth and all other natural things either invariably or normally come about in a given way; but of not one of the results of chance or spontaneity is this true. We do not ascribe to chance or mere coincidence the frequency of rain in winter, but frequent rain in summer we do; nor heat in the dog-days, but only if we have it in winter. If then, it is agreed that things are either the result of coincidence or for an end, and these cannot be the result of coincidence or spontaneity, it follows that they must be for an end; and that such things are all due to nature even the champions of the theory which is before us would agree. Therefore action for an end is present in things which come to be and are by nature.

Further, where a series has a completion, all the preceding steps are for the sake of that. Now surely as in intelligent action, so in nature; and as in nature, so it is in each action, if nothing interferes. Now intelligent action is for the sake of an end; therefore the nature of things also is so. Thus if a house, e.g. had been a thing made by nature, it would have been made in the same way as it is now by art; and if things made by nature were made also by art, they would come to be in the same way as by nature. Each step then in the series is for the sake of the next; and generally art partly completes what nature cannot bring to a finish, and partly imitates her. If, therefore, artificial products are for the sake of an end, so clearly also are natural products. The relation of the later to the earlier terms of the series is the same in both.

The infinite

64

Belief in the existence of the infinite comes mainly from five considerations:

1. From the nature of time—for it is infinite.

2. From the division of magnitudes—for the mathematicians also use the notion of the infinite.

3. If coming to be and passing away do not give out, it is only because that from which things come to be is infinite.

4. Because the limited always finds its limit in something, so that there must be *no* limit, if everything is *always* limited by something different from itself.

5. Most of all, a reason which is peculiarly appropriate and presents the difficulty that is felt by everybody—not only number but also mathematical magnitudes and what is outside the heaven are supposed to be infinite because they never give out in our *thought*.

The last fact (that what is outside is infinite) leads people to suppose that body also is infinite, and that there is an infinite number of worlds. Why should there be body in one part of the void rather than in another? Grant only that mass is anywhere and it follows that it must be everywhere. Also, if void and place are infinite, there must be infinite body too, for in the case of eternal things what may be must be.

But the problem of the infinite is difficult: many contradictions result whether we suppose it to exist or not to exist. If it exists, we have still to ask *how* it exists; as a substance or as the essential attribute of some entity? Or in neither way, yet nonetheless is there something which is infinite or some things which are infinitely many?

65

In general, the view that there is an infinite body is plainly incompatible with the doctrine that there is necessarily a proper place for each kind of body, if every sensible body has either weight or lightness, and if a body has a natural locomotion towards the centre if it is heavy and upwards if it is light. This would need to be true of the infinite also. But neither character can belong to it: it cannot be either as a whole, nor can it be half the one and half the other. For how should you divide it? or how can the infinite have the one part up and the other down, or an extremity and a centre?

Further, every sensible body is in place, and the kinds or differences of place are up-down, before-behind, right-left; and these distinctions hold not only in relation to us and by arbitrary agreement, but also in the whole itself. But in the infinite body they cannot exist. In general, if it is impossible that there should be an infinite place, and if every body is in place, there cannot be an infinite body.

Surely what is in a special place is in place, and what is in place is in a special place. Just, then, as the infinite cannot be quantity—that would imply that it has a particular quantity, e.g. two or three cubits; quantity just means these—so a thing's being in place means that it is *some*where, and that is either up or down or in some other of the six differences of position; but each of these is a limit.

It is plain from these arguments that there is no body which is *actually* infinite.

Place, natural place and natural motion

66

The question, what is place? presents many difficulties. An examination of all the relevant facts seems to lead to divergent conclusions. Moreover, we have inherited nothing from previous thinkers, whether in the way of a statement of difficulties or of a solution.

The existence of place is held to be obvious from the fact of mutual replacement. Where water now is, there in turn, when the water has gone out as from a vessel, air is present. When therefore another body occupies this same place, the place is thought to be different from all the bodies which come to be in it and replace one another. What now contains air formerly contained water, so that clearly the place or space into which and out of which they passed was something different from both.

Further, the typical locomotions of the elementary natural bodies—namely, fire, earth and the like—show not only that place is something, but also that it exerts a certain influence. Each is carried to its own place, if it is not hindered, the one up, the other down. Now these are regions or kinds of place—up and down and the rest of the six directions. Nor do such distinctions (up and down and right and left, etc.) hold only in relation to us. To *us* they are not always the same but change with the direction in which we are turned: that is why the same thing may be both right *and* left, up *and* down, before *and* behind. But in *nature* each is distinct, taken apart by itself. It is not every chance direction which is 'up', but where fire and what is light are carried; similarly too 'down' is not any chance direction but where what has weight and what is made of earth are carried—the implication being that these places do not differ merely in relative position, but also as possessing distinct potencies. This is made plain also by the objects studied by mathematics. Though they have no real place, they nevertheless, in respect of their position relatively to us, have a right and left as attributes ascribed to them only in consequence of their relative position, not having by nature these various characteristics.

67

What then after all is place? The answer to this question may be elucidated as follows.

Let us take for granted about it the various characteristics which are supposed correctly to belong to it essentially. We assume then—

1. Place is what contains that of which it is the place.
2. Place is no part of the thing.
3. The immediate place of a thing is neither less nor greater than the thing.
4. Place can be left behind by the thing and is separable.

In addition:

5. All place admits of the distinction of up and down, and each of the bodies is naturally carried to its appropriate place and rests there, and this makes the place either up or down.

68

There are just four things of which place must be one—the shape, or the matter, or some sort of extension between the bounding surfaces of the containing body, or this boundary itself if it contains no extension over and above the bulk of the body which comes to be in it.

Three of these it obviously cannot be:

1. The shape is supposed to be place because it surrounds, for the extremities of what contains and of what is contained are coincident. Both the shape and the place, it is true, are boundaries. But not of the same thing: the form is the boundary of the thing, the place is the boundary of the body which contains it.

2. The extension between the extremities is thought to be something, because what is contained and separate may often be changed while the container remains the same (as water may be poured from a vessel)—the assumption being that the extension is something over and above the body displaced. But there is no such extension. One of the bodies which change places and are naturally capable of being in contact with the container falls in—whichever it may chance to be.

If there were an extension which were such as to exist independently and—permanent, there would be an infinity of places in the same thing. [. . .]

3. The matter, too, might seem to be place, at least if we consider it in what is at rest and is thus separate but in continuity. For just as in change of quality there is something which was formerly black and is now white, or formerly soft and now hard—this is just why we say that the matter exists—so place, because it presents a similar phenomenon, is thought to exist—only in the one case we say so because *what* was air is now water, in the other because *where* air formerly was there is now water. But the matter, as we said before, is neither separable from the thing nor contains it, whereas place has both characteristics.

Well, then, if place is none of the three—neither the form nor the matter nor an extension which is always there, different from, and over and above, the extension of the thing which is displaced—place necessarily is the one of the four which is left, namely, the boundary of the containing body at which it is in contact with the contained body. (By the contained body is meant what can be moved by way of locomotion.)

Place is thought to be something important and hard to grasp, both because the matter and the shape present themselves along with it, and because the displacement of the body that is moved takes place in a stationary container, for it seems possible that there should be an interval which is other than the bodies which are moved. The air, too, which is thought to be incorporeal, contributes something to the belief: it is not only the boundaries of the vessel which seem to be place, but also what is between them, regarded as empty. Just, in fact, as the vessel is transportable place, so place is a non-portable vessel. So when what is within a thing which is moved, is moved and changes its place, as a boat on a river, what contains plays the part of a vessel rather than that of place. Place on the other hand is rather what is motionless: so it is rather the whole river that is place, because as a whole it is motionless.

Hence we conclude that *the innermost motionless boundary of what contains is place.*

Motion and the void

69

The investigation of similar questions about the void, also, must be held to belong to the physicist—namely, whether it exists or not, and how it exists or what it is—just as about place. The views taken of it involve arguments both for and against, in much the same sort of way. For those who hold that the void exists regard it as a sort of place or vessel which is supposed to be 'full' when it holds the bulk which it is capable of containing, 'void' when it is deprived of that—as if 'void' and 'full' and 'place' denoted the same thing, though the essence of the three is different. [. . .]

If people say that the void must exist, as being necessary if there is to be movement, what rather turns out to be the case, if one studies the matter, is the opposite, that not a single thing can be moved if there *is* a void; for as with those who for a like reason say the earth is at rest, so, too, in the void things must be at rest; for there is no place to which things can move more or less than to another; since the void insofar as it is void admits no difference.

The second reason is this: all movement is either compulsory or according to nature, and if there is compulsory movement there must also be natural (for compulsory movement is contrary to nature, and movement contrary to nature is posterior to that according to nature, so that if each of the natural bodies has not a natural movement, none of the other movements can exist); but how can there be natural movement if there is no difference throughout the void or the infinite? For insofar as it is infinite, there will be no up or down or middle, and insofar as it is a void, up differs no whit from down; for as there is no difference in what is nothing, there is none in the void (for the void seems to be non-existent and a privation of being), but natural locomotion seems to be differentiated, so that the things that exist by nature must be differentiated. Either, then, nothing has a natural locomotion, or else there is no void.

Further, in point of fact things that are thrown move though that which gave them their impulse is not touching them, either by reason of mutual replacement, as some maintain,[1] or because the air that has been pushed pushes them with a movement quicker than the natural locomotion of the projectile wherewith it moves to its proper place. But in a void none of these things can take place, nor can anything be moved save as that which is carried is moved.

Further, no one could say why a thing once set in motion should stop anywhere; for why should it stop *here* rather than *here*? So that a thing will either be at rest or must be moved *ad infinitum*, unless something more powerful gets in its way.

Further, things are now thought to move into the void because it yields; but in a void this quality is present equally everywhere, so that things should move in all directions.

Further, the truth of what we assert is plain from the following considerations. We see the same weight or body moving faster than another for two reasons, either because there is a difference in what it moves through, as between water, air and earth, or because, other things being equal, the moving body differs from the other owing to excess of weight or of lightness.

Now the medium causes a difference because it impedes the moving thing, most of all if it is moving in the opposite direction, but in a secondary degree even if it is at rest; and especially a medium that is not exactly divided, i.e. a medium that is somewhat dense.

A, then, will move through B in time C, and through D, which is thinner, in time E (if the length of B is equal to D), in proportion to the density of the hindering body. For let B be water and D air; then by so much as air is thinner

---

1 The hypothesis of some of the ancients according to which the air displaced in front of a moving projectile fills the place left behind it and thus acts as a pushing force. (Ed.)

and more incorporeal than water, *A* will move through *D* faster than through *B*. Let the speed have the same ratio to the speed, then, that air has to water. Then if air is twice as thin, the body will traverse *B* in twice the time that it does *D*, and the time *C* will be twice the time *E*. And always, by so much as the medium is more incorporeal and less resistant and more easily divided, the faster will be the movement.

Now there is no ratio in which the void is exceeded by body, as there is no ratio of zero to a number. For if 4 exceeds 3 by 1, and 2 by more than 1, and 1 by still more than it exceeds 2, still there is no ratio by which it exceeds zero; for that which exceeds must be divisible into the excess + that which is exceeded, so that 4 will be what it exceeds zero by zero. For this reason, too, a line does not exceed a point—unless it is composed of points. Similarly the void can bear no ratio to the full, and therefore neither can movement through the one to movement through the other, but if a thing moves through the thinnest medium such and such a distance in such and such a time, it moves through the void with a speed beyond any ratio. [. . .][1]

These are the consequences that result from a difference in the media; the following depend upon an excess of one moving body over another. We see that bodies which have a greater impulse either of weight or of lightness, if they are alike in other respects, move faster over an equal space, and in the ratio which their magnitudes bear to each other. Therefore they will also move through the void with this ratio of speed. But that is impossible; for why should one move faster? (In moving through *plena* it must be so; for the greater divides them faster by its force. For a moving thing cleaves the medium either by its shape, or by the impulse which the body that is carried along or is projected possesses.) Therefore all will possess equal velocity. But this is impossible.

Time

70

Next for discussion after the subjects mentioned is Time.

The best plan will be to begin by working out the difficulties connected with it, making use of the current arguments. First, does it belong to the class of things that exist or to that of things that do not exist? Then secondly, what is its nature? To start then: the following considerations would make one suspect that it either does not exist at all or barely, and in an obscure way. One part of it has been and is not, while the other is going to be and is not yet. Yet time— both infinite time and any time you like to take—is made up of these. One would naturally suppose that what is made up of things which do not exist could have no share in reality.

Further, if a divisible thing is to exist, it is necessary that, when it exists, all or

---

1. Since zero is not considered a number, and the void is a medium of zero density, there cannot be any ratio between the speed in a medium and in the void. (Ed.)

some of its parts must exist. But of time some parts have been while others have to be, and no part of it *is*, though it is divisible. For what is 'now' is not a part: a part is a measure of the whole, which must be made up of parts. Time, on the other hand, is not held to be made up of 'nows'.

71

But as time is most usually supposed to be motion and a kind of change, we must consider this view.

Now (a) the change or movement of each thing is only *in* the thing which changes or *where* the thing itself which moves or changes may chance to be. But time is present equally everywhere and with all things.

Again, (b) change is always faster or slower, whereas time is not: for 'fast' and 'slow' are defined by time—'fast' is what moves much in a short time, 'slow' what moves little in a long time; but time is not defined by time, by being either a certain amount or a certain kind of it.

Clearly then it is not movement. (We need not distinguish at present between 'movement' and 'change'.) [. . .]

It is evident, then, that time is neither movement nor independent of movement.

We must take this as our starting-point and try to discover—since we wish to know what time is—what exactly it has to do with movement.

Now we perceive movement and time together; for even when it is dark and we are not being affected through the body, if any movement takes place in the mind we at once suppose that some time also has elapsed; and not only that but also, when some time is thought to have passed, some movement also along with it seems to have taken place. Hence time is either movement or something that belongs to movement. Since then it is not movement, it must be the other.

But what is moved is moved from something to something, and all magnitude is continuous. Therefore the movement goes with the magnitude. Because the magnitude is continuous, the movement too must be continuous, and if the movement, then the time; for the time that has passed is always thought to be in proportion to the movement.

The distinction of 'before' and 'after' holds primarily, then, in place; and there in virtue of relative position. Since then 'before' and 'after' hold in magnitude, they must hold also in movement, these corresponding to those. But also in time the distinction of 'before' and 'after' must hold, for time and movement always correspond with each other. The 'before' and 'after' in motion identical in substratum with motion yet differs from it in definition, and is not identical with motion.

But we apprehend time only when we have marked motion, marking it by 'before' and 'after'; and it is only when we have perceived 'before' and 'after' in motion that we say that time has elapsed. Now we mark them by judging that *A* and *B* are different, and that some third thing is intermediate to them. When we think of the extremes as different from the middle and the mind pronounces that the 'nows' are two, one before and one after, it is then that we say that there is time, and this that we say is time. For what is bounded by the 'now' is thought to be time—we may assume this.

When, therefore, we perceive the 'now' as one, and neither as before and after in a motion nor as an identity but in relation to a 'before' and an 'after', no time is thought to have elapsed, because there has been no motion either. On the other hand, when we do perceive a 'before' and an 'after', then we say that there is time. For time is just this—number of motion in respect of 'before' and 'after'.

Hence time is not movement, but only movement insofar as it admits of enumeration. A proof of this: we discriminate the more or the less by number, but more or less movement by time. Time then is a kind of number.

72

Since what is 'in time' is so in the same sense as what is in number is so, a time greater than everything in time can be found. So it is necessary that all the things in time should be contained by time, just like other things also which are 'in anything', e.g. the things 'in place' by place.

A thing, then, will be affected by time, just as we are accustomed to say that time wastes things away, and that all things grow old through time, and that there is oblivion owing to the lapse of time, but we do not say the same of getting to know or of becoming young or fair. For time is by its nature the cause rather of decay, since it is the number of change, and change removes what is. [. . .]
A sufficient evidence of this is that nothing comes into being without itself moving somehow and acting, but a thing can be destroyed even if it does not move at all. And this is what, as a rule, we chiefly mean by a thing's being destroyed by time. Still, time does not work even this change; even this sort of change takes place *incidentally* in time.

Forced motion and the law of force

73

Since everything that is in motion must be moved by something, let us take the case in which a thing is in locomotion and is moved by something that is itself in motion, and that again is moved by something else that is in motion, and that by something else, and so on continually: then the series cannot go on to infinity, but there must be some first movent. [. . .]
That which is the first movent of a thing—in the sense that it supplies not 'that for the sake of which' but the source of the motion—is always together

with that which is moved by it (by 'together' I mean that there is nothing intermediate between them). This is universally true wherever one thing is moved by another. And since there are three kinds of motion, local, qualitative, and quantitative, there must also be three kinds of movent, that which causes locomotion, that which causes alteration, and that which causes increase or decrease.

Let us begin with locomotion, for this is the primary motion. Everything that is in locomotion is moved either by itself or by something else. In the case of things that are moved by themselves it is evident that the moved and the movent are together: for they contain within themselves their first movent, so that there is nothing in between. The motion of things that are moved by something else must proceed in one of four ways: for there are four kinds of locomotion caused by something other than that which is in motion, viz. pulling, pushing, carrying, and twirling. All forms of locomotion are reducible to these.

74

Now since wherever there is a movent, its motion always acts upon something, is always in something, and always extends to something (by 'is always in something' I mean that it occupies a time: and by 'extends to something' I mean that it involves the traversing of a certain amount of distance: for at any moment when a thing is causing motion, it also has caused motion, so that there must always be a certain amount of distance that has been traversed and a certain amount of time that has been occupied). If, then, A the movent has moved B a distance C in a time D, then in the same time the same force A will move $\frac{1}{2}B$ twice the distance C, and in $\frac{1}{2}D$ it will move $\frac{1}{2}B$ the whole distance C: for thus the rules of proportion will be observed. Again if a given force moves a given weight a certain distance in a certain time and half the distance in half the time, half the motive power will move half the weight the same distance in the same time. Let E represent half the motive power A and F half the weight B: then the ratio between the motive power and the weight in the one case is similar and proportionate to the ratio in the other, so that each force will cause the same distance to be traversed in the same time.

But if E moves F a distance C in a time D, it does not necessarily follow that E can move twice F half the distance C in the same time. If, then, A moves B a distance C in a time D, it does not follow that E, being half of A, will in the time D or in any fraction of it cause B to traverse a part of C the ratio between which and the whole of C is proportionate to that between A and E (whatever fraction of $A E$ may be): in fact it might well be that it will cause no motion at all; for it does not follow that, if a given motive power causes a certain amount of motion, half that power will cause motion either of any particular amount or in any length of time: otherwise one man might move a ship, since both the motive power of the ship-haulers and the distance that they all cause the ship to traverse are divisible into as many parts as there are men.

75

If everything that is in motion with the exception of things that move themselves is moved by something else, how is it that some things, e.g. things thrown, continue to be in motion when their movent is no longer in contact with them? If we say that the movent in such cases moves something else at the same time, that the thrower, e.g. also moves the air, and that this in being moved is also a movent, then it would be no more possible for this second thing than for the original thing to be in motion when the original movent is not in contact with it or moving it: all the things moved would have to be in motion simultaneously and also to have ceased simultaneously to be in motion when the original movent ceases to move them, even if, like the magnet, it makes that which it has moved capable of being a movent. Therefore, while we must accept this explanation to the extent of saying that the original movent gives the power of being a movent either to air or to water or to something else of the kind, naturally adapted for imparting and undergoing motion, we must say further that this thing does not cease simultaneously to impart motion and to undergo motion: it ceases to be in motion at the moment when its movent ceases to move it, but it still remains a movent, and so it causes something else consecutive with it to be in motion, and of this again the same may be said. The motion begins to cease when the motive force produced in one member of the consecutive series is at each stage less than that possessed by the preceding member, and finally ceases when one member no longer causes the next member to be a movent but only causes it to be in motion. The motion of these last two—of the one as movent and of the other as moved—must cease simultaneously, and with this the whole motion ceases. Now the things in which this motion is produced are things that admit of being sometimes in motion and sometimes at rest, and the motion is not continuous but only appears so: for it is motion of things that are either successive or in contact, there being not one movent but a number of movents consecutive with one another: and so motion of this kind takes place in air and water. Some say that it is 'mutual replacement': but we must recognize that the difficulty raised cannot be solved otherwise than in the way we have described.

The prime mover

76

Now that these points are settled, it is clear that the first unmoved movent cannot have any magnitude. For if it has magnitude, this must be either a finite or an infinite magnitude. Now we have already proved in our course on Physics that there cannot be an infinite magnitude: and we have now proved that it is impossible for a finite magnitude to have an infinite force, and also that it is impossible for a thing to be moved by a finite magnitude during an infinite time. But the first movent causes a motion that is eternal and does cause it during an infinite time. It is clear, therefore, that the first movent is indivisible and is without parts and without magnitude.

The prime form of movement is circular motion

77

It can now be shown plainly that rotation is the primary locomotion. Every locomotion, as we said before, is either rotatory or rectilinear or a compound of the two: and the two former must be prior to the last, since they are the elements of which the latter consists. Moreover rotatory locomotion is prior to rectilinear locomotion, because it is more simple and complete, which may be shown as follows. The straight line traversed in rectilinear motion cannot be infinite: for there is no such thing as an infinite straight line; and even if there were, it would not be traversed by anything in motion: for the impossible does not happen and it is impossible to traverse an infinite distance. On the other hand rectilinear motion on a finite straight line is if it turns back a composite motion, in fact two motions, while if it does not turn back it is incomplete and perishable: and in the order of nature, of definition, and of time alike the complete is prior to the incomplete and the imperishable to the perishable. Again, a motion that admits of being eternal is prior to one that does not. Now rotatory motion can be eternal: but no other motion, whether locomotion or motion of any other kind, can be so, since in all of them rest must occur, and with the occurrence of rest the motion has perished. Moreover the result at which we have arrived, that rotatory motion is single and continuous, and rectilinear motion is not, is a reasonable one. In rectilinear motion we have a definite starting-point, finishing-point, and middle-point, which all have their place in it in such a way that there is a point from which that which is in motion can be said to start and a point at which it can be said to finish its course (for when anything is at the limits of its course, whether at the starting-point or at the finishing-point, it must be in a state of rest). On the other hand in circular motion there are no such definite points: for why should any one point on the line be a limit rather than any other? Any one point as much as any other is alike starting-point, middle-point, and finishing-point.

The five elements, their natural movements and the four elementary qualities

78

Bodies are either simple or compounded of such; and by simple bodies I mean those which possess a principle of movement in their own nature, such as fire and earth with their kinds, and whatever is akin to them. Necessarily, then, movements will also be either simple or in some sort compound—simple in the case of simple bodies, compound in that of the composite—and in the latter case the motion will be that of the simple body which prevails in the composition. Supposing, then, that there is such a thing as simple movement, and that circular movement is an instance of it, and that both movement of a simple body is simple and simple movement is of a simple body (for if it is movement of a compound it will be in virtue of a prevailing simple element), then there must necessarily be some simple body which revolves naturally and in virtue of its own nature with a circular movement. By constraint, of course, it may be brought to move with the motion of something else different from itself, but it cannot so move naturally, since there is one sort of movement natural to each of the simple bodies. Again, if the unnatural movement is the contrary of the natural and a thing can have no more than one contrary, it will follow that circular movement, being a simple motion, must be unnatural, if it is not natural, to the body moved. If then (1) the body, whose movement is circular, is fire or some other element, its natural motion must be the contrary of the circular motion. But a single thing has a single contrary; and upward and downward motion are the contraries of one another. If, on the other hand, (2) the body moving with this circular motion which is unnatural to it is something different from the elements, there will be some other motion which is natural to it. But this cannot be. For if the natural motion is upward, it will be fire or air, and if downward, water or earth. Further, this circular motion is necessarily primary. For the perfect is naturally prior to the imperfect, and the circle is a perfect thing. This cannot be said of any straight line:—not of an infinite line; for, if it were perfect, it would have a limit and an end: nor of any finite line; for in every case there is something beyond it, since any finite line can be extended. And so, since the prior movement belongs to the body which is naturally prior, and circular movement is prior to straight, and movement in a straight line belongs to simple bodies—fire moving straight upward and earthy bodies straight downward towards the centre—since this is so, it follows that circular movement also must be the movement of some simple body. For the movement of composite bodies is, as we said, determined by that simple body which preponderates in the composition. These premises clearly give the conclusion that there is in nature some bodily substance other than the formations we know, prior to them all and more divine than they. [. . .]

In consequence of what has been said, in part by way of assumption and in part by way of proof, it is clear that not every body either possesses lightness or heaviness. As a preliminary we must explain in what sense we are using the words 'heavy' and 'light', sufficiently, at least, for our present purpose: we can examine the terms more closely later, when we come to consider their essential nature. Let us then apply the term 'heavy' to that which naturally moves towards the centre, and 'light' to that which moves naturally away from the centre. The heaviest thing will be that which sinks to the bottom of all things that move downward, and the lightest that which rises to the surface of everything that moves upward. Now, necessarily, everything which moves either up or down possesses lightness or heaviness or both—but not both relatively to the same thing: for things are heavy and light relatively to one another; air, for instance, is light relatively to water, and water light relatively to earth. The body, then, which moves in a circle cannot possibly possess either heaviness or lightness. For neither naturally nor unnaturally can it move either towards or away from the centre. Movement in a straight line certainly does not belong to it *naturally*, since one sort of movement is, as we saw, appropriate to each simple body, and so we should be compelled to identify it with one of the bodies which move in this way. Suppose, then, that the movement is *unnatural*. In that case, if it is the downward movement which is unnatural, the upward movement will be natural; and if it is the upward which is unnatural, the downward will be natural. [. . .]

It is equally reasonable to assume that this body will be ungenerated and indestructible and exempt from increase and alteration, since everything that comes to be comes into being from its contrary and in some substrate, and passes away likewise in a substrate by the action of the contrary into the contrary, as we explained in our opening discussions. Now the motions of contraries are contrary. If then this body can have no contrary, because there can be no contrary motion to the circular, nature seems justly to have exempted from contraries the body which was to be ungenerated and indestructible. For it is in contraries that generation and decay subsist. [. . .]

The reasons why the primary body is eternal and not subject to increase or diminution, but unageing and unalterable and unmodified, will be clear from what has been said to any one who believes in our assumptions. Our theory seems to confirm experience and to be confirmed by it. [. . .] The common name, too, which has been handed down from our distant ancestors even to our own day, seems to show that they conceived of it in the fashion which we have been expressing. The same ideas, one must believe, recur in men's minds not once or twice but again and again. And so, implying that the primary body is something else beyond earth, fire, air, and water, they gave the highest place a name of its own, *either*, derived from the fact that it 'runs always' for an eternity of time. [. . .]

It is also clear from what has been said why the number of what we call simple bodies cannot be greater than it is. The motion of a simple body must itself be simple, and we assert that there are only these two simple motions, the circular and the straight, the latter being subdivided into motion away from and motion towards the centre.

79

Next in order we must discuss 'action' and 'passion'. The traditional theories on the subject are conflicting. For (i) most thinkers are unanimous in maintaining (a) that 'like' is always unaffected by 'like', because (as they argue) neither of two 'likes' is more apt than the other either to act or to suffer action, since all the properties which belong to the one belong identically and in the same degree to the other; and (b) that 'unlikes', i.e. 'differents', are by nature such as to act and suffer action reciprocally. For even when the smaller fire is destroyed by the greater, it suffers this effect (they say) owing to its 'contrariety'—since the great is contrary to the small. But (ii) Democritus dissented from all the other thinkers and maintained a theory peculiar to himself. He asserts that agent and patient are identical, i.e. 'like'. It is not possible (he says) that 'others', i.e. 'differents', should suffer action from one another: on the contrary, even if two things, being 'others', do act in some way on one another, this happens to them not *qua* 'others' but *qua* possessing an identical property.

Such, then, are the traditional theories, and it looks as if the statements of their advocates were in manifest conflict. But the reason of this conflict is that each group is in fact stating *a part*, whereas they ought to have taken a comprehensive view of the subject *as a whole*. For (i) if *A* and *B* are 'like'—absolutely and in all respects without difference from one another—it is reasonable to infer that neither is in any way affected by the other. Why, indeed, should either of them tend to act any more than the other? Moreover, if 'like' can be affected by 'like', a thing can also be affected by itself: and yet if that were so—if 'like' tended in fact to act *qua* 'like'—there would be nothing indestructible or immovable, for everything would move itself. And (ii) the same consequence follows if *A* and *B* are absolutely 'other', i.e. in no respect identical. *Whiteness* could not be affected in any way by *line* nor *line* by *whiteness*—except perhaps 'coincidentally', viz. if the line happened to be white or black: for unless two things either are, or are composed of, 'contraries', neither drives the other out of its natural condition. But (iii) since only those things which either involve a 'contrariety' or are 'contraries'—and not any things selected at random—are such as to suffer action and to act, agent and patient must be 'like' (i.e. identical) in kind and yet 'unlike' (i.e. contrary) in species. (For it is a law of nature that body is affected by body, flavour by flavour, colour by colour, and so in general what belongs to any kind by a member of the same kind—the reason being that 'contraries' are in every case within a single identical kind, and it is 'contraries' which reciprocally act and suffer action.) Hence agent and patient must be in one sense identical, but in another sense other than (i.e. 'unlike') one another. And since (a) patient and agent are generically identical (i.e. 'like') but specifically 'unlike', while (b) it is 'contraries' that exhibit this character: it is clear that 'contraries' and their 'intermediates' are such as to suffer action and to act reciprocally—for indeed it is these that constitute the entire sphere of passing-away and coming-to-be.

80

Accordingly, we must segregate the tangible differences and contrarieties, and distinguish which amongst them are primary. Contrarieties correlative to touch are the following: hot-cold, dry-moist, heavy-light, hard-soft, viscous-brittle, rough-smooth, coarse-fine. Of these (i) heavy and light are neither active nor susceptible. Things are not called 'heavy' and 'light' because they act upon, or suffer action from, other things. But the 'elements' must be reciprocally active and susceptible, since they 'combine' and are transformed into one another. On the other hand (ii) hot and cold, and dry and moist, are terms of which the first pair implies *power to act* and the second pair *susceptibility*. 'Hot' is that which 'associates' things of the same kind (for 'dissociating', which people attribute to Fire as its function, *is* 'associating' things of the same class, since its effect is to eliminate what is foreign), while 'cold' is that which brings together, i.e. 'associates' homogeneous and heterogeneous things alike. And 'moist' is that which, being readily adaptable in shape, is not determinable by any limit of its own: while 'dry' is that which is readily determinable by its own limit, but not readily adaptable in shape. [. . .]

The elementary qualities are four, and any four terms can be combined in six couples. Contraries, however, refuse to be coupled: for it is impossible for the same thing to be hot and cold, or moist and dry. Hence it is evident that the 'couplings' of the elementary qualities will be four: hot with dry and moist with hot, and again cold with dry and cold with moist. And these four couples have attached themselves to the *apparently* 'simple' bodies (Fire, Air, Water, and Earth) in a manner consonant with theory. For Fire is hot and dry, whereas Air is hot and moist (Air being a sort of aqueous vapour); and Water is cold and moist, while Earth is cold and dry. Thus the differences are reasonably distributed among the primary bodies, and the number of the latter is consonant with theory.

81

It has been established before that the coming-to-be of the 'simple' bodies is reciprocal. At the same time, it is manifest, even on the evidence of perception, that they *do* come-to-be: for otherwise there would not have been 'alteration', since 'alteration' is change in respect to the qualities of the objects of touch. Consequently, we must explain (i) what is the manner of their reciprocal transformation, and (ii) whether every one of them can come-to-be out of every one—or whether some can do so, but not others.

Now it is evident that all of them are by nature such as to change into one another: for coming-to-be is a change into contraries and out of contraries, and the 'elements' all involve a contrariety in their mutual relations because their distinctive qualities are contrary. For in some of them *both* qualities are contrary —e.g. in Fire and Water, the first of these being dry and hot, and the second moist and cold; while in others *one* of the qualities (though only one) is contrary —e.g. in Air and Water, the first being moist and hot, and the second moist and cold. It is evident, therefore, if we consider them in general, that every one is by nature such as to come-to-be out of every one: and when we come to consider them severally, it is not difficult to see the manner in which their transformation is effected. For, though all will result from all, both the speed and the facility of their conversion will differ in degree.

Thus (i) the process of conversion will be quick between those which have interchangeable 'complementary factors', but slow between those which have none. The reason is that it is easier for a single thing to change than for many. Air, e.g. will result from Fire if a single quality changes: for Fire, as we saw, is hot and dry while Air is hot and moist, so that there will be Air if the dry is overcome by the moist. Again, Water will result from Air if the hot is overcome by the cold: for Air, as we saw, is hot and moist while Water is cold and moist, so that, if the hot changes, there will be Water. So too, in the same manner, Earth will result from Water and Fire from Earth, since the two 'elements' in both these couples have interchangeable 'complementary factors'. For Water is moist and cold while Earth is cold and dry—so that, if the moist be overcome, there will be Earth: and again, since Fire is dry and hot while Earth is cold and dry, Fire will result from Earth if the cold passes away.

The physical doctrine of concentric spheres

82

Eudoxus supposed that the motion of the sun or of the moon involves, in either case, three spheres, of which the first is the sphere of the fixed stars, and the second moves in the circle which runs along the middle of the zodiac, and the third in the circle which is inclined across the breadth of the zodiac; but the circle in which the moon moves is inclined at a greater angle than that in which the sun moves. And the motion of the planets involves, in each case, four spheres, and of these also the first and second are the same as the first two mentioned above (for the sphere of the fixed stars is that which moves all the other spheres, and that which is placed beneath this and has its movement in the circle which bisects the zodiac is common to all), but the *poles* of the third sphere of each planet are in the circle which bisects the zodiac, and the motion of the fourth sphere is in the circle which is inclined at an angle to the equator of the third sphere; and the poles of the third sphere are different for each of the other planets, but those of Venus and Mercury are the same.

Callippus made the position of the spheres the same as Eudoxus did, but while he assigned the same number as Eudoxus did to Jupiter and to Saturn, he thought two more spheres should be added to the sun and two to the moon, if one is to explain the observed facts; and one more to each of the other planets.

But it is necessary, if all the spheres combined are to explain the observed facts, that for each of the planets there should be other spheres (one fewer than those hitherto assigned) which counteract those already mentioned and bring back to the same position the outermost sphere of the star which in each case is situated below the star in question; for only thus can all the forces at work produce the observed motion of the planets. Since, then, the spheres involved in the movements of the planets themselves are—eight for Saturn and Jupiter and twenty-five for the others, and of these only those involved in the movement of the lowest situated planet need not be counteracted, the spheres which counteract those of the outermost two planets will be six in number, and the spheres which counteract those of the next four planets will be sixteen; therefore the number of all the spheres—both those which move the planets and those which counteract these—will be fifty-five.

THE HELLENISTIC AGE

EPICURUS

Atoms and the void

83

Further, the whole of being consists of bodies and space. For the existence of bodies is everywhere attested by sense itself, and it is upon sensation that reason must rely when it attempts to infer the unknown from the known. And if there were no space, bodies would have nothing in which to be and through which to move, as they are plainly seen to move. Beyond bodies and space there is nothing which by mental apprehension or on its analogy we can conceive to exist. When we speak of bodies and space, both are regarded as wholes or separate things, not as the properties or accidents of separate things.

Again, of bodies some are composite, others the elements of which these composite bodies are made. These elements are indivisible and unchangeable, and necessarily so, if things are not all to be destroyed and pass into non-existence, but are to be strong enough to endure when the composite bodies are broken up, because they possess a solid nature and are incapable of being anywhere or anyhow dissolved. It follows that the first beginnings must be indivisible, corporeal entities.

Again, the sum of things is infinite. For what is finite has an extremity, and the extremity of anything is discerned only by comparison with something else. Now the sum of things is not discerned by comparison with anything else: hence, since it has no extremity, it has no limit; and, since it has no limit, it must be unlimited or infinite.

Moreover, the sum of things is unlimited both by reason of the multitude of the atoms and the extent of the void. For if the void were infinite and bodies finite, the bodies would not have stayed anywhere but would have been dispersed in their course through the infinite void, not having any supports or counter-checks to send them back on their upward rebound. Again, if the void were finite, the infinity of bodies would not have anywhere to be.

Furthermore, the atoms, which have no void in them—out of which composite bodies arise and into which they are dissolved—vary indefinitely in their shapes; for so many varieties of things as we see could never have arisen out of a recurrence of a definite number of the same shapes. The like atoms of each shape are absolutely infinite; but the variety of shapes, though indefinitely large, is not absolutely infinite unless one is prepared to keep enlarging their magnitudes also simply *ad infinitum*.

The atoms are in continual motion through all eternity. Some of them rebound to a considerable distance from each other, while others merely oscillate in one place when they chance to have got entangled or to be enclosed by a mass of other atoms shaped for entangling.

This is because each atom is separated from the rest by void, which is incapable of offering any resistance to the rebound; while it is the solidity of the atom which makes it rebound after a collision, however short the distance to which it rebounds, when it finds itself imprisoned in a mass of entangling atoms. Of all this there is no beginning, since both atoms and void exist from everlasting.

The repetition at such length of all that we are now recalling to mind furnishes an adequate outline for our conception of the nature of things.

84

Moreover, there is an infinite number of worlds, some like this world, others unlike it. For the atoms being infinite in number, as has just been proved, are borne ever further in their course. For the atoms out of which a world might arise, or by which a world might be formed, have not all been expended on one world or a finite number of worlds, whether like or unlike this one. Hence there will be nothing to hinder an infinity of worlds.

Sense perception and the doctrine of 'images'

85

Again, there are outlines or films, which are of the same shape as solid bodies, but of a thinness far exceeding that of any object that we see. For it is not impossible that there should be found in the surrounding air combinations of this kind, materials adapted for expressing the hollowness and thinness of surfaces, and effluxes preserving the same relative position and motion which they had in the solid objects from which they come. To these films we give the name of 'images' or 'idols'. Furthermore, so long as nothing comes in the way to offer resistance, motion through the void accomplishes any imaginable distance in an inconceivably short time. For resistance encountered is the equivalent of slowness, its absence the equivalent of speed.

86

Besides this, remember that the production of the images is as quick as thought. For particles are continually streaming off from the surface of bodies, though no diminution of the bodies is observed, because other particles take their place. And those given off for a long time retain the position and arrangement which their atoms had when they formed part of the solid bodies, although occasionally they are thrown into confusion. Sometimes such films are formed very rapidly in the air, because they need not have any solid content; and there are other modes in which they may be formed. For there is nothing in all this which is contradicted by sensation, if we in some sort look at the clear evidence of sense, to which we should also refer the continuity of particles in the objects external to ourselves.

We must also consider that it is by the entrance of something coming from external objects that we see their shapes and think of them. For external things would not stamp on us their own nature of colour and form through the medium of the air which is between them and us, or by means of rays of light or currents of any sort going from us to them, so well as by the entrance into our eyes or minds, to whichever their size is suitable, of certain films coming from the things themselves, these films or outlines being of the same colour and shape as the external things themselves. They move with rapid motion; and this again explains why they present the appearance of the single continuous object, and retain the mutual interconnexion which they had in the object, when they impinge upon the sense, such impact being due to the oscillation of the atoms in the interior of the solid object from which they come. And whatever presentation we derive by direct contact, whether it be with the mind or with the sense-organs, be it shape that is presented or other properties, this shape as presented is the shape of the solid thing, and it is due either to a close coherence of the image as a whole or to a mere remnant of its parts.

The numbers and properties of atoms

87

Moreover, we must hold that the atoms in fact possess none of the qualities belonging to things which come under our observation, except shape, weight, and size, and the properties necessarily conjoined with shape. For every quality changes, but the atoms do not change, since, when the composite bodies are dissolved, there must needs be a permanent something, solid and indissoluble, left behind, which makes change possible: not changes into or from the non-existent, but often through differences of arrangement, and sometimes through additions and subtractions of the atoms. Hence these somethings capable of being diversely arranged must be indestructible, exempt from change, but possessed each of its own distinctive mass and configuration. This must remain.

For in the case of changes of configuration within our experience the figure is supposed to be inherent when other qualities are stripped off, but the qualities are not supposed, like the shape which is left behind, to inhere in the subject of change, but to vanish altogether from the body. Thus, then, what is left behind is sufficient to account for the differences in composite bodies, since something at least must necessarily be left remaining and be immune from annihilation.

Again, you should not suppose that the atoms have any and every size, lest you be contradicted by facts; but differences of size must be admitted; for this addition renders the facts of feeling and sensation easier of explanation. But to attribute any and every magnitude to the atoms does not help to explain the differences of quality in things; moreover, in that case atoms large enough to be seen ought to have reached us, which is never observed to occur; nor can we conceive how its occurrence should be possible, i.e. that an atom should become visible.

Besides, you must not suppose that there are parts unlimited in number, be they ever so small, in any finite body. Hence not only must we reject as impossible subdivision *ad infinitum* into smaller and smaller parts, lest we make all things too weak and, in our conceptions of the aggregates, be driven to pulverize the things that exist, i.e. the atoms, and annihilate them; but in dealing with finite things we must also reject as impossible the progression *ad infinitum* by less and less increments.

For when once we have said that an infinite number of particles, however small, are contained in anything, it is not possible to conceive how it could any longer be limited or finite in size. For clearly our infinite number of particles must have some size; and then, of whatever size they were, the aggregate they made would be infinite.

Cosmogony

88

After the foregoing we have next to consider that the worlds and every finite aggregate which bears a strong resemblance to things we commonly see have arisen out of the infinite. For all these, whether small or great, have been separated off from special conglomerations of atoms; and all things are again dissolved, some faster, some slower, some through the action of one set of causes, others through the action of another.

And further, we must not suppose that the worlds have necessarily one and the same shape. For nobody can prove that in one sort of world there might not be contained, whereas in another sort of world there could not possibly be, the seeds out of which animals and plants arise and all the rest of the things we see.

Man and his attitude towards celestial phenomena

89

Nay more: we are bound to believe that in the sky revolutions, solstices, eclipses, risings and settings, and the like, take place without the ministration or command, either now or in the future, of any being who at the same time enjoys perfect bliss along with immortality. For troubles and anxieties and feelings of anger and partiality do not accord with bliss, but always imply weakness and fear and dependence upon one's neighbours. Nor, again, must we hold that things which are no more than globular masses of fire, being at the same time endowed with bliss, assume these motions at will. Nay, in every term we use we must hold fast to all the majesty which attaches to such notions as bliss and immortality, lest the terms should generate opinions inconsistent with this majesty. Otherwise such inconsistency will of itself suffice to produce the worst disturbance in our minds. Hence, where we find phenomena invariably recurring, the invariableness of the recurrence must be ascribed to the original interception and conglomeration of atoms whereby the world was formed.

90

In the first place, remember that, like everything else, knowledge of celestial phenomena, whether taken along with other things or in isolation, has no other end in view than peace of mind and firm conviction. We do not seek to wrest by force what is impossible, nor to understand all matters equally well, nor make our treatment always as clear as when we discuss human life or explain the principles of physics in general—for instance, that the whole of being consists of bodies and intangible nature, or that the ultimate elements of things are indivisible, or any other proposition which admits only one explanation of the phenomena to be possible. But this is not the case with celestial phenomena: these at any rate admit of manifold causes for their occurrence and manifold accounts, none of them contradictory of sensation, of their nature.

For in the study of nature we must not conform to empty assumptions and arbitrary laws, but follow the promptings of the facts; for our life has no need now of unreason and false opinion; our one need is untroubled existence. All things go on uninterruptedly, if all be explained by the method of plurality of causes in conformity with facts, so soon as we duly understand what may be plausibly alleged respecting them. But when we pick and choose among them, rejecting one equally consistent with the phenomena, we clearly fall away from the study of nature altogether and tumble into myth. Some phenomena within our experience afford evidence by which we may interpret what goes on in the heavens. We see how the former really take place, but not how the celestial phenomena take place, for their occurrence may possibly be due to a variety of causes. However, we must observe each fact as presented, and further separate from it all the facts presented along with it, the occurrence of which from various causes is not contradicted by facts within our experience.

LUCRETIUS

The visible effects of invisible particles

91

Now mark me: since I have taught that things cannot be created from nothing and when brought forth cannot be brought back to nothing, that you may not by any chance begin nevertheless to distrust my words because the first-beginnings of things cannot be distinguished by the eye, learn in addition of bodies which you must yourself of necessity confess to be numbered amongst things and yet impossible to be seen. First the mighty wind when stirred up beats upon harbours and overwhelms huge ships and scatters the clouds, at times sweeping over the plains with rapid hurricane strews them with great trees and flogs the topmost mountains with tree-crashing blasts: so furious and fierce its howling, so savage and threatening the wind's roar. Therefore undoubtedly the winds are bodies that unseen sweep the sea, that sweep the earth, sweep the clouds of the sky also, beating them suddenly and catching them up in a hurricane, and they flow and they deal devastation in the same way as water: which soft as it is, suddenly rolls in overswelling stream when a great deluge of water from the high mountains swells the flood with torrents of rain, dashing together wreckage of forests and whole trees, nor can strong bridges withstand the sudden force of the coming water, with so mighty a force does the river boiling with rain-torrents rush against the solid piers; it works devastation with loud uproar and rolls huge rocks under its waves, and sweeps away whatever stands in its path. Thus therefore the blasts of the wind also needs must be borne along, which like a strong river, when they have borne down in any direction, thrust all before them and sweep all away with frequent attacks, at times catch things up in a whirling eddy and carry them in a rapidly twisting tornado. Wherefore I say again and again, the winds are bodies unseen, since in deeds and ways they are found to rival great rivers, which possess a body which can be seen. Then further, we smell the various odours of things and yet we never see them approaching our nostrils, nor do we behold scorching heat, nor can we set eyes on cold, nor are we wont to see sounds: yet all these must of necessity consist of bodily structure, since they can act upon our senses. For nothing can touch or be touched, save body. Again, garments hung up on a surf-beaten shore grow damp, the same spread in the sun grow dry; yet none has seen either how the damp of the water pervaded them, or again how it departed in the heat. Therefore the water is dispersed into small particles, which the eye cannot in any way see. Moreover, with many revolutions of the sun's year, a ring on the finger is thinned underneath by wear, the fall of drippings hollows a stone, the curved ploughshare of iron imperceptibly dwindles away in the fields, and the stony pavement of the roads we see already to be rubbed away by men's feet; again bronze statues set by gateways display the right hands thinned away by the often touching of those who kiss and those who pass by. These therefore we observe to be growing less because they are rubbed away; but what particles are separated on each occasion, our niggardly faculty of sight has debarred us from seeing. Lastly, whatever time and nature little by little adds to things, compelling them to grow greater by degrees, no keenness of sight, however strained, can perceive; nor further when things grow old by age and wasting, nor when rocks hanging over the sea are eaten away by the fine salt, could you discern what they lose upon each occasion. Therefore nature works by means of bodies unseen.

The constant movement of atoms

92

Listen now, and I will set forth by what motion the generative bodies of matter beget the various things and once begotten dissolve them, and by what force they are compelled to do it, and what swiftness has been given them to travel through the great void: do you remember to give heed to my words. For certainly matter is not one packed and coherent mass, since we see each thing decreasing, and we perceive all things as it were ebbing through length of time, and age withdrawing them from our eyes; although nevertheless the sum is seen to remain unimpaired for this reason, that whenever bodies pass away from a thing, they diminish that from which they pass and increase that to which they have come, they compel the first to fade and the second on the contrary to bloom, yet do not linger there. Thus the sum of things is ever being renewed, and mortal creatures live dependent one upon another. Some nations increase, others diminish, and in a short space the generations of living creatures are changed and like runners pass on the torch of life.

If you think the first-beginnings of things can stand still, and by standing still can beget new motions amongst things, you are astray and wander far from true reasoning. For since the first-beginnings of things wander through the void, they must needs all be carried on either by their own weight or by a chance blow from one or other. For when in quick motion they have often met and collided, it follows that they leap apart suddenly in different directions; and no wonder, since they are perfectly hard in their solid weight and nothing obstructs them from behind. And to show you more clearly that all the bodies of matter are constantly being tossed about, remember that there is no bottom in the sum of things and the first bodies have nowhere to rest, since space is without end or limit, and I have shown at large and proved by irrefragable reasoning that it extends immeasurable from all sides in all directions. Since this stands firm, beyond doubt no rest is granted to the first bodies throughout the profound void, but rather driven by incessant and varied motions, some after being pressed together then leap back with wide intervals, some again after the blow are tossed about within a narrow compass. And those which being held in combination more closely condensed collide and leap back through tiny intervals, caught fast in the complexity of their own shapes, these constitute the strong roots of stone and the bodies of fierce iron and the others of their kind. Those are few, moreover, which travel through the great void. The rest leap far apart and pass far back with long intervals between: these supply thin air for us and the gleaming light of the sun. And many besides travel through the great void which have been rejected from combination with things, and have nowhere been able to conjoin their motions even when received. Of this as I now describe it there is an image and similitude always moving and present before our eyes. Do but apply your scrutiny when the sun's light and his rays penetrate and spread through a dark room: you will see many minute specks mingling in many ways throughout the void in the light itself of the rays, and as it were in everlasting conflict struggling, fighting, battling in troops without any pause, driven about with frequent meetings and partings; so that you may conjecture from this what it is for the first-beginnings of things to be ever tossed about in the great void. So far as it goes, a small thing may give an analogy of great things, and show tracks of knowledge. Even more for another reason, it is proper that you give attention to these bodies which are seen to be in turmoil within the sun's rays, because such turmoil indicates that there are secret and blind motions also hidden in matter. For there you will see how many things set in motion by blind blows change their course and beaten back return back again, now this way now that way, in all directions. You may be sure that all take their restlessness from the first-beginnings. For first the first-beginnings of things move of themselves; then the bodies that form a small combination and as one may say are nearest to the powers of the beginnings, are set moving, driven by the blind blows of these, while they in their turn attack those that are a little larger. Thus the movement ascends from the beginnings and by successive degrees emerges upon our senses, so that those bodies also are moved which we are able to perceive in the sun's light, yet it does not openly appear by what blows they are made to do so.

The swerving of atoms

93

One further point in this matter I desire you to understand: that while the first bodies are being carried downwards by their own weight in a straight line through the void, at times quite uncertain and in uncertain places, they swerve a little from their course, just so much as you might call a change of motion. For if they were not apt to incline, all would fall downwards like raindrops through the profound void, no collision would take place and no blow would be caused amongst the first-beginnings: thus nature would never have produced anything.

But if peradventure anyone believes it to be possible that heavier elements, being carried more quickly straight through the void, fall from above on the lighter, and so deal blows which can produce generative motions, he is astray and departs far from true reasoning. For whatever things fall through water and through fine air, these must speed their fall in accordance with their weights, because the body of water and the thin nature of air cannot delay each thing equally, but yield sooner overcome by the heavier; but contrariwise empty void cannot offer any support to anything anywhere or at any time, but it must give way continually as its nature demands: wherefore they must all be carried with equal speed, although not of equal weight, through the unresisting void. Therefore the heavier bodies will never be able to fall from above on the lighter, nor deal blows of themselves so as to produce the various motions by which nature carries on her processes. Wherefore again and again I say, the bodies must incline a little; and not more than the least possible, or we shall seem to assume oblique movements, and thus be refuted by the facts. For this we see to be manifest and plain, that weights as far as in them lies cannot travel obliquely, when they drop straight from above, as far as one can perceive; but who is there who can perceive that they never swerve ever so little from the straight undeviating course?

Again, if all motion is always one long chain, and new motion always arises out of the old in order invariable, and if the first-beginnings do not make by swerving a beginning of motion such as to break the decrees of fate, that cause may not follow cause to infinity, whence comes this free will in living creatures all over the earth, whence I say is this will wrested from the fates by which we proceed whither pleasure leads each, swerving also our motions not at fixed times and fixed places, but just where our mind has taken us? For undoubtedly it is his own will in each that begins these things, and therefrom go movements rippling through the limbs.

Mind and body

94

Now I shall go on to explain to you, of what kind of body this mind is, and whence it is formed. First I say that it is exceedingly delicate and formed of exceedingly minute particles. That this is so, you may consider the following points to convince you. Nothing is seen to be done so swiftly as the mind determines it to be done and does its own first act: therefore the mind bestirs itself more quickly than any other of these things which are seen plain before our eyes. But that which is so readily moved must consist of seeds exceedingly rounded and exceedingly minute, that they may be moved when touched by a small moving power. For water moves and flows with so very small a moving power because it is made of small rolling shapes. But on the other hand the nature of honey has more cohesion, its fluid is more sluggish, and its movement more tardy; for the whole mass of its matter coheres more closely, assuredly because it is not made of bodies so smooth or so delicate and round. For a light breath of air can make you a high heap of poppy-seed slip down from the top; but contrariwise it cannot stir a pile of pebbles or wheat-ears. So, according as bodies are extremely small and smooth, they have power of motion; but contrariwise, whatever is found to be most weighty and rough is by so much the more stable. Now, therefore, since the nature of the mind has been found to be moved with unusual ease, it must consist of bodies exceedingly small and smooth and round. If this be known to you, my good friend, it will be found of advantage in many ways, and you will call it useful. Another thing also makes clear of how fine a texture it is, and in how small a space it is contained if it could be gathered together; namely that as soon as death's peaceful calm has taken possession of a man, when mind and spirit have departed, you could not perceive any jot or tittle to be diminished from the body whether in look or in weight: death presents all, except vital sense and warming heat. Accordingly the whole spirit must consist of very small seeds, being interlaced through veins, flesh, and sinews; wherefore, when the whole has already departed from all the body, nevertheless the outward contour of the limbs presents itself undiminished, nor is one jot of weight lacking; even as happens when the bouquet of wine has vanished, or when the sweet breath of ointment has dispersed into air, or when the flavour has passed from a substance, and yet the thing itself does not seem any smaller to the eye for all that, nor is anything lost in the weight, because assuredly many minute seeds compose the flavour and the smell in the whole substance of mind and spirit to be made of very minute seeds, since in departing it takes nothing from the weight.

95

This nature then is contained in every body, being itself the body's guardian and source of its existence: for they cling together with common roots, and manifestly they cannot be torn asunder without destruction. Even as it is not easy to tear out the scent from lumps of frankincense, without its very nature being destroyed: so it is not easy to draw out mind and spirit from the whole body without the dissolution of all. So interwoven are their elements from their first origin in the life which they live together; and we see that neither body nor mind has the power to feel singly without the other's help, but by common notions proceeding from both conjointly sensation is kindled for us in our flesh. Besides, a body is never born by itself, nor grows, nor is it seen to last long after death. For it is not as when the liquid of water often throws off the heat which has been given to it, and yet is not itself torn to pieces for that reason but remains uninjured; not thus I say can the frame endure disruption apart from the spirit which has left it; but it is utterly undone, torn to pieces along with it, and along with it rots away. From the first moment of life, the interdependent contacts of body and spirit while yet laid away in the mother's body, and the womb, so learn the vital motions, that disruption apart cannot be without their ruin and damage; so that you may see that since conjunction is necessary to their existence, so also theirs must be a joint nature.

THE STOICS: ZENO AND CHRYSIPPUS

The universe as a dynamical continuum

96

The physical states are nothing else but spirits, because the bodies are made cohesive by them. And the binding air is the cause for those bound into such a state being imbued with a certain property which is called hardness in iron, solidity in stone, brightness in silver.

97

The Stoics say that earth and water have no binding force of their own, nor can they bind other substances together. They maintain their unity by partaking of the power of *pneuma* and fire. Air and fire on the other hand, through their inner tension and through being mixed with the other two, provide these with tension, permanence and substantiality.

98

Some of the bodies are unified, some formed of parts conjoined, some of separate parts. Unified bodies are such as are controlled by a single hexis similar to that of plants and animals; those formed of conjoined parts are such as are composed of adjacent elements which tend to combine into one main structure, like cables and turrets and ships; those formed of separate parts are compounded of things which are disjointed and isolated and existing by themselves, like armies and flocks and choruses.

Seeing, then, that the universe also is a body, it is either unified or of conjoined or separate parts. But it is neither of conjoined nor of separate parts, as we prove from the sympathies it exhibits. For in accordance with the waxings and wanings of the moon many sea and land animals wane and wax, and ebb-tides occur in some parts of the sea. And in the same way, too, in accordance with certain risings and settings of the stars alterations in the surrounding atmosphere and all varieties of change in the air take place, sometimes for the better, but sometimes fraught with pestilence. And from these facts it is obvious that the universe is a unified body. For in the case of bodies formed from conjoined or separate elements the parts do not sympathize with one another, since if all soldiers, say, in an army have perished the one who survives is not seen to suffer at all through transmission; but in the case of unified bodies there exists a certain sympathy, since, when the finger is cut, the whole body shares in its condition. So then, the universe also is a unified body.

99

He endowed some substances with physical structure, some with the power of growth, some with soul and some with an intelligent soul. In stones and trees which are divorced from the organic creation, he formed the physical nexus as a very strong bond. This is the *pneuma* that returns upon itself. It begins in the centre of the substance and stretches outwards to its edges [. . .] and returns back again to the place from which it started.

100

Chrysippus' theory of mixture is as follows: he assumes that the whole material world is unified by a *pneuma* which wholly pervades it, and by which the universe is made coherent and kept together and is made intercommunicating. And of the compounded bodies some become mixed by juxtaposition when two or more substances are put together in the same place and placed side by side, joining each other, as he says, and preserving in this juxtaposition their proper essence and quality according to their individuality, as happens for instance when beans and grains of wheat get mixed with one another. Some bodies are destroyed together through a complete fusion of their substances and their respective qualities as is the case with drugs whose components undergo simultaneous destruction and as a result of it another body emerges.

Certain mixtures, he says, result in a total interpenetration of substances and their qualities, the original substances and qualities being preserved in this mixture; this he calls specifically *krasis* of the mixed components. It is characteristic of the mixed substances that they can again be separated, which is only possible if the components preserve their properties in the mixture [. . .]. This interpenetration of the components he assumes to happen in that the substances mixed together interpenetrate each other such that there is not a particle among them that does not contain a share of all the rest. If this were not the case, the result would not be *krasis* but juxtaposition. The supporters of this theory adduce evidence for their belief from the fact that many bodies preserve their qualities whether they are present in smaller or in larger quantities, as can be seen in the case of frankincense. When burnt, it becomes greatly rarefied, but for all that preserves its quality. Further, there are many substances which, when assisted by others, expand to an extent which they could not do by themselves. Gold, for instance, when mixed with certain drugs, can be spread and rarefied to an extent which is not possible if it is beaten out by itself. Similarly, there are cases where we can be effective together with others, while we cannot when we are alone. We can cross rivers if we form a chain, a thing we could not do by ourselves; and together with others we can bear weights which we could not bear if the whole load were to fall on us alone. And the tendrils of grape-vine which could not stand up by themselves can do so if they are entangled with each other. Thus, he says, we should not be surprised that certain substances assist each other by forming a complete union such as to preserve their own qualities while totally interpermeating each other, even if the mass of one is so slight that by itself it could not preserve its quality if spread to such an extent. Thus a ladle of wine mixes with a large amount of water, being assisted by the latter to spread throughout a great volume. In order to prove this assertion, the Stoics adduce as clear evidence the fact that the soul has substantiality of its own as has the body containing it. By totally pervading the body it preserves in this mixture its own essence (there is no part of the living body which does not have its share in the soul), and the same holds for the nature of plants and for the physical structure of those things held together by hexis.

101

Outside the cosmos there is the void, extending from all sides into the infinite. That which is occupied by a body is called place, and that which is not occupied is void. Every body must necessarily be in something. That in which it is must be different from the body occupying and filling it, and thus is incorporeal and so to speak insipid. That reality which can receive the body and be occupied by it we call the void. [. . .]

We can indeed imagine the cosmos moving from the place it occupies now. By this transition we must understand that the place it left becomes void and that which it enters becomes occupied by it; this would be a void which has been filled up. Moreover, if the whole material world is dissolved into fire, as most accomplished physicists believe, the cosmos must necessarily occupy a place which is more than a thousand times greater, like the solid bodies which get vaporized. The place, therefore, which in the state of conflagration will be occupied by the spreading matter is now void and not filled up by any body. [. . .] When the material world contracts again and is forced into a smaller size, a void will be created again. [. . .] That which can be filled or left by a body is a void. Therefore, the void must necessarily have some sort of reality.

Such a void, however, does absolutely not exist within the cosmos. This is evident from the phenomena. For if the whole material world were not completely grown together, the cosmos as it is could not be kept together and administered by nature, nor would there exist a mutual sympathy of its parts; nor could we see and hear if the cosmos were not held together by one tension and if the *pneuma* were not cohesive throughout the whole being. For the empty spaces in between would impede sense perception.

102

But, they [the Peripatetics] say, if outside the cosmos were a void, the cosmos would move through it as nothing would be able to hold it together and to support it. To this we reply that this would not happen because the cosmos tends towards its own centre, and this centre is situated below, wherever the cosmos tends. For if the centre and the 'below' of the cosmos were not the same, the cosmos would move downwards through the void, as will be shown in the discourse on the centripetal motion.

They also argue that if there existed a void outside the cosmos, matter would be poured into it and dispersed and scattered into infinity. Our reply is that this could never happen, because the cosmos has a hexis which holds it together and protects it, and the surrounding void cannot affect it. It maintains itself by the rule of an immense force, contracting and again expanding into the void according to its natural transmutation, alternately dissolving into fire and starting creation again.

That a void outside the cosmos is necessary is evident by what has been shown above. However, that it extends from all sides to infinity is of the utmost necessity, as can be seen from that which follows. Everything finite is bordered by something of a different kind. [. . .] It would, therefore, be necessary, if the void surrounding the cosmos were finite, for it to terminate into something of a different kind. But there does not exist anything differing in kind from the void into which it terminates and therefore it is infinite.

103

Man does not consist of more parts than his finger, nor the cosmos of more parts than man. For the division of bodies goes on infinitely, and among the infinites there is no greater and smaller nor generally any quantity which exceeds the other. Nor cease the parts of the remainder to split up and to supply quantity out of themselves.

104

There is no extreme body in nature, neither first nor last, into which the size of a body terminates. But there always appears something beyond the assumed, and the body in question is thrown into the infinite and boundless.

105

The Stoics do not admit the existence of a shortest element of time, nor do they concede that the Now is indivisible, but that which someone might assume and think of as present is according to them partly future and partly past. Thus nothing remains of the Now, nor is there left any part of the present, but what is said to exist now is partly spread over the future and partly over the past.

Causality and determinism

106

They say that this cosmos is one and comprises in it all that exists and is administered by a vital, reasonable and intellectual principle of growth. It holds the eternal administration of all that exists which advances by a certain sequence and order. The prior events are causes of those following them, and in this manner all things are bound together with one another, and thus nothing can happen in the cosmos which is not a cause to something else following it and linked with it. Nor can any one of those events happening later be separated from what happened earlier and not be tied up with one of those earlier happenings. But from everything that happens something else follows depending on it by necessity as cause. [. . .] For there neither exists nor happens anything uncaused in the cosmos, because there is nothing in it which is separated and divorced from all that happened before. The cosmos would break up and be scattered and could no longer remain a unity administered by one order and plan, if some uncaused movement were to be introduced into it. This would be the case if there were not some cause preceding it for all that exists and happens, from which it results by necessity. According to them a causeless event is equal in essence to and equally impossible as a creation out of nothing. Such is the administration of the All going on from infinity to infinity manifestly and unceasingly. [. . .]

In view of the multiplicity of causes, the Stoics equally postulate about all of them that, wherever the same circumstances prevail with regard to the cause and the things affected by the cause, it is impossible that sometimes the result should be this and sometimes that; otherwise there would exist some uncaused motion.

107

The Stoics say that '*in* our power' is that which happens '*through* our power', thus denying that man has the power to choose and to do one of opposites. For, they argue, as the nature of things that are and happen is various and different, [. . .] all that happens to any thing will happen in accordance with its specific nature. What happens to a stone will happen in accordance with the nature of the stone, and so with fire and with a living being. However, none of these happenings in accordance with their specific nature could happen otherwise, but they all happen so by a necessity and not by force, out of the impossibility of what is natural to move in one case thus and in another differently. For if one lets a stone fall from a certain height it must move down if there is no impediment, because it contains gravity in itself which is the cause of its natural movement. And if there is also present the external cause which assists the stone in its natural movement, the stone must by necessity move according to its nature. [. . .] The same law applies to other things, and what holds for inanimate things holds also for living beings. For there exists a certain natural movement in living beings—the movement by impulse. For every living being which moves as a living being carries out a movement originating in it according to fate as a result of an impulse. Under these circumstances movements and actions go on in the cosmos according to fate, some for instance through the medium of earth, some through air, some through fire and some through something else, and in the same way some are effected by living beings [. . .] and these, happening according to fate, are said by the Stoics to happen by free will of those living beings.

108

If you roll a cylindrical stone over a sloping steep piece of ground, you do indeed furnish the beginning and cause of its rapid descent, yet soon it speeds onward, not because you make it do so, but because of its nature and because of the ability of its form to roll. Just so the order, the law and the necessity of fate set in motion the classes and beginnings of causes, but the impulse of our design and thought and the actions themselves are determined by each individual's own will and the characteristics of his mind.

109

The assertion that chance is a cause obscure to the human mind is not a statement about the nature of chance but means that chance is a specific relation of men towards cause, and thus the same event appears to one as chance and to another not, depending on whether one knows the cause or does not know it.

110

Chrysippus confuted those who would impose lack of causality upon nature by repeatedly referring to the astragalos and the balance and many other things which cannot alter their falling motions or inclinations without some cause and variance occurring in themselves or outside them. [. . .] For there is no such thing as lack of cause, or spontaneity. In the so-called accidental impulses which some have invented, there are causes hidden from our sight which determine the impulse in a definite direction.

ARCHIMEDES

Aristarchus' hypothesis of a heliocentric universe

111

There are some, king Gelon, who think that the number of the sand is infinite in multitude; and I mean by the sand not only that which exists about Syracuse and the rest of Sicily but also that which is found in every region whether inhabited or uninhabited. Again there are some who, without regarding it as infinite, yet think that no number has been named which is great enough to exceed its multitude. And it is clear that they who hold this view, if they imagined a mass made up of sand in other respects as large as the mass of the earth, including in it all the seas and the hollows of the earth filled up to a height equal to that of the highest of the mountains, would be many times further still from recognizing that any number could be expressed which exceeded the multitude of the sand so taken. But I will try to show you by means of geometrical proofs, which you will be able to follow, that, of the numbers named by me and given in the work which I sent to Zeuxippus, some exceed not only the number of the mass of sand equal in magnitude to the earth filled up in the way described, but also that of a mass equal in magnitude to the universe. Now you are aware that 'universe' is the name given by most astronomers to the sphere whose centre is the centre of the earth and whose radius is equal to the straight line between the centre of the sun and the centre of the earth. This is the common account, as you have heard from astronomers. But Aristarchus of Samos brought out a book consisting of some hypotheses, in which the premises lead to the result that the universe is many times greater than that now so called. His hypotheses are that the fixed stars and the sun remain unmoved, that the earth revolves about the sun in the circumference of a circle, the sun lying in the middle of the orbit, and that the sphere of the fixed stars, situated about the same centre as the sun, is so great that the circle in which he supposes the earth to revolve bears such a proportion to the distance of the fixed stars as the centre of the sphere bears to its surface. Now it is easy to see that this is impossible; for, since the centre of the sphere has no magnitude, we cannot conceive it to bear any ratio whatever to the surface of the sphere. We must, however, take Aristarchus to mean this: since we conceive the earth to be, as it were, the centre of the universe, the ratio which the earth bears to what we describe as the 'universe' is the same as the ratio which the sphere containing the circle in which he supposes the earth to revolve bears to the sphere of the fixed stars. For he adapts the proofs of his results to a hypothesis of this kind, and in particular he appears to suppose the magnitude of the sphere in which he represents the earth as moving to be equal to what we call the 'universe'.

I say then that, even if a sphere were made up of the sand, as great as Aristarchus supposes the sphere of the fixed stars to be, I shall still prove that, of the numbers named in the *Principles*, some exceed in multitude the number of the sand which is equal in magnitude to the sphere referred to.

A method of discovering mathematical theorems

112

Seeing moreover in you, as I say, an earnest student, a man of considerable eminence in philosophy, and an admirer of mathematical inquiry, I thought fit to write out for you and explain in detail in the same book the peculiarity of a certain method, by which it will be possible for you to get a start to enable you to investigate some of the problems in mathematics by means of mechanics. This procedure is, I am persuaded, no less useful even for the proof of the theorems themselves; for certain things first became clear to me by a mechanical method, although they had to be demonstrated by geometry afterwards because their investigation by the said method did not furnish an actual demonstration. But it is of course easier, when we have previously acquired, by the method, some knowledge of the questions, to supply the proof than it is to find it without any previous knowledge. This is a reason why, in the case of the theorems the proof of which Eudoxus was the first to discover, namely that the cone is a third part of the cylinder, and the pyramid of the prism, having the same base and equal height, we should give no small share of the credit to Democritus who was the first to make the assertion with regard to the said figure though he did not prove it. I am myself in the position of having first made the discovery of the theorem now to be published by the method indicated, and I deem it necessary to expound the method partly because I have already spoken of it and I do not want to be thought to have uttered vain words, but equally because I am persuaded that it will be of no little service to mathematics; for I apprehend that some, either of my contemporaries or of my successors, will, by means of the method when once established, be able to discover other theorems in addition, which have not yet occurred to me.

SENECA

Progress in science

113

Why should we be surprised, then, that comets, so rare a sight in the universe, are not embraced under definite laws, or that their beginning and end are not known, seeing that their return is at long intervals? It is not yet fifteen hundred years since Greece

> Counted the number of stars and named them every one.

And there are many nations at the present hour who merely know the face of the sky and do not yet understand why the moon is obscured in an eclipse. It is but recently indeed that science brought home to ourselves certain knowledge on the subject. The day will yet come when the progress of research through long ages will reveal to sight the mysteries of nature that are now concealed. A single lifetime, though it were wholly devoted to the study of the sky, does not suffice for the investigation of problems of such complexity. And then we never make a fair division of the few brief years of life as between study and vice. It must, therefore, require long successive ages to unfold all. The day will yet come when posterity will be amazed that we remained ignorant of things that will seem to them so plain. The five planets are constantly thrusting themselves on our notice; they meet us in all the different quarters of the sky with a positive challenge to our curiosity. Yet it is but lately we have begun to understand their motions, to realize what their morning and evening settings mean, what their turnings, when they move straight ahead, why they are driven back. We have learned but a few years ago whether Jupiter would rise or set, or whether he would retrograde—the term that has been applied to his regression. [. . .] Men will some day be able to demonstrate in what regions comets have their paths, why their course is so far removed from the other stars, what is their size and constitution. Let us be satisfied with what we have discovered, and leave a little truth for our descendants to find out.

PLUTARCH

The attracting force of celestial bodies

114

Thereupon Lucius laughed and said: 'Oh, sir, just don't bring suit against us for impiety as Cleanthes thought that the Greeks ought to lay an action for impiety against Aristarchus the Samian on the ground that he was disturbing the hearth of the universe because he sought to save the phenomena by assuming that the heaven is at rest while the earth is revolving along the ecliptic and at the same time is rotating about its own axis. We express no opinion of our own now; but those who suppose that the moon is earth, why do they, my dear sir, turn things upside down any more than you do who station the earth here suspended in the air? Yet the earth is a great deal larger than the moon according to the mathematicians who during the occurrences of eclipses and the transits of the moon through the shadow calculate her magnitude by the length of time that she is obscured. For the shadow of the earth grows smaller the further it extends, because the body that casts the light is larger than the earth; and that the upper part of the shadow itself is tapered and narrow was recognized, as they say, even by Homer, who called night "nimble" because of the "sharpness" of the shadow. Yet captured by this part in eclipses the moon barely escapes from it in a space thrice her own magnitude. Consider then how many times as large as the moon the earth is, if the earth casts a shadow which at its narrowest is thrice as broad as the moon. All the same, you fear for the moon lest it fall; whereas concerning the earth perhaps Aeschylus has persuaded you that Atlas

Stands, staying on his back the prop of earth
And sky, no tender burden to embrace.

Or, while under the moon there stretches air unsubstantial and incapable of supporting a solid mass, the earth, as Pindar says, is encompassed by "steel-shod pillars"; and therefore Pharnaces is himself without any fear that the earth may fall but is sorry for the Ethiopians or Taprobanians, who are situated under the circuit of the moon, lest such a great weight fall upon them. Yet the moon is saved from falling by its very motion and the rapidity of its revolution, just as missiles placed in slings are kept from falling by being whirled around in a circle. For each thing is governed by its natural motion unless it be diverted by something else. That is why the moon is not governed by its weight: the weight has its influence frustrated by the rotatory motion. Nay, there would be more reason perhaps to wonder if she were absolutely unmoved and stationary like the earth. As it is, while the moon has good cause for not moving in this direction, the influence of weight alone might reasonably move the earth, since it has no part in any other motion; and the earth is heavier than the moon not merely in proportion to its greater size but still more, inasmuch as the moon has, of course, become light through the action of heat and fire. In short, your own statements seem to make the moon, if it is fire, stand in greater need of earth, that is of matter to serve it as a foundation, as something to which to adhere, as something to lend it coherence, and as something that can be ignited by it, for it is impossible to imagine fire being maintained without fuel, but you people say that earth does abide without root or foundation.' 'Certainly it does,' said Pharnaces, 'in occupying the proper and natural place that belongs to it, the middle, for this is the place about which all weights in their natural inclination press against one another and towards which they move and converge from every direction, whereas all the upper space, even if it receives something earthy which has been forcibly hurled up into it, straightway extrudes it into our region or rather lets it go where its proper inclination causes it naturally to descend.'

At this—for I wished Lucius to have time to collect his thoughts—I called to Theon. 'Which of the tragic poets was it, Theon,' I asked, 'who said that physicians

With bitter drugs the bitter bile purge?'

Theon replied that it was Sophocles. 'Yes,' I said, 'and we have of necessity to allow them this procedure; but to philosophers one should not listen if they desire to repulse paradoxes with paradoxes and in struggling against opinions that are amazing fabricate others that are more amazing and outlandish as these people do in introducing their "motion to the centre". What paradox is not involved in this doctrine? Not the one that the earth is a sphere although it contains such great depths and heights and irregularities? Not that people live on the opposite hemisphere clinging to the earth like woodworms or geckos turned bottomside up?—and that we ourselves in standing remain not at right angles to the earth but at an oblique angle, leaning from the perpendicular like drunken men? Not that incandescent masses of forty tons falling through the depth of the earth stop when they arrive at the centre, though nothing encounters or supports them; and, if in their downward motion the impetus should carry them past the centre, they swing back again and return of themselves? Not that pieces of meteors burnt out on either side of the earth do not move downwards continually but falling upon the surface of the earth force their way into it from the outside and conceal themselves about the centre? Not that a turbulent stream of water, if in flowing downwards it should reach the middle point, which they themselves call incorporeal, stops suspended or moves round about it, oscillating in an incessant and perpetual see-saw? Some of these a man could not even mistakenly force himself to conceive as possible. For this amounts to "upside down" and "all things topsy-turvy", everything as far as the centre being "down" and everything under the centre in turn being "up". The result is that, if a man should so coalesce with the earth that its centre is at his navel, the same person at the same time has his head up and his feet up too. Moreover, if he digs through the further side, his bottom in emerging is up, and the man digging himself "up" is pulling himself "down" from "above"; and, if someone should then be imagined to have gone in the opposite direction to this man, the feet of both of them at the same time turn out to be "up" and are so called.

'Nevertheless, though of tall tales of such a kind and number they have shouldered and lugged in—not a wallet-full, by heaven, but some juggler's pack and hotchpotch, still they say that others are playing the buffoon by placing the moon, though it is earth, on high and not where the centre is. Yet if all heavy body converges to the same point and is compressed in all its parts upon its own centre, it is no more as centre of the sum of things than as a whole that the earth would appropriate to herself the heavy bodies that are parts of herself; and the downward tendency of falling bodies proves not that the earth is in the centre of the cosmos but that those bodies which when thrust away from the earth fall back to her again have some affinity and cohesion with her. For as the sun attracts to itself the parts of which it consists so the earth too accepts as her own the stone that has properly a downward tendency, and consequently every such thing ultimately unites and coheres with her. If there is a body, however, that was not originally allotted to the earth or detached from it but has somewhere independently a constitution and nature of its own, as those men would say of the moon, what is to hinder it from being permanently separate in its own place, compressed and bound together by its own parts? For it has not been proved that the earth is the centre of the sum of things, and the way in which things in our region press together and concentrate upon the earth suggests how in all probability things in that region converge upon the moon and remain there. The man who drives together into a single region all earthy and heavy things and makes them part of a single body—I do not see for what reason he does not apply the same compulsion to light objects in their turn but allows so many separate concentrations of fire and, since he does not collect all the stars together, clearly does not think that there must also be a body common to all things that are fiery and have an upward tendency.'

PTOLEMY

ALMAGEST

Book I, Chapter 1 Preface

115

Real philosophers have rightly, I think, distinguished the theoretical part of philosophy from the practical. It is true that, before this distinction was made, theory and practice were intimately connected; nevertheless, there is a considerable difference between them, not only because men can have moral virtues even without instruction, whereas it is impossible to acquire an understanding of the universe without study, but also, and especially, because in the case of the former, improvement is the result of continuous practical activity, while in the latter progress springs from an increase in theoretical knowledge. It is for this reason that, where practical conduct is concerned, I think it our duty to regulate it in accordance with intuitively acquired principles, in order that we may never lose sight, even in incidental and unimportant matters, of those things which contribute to the shaping of an honourable and disciplined character. As for intellectual study, we ought to devote ourselves to the teaching of theoretical science (which has many interesting and important branches) and particularly to the teaching of that branch which is properly called mathematics.

Aristotle, again, rightly distinguishes three main kinds of theoretical philosophy: physics, mathematics and theology. Whatever exists is what it is because of matter, form and movement; none of these three can be *seen*, without the other two, in the object to which it belongs; by itself it can only be *thought of*. Hence, if one takes a simple and straightforward view, one would think that an invisible and immovable god is the first cause of the first movement of the universe; and the science that deals with the investigation of these matters is theology. Such a power can be thought of only in the most sublime regions of the world, entirely separate from those things which are perceived by our senses.

The branch of science that investigates the qualities of ever-changing matter (i.e. such things as the white and the warm, the sweet and the soft, etc.) one would call physics, since qualitative change takes place, on the whole, in the sublunary world which is the world of change and corruption.

Again, that branch of science which (a) deals with forms and local motion, i.e. those movements which lead to changes in location, and which (b) investigates shape, quantity, magnitude, space, time and similar concepts, is to be called mathematics. Its subject matter lies, as it were, between that of theology and that of physics; it can be conceived of both by means of sense perception and without it; and, thirdly, it belongs to all beings, mortal as well as immortal.

It shares in the variability of those things that constantly change in respect of their form which is inseparable from them; it is equally constant in the invariability of form in respect of eternal things which are of ethereal nature.

Now, one would be right in calling the other two branches of theoretical philosophy speculative guesswork rather than scientific knowledge; in the case of theology one would point to the fact that its subject matter is invisible and incapable of being critically examined; in the case of physics to the uncertainty and instability of matter. Hence there is no hope that philosophers will ever agree on these things. Only mathematics, if approached with an inquiring mind, can offer reliable and stable knowledge; for proof there is unambiguous and indisputable, since it uses the methods of arithmetic and geometry.

This is the reason why I have devoted my best endeavours to this science and particularly to that branch of it which is concerned with the attempt to observe and understand divine and heavenly bodies. For this science alone investigates things which are eternal and unchanging; and therefore it is itself eternal and unchanging in respect of the clear and well-ordered nature of the understanding which it provides. Such stability is the characteristic of true Science. And further: mathematics can be helpful to other sciences, and, in its assistance to them, its contribution can be as effective as their own.

It can be particularly helpful towards progress in theology since it alone can draw valid conclusions about the separate and umoved force on the basis of the phenomena of heavenly motions and their systematic order. These phenomena are, it is true, phenomena of bodies which are perceived by the senses, of bodies which are moving and being moved; but they are eternal and unchangeable.

Its contribution to physics is equally important. The general properties of matter become apparent as a result of specific behaviour in the case of local motion; thus, rectilinear movement is a specific characteristic of the corruptible and perishable, circular movement of the incorruptible and eternal, centripetal movement of the heavy or passive, and centrifugal movement of the light and active.

Again, this science, more than any other, is capable of providing men with discriminating standards of moral conduct and gentlemanly character; through observing the uniformity, order, symmetry and regularity of the movements of the heavenly bodies its followers become lovers of this divine beauty. Through habituation it makes a like disposition of order and discipline in their souls into, as it were, a second nature.

This love for the science of things which are eternal and unchanging we shall constantly try to foster. We shall do so by studying what has been already achieved in understanding these matters by those men of the past who have approached them in a genuinely scientific spirit and also by resolving to make such a contribution ourselves as has been made possible by the accumulation of material since their time.

We shall try to explain as briefly as possible (and in such a way as will enable those who have already made some progress to follow our explanations) everything that has been discovered up to the present time. For the sake of completeness we shall set out in their proper order all those matters which are useful for the theory of the heavenly bodies. On the other hand, in order not to make this account too long, we shall treat only summarily those matters in which certainty has been attained by the ancients, while dealing in much greater detail with matters which, either, have not yet been understood and explained at all, or, at any rate, not as fully as they might have been.

Chapter 2 On the order of the theorems

We shall begin by considering, in general, the relation of the Earth as a whole to the Heaven as a whole and then, in particular, we shall, first, give an account of the position of the ecliptic, of the places of the inhabited part of the Earth, of the differences which arise between them from one horizon to the other as a consequence of the inclination of the sphere. A preliminary discussion of these matters will facilitate the investigation of the rest.

It will be our second task to explain the motions of the Sun and of the Moon and the phenomena connected with them; for without a prior understanding of these things it would be impossible to work out in detail a theory of the stars. Lastly, in offering this theory, we begin with an account of the sphere of the fixed stars, and, after this, there follows the theory of the five so-called planets.

In our attempt to establish these theories we shall use as principles and foundations for our arguments the visible phenomena and the undisputed observations both of the ancients and of our contemporaries. We shall then fit together the successive and, up to that point, theoretically unconnected pieces of observation by means of demonstrations based on geometrical method.

Our general discussion will treat the following preliminary points:

1. The heaven is spherical and has a spherical motion.
2. Taken as a whole, the Earth, too, is spherical in shape.
3. The Earth is situated in the middle of the whole heaven like a centre.
4. The size of the Earth and its distance from the sphere of the fixed stars are such, that it has to the sphere of the fixed stars the ratio of a point to a sphere.
5. The Earth itself has no local motion.

Each of these points we shall discuss briefly so that they may all be present to our memory when needed.

Chapter 3 The Heavens revolve as a sphere

It is reasonable to assume that the ancients formed their first notions on these matters as a consequence of observations such as the following:

They saw the Sun, the Moon and the other stars moving always on parallel circles from east to west. They saw that these bodies begin by moving upwards from below, and, as it were, out of the Earth itself, that they rise gradually to a high point and that from there they sink down again in the same way until, in the end, they fall, if I may so put it, back into the Earth and vanish from sight.

Having remained invisible for some time they then rise and set again. They also noticed that there is a regular correspondence between the times and the places of risings and settings.

They were led to the notion of a spherical pattern by the observation of the circular revolution of those stars that are always visible; this revolution takes place around one and the same centre. Now, clearly, this point is the pole of the heavenly sphere, for the stars nearer to it move in smaller circles, whereas those further away describe circles that are greater proportionately as their distance is greater. Stars even further distant become invisible for a time. In the case of these it was observed that those nearer the stars that are always visible remain invisible for a short time only, but those further away for a proportionately longer time. Thus, to begin with, they based their notion of sphericity on such observations as these; subsequent consideration enabled them to draw further conclusions from them since all phenomena bear witness against any theory of non-spherical motion.

Let us suppose (as some thinkers did) that the stars move in a straight line to infinity. How could one then account for the fact that all stars seem to move every day from the same starting-point? How could the stars turn round, having set out on their path into infinity? Or, if they do turn round, why is this turning round not observed? Why is it that their size does not gradually become smaller and smaller until they finally disappear altogether? What really happens is this: they actually seem to be bigger when they are just on the point of disappearing; and when they do disappear it is because the surface of the Earth is in the way and, so to speak, cuts off our line of vision. [. . .]

Briefly, if one assumes anything else for the motion of the heavens, the distances from the Earth to the heavenly bodies will inevitably turn out to be unequal, no matter what hypothesis is adopted concerning the position of the Earth. Hence, at every revolution the magnitudes of the stars and the distances of these from each other would appear to the same observers on the Earth unequal since they would be seen from different distances at different times. This, however, is evidently not the case. It is true that, just above the horizon, magnitudes of heavenly bodies seem to be greater. This, however, is not due to a diminution in their distance but to vaporization of the moisture surrounding the Earth. This process of refraction takes place in the space between our eyes and the stars just as objects thrown into water seem to be greater in proportion as they go lower down.

We are led to the notion of a spherical pattern by other considerations too. There is no other hypothesis on the basis of which time-measuring instruments would agree with each other. Secondly, heavenly bodies have a movement more unimpeded and smoother than any other: now, smooth and unimpeded movement is a characteristic of the circle (in the case of plane figures) and of the sphere (in that of solids). And further: of different isoperimetric figures those that have more angles are greater; hence the circle has an area greater than any other figure of equal perimeter, and the sphere a greater volume than any other solid; but the heaven is greater than any other body.

Physical considerations lead to the same conclusion: aether is the least dense and the most homogeneous of all bodies; the surfaces of all homogeneuos bodies are homogeneous; the only homogeneous surfaces are (a) the circular, among plane figures and (b) the surface of the sphere among solids; but the aether is not a plane; it is a solid; hence it must be spherical.

Similarly, nature has formed all earthly and corruptible bodies out of round (though not homogeneous) figures; but divine bodies in the aether are formed out of constituents that are both homogeneous and spherical, since, if they were plane or disc-shaped they would not appear to have a circular outline to all those who observe them at the same time from different places. It is therefore reasonable to assume that the aether surrounding these bodies (a) being of a like nature should also be spherical, and (b) being homogeneous, should have a motion that is circular and uniform.

Chapter 4 The earth taken as a whole is sensibly spherical

That the Earth, too, taken as a whole, is sensibly spherical can best be understood in the following way: the Sun and the Moon and the other stars do not rise and set at the same time for all men on the Earth. They rise earlier for observers in the east and later for those in the west. We find that the simultaneous appearances of eclipses, particularly lunar eclipses, are not registered as taking place at the same hours, that is to say, at the same time interval from mid-day, in all places; the times registered by observers further east are consistently later than those further west. Now, we find that the difference in time is proportional to the distance between the places in an east–west direction; hence we conclude that the surface of the Earth is spherical; for, in respect of its curvature, the Earth, taken as a whole, is homogeneous; and thus the times of successive occultations by the earth of heavenly bodies are always proportional to each other. If the shape of the Earth were other than spherical this would not be the case, as can be shown by the following considerations: if the Earth were concave observers in the west would see the stars rise before they are seen by observers in the east; if the Earth were flat all observers anywhere on the Earth would simultaneously see the stars rise or set; if the Earth were a solid with triangular or square surfaces (such as a pyramid or a cube) or, for that matter, a solid with surfaces of any other polygonal shape, the stars would rise or set simultaneously for all observers on the same plane surfaces of these solids.

None of those things, however, is true. [. . .]

Yet, the fact is that the further north we go the greater is the number of stars in the southern sky which become invisible while the number of stars visible in the northern sky increases. It is clear, therefore, that here, too, *all* the facts connected with the curvature of the Earth point to it being a special kind of curvature, namely sphericity; we have seen this already when we argued earlier that the curvature of the Earth is responsible for the proportionality of occultation phenomena in the crosswise direction.

A further argument for the sphericity of the Earth is the fact that if we sail towards a mountain or other high-lying place (from whatever angle and in whatever direction) we notice that its height gradually increases as if it were emerging gradually from the sea, while earlier it was hidden because of the curvature of the water's surface.

Chapter 6 The earth has the ratio of a point to the sphere of the heavenly bodies
That the Earth has sensibly the ratio of a point to its distance from the sphere of the so-called fixed stars is clearly shown by the fact that the magnitudes of the stars and their distances from each other as observed at the same time from any part of the Earth, are in every way the same; no discrepancies in these respects are ever found between observations of the same stars from different geographical latitudes.

We may add this further point: the end-points of gnomons erected on whatever point of the Earth, and, similarly, the centres of armillary spheres, perform the same function as the real centre of the Earth; that is to say, they validate the sightings and the rotations of the shadows as much in conformity with the hypotheses concerning the phenomena, as if the line of observation really passed through the centre of the Earth.

That this is so is clearly indicated also by the fact that the planes through our eyes, which we call horizons, cut the whole sphere of the heaven everywhere into halves. This would not be the case if the size of the Earth were perceptibly significant in relation to the distance of the heavenly bodies; only the plane through the centre of the Earth could cut the sphere into two equal halves, whereas planes through any point on the surface of the Earth would always cut the sphere in such a way that the section below the Earth would be bigger than that above it.

Chapter 7 The earth has no local motion
In the same way we can show that it is impossible for the Earth to have any movement in the crosswise direction we discussed above or, indeed, to move at all from its location at the centre. For, if it could so move, the consequences would be the same as those which would follow if the Earth were situated at a place other than the centre.

I think it superfluous to inquire into the causes of centripetal motion once it has become clear from the phenomena themselves that (a) the Earth occupies the centre of the Universe and (b) all weights fall towards it.

There is one simple way in which we can understand centripetal fall; on any point on the Earth which, as has been demonstrated, is spherical and occupies the centre of the universe, the line of fall (I mean free fall) of heavy bodies is always and everywhere perpendicular to the tangential plane through the point at which the falling body hits the ground; from this it follows clearly that falling bodies if they were not held up by the surface of the Earth would be carried in their fall all the way down to the centre, since the straight line leading to the centre is always perpendicular to the tangential plane through the point of intersection of that line and the surface of the sphere.

Those who regard it as paradoxical that a mass such as that of the Earth should stand firmly fixed in one place without moving from it seem to me to be mistaken in that they are not comparing like with like: they employ reasoning appropriate to what one would expect to happen in the case of small bodies such as their own, but not to the behaviour peculiar to the Universe. I think they would no longer regard this as so strange if they understood that the size of the Earth which seems to them so great has to the whole body surrounding it the ratio of a point. They would then understand that it is possible for a relatively very small body to be supported and buttressed by the absolutely greatest and homogeneous body exerting force on it equally and uniformly from all directions. There is no 'up' or 'down' in the Universe with respect to the Earth, precisely as one would not speak of 'up' or 'down' in the case of an ordinary sphere; and, as regards composite substances in the Universe, those that are light and composed of fine particles will rush outwards to the periphery while appearing to move in what, in each case, looks like an 'upward' direction, precisely as in our case that which is above our heads (we use the word 'up' here) points in the direction of the periphery; heavy substances consisting of coarse particles move towards the centre, while appearing to fall 'downwards'; precisely as, to use the example of men again, that which is in the direction of our feet (we call it 'down') points towards the centre of the Earth.

Their common tendency towards the centre is, of course, due to the pressures and counterpressures working on each other equally and uniformly from all sides. It is thus not difficult to understand that the whole solid mass of the Earth which is so very big in proportion to the bodies falling on to it absorbs their fall and is unmoved by the impact of their very small weights (particularly since these weights exert their pressure from all sides equally). In any case, if the Earth had the same tendency to fall as other heavy bodies, it would clearly overtake them all in its downward movement because of its size, which is so greatly in excess of theirs; and the animals and other heavy bodies would be left behind floating on the air, and in the end the Earth would soon enough fall out of the heaven altogether. Now this is too ridiculous even to think about.

In spite of the fact that they have not been able to answer these arguments some thinkers have in the past worked out a theory which seems plausible to them, and which, they think, is in agreement with all the phenomena. They argue that we could assume, for example, that the heaven is immobile while the Earth turns in a west–east direction around the same axis very nearly one complete rotation each day; or, alternatively, one could attribute a certain amount of movement to both heaven and Earth as long as both these movements take place around the same axis and at speeds which are so calculated as to make it possible to account for the one overtaking the other.

Now, as far as the *heavenly phenomena* are concerned, such a theory might well be acceptable—indeed it has the virtue of greater simplicity. What its proponents, however, do not take into account is the fact that the absurdity of such a theory would become apparent on consideration of the characteristic properties of *things here on Earth and in the air*. Suppose we conceded to them that the finest and lightest substances, against their natural tendencies, either do not move at all or that they move in the same way as substances of exactly opposite tendencies, that is to say, heavy, coarse-particled substances (whereas, in fact, of course, it is clear that airy and fine-particled substances move faster than more earthy ones); suppose we conceded further that the more coarse-particled and heaviest substances have a movement proper, fast and uniform (whereas in fact again it is generally agreed that they sometimes do not even readily respond to movement initiated by other bodies); if we conceded all this, they would still have to admit that the rotation of the Earth would be more violent and rapid than any of the movements that take place around it, since it would return to its original position in such a very short time. As a consequence of such a rapid movement of the Earth all things on it that are not firmly fixed would appear to be moving in a direction opposite to that in which the Earth rotates: no cloud would ever be seen passing towards the east, nor would anything else, whether it be a flying animal or an object thrown up; for the Earth would always overtake everything in its eastward movement so that all other things would appear to be left behind and therefore to be moving westwards.

If it be argued that the air may be carried round together with the Earth in the same direction and at the same speed it would still be true that other bodies in the air would always seem to be left behind the Earth and the air in their common rotation; again, if these bodies were to be assumed to be, as it were, attached to the air and thus to be carried around with it then it would no longer be possible for any objects to appear to move forward or to be left behind; they would always remain fixed in one place, they would never change their position from one location to another, no matter whether they are flying or have been thrown (this, of course, is contrary to what is clearly seen happening); thus, it would follow from the hypothesis that the Earth moves that these bodies could never move, either slowly or fast, at all.

Book XIII, Chapter 2 On the hypotheses concerning heavenly movements

116

One must not think that these hypotheses are too elaborate just because our technical devices are inadequate. It would be wrong to compare human contrivances with things divine, nor would it be right to derive arguments concerning the latter from examples that are altogether different. For what greater difference could there be than that which exists between, on the one hand, beings that always remain the same and, on the other, those that always change? What could be more different from things affected by any change in their environment than things which do not change even at their own internal impulse? It is, of course, right to try to adopt the simplest possible hypotheses consonant with the movements in the heaven; but where this is not possible one must adopt those hypotheses that lend themselves to such harmonization. For, once all the phenomena have been systematically accounted for on the basis of the hypotheses, why should it then still be thought strange that the heavenly movements are so complex? There is, after all, nothing in the nature of heavenly bodies which would prevent such complexity; what is, however, in their nature is that they should be able to yield and make way for the natural movements of each of the other heavenly bodies, even though these movements happen to be in a direction opposite to their own. Thus, they can pass through literally all heavenly matter and penetrate it with their light. They travel freely not only on their own particular circles but also around the spheres themselves and around the axes of the revolutions.

The complexity of the different movements and the fact that they are so closely inter-related make it very difficult to represent their free and easy movement by means of artificial contrivances. On the sky, however, this complexity is not in any way affected by the difficulties inherent in the mixture of these movements.

There is this further point: we must not judge the 'simplicity' of heavenly movement by the criteria of simplicity derived from human experience, since even these criteria are not the same for all men. Looked at purely from a human point of view, nothing that happens in the heaven would appear simple, not even the immutability of the first movement; for it is precisely this eternal constancy which presents us men not just with a difficulty but with an impossibility. One must start instead from the immutability of heavenly natures and their movements. They will then all appear 'simple', indeed more so than the things here on Earth which seem simple, since one cannot imagine any disturbance or difficulty in their circuits.

PROCLUS

Doubts concerning astronomical hypotheses

117

My dear friend: The great Plato thinks that the real philosopher ought to study the sort of Astronomy that deals with entities more abstract than the visible heaven, without reference to either sense perception or ever-changing matter. In that world of abstract entities he will come to know slowness itself and speed itself in their true numerical relationships. Now, I think, you wish to bring us down from that contemplation of abstract truth to consideration of the orbits on the visible heaven, to the observations of professional astronomers and to the hypotheses which they have devised from these observations, hypotheses which people like Aristarchus, Hipparchus, Ptolemy and others like them are always writing about. I suppose you want to become acquainted with their theories because you wish to examine carefully all the theories, as far as that is possible, with which the ancients, in their speculations about the universe, have abundantly supplied us.

Last year, when I was staying with you in Central Lydia, I promised you that when I had time, I would work with you on these matters in my accustomed way. Now that I have arrived in Athens and heaven has freed me from those many unending troubles, I keep my promise to you and will, in what follows, explain to you the real truth which those who are so eager to contemplate the heavenly bodies have come to believe by means of long and, indeed, endless chains of reasoning. In doing so I must, of course, pretend to myself to forget, for the moment at any rate, Plato's exhortations and the theoretical explanations which he taught us to maintain. Even so, I shall not be able to refrain from applying, as is my habit, a critical mind to their doctrines, though I shall do so sparingly, since I am convinced that the exposition of their doctrines will suggest to you quite clearly what the weaknesses in their hypotheses are, hypotheses of which they are so proud when developing their theories.

I think I ought to begin with the following question: what were the phenomena that led them to an investigation of the causes of each of them, because they distrusted the obvious and apparent? They started from the perfectly proper assumption that the movement of divine bodies must be circular and ordered, even though 'circular' does not mean quite the same in all cases, and even though it is sometimes mixed with a movement of a different kind; but, though different in kind from that which is circular in the ordinary sense of the word, it is nevertheless completely ordered and regular.

An unchanging movement in accordance with one unchanging principle and an order always in agreement with itself surely belongs to the most divine of the visible bodies, particularly in the theories of those who assume that all the heavenly bodies revolve in a way that conforms to reason. For reason always establishes order in all matters with which it is concerned. Holding fast to this notion as a secure guide-line they were rightly dissatisfied with the apparent disorder. They therefore asked themselves which hypotheses could represent the paths of heavenly bodies on those circles as rational instead of as irrational, and as determined by numerical relationships proper to each of them, instead of as indeterminate and irregular.

To show you clearly which observed phenomena disturbed them as being unworthy of divine nature and stimulated them to the formulation of these hypotheses (and thus, incidentally, to the creation of astronomy) I shall discuss each of them from the beginning.

We have now completed the outline of astronomical hypotheses. However, before I end, I wish to add this: in their endeavour to demonstrate that the movements of the heavenly bodies are uniform the astronomers have unwittingly shown the nature of these movements to be lacking in uniformity and to be subject to outside influences. What shall we say of the eccentrics and the epicycles of which they speak so much? Are they only conceptual notions or do they have a substantial existence in the spheres with which they are connected? If they exist only as concepts, then the astronomers have passed, without noticing it, from bodies really existing in nature to mathematical notions and, again without noticing it, have derived the causes of natural movements from something that does not exist in nature. I will add further that there is absurdity also in the way in which they attribute particular kinds of movement to heavenly bodies. That we conceive of these movements, that is not proof that the stars which we conceive of as moving in these circles really move anomalously.

On the other hand, if the astronomers say that the circles have a real, substantial existence, then they destroy the coherence of the spheres themselves on which the circles are situated. They attribute a separate movement to the circles and another to the spheres, and again, the movement they attribute to the circles is not the same for all of them; indeed, sometimes these movements take place in opposite directions. They vary the distances between them in a confused way, sometimes the circles come together in one plane, at other times they stand apart, and cut each other. There will, therefore, be all sorts of divisions, foldings and separations.

I want to make this further observation: the astronomers exhibit a very casual attitude in their exposition of these hypothetical devices: why is it that, on any given hypothesis, the eccentric or, for that matter, the epicycle moves (or is stationary) in such and such a way while the star moves either in direct or retrograde motion? And what are the explanations (I mean the real explanations) of those planes and their separations? That they never explain in a way that would satisfy our yearning for complete understanding. They really go backwards: they do not derive their conclusions deductively from their hypotheses, as one does in the other sciences; instead, they attempt to formulate the hypotheses starting from the conclusions, which they ought to derive from the hypotheses; it is clear that they do not even solve such problems as could well be solved.

One must, however, admit that these are the simplest hypotheses and the most fitting for divine bodies, and that they have been constructed with a view to discovering the characteristic movements of the planets (which, in real truth, move in exactly the same way as they *seem* to move) and to formulating the quantitative measures applicable to them.

JOHANNES PHILOPONUS

Irregularities and harmony

119

Perhaps there are no things which are contrary to nature in an absolute sense, but one has to distinguish between things of a partial character which are not natural and are contrary to nature, and things which are a whole and are natural as well as in accordance with nature. [. . .] A deformity from birth is not natural for man nor is it according to nature. But taking Nature as a whole in which nothing is contrary to nature (because there is no evil in the whole), a deformity from birth is natural and according to nature. For it happens when total nature transforms the underlying substance and makes it unfit to receive the proper form of partial nature. What I mean is this: If the environment is disposed in a certain way by the revolution of the heavens and does something to the substance of man at his birth so that he becomes unfit to receive the form normally put into him by Nature, his nature will not attain its purpose because of the unfitness of his substance, and another form will emerge, contrary to nature with respect to its partial character, but in accordance with nature and natural as seen within the totality of Nature. It will not be contrary to nature with regard to this totality, because corruption is not contrary to nature if indeed generation is in accordance with nature. This must be the case, because generation of one part means destruction of another.

I will give you an illustration that will explain what happens with things contrary to nature: Suppose that a lyre player tunes his instrument according to one of the musical scales and is then ready to begin his music. Suppose, however, that someone else loosens the tension of some of the strings or all of them, or rather let us assume for the sake of this illustration that the strings are affected by the state of humidity of the environment and thus get out of tune. Now the player's fingers move the strings so that a perfect melody would result if the strings were still properly tuned; when the player strikes the lyre thus, the substance of the strings does not perform the melody that he had in mind, but instead an unmusical, distorted and indefinite sound is produced. The same happens with organic nature. When the substance underlying the human form, or that of another being, becomes unfit through the constellation of the revolving heavens, it comes to pass that its actualization fails. And just as we do not say that the sound of the untuned lyre is artistic or in accordance with art, although an artist produces it, neither do we say in the case of organic nature that it was a natural event, because it did not happen according to the well-defined laws of nature. However, with regard to the nature of the whole we do say that it was a natural event, for it is in accordance with the nature of the whole that it destroys one thing when it creates another.

Fitness and actualization

120

Some people claim that the soul moves the body so to say by a mechanical device, as if the body were pushed by the motion of the soul, as when children in their play make small, very thin hollow balls of wax and enclose in them some bluebottles or beetles so that, when these move, the ball is set in motion. Or like an animal which is enclosed in a cage moving the cage by its own pushing movements. [. . .] But if the soul moves the body and pushes it so to say mechanically, Aristotle says that it could leave the body and then re-enter and move it again, and thus cause the resurrection of dead animals. For what could prevent this if the motion is produced by purely mechanical means? However, one can object to this and say that Aristotle was wrong when he maintained that if the soul moves the body by moving itself, it could re-enter and bring the dead to life. For instance, take a pillar whose action as a lever lifts up a wall or something similar. When the pillar slips the wall collapses and its joints break up and nobody is able to lift it again by applying the lever. One can find other similar examples. The argument is that in such cases mechanical devices alone are not enough, but there has to be a fitness of the object to be lifted. [. . .] Those explaining motion by some sort of mechanism alone do not attribute to the body a fitness nor any natural capacity for motion. We, however, assume that through the presence of the soul some vital force is implanted in the body, and accordingly its absence leads to a collapse of the body. It is therefore probable that the soul cannot re-enter the body after the removal of that fitness which had been implanted in the body at the beginning. [. . .] Similarly a stick pushed against a door cannot move the door when it has not the fitness necessary for being moved but, having this fitness, it will move when pushed by the stick. But it will not do so when fastened by nails or when the hinges are loose. Everything set in motion by something else generally needs a certain specific fitness. [. . .] Democritus, too, among all the others could have said that assuming that the soul moves the body by moving itself, one must presuppose a certain fitness of the body to receive the motions of the soul, for instance such and such an order and position of such and such atoms. Others again would have to presuppose a harmony of certain elements whose dissolution would make it impossible to push the body, exactly as when the shape or quality of the wax is lost, for instance when it becomes soft or undergoes some other change, the animal enclosed in it cannot move it any more.

Natural motion, forced motion and impetus

121

First we must ask those who hold this view the following question: If somebody

throws a stone by force, is it by pushing the air behind the stone that he forces
it into motion against its natural direction? Or does the thrower impart to the
stone a kinetic power? If he does not impart any power to the stone, but moves
it merely by pushing the air, and similarly the bowstring the arrow, what is the
use of the contact between the hand and the stone or the string and the notched
end of the arrow? [. . .] And further, if string and arrow or hand and stone are in
direct contact, and there is nothing between them, what air behind the missile is
being moved? And if the air at the side is moved, what connection has it with
the thrown body, for it is not within its path? From these and many other
considerations one can see that it is impossible that things moved by force should
move in this way. It must rather be that some incorporeal kinetic power is
imparted by the thrower to the object thrown and that the pushed air contributes either nothing or very little to this motion. If, then, objects are moved
by force in this manner, it is evident that if an arrow or a stone is projected by
force and contrary to nature in a vacuum, the same thing will happen much
more easily, nothing being necessary except the thrower. [. . .] And doubtless
this theory, proved by the facts—namely that some incorporeal kinetic force
is imparted by the thrower to the body which is thrown during the time the
thrower is in contact with the missile—is no more difficult to accept than that
certain forces reach the eyes from objects which are seen, as Aristotle thinks.
Indeed, we can see from the colours which stain solid bodies exposed to them,
that certain forces of an incorporeal form are emitted when the sun's rays pass
through a transparent coloured object. [. . .] It is thus evident that certain forces
can reach bodies in an incorporeal way from other bodies.

122

Air does not only possess the principle of upward motion but also that of
downward motion. For if some of the earth or the water beneath it is removed,
air immediately fills its place and likewise when something above is removed,
air moves upward. If one may attribute downward motion to the force of the
vacuum and not to a natural principle, what prevents us from saying that
upward motion too has the same cause? Air may be moving up if there happens
to be an empty space, and otherwise not.

The unity of heaven and earth

123

'One star differeth from another star in glory,' says Paulus. Indeed, there is much
difference among them in magnitude, colour and brightness, and I think that the
reason for this is to be found in nothing else than the composition of the matter
of which the stars are constituted. They cannot be simple bodies, for how could
they differ but for their different constitution? This also causes a very great

variation in sublunar fires, in thunderbolts, comets, meteors, shooting stars and lightning. Each of these fires is produced in principle when more or less dense matter is penetrated and inflamed. Terrestrial fires lit for human purposes also differ according to the fuel, be it oil or pitch, reed, papyrus or different kinds of wood, either in a humid or in a dry state.

124

The fact that in all past time, heaven seems not to have changed either as a whole or in its parts should not be taken as a proof of its being imperishable and un-created. There are animals which live longer than others, and parts of the earth, such as mountains, stones and hard metals, are all roughly speaking as old as time, and there is no record of Mount Olympus having a beginning, or growing or diminishing. Moreover, for the survival of mortal beings it is necessary that their principal parts should persist in their own natural state. Thus, as long as God wants the universe to exist its principal parts have to be preserved, and admittedly the heaven as a whole and in its parts is the principal and most essential part of the universe.

125

It can be proved that the totality of the elements does not possess infinite power. For experience shows that the smaller a quantity of matter the more rapid its decay, and the greater it is the slower the process. Let us assume that a ladleful of water could last a year, and that each equal quantity of water will last the same time. The existing mass of water being finite it could be measured out by ladles and divided into a finite number of them. As each of these parts will be of finite power the same will hold for the whole which is the sum of all the parts; and similarly with the other elements. It is thus proved that each of the four elements in its totality possesses only finite power.

126

If the celestial bodies are composite, and composite things imply decomposition, and things implying decomposition imply decay (for the decomposition of the elements is a decay of the composite, and what implies decay has no omnipo-tence), it follows that the things in heaven, by their own nature, have no omnipotence. [. . .] Moreover, those who declare heaven to consist not of the four elements but of the fifth essence do assume it to be a composition of the underlying fifth matter and of the solar or lunar form. However, if one abstracts the forms of all things, there obviously remains their tri-dimensional extension only, in which respect there is no difference between any of the celestial and the terrestrial bodies.

127

Matter may become fire out of another fire preceding it, and this out of yet another one, but this chain will finally lead to a fire which was not produced in this way but by friction or some cause different from fire. Similarly it is not impossible that in all processes of generation of one thing out of another something analogous could be assumed. All things generated by nature one out of another are equally subject to a beginning of their existence. Each species—and this particularly holds for the primary elements—must have a first member which derived its origin not from a similar or dissimilar one preceding it, but was created by God together with the formation of the universe.

128

If people assign the place above to the Divinity, this is not yet to be taken as a proof that heaven is imperishable. Because those too who believe the holy places and temples to be full of gods and raise their hands towards them do not assume these dwellings to be without beginning or imperishable but regard them only as a place more fit than others to be inhabited by God.

SECTION II

The Middle Ages

SELECTED, INTRODUCED AND EDITED BY
PROFESSOR SHLUMO PINES

Section II: The Middle Ages

INTRODUCTION

Two distinct periods can be discerned as far as the history of scientific thought in the Middle Ages is concerned.

Before the thirteenth century Arabic science—whose protagonists were Moslems, Christians, and Jews—was dominant, and its supremacy was recognized in the Christian West. Many of the principal Greek scientific texts were available in reasonably good Arabic translations. There were also Arabic versions of Sanskrit works.

For the first time in the history of Western civilization, Indian mathematical, astronomical and other scientific notions and theories were taken over (by the Arabic scientists, and—immediately before or at the time of the Moslem conquests—by their Syriac predecessors), then developed and incorporated into the system of the sciences.

Lively scientific debates were taking place all over the Moslem world, and in certain disciplines the Moslems progressed beyond the Greeks. Optics, as developed by Ibn al-Haytham, is a case in point.

This state of affairs existed throughout the thirteenth century and after in the countries of the Moslem East, particularly in Iran. Scientific discussions continued with almost unreduced vigour, and occasionally important new ideas and theories were put forward. Yet from the point of Latin Europe—which was coming to terms with the abundance of philosophical and scientific texts provided by translations from Greek and Arabic—this century marked the end of the unquestioning acceptance of Arabic intellectual superiority. Samuel Ibn Tibbon, a Jewish philosopher and translator, who lived in the thirteenth century in the south of France, in a Hebrew treatise entitled *Let the Waters be gathered together*, put it thus (Pressburg edition of 1837, page 175): 'I have seen that the true sciences have become very widely known amongst the nations in whose countries and under whose dominion I live, in fact more widely known than in the countries of Ishmael, [i.e. Islām].'

From another point of view, the development from their first beginnings up to the thirteenth century of physical theories in the world of Islām presents a certain parallel to the history of such theories in Latin Europe from the twelfth or thirteenth century until the emergence of the classical physics of Galileo.

In both civilizations, Aristotelian orthodoxy dominated for a time, in spite of its incompatibility with the religious teachings of Islam, Christianity and Judaism. It was the thought of a considerable portion of the intellectual élite. But in both civilizations this orthodoxy eventually underwent a process of

123

disintegration, due in part to built-in difficulties inherent in the systematization of Greek science attempted in the Middle Ages.

One of the main difficulties arose from the contradictions between the laws of physics laid down by Aristotle and the Ptolemaic system. A text of Averroes figuring in the present anthology deals with this problem from the point of view of Aristotelian orthodoxy, while one of Maimonides illustrates another tendency. While Maimonides has an implicit belief in the correctness of Aristotelian physics as an explanation of the phenomena of the sublunar world, he wishes to point out that Aristotle's views concerning celestial physics are subject to doubt. Like Kant, he is interested in showing that some of the so-called verities and certainties of science are unfounded; as a consequence room can be left for faith.

Nevertheless the difficulties which Maimonides emphasizes were real ones. The problem posed by the irreconcilability of Aristotelian theory and the Ptolemaic system continued to bedevil scientific debates up to the time of Copernicus. The theory of impetus is indicative of another major difficulty which Aristotelians had to face both in Islam and Christian Europe.

Religious opposition to the Aristotelian philosophy was not confined to the Moslems and Jews who lived in the centres of Islamic civilization. It was also a potent factor in the Christian West. In the course of the thirteenth century, belief in some of the main tenets of Aristotelian physics was on several occasions declared heretical by ecclesiastical edicts. (Even the system of Saint Thomas Aquinas, who acknowledged the supremacy of the Church's doctrine in a much less ambiguous way than the Latin followers of Averroes, occasionally aroused strong misgivings in the ecclesiastical hierarchy; here was an attempt to integrate Aristotelianism within the framework of orthodox religious belief.) It has been maintained—perhaps wrongly—that these church prohibitions furthered the emergence in the fourteenth century of largely anti-Aristotelian physical theories—those of Buridan, Nicole Oresme, and others—at the European universities.

In Islam a peculiar kind of atomism was elaborated in the ninth and tenth centuries by the Moslem theologians known as the Mutakallimūn, who were generally suspicious of, and hostile to, Greek learning. This atomism is very unlike the antique one, as we know it; and some of its fundamental assumptions are reminiscent of conceptions found in Indian atomistic theories, both Buddhist and non-Buddhist. A coherent exposition of this doctrine of the Mutakallimūn is given by Maimonides, one of their opponents, in the *Guide of the Perplexed*.

According to this doctrine, not only bodies, but also time, space, and motion, and all the qualities and properties pertaining to the atoms (for instance their colours) have an atomistic structure. In other words, everything is discontinuous. All atoms or quanta belonging to one particular category are similar. To take one example, we note no difference whatever between the quanta of motion from which every movement may be synthesized. A given movement is slower than another only because more quanta of rest accompany the quanta of motion.

Many of the Mutakallimūn absolutely deny causality. The fact that one phenomenon generally, or even invariably, follows upon another is not held to be due to a natural law; it is merely a 'habit' which can be 'broken' by the divine Will. This Will, in the view of many Mutakallimūn, recreates the discontinuous world anew in every successive instant of time.

A very different alternative to Aristotelian physics was offered in Islam by the adherents of a tradition which was called with only partial justification the Platonic one. An early representative of this tradition—and a more radical one than most—is Abū Bakr al-Rāzī, who took issue with Aristotelianism on practically all the fundamental problems of physics.

He substituted for the Aristotelian two-dimensional place, whose existence is a function of that of the bodies, a three-dimensional infinite and 'absolute' space. He also maintained the conception of an 'absolute' time, regarded by him as a never-ending flow; its existence, contrary to Aristotle's view, was wholly independent of movement or indeed of the existence of the cosmos. Moreover, he was also a partisan of an atomistic theory which has very little in common with that of the Mutakallimūn; though differing from the Greek atomistic doctrines, it is probably related to them.

Al-Rāzī believed in the unending progress of human science. Because of the gradual—but continuous—accumulation of knowledge, the moderns were in his opinion justified in criticizing even the greatest of the ancients (for instance Galen). He held no brief for orthodoxy in any of its manifestations.

A notion of experimental methods may be found in Islam in some passages of Avicenna, and in Christian Europe in certain texts of Robert Grosseteste; some of the latter figure in this Anthology.

The notion that science means (*ought* to mean) power over nature is not at all a common one in classical antiquity. *Inter alia*, it may be adumbrated in the conception, occasionally formulated by Greek Neoplatonists, that it ought to be part of Man's imitation of God to endeavour within the limits of his capacity to be powerful as God is powerful. In Islām the alchemists sometimes claimed to be able to determine by means of their science the issues of wars and of revolutions and to create men and even prophets in the laboratory. In Christendom Roger Bacon has a very vivid albeit fanciful vision of the potentialities of technology.

The so-called theory of impetus played—both in the Moslem world and in the Christian West—an important part in the gradual disintegration of Aristotelian orthodoxy. This theory of motion runs counter to the Aristotelian doctrine. According to the latter, a body can move only if a mover exercises upon it an action at every instant of the motion; if left alone, it will not continue to move. This doctrine is clearly antithetical to the modern principle of inertia. The fact that missiles—for instance, stones or arrows—go on moving for some time after having been projected is supposed by many Aristotelians to be due to the action of successive layers of air, the first of which was moved at the same time as the projectile. In the sixth century the Christian philosopher Johannes Philoponus, who wrote in Greek and was steeped in Greek philosophical tradition as well as in biblical lore, took issue upon this point with Aristotle. He outlined an hypothesis which was later known as the theory of impetus. We may add that Philoponus' thesis was to some extent anticipated by the astronomer Hipparchus (second century B.C.) and, very probably, also by an unidentified Neoplatonic philosopher.

Opposing Aristotle, Johannes Philoponus propounds the idea that the continuance of the motion of a projectile is due to the action of a 'force'—called by the schoolmen 'impetus'—which is communicated to the projectile by the mover. In other words, motion can go on, even if no cause of the motion (in the Aristotelian sense) is present.

Philoponus' account of this thesis, translated into Arabic, had a considerable impact on physical doctrine in the Islamic world. Under its influence, and perhaps, under that of other sources, Arabic philosophers constructed a highly complex theory of 'violent inclination', this being their term for impetus.

The theory was elaborated by Avicenna, who formulated the belief that, if there were a vacuum (he agreed with Aristotle in regarding this as an impossibility), a projectile moving in it as a result of an 'inclination' would never stop. In other words, movement, if not impeded by external factors, can go on indefinitely. In spite of the difference of basic notions and vocabulary, this conception calls to mind the law of inertia.

Abū'l-Barakāt al-Baghdādī discarded this Avicennian conception, though not the theory of 'violent inclination' as a whole. His doctrine anticipates classical physics with regard to another point. Aristotle and the Aristotelians teach that the continued application of a constant force to a moveable body produces uniform motion. On the other hand, according to Newtonian physics, such a force produces a uniformly accelerated motion. This conclusion, which within the framework of Newtonian physics is closely connected with the law of inertia, was also reached by Abū'l-Barakāt in his study of free fall—part of his exposition of natural and violent inclination.

The Moslem variety of the notion of impetus made a decisive contribution to the corresponding theories of the Latin scholastics. In the fourteenth century these theories as well as other anti-Aristotelian doctrines were being taught by Buridan, Oresme, Albert of Saxony, and many others. The Aristotelian view of the cosmos was disintegrating, as may be seen from the text of Oresme quoted in the present Anthology; various authors refer to the new—sometimes called Parisian—physics.

However, many Aristotelian conceptions were still considered as valid; there was no immediate breakthrough. This had to wait for Galileo—after the inspired guesses of Renaissance science and philosophy had had their day. Nevertheless it is evident that the physics of fourteenth-century Christendom prepared the way for this new development. It is an interesting fact that, in many ways, the physical doctrines held in the twelfth or thirteenth century by Arabic philosophers who rejected orthodox Aristotelianism forcibly call to mind those of fourteenth-century schoolmen who taught a new kind of physics. Yet for some reason no such breakthrough ever occurred in Islam.

The relatively precise, often quantified, theories of these schoolmen are indicative of the progressive disintegration of the certainties of Aristotelianism within the domain of the various natural sciences. The impact of the conceptions of Nicholas of Cusa is of a different order—these ideas are of a metaphysical nature. It is largely for metaphysical reasons that Nicholas dismisses the orderly Aristotelian conception of the universe, with the earth in its centre, as inadequate.

Later on some authors of the Renaissance combined metaphysical considerations of a similar, and perhaps more radical, nature with physical theories more revolutionary than those of the fourteenth-century Scholastics—to create a philosophy in a new key.

II: THE MIDDLE AGES

RHAZES

The four elements and the void

129

The hardest and densest of these four elements of bodies is earth, which is in the centre. Water is more rarefied than earth, because it is higher. The wind is more rarefied than water, because it is higher than water. And then, fire is even more rarefied than air, since it is higher than air.

Muhammad ibn Zakariyā al-Rāzī says in his book, which he names *The Commentary of Divine Knowledge*, that these substances received their forms by a combination of the absolute matter with the substance of vacuum.

In fire the substance of matter is mixed with the substance of vacuum. But there is more vacuum in it than matter. He says that in air there is less vacuum than matter. And he says that in water there is less of it than in the substance of air. In earth there is even less vacuum than in water.

He also says that when one strikes a stone on iron, fire appears in the air, because both the stone and the iron rarefy the air until it becomes fire.

Time: absolute and relative

130

ABŪ HĀTIM: We know time by the motion of the spheres, by the passing of days and nights, by the number of years and months, and by the elapsing of periods. Are these co-eternal with time, or are they originated?

ABŪ BAKR: It is not possible that these be eternal, since all of them are determined according to the motions of the celestial sphere, and counted by the rising and the setting of the sun. And the sphere with what it contains is originated.

This is Aristotle's opinion on time. Others disagreed with him and expressed various opinions on it.

I hold that there is absolute time and restricted time.

Absolute time is duration, eternity. It is eternal. It is moving perpetually.

Restricted time is that which results from the motion of the spheres and the course of the sun and the stars.

When you make this distinction and think of the motion of eternity, you will think of absolute time. This is eternity.

But when you think of the motion of the celestial sphere, you will think of restricted time.

ABŪ HĀTIM: Let me have the real meaning of absolute time so that I should be able to think of it. For when we eliminate from thought the motions of the celestial spheres, the passing of days and nights, and the elapsing of hours—time is eliminated from thought. Thus we do not know its real meaning.

Let me have the motion of eternity, of which you say that it is absolute time.

ABŪ BAKR: Don't you see how this world passes as time elapses: taf, taf, taf, whereas time is something which neither passes nor vanishes. And such is the motion of eternity, when you imagine absolute time.

ABŪ HĀTIM: It is by the passage of time which is by the motion of the celestial sphere that this world draws nearer towards its end. And the world is originated and the celestial sphere is originated. You acknowledge this. But time is connected with the world. Hence it is originated together with it.

Time passes and comes nearer to an end, when the world draws nearer towards its end—just as time was originated when the world was originated. The only real meaning of time which we know is that which we have mentioned, namely the motions of the sphere and the sun, and the number of years, months, days and hours. And when you eliminate these from thought, there remains no time as we have stated.

Either you consider these things eternal together with time. Thus the number of eternal things would increase: the celestial sphere and whatever is moved by it would be included in this number. This would lead to the view that the world is eternal.

Or, you admit that time is originated as these things are.

Or, let me have another meaning of time, so that it could be thought of, just as it can be thought of now whenever these (i.e. the motions of the sphere and the passage of days) are thought of. These sounds which you have mentioned: 'taf, taf, taf', are also something which may be counted. And pronunciation and number are originated.

Since this is so, you have not brought forward anything, when you brought forward these sounds of which an intelligent person would be ashamed. Provide us therefore with something which has a real meaning so that it may be thought of.

ABŪ BAKR: This is not yet the end of it. I have told you that Aristotle believed that which you hold. But people disagreed with him.

Yet the view of Plato scarcely disagrees with what I believe about time. And this is to my mind the most correct of opinions.

ABŪ HĀTIM: You have thus returned to following authority upon trust, and to that disagreement of which you disapproved.

And you follow Plato on this point and take from him on trust, and leave the view of Aristotle and disagree with him, while I accept his view.

And what you have to admit as regards time (i.e. that it is originated and not eternal) you will have to admit as regards place.[1]

ABŪ BAKR: How is this?

1. In its Platonic sense, i.e. space. (Ed.)

ABŪ HĀTIM : Tell me, does place surround the extensions,[1] or do the extensions surround place?

ABŪ BAKR : Nay, the extensions surround place.

ABŪ HĀTIM : Why then are not the extensions counted together with the five principles which you hold to be eternal?[2] For if place is eternal, it follows that the extensions are eternal together with it.

ABŪ BAKR : The extensions are place, and place is the extensions. Both are one thing. There is no difference between them.

ABŪ HĀTIM : How can it be that there should be no difference between them, and how should both be one thing, when you have claimed that the extensions surround place, and place does not surround the extensions?

Have you not, in saying this, made a distinction between place and the extensions?

Upon my life! It is right that you should distinguish between them. Yet the case has made it necessary for you to lie—flagrantly and say that they are one thing, since you contradict yourself when you hold that place is eternal and the extensions not. You must either consider the six extensions to be eternal together with place, the number of eternal things thus being eleven. Or, you must relinquish the tenet that place is eternal.

ABŪ BAKR : The philosophers hold various opinions on the extensions. Some deny that they are six, and hold many opinions on this.

ABŪ HĀTIM [*from whose book the whole discussion is reproduced, adds at this point*]: When I saw that he tried to find refuge in these words and change the subject, I said:

We do not bother whether they disagreed or agreed about their number, whether they increased or decreased; or whether they held the number of extensions to be one or many. These extensions, or this extension, are together with place. If place is eternal, then the extension is eternal. And if the extension is originated, then place is originated. And place cannot do without extensions, because if there are no extensions there is no place.

Place: absolute and relative

131

ABŪ BAKR : I hold about place too that there is absolute place and relative place. Absolute place is like a vessel which contains bodies. When you eliminate the bodies from thought, the vessel does not vanish. Just as when you eliminate the celestial sphere from thought, the thing in which it is does not vanish but remains in the thought.

Or take a jug from which the wine has been poured out: wine is eliminated from thought, but the jug is not eliminated at all.

1. The six sides, directions, or dimensions: front, behind, left, right, above, beneath. (Ed.)
2. i.e. the Creator; the soul; matter; time; place. (Ed.)

But relative place is relative to that which occupies it. And when there is nothing which occupies, there is no place. This is comparable to the fact that when you eliminate width from thought the body is eliminated, and if you eliminate the line from thought—the surface is eliminated from thought.

ABŪ HĀTIM: The surface comes from the line. This is not like the relation between place and what occupies place, which is rather to be compared to what you have said about the celestial sphere. But the case is not such as you have stated it, namely, that if one eliminates the celestial sphere from thought, place is not thereby eliminated from thought. Nay, it is rather that place is eliminated from thought when you eliminate the celestial sphere from it. What you have said about the jug and the wine is also like the line and the surface, for both are bodies and they are not like place and that which occupies place.

ABŪ BAKR: Give then a definition of the extensions, so that one should be able to refer to this definition.

ABŪ HĀTIM: Answer me: Are we in a place?

ABŪ BAKR: Yes.

ABŪ HĀTIM: So point to the place in which we are. Will anybody dispute it?

ABŪ BAKR: Nobody will dispute it.

ABŪ HĀTIM: When you point towards the earth, we say, 'This is earth and it has extensions.' And when you point towards the air, we say, 'This is air and it has extensions.' And when you point towards heaven, we say, 'This is heaven and it has extensions.'

ABŪ BAKR: All these occupy place, and place has no body at which one could point. Rather, it is known by imagination.

ABŪ HĀTIM: Likewise the extensions which encompass place are perceived by imagination. When these extensions are eliminated from thought, place is eliminated. And then there is no place and no extensions. Both can be thought of together only. This problem is like that of time.

ABŪ BAKR: True, upon my life! As to place too I agree with Plato. And what you cling to is Aristotle's opinion.

The necessity for a critical attitude in science

132

I am well aware that the fact that I have written this book will provoke many to tax me with ignorance or indeed to censure me and to judge me harshly. For there is every indication that it tends to assign the leading role to the one who goes about setting himself up (deriving profit and pleasure from so doing) as the gainsayer of that eminent man Galen, disregarding the great man's genius and learning, the fact that he made a careful study of all branches of philosophy, and the position that he holds in this field of knowledge. Now, Heaven knows, I fully realized that writing this book was a necessity that was forced upon me.

For it was painful to me to set myself up against the one who, of all men, has showered the most benefits upon me and has been the greatest source of help to me, by whom I have let myself be guided and in whose footsteps I have followed, like a well of knowledge from which I drank deeply, and to stand up to him in a manner that ill becomes a slave face to face with his master, a pupil in the presence of his teacher, or indeed one who meets with kindness in his dealings with a benefactor. [. . .] The art of medicine, however, is a philosophy and admits of no mute acquiescence in the pronouncements of the leaders of a school of thought, no blind acceptance of their views, nor any preferential treatment that they might enjoy; nor, indeed, any hesitation in subjecting their pronouncements to careful scrutiny. Such an attitude is, furthermore, something that no philosopher likes to see in his disciples and pupils, as Galen himself says in his book *De usu partium*, in the passage in which he denounces those who demand blind acceptance of their pronouncements from their followers and supporters, without any proof of the validity of these sayings being given. It is as if he were lavishing encouragement upon me and making my task easier; and if he were living among us now, this man of genius would not have censured me for having written this book and would not have taken exception to it, because of his passion for the truth and because he thought that studies should be pursued farther and farther until they can be brought to their conclusion. No, far from taking exception to it, he would have been all eagerness to examine this book at close hand and to reflect upon it. [. . .]

As for anyone who would censure me and tax me with ignorance because I have let my thoughts dwell on these doubts and given voice to them here, I shall pay no attention to him, since I do not consider him to be a philosopher. For it would be as if he were rejecting the tradition of philosophers in favour of that of the common herd, which demands blind acceptance of opinions and renouncement of any criticism of them. Whereas the philosophical tradition has always embraced simultaneously both glorification of its leaders and considerable harshness in the criticisms that are directed against them; they enjoy no preferential treatment at all. Thus it was that Aristotle was able to say: 'The truth and Plato are at variance with each other, and we hold both dear; the truth, however, is dearer to us than Plato is.' And he did in fact set himself up against Plato, refuting his principal views. Theophrastus, in his turn, refuted Aristotle (and rightly so) with regard to a branch of philosophy which, obviously, is second only to geometry, namely logic. Themistius, in the same way, demonstrates errors committed by Aristotle in many passages, and frequently expresses true astonishment at them, sometimes going so far as to say: 'I fail to see how it can have happened that this idea did not occur to the Sage; it is, however, very close to being obvious.' As for Galen himself, there would be no point in my enlarging upon the many criticisms that he directs against the ancients and against the most eminent of his contemporaries, the pertinacity and vigour that he brings to doing this, and the amount of space that he devotes to it. For these criticisms are so numerous that I should not be able to treat them in detail, and it is obvious to readers of his books that it is with these criticisms that he is most greatly concerned. Indeed, I do not believe that any philosopher or man of medicine escapes being much abused by him. Now, the majority of these criticisms are just ones; the same may possibly be said even of all of them. And this indicates the extent of his knowledge, his native intelligence, and the vast number of matters in which he was well versed.

Now, if the question were put to me of why it is that those who come after are able to introduce corrections of this kind to the sayings of the most eminent of the ancients, I should reply that it can come about as a result of a variety of reasons: among them are carelessness and shallowness of judgement, which are sources of error, and also the violence that is done to sound reasoning by passion. [. . .]

An additional reason lies in the fact that fields of knowledge develop continuously with the passage of time, drawing ever nearer to a state of perfection. This is why a man living in a later age takes very little time to make discoveries that demanded a great deal of time from the ancient who originally made them. Consequently, he is able to elucidate them farther, and this means that he will have a greater aptitude for going on to make additional discoveries.

In this respect an analogy can be drawn between the ancients and those who make a fortune and, likewise, between those who come after these ancients and the heirs of the fortune in question, to whom the very fact of this inheritance gives the means to add to it ever-increasing acquisitions.

Now, if it were said to me that this viewpoint invites the conclusion that the learned men who come after are superior to the ancients as far as their knowledge is concerned, I should reply that I do not consider that an absolute judgement of this kind can be put forward without its being made subject to a condition bearing upon the characteristics of the learned man who comes after: I should say, namely, that he will be the superior of the ancient provided that he possesses a perfect knowledge of the latter's work.

AL-BIRUNI

On the hypothesis of the axial rotation of the earth

133

Let us now come back to the second part of the earth's motion, namely the one round itself, eastwards, without leaving its place.

The followers of Aryabhatta, an Indian sage, believed in it. We think that what induced people to believe in this was the desire to attribute to the sky the motions of the stars in it in the second, eastward, motion, and to attribute to the earth the effects of the first, westward, motion, so that it should not be necessary to attribute to the sky two different motions.

Although this does not undermine the foundations of the science of astronomy, we say that the first motion has no impact on the ether since it makes the whole of it revolve in one and the same rotation.

A scientific system should not be adopted when it is contradicted in some respects. As long as no decision concerning this emerges, research as to whether it is true should not be delayed.

Ptolemy accused those who held this view, i.e. that the earth turns round itself, of ignorance, because they attribute quick motion to heavy dense objects, and slow motion, or lack of any motion, to light rare ones. But this argument is suitable to physical rather than to geometrical research. In fact it is an argument of persuasion and not a scientific proof. [. . .]

Aristotle and his followers, the foremost amongst the natural philosophers, refused to attribute any lightness or heaviness to the ether. One of them was asked what would happen to a piece of ether, if one were to suppose that it had been put on the surface of the earth. He answered that it would rest and not move, being different from the things which move straight and move towards their natural regions and places, when they have been taken out of them to other regions. Thus according to Ptolemy the lack of motion in the light and rare would arouse amazement.

From the geometrical point of view one has to consider that if the earth were to move round itself, anything that had become separated from it would lag behind, e.g. soaring birds, things thrown in the direction of the sky, clouds standing in the air. One would see these things always moving to the west. If on the other hand they too were to have this motion, like the earth, it would follow that they would be seen resting because they would move in a parallel way. But in fact we see them moving in all directions. Hence neither they nor the earth move in this rotation which brings about night and day.

I have personally come across a most distinguished astronomer who tended to support this view. He was of the opinion that heavy objects do not fall on the earth in a line with its diameter and perpendicularly to its surface, but obliquely at various angles which we do not apprehend. Yet we see them as if they were falling vertically. For that man held that a heavy object separated from the earth has two motions. One is circular because it is in the nature of a part to share the properties of the whole. The other motion is straight because the heavy object is attracted to its source. Thus when a heavy object is separated from the earth it moves owing to the first motion so that it keeps to the perpendicular opposition. The second motion on the other hand, if it were isolated would make the object fall to the west of the perpendicular opposition. But the fall of the object is composed of both motions. Therefore it does not leave the perpendicular opposition. The line in which it falls on the earth is in reality not perpendicular to the earth, but inclining to the east. Its trace in the air is not observable or apprehensible by the senses. Hence one cannot make out whether it is vertical or inclining. One imagines it to be vertical because the picture of its perpendicular opposition sticks in the mind. [. . .]

Let it now be known that if the earth were to move in the above-mentioned way, i.e. turn round its own axis, its equatorial belt would cover the above-mentioned number of miles[1] three hundred and sixty times in twenty-four hours. One nine-hundredth of an hour, which is a second of the celestial sphere, would correspond to three thousand seven hundred and seventy-eight cubits.[2] According to the estimate of the Indians the time necessary for the earth to turn [an angle of] one minute is one breath of a human being. If during this span of time its motion were to be nearly one mile, this would be perceptible and measurable.

If the things separated from the earth keep to the perpendicular opposition because of the motion they have in common with the earth, it is known that when an added violent force operates on them, it makes them leave that imagined rest. The effect of this force will be manifest in producing their different directions. For a violent force towards the East will co-operate with nature, while a westward force will oppose nature, and repel it. Thus the jump of one who jumps will be different in either direction, and so the passage of an arrow in one of the two directions, and the flight of the bird flying in one direction will be different from that flying in the opposite direction.

They will be also different in the North and the South because of the widening in the one and the narrowing in the other.[3] Nothing of this exists. Hence the earth, in its place, does not rotate round its centre.

AL HAZEN

The science of optics and the propagation of light

134

The discussion what light is, is a part of the science of physics. But in order to discuss how light shines one requires the science of mathematics, because of the lines according to which light travels. Likewise, it is a part of physics to discuss what a ray is, and a part of mathematics to discuss its form. Thus the question what it means that bodies are transparent, so that light shines through them, is a part of physics. But the discussion of how light travels in them is a part of mathematics. Hence the discussion of light, rays and transparent matter must consist of both physics and mathematics.

135

The bodies which shine of themselves, apprehended by sense-perception, are of two kinds, namely the stars and fire. The light of these bodies shines on all adjacent bodies, and it is this which is apprehended by sense-perception. We

1. 56 2/3 Arabian miles, which were supposed to be equal to the length of one degree on the equator. (Ed.)
2. One Arabian mile equals 4000 cubits. (Ed.)
3. The linear velocity of a point on a rotating sphere is the greater the nearer it is to the equator. (Ed.)

have explained in the first chapter of our book on optics that any light in any body shines on all confronting bodies, be this light essential or accidental. We have explained this exhaustively there. Nevertheless, induction suffices to convince in this matter, since it is impossible for an opaque body to be opposite a shining one without the light of the latter being seen upon it, as long as there is nothing to intervene between them, as long as there is not too great a distance between them and as long as the light of the shining body is not too weak.

All natural bodies, the transparent ones as well as the opaque ones, have the power to receive light. Thus they receive the light of shining bodies. Transparent bodies have, apart from the power to receive light, the power to transmit light. This is transparency. The bodies called transparent are those through which light shines, and the eye perceives what is behind them. [. . .]

How does light travel through transparent bodies? Light travels in transparent bodies along straight lines only. It travels from every point in the shining body along every straight line in the transparent body along which it is possible for it to travel. We have explained this exhaustively in our book on optics. But let us now mention something to prove this convincingly: the fact that light travels in straight lines is clearly observed in the lights which enter into dark rooms through holes. For, when sunlight, moonlight, or the light of fire, enter through a medium-sized hole into a dark room in which there is dust or dust has been raised, the entering light will be clearly observable in the dust which fills the air. It will also be seen on the surface of the floor or on the wall opposite the hole. And one will find that the light travels in straight lines from the hole to the floor or to the wall opposite the hole. If one examines this light with a straight stick one finds that the light travels as straight as the stick.

If there is no dust in the room the light is to be seen on the floor or on the wall opposite the hole. If one puts between the light seen and the hole a straight stick, or if one tightly stretches a thread between them, and one puts between the light and the hole an opaque body, the light will be seen on this opaque body and will disappear from the spot at which it was seen. If then the opaque body is moved along the straight way of the stick, the light will always be found to be observable upon the opaque body.

Thus it becomes clear that the light travels in straight lines from the hole to the spot at which the light is seen. We have already explained in our *Optics* how the spreading of light can be examined in each of the kinds of transparent bodies. What we have mentioned here suffices. [. . .]

The author of *Logic* [i.e. Aristotle] was of the opinion that the transparency of the celestial sphere is clearer than the transparency of all transparent bodies and that its transparency is the utmost, and that there can be no body more transparent than the celestial sphere. But the mathematicians think that there is no end to transparency and that for every transparent body there may be another body which is even more transparent. One of the later mathematicians has made this point clear, namely Abū Saʿd al-ʿAlāʾ ibn Suhayl. For he has written a treatise in which he has explained this by means of a geometrical demonstration. We shall mention the demonstration of this, but more succinctly than al-ʿAlāʾ ibn Suhayl, and explain it more clearly than he has. Thus we say: every light which shines upon a transparent body travels through this transparent body in straight lines. Factual findings prove this. If then light travels through a transparent body and arrives at another transparent body, the transparency of which differs from that of the first body in which it travelled, and it is oblique to the surface of the second body, the light is refracted and does not travel straight into the second body. We have made this point clear in the seventh treatise of our book on optics, and we have there shown a way to establish this in any transparent body. We have explained there that the refraction is according to special angles. If the refraction is from a rare body to a denser one, the refraction is towards the perpendicular line which emerges—at right angles to the surface of the denser body—from the point at which the refraction occurs. If the refraction is from the denser body to the rarer one, the refraction is away from the perpendicular. When light travels in the rarer body and is refracted into the denser one, a certain angle is formed at the point of refraction. If light travels in the denser body and is refracted into the rarer one, the light which travels in the denser body along the refracted line will be refracted into the rarer body according to that very same angle, which was formed between the first ray and the refracted ray.

When light is refracted from a rare transparent body into two other bodies which are denser than the first body, but different in density from one another, the refraction of light in the denser body will be greater. I mean that when the light is refracted in the denser body it will be nearest to the perpendicular which emerges from the point of refraction. When light is refracted from a dense transparent body into two rare bodies, different in rarity from one another, the refraction of light in the rarest body will be the furthest from the perpendicular emerging from the point of refraction.

Ptolemy has explained this point also as regards the ray of the eye in the fifth chapter of his book on optics. I mean that he has explained that when the ray of the eye travels in a transparent body and then meets another transparent body, which differs as to the degree of its transparency from the first body, and it (i.e. the ray) is oblique to the surface of the second body, it is refracted and does not travel straight. He explained that the refraction of the eye-ray from air into glass is greater than from air into water. And glass is denser than water. He also explained that when the eye is in the rarer body and the ray is refracted in a denser body according to a certain angle, and then the eye passes into the denser body along the refracted ray, the ray will be refracted according to the same angle.

From all this it follows that every ray which travels through a transparent body and then meets another transparent body, and the transparency of the second body is denser than that of the first in which it travelled, will be refracted in the second body. Its refraction in the second body will be in accordance with the density of the second body. The denser it is the larger is the angle of refraction. And any ray which travels in a transparent body and then comes across another transparent body, the transparency of which is rarer than that of the first body, will be refracted in the second body. Its refraction in the second body will be in proportion to the rarity of the second body.[. . .] So far we have described the view of the mathematicians, namely that transparence in transparent bodies may become rarer and clearer *ad infinitum,* i.e. that for any transparence in a transparent body it is possible to imagine a clearer transparence.

But the physicists hold that everything pertaining to natural bodies is limited and finite, and not infinite. The angles divisible *ad infinitum* are imagined angles between imagined lines. But the angles in natural bodies as well as those imagined in natural bodies are not infinitely divisible, while the body in which the angles are remains as it is. For the body in which the imagined angles are cannot be divided *ad infinitum.* For every natural body is divisible up to a certain limit, up to which the body keeps its form. If it is divided beyond that limit it takes off its form and takes on another form.

Take for instance water. When one divides it to the utmost one reaches a limit which is the very smallest particle of water. When it is divided beyond that it sheds the form of water and takes on the form of air. Air in its turn is divisible up to the very smallest particle of air. When it is divided beyond that it sheds the form of air and puts on the form of fire. Fire in its turn is divisible up to the very smallest particle of fire. Beyond that it cannot be divided, for there does not exist any form rarer than the form of fire. If the form of the celestial sphere is rarer than the form of fire, and if it is possible for fire to become of the kind of the celestial sphere, the very smallest particle of fire can be divided and turn into the substance of the celestial sphere. The body of the celestial sphere is indivisible. Were it to be imagined divided, one would arrive at the very smallest of its parts. Beyond this it would be indivisible. For there exists no rarer form than the form of the celestial sphere. If it were imagined as divided beyond its very smallest parts, if such a division were possible, what one would imagine would be a division of the body's dimensions and not of its substance. If, nevertheless, the body were imagined as divided this would be an imaginary, not a real, division.

When the author of *Logic* says that the celestial sphere is transparent to the utmost he means that there are no natural bodies more transparent than the sphere, and that it is impossible for such bodies to exist. For according to his view every species which can possibly exist has already come into existence.

The role of criticism in science

136

Truth is sought for its own sake. And those who are engaged upon the quest for anything that is sought for its own sake are not interested in other things. Finding the truth is difficult, and the road to it is rough. For the truths are plunged in obscurity. It is natural to everyone to regard scientists favourably. Consequently a person who studies their books, giving a free rein to his natural disposition and making it his object to understand what they say and to possess himself of what they put forward, comes to consider as truth the notions which they had in mind and the ends they indicate. God, however, has not preserved the scientist from error and has not safeguarded science from shortcomings and faults. If this had been the case, scientists would not have disagreed upon any point of science, and their opinions upon any question concerning the truth of things would not have diverged. The real state of affairs is, however, quite different. Accordingly, it is not the person who studies the books of his predecessors and gives a free rein to his natural disposition to regard them favourably who is the real seeker after truth. But rather the person who in thinking about them is filled with doubts, who holds back with his judgement with respect to what he has understood of what they say, who follows proof and demonstration rather than the assertions of a man whose natural disposition is characterized by all kinds of defects and shortcomings. A person who studies scientific books with a view to knowing the truth ought to turn himself into a hostile critic of everything that he studies. [. . .] He should criticize it from every point of view and in all its aspects. And while thus engaged in criticism he should also be suspicious of himself and not allow himself to be easy-going and indulgent with regard to the object of his criticism. If he takes this course, the truth will be revealed to him and the flaws in the writings of his predecessors will stand out clearly. [. . .]

If we should study the works of a man who, famed for his excellence, shows great ingenuity in his mathematical ideas, and who is always being cited in the true sciences, I refer to Claudius Ptolemy, we should find in them any scientific doctrines and precious, most instructive and useful ideas. [. . .] But when we criticize them we shall find in them obscure passages, improper terms and contradictory notions. There are, however, few of them, if they are set beside the correct notions which he hit upon.

AVICENNA

The theory of impetus as against Aristotle's explanation of forced motion

137

But when we investigate this we find that the most correct theory is that what moves receives an inclination[1] from that which makes it move. This inclination

1. The Arabic equivalent of impetus. (Ed.)

is what one feels when one tries to stop natural motion by violence, or violent motion by another violence. In that case one feels a power of resistance which is liable to increase and decrease. Sometimes it is stronger, sometimes it is weaker. There is no doubt that it exists in the body, even when the body is violently made to rest.

The theory of those who hold that air is pushed and pushes is also not correct. How should it be correct when the case of air is like the case of something thrown? For after what has made it move rests, this (allegedly) pushed air either continues moving or does not continue. If it does not continue, how could it make the thrown object penetrate a wall while it carries it? But if it continues the following is clear: if it moved more quickly than an arrow it would penetrate a wall more powerfully than an arrow does. For according to them an arrow penetrates by virtue of a penetrating force which is derived from the quicker motion of the air.

But in fact air is blocked and repelled by the things which stand in its way. Why then is the arrow not blocked and repelled? If the reason for this were that what is near to the arrowhead is blocked whereas what is near to the upper part of the arrow retains its power, it would follow that the arrow precedes the air. But they made the air precede. But if the arrow precedes, it follows necessarily that the air adjacent to the arrow has not got the impetus to make the arrow, which is blocked by a wall, penetrate it without being pushed from behind. For it is wrong to say that the arrow penetrates the air. For according to them the air carries it and pushes it by its own impetus.

But if the arrow penetrates the wall because it attracts the air behind it, so that this attraction turns into a push for the attractor, then the attracted is attracted more forcibly than the adherent attractor. If this excess of force is due to a power and an inclination it has thus been proved that the arrow moves by virtue of an inclination.

If on the other hand this were only a 'following up', it would have to cease when the cause ceases. But if it continues to move when the mover does not move any more, then this is due to a power and an inclination.

Why do things drop like an arrow, which happen to be in the adherent air? Why does the air not carry them while, according to the opponents, the air prevents heavy bodies carried in it from dropping by a powerful motion and by this motion it resists the penetration due to heaviness.

When the wind blows on the branches of a tree it may smash them, and yet if an arrow is put into the winds they would not carry it. If the air can carry a big stone it should follow that when small bodies come in its way it would crush them.

These people think that they say anything at all when they say the following: 'The air moves more quickly and causes successive forward movements in the parts of the air. And the arrow is put into them.' But it is not so. Because if these parts of the air move forward one after the other, then those of them which move, move when those which moved them are already at rest. This would contradict what they claim.

If on the other hand the parts of the air move together, there are two possibilities: either their first mover moves with them or it stands still. If the motion of the air is together with the motion of the first mover, the arrow should halt when the first mover halts. But since the arrow does go on, the doubt remains, namely that there is a motion, and a cause by which the motion continues other than the first mover.

And when one seeks an explanation why what moves violently accelerates in the middle, the supposition that there is a power (i.e. the inclination) does no harm, just as the alternative theory, namely that the air carries, does not help. For here the difficulty remains. For he who first raises doubts may say: why does this air become quicker in the middle of the time of the motion? For if this is so because the motion rarefies the air, the arrow, which is carried in it, is more likely not to be affected by this acceleration. For the air becomes more voluminous and less substantial. And what is more voluminous and less substantial moves more slowly than what is not so when both are equally driven.

But if the said rarefication applies to the air penetrated and not the penetrating air, why then is the decomposing and rarefying action of the friction stronger in the middle of the motion than at its beginning.

Indeed, if this friction were continually to apply to one thing which it touches, i.e. either the rubbing or the rubbed, this would make sense. What rubs is like what pierces. Through long exertion it becomes hotter and more capable of rarefying. On the other hand what is continuously rubbed is more and more affected.

But in this case neither what rubs nor what is rubbed remain one and the same. It is rather that according to them, and as a consequence of what they say, the air must move like a chain which is pushed forward. Every specific part is assumed to rub another specific part.

It may be that this is the most obvious cause for the increase in power. It may be that the more the thrown object is rubbed the hotter it becomes. Its heat through rubbing goes on increasing and the intensity goes on decreasing. Yet the rarefication which results from heating compensates and makes up for what is lost in intensity as long as the power is steady.

But after the power has pushed for a considerable time and has as a result abated, the friction too becomes weaker until it no longer compensates and makes up for the effect of the push.

But we do not consider this to be the only cause, although it is possible that it is one of the supporting causes which are added in the middle of the motion.

Thus the quality of the violent power and its parts have become clear. It has also become clear that every motion occurs through a power in the moving object by which it is impelled. This power is either violent, or accidental, or natural.

ABŪ 'L-BARAKĀT

Impetus

138

Some[1] argue that the fact that the natural inclination becomes stronger towards the end of the motion is due, not to the cessation of the violent inclination, but to something specific to the natural inclination itself. To prove this they give the example of a stone thrown from above which has not previously ascended in a violent motion, and has no violent inclination. For one sees that the farther away the point from which it is thrown, the quicker is its final motion and the stronger its inclination. As a result of this it breaks and is crushed. This does not happen to it when it is thrown from a shorter distance. It becomes clear that this differs in proportion to the length of the distance the stone has fallen. For[2] initially the natural inclination was opposed by the violent inclination, but this was weakened by the resistance of what the stone passed through on its way. This is so because the cause and originator of the violent inclination became disjointed from the stone, so that no successive inclinations were brought about in it to make up for the weakening due to the resistance. But the source of the natural inclination is in the stone, and provides it with one inclination after the other.

Because of this one sees that when a ball's motion, due to a stroke, ceases, the striker tries to add another stroke to make the motion reach its utmost speed. If a thrower were able to do this, he would do it also. But the source of the natural inclination is not separated from the stone. It goes on producing the natural inclination until the stone arrives through it at its natural region. The farther the power moves the stone away from its natural region, the more natural inclinations are produced. Hence, the longer the motion, the more increases the power of inclination.

AVERROES

A critique of Ptolemy's system

139

The opinion that there is an eccentric sphere or an epicyclic sphere is alien to nature. An epicyclic sphere is absolutely impossible, for a body which moves in a circle moves round the centre of the universe, not outside it. For it is what moves in a circle that makes the centre. If there were any circular motion outside this centre, there would be another centre outside this centre. Then there would

1. The passage occurs in the course of a discussion of the view that a thrown object has two inclinations, a natural one towards the earth, and a violent one given to it by the thrower. (Ed.)
2. It appears that at this point the discussion returns to deal with the object falling after having been thrown upwards. (Ed.)

be another earth outside this earth. And it is manifest that according to the science of physics all this is impossible.

The same seems to be true of the eccentric sphere which Ptolemy assumes. For if there were many centres, there would be heavy bodies outside the place of the earth, and there would be more than one middle, and it would have a width and be divisible. All this is impossible.

Furthermore, if there were eccentric spheres, there would be redundant bodies among the celestial bodies. They would be of no advantage, but be a sort of stuffing of the kind that is thought to exist in the bodies of animals. There is nothing in the apparent motions of the stars which inevitably necessitates one to assume either an epicyclic or an eccentric sphere.

Perhaps these two things are made unnecessary by the spiral motions which were assumed by Aristotle in astronomy, following his predecessors. For the astronomers before Hipparchus and Ptolemy assumed neither an epicyclic nor an eccentric sphere. Ptolemy says this explicitly in his book called *Hypotheses ton planomenon*. He maintains that Aristotle and his predecessors assumed, instead of this, spiral motions. They held, according to his report, that there are many motions. Ptolemy maintains that those who came after Aristotle found a simpler method than that, i.e. they were able to explain the phenomena by assuming a smaller number of bodies. By this he refers to the epicyclic sphere and the eccentric sphere. He maintains that this method is better because of the accepted maxim that nature makes nothing which is superfluous, and if it is possible for nature to move something by few media, it will not move it by many. But Ptolemy forgot what compelled the early thinkers to resort to spiral motions, namely the fact that epicyclic spheres as well as eccentric ones are impossible.

And since people thought this astronomy to be simpler and easier, because of the repetition of the motions, namely those written down nowadays in Ptolemy's book, people abandoned this ancient astronomy so that its knowledge has passed away. Thus people do not understand nowadays what Aristotle says about those early thinkers. Alexander and Themistius have confessed this. But they did not grasp the reason for it which we have mentioned.

Thus it becomes necessary to start investigating this ancient astronomy from the beginning, because it is the true astronomy which is correct according to the principles of nature. In my opinion it is based upon the motion of one and the same sphere round one and the same centre and two or more axes in accordance with what is apparent. For it is possible that through such motions the stars acquire speed or slowness, advance or retreat, and other movements for which Ptolemy has no geometrical scheme. They may also—through this—be seen to draw nearer or go farther away, as it happens with the moon.

When I was young I hoped to carry out this investigation. In my old age I have given up this hope, because obstacles hindered me from accomplishing it. But perhaps these words will stimulate somebody to investigate these things later on. For in our day there is no science of astronomy in existence. The astronomy which exists in our time agrees with calculation rather than with reality.

MOSES MAIMONIDES

Aristotelian physics versus Ptolemaic astronomy

140

You know of astronomical matters what you have read under my guidance and understood from the contents of the *Almagest*. But there was not enough time to begin another speculative study with you. What you know already is that as far as the action of ordering the motions and making the course of the stars conform to what is seen is concerned, everything depends on two principles: either that of the epicycles or that of the eccentric spheres or on both of them. Now I shall draw your attention to the fact that both those principles are entirely outside the bounds of reasoning and opposed to all that has been made clear in natural science. In the first place, if one affirms as true the existence of an epicycle revolving round a certain sphere, positing at the same time that that revolution is not around the centre of the sphere carrying the epicycles—and this has been supposed with regard to the moon and to the five planets—it follows necessarily that there is rolling, that is, that the epicycle rolls and changes its place completely. Now this is the impossibility that was to be avoided, namely, the assumption that there should be something in the heavens that changes its place. For this reason Abū Bakr ibn al-Sā'igh[1] states in his extant discourse on astronomy that the existence of epicycles is impossible. He points out the necessary inference already mentioned. In addition to this impossibility necessarily following from the assumption of the existence of epicycles, he sets forth there other impossibilities that also follow from that assumption. I shall explain them to you now.

The revolution of the epicycles is not around the centre of the world. Now it is a fundamental principle of this world that there are three motions: a motion from the midmost point of the world, a motion towards that point, and a motion around that point. But if an epicycle existed, its motion would be neither from that point nor towards it nor around it.

Furthermore, it is one of the preliminary assumptions of Aristotle in natural science that there must necessarily be some immobile thing around which circular motion takes place. Hence it is necessary that the earth should be immobile. Now if epicycles exist, theirs would be a circular motion that would not revolve round an immobile thing. I have heard that Abū Bakr has stated that he had invented an astronomical system in which no epicycles figured, but only eccentric circles. However, I have not heard this from his pupils. And even if this were truly accomplished by him, he would not gain much thereby. For

1. Better known by the name of Ibn Bājja or Avempace; died 1138. (Ed.)

eccentricity also necessitates going outside the limits posed by the principles established by Aristotle, those principles to which nothing can be added. It was by me that attention was drawn to this point. In the case of eccentricity, we likewise find that the circular motion of the spheres does not take place around the midmost point of the world, but around an imaginary point that is other than the centre of the world. Accordingly, that motion is likewise not a motion taking place around an immobile thing. If, however, someone having no knowledge of astronomy thinks that eccentricity with respect to these imaginary points may be considered—when these points are situated inside the sphere of the moon, as they appear to be at the outset—as equivalent to motion round the midmost point of the world, we would agree to concede this to him if that motion took place round a point in the zone of fire or of air, though in that case that motion would not be around an immobile thing. We will, however, make it clear to him that the measures of eccentricity have been demonstrated in the *Almagest* according to what is assumed there. And the latter-day scientists have given a correct demonstration, regarding which there is no doubt, of how great the measure of these eccentricities is compared with half the diameter of the earth, just as they have set forth all the other distances and dimensions. It has consequently become clear that the eccentric point around which the sun revolves must of necessity be outside the concavity of the sphere of the moon and beneath the convexity of the sphere of Mercury. Similarly the point around which Mars revolves, I mean to say the centre of its eccentric sphere, is outside the concavity of the sphere of Mercury and beneath the convexity of the sphere of Venus. Again the centre of the eccentric sphere of Jupiter is at the same distance—I mean between the sphere of Mercury and Venus. As for Saturn, the centre of its eccentric sphere is between the spheres of Mars and Jupiter. See now how all these things are remote from natural speculation! All this will become clear to you if you consider the distances and dimensions, known to you, of every sphere and star, as well as the evaluation of all of them by means of half the diameter of the earth so that everything is calculated according to one and the same proportion and the eccentricity of every sphere is not evaluated in relation to the sphere itself.

Even more incongruous and dubious is the fact that in all cases in which one of two spheres is inside the other and adheres to it on every side, while the centres of the two are different, the smaller sphere can move inside the bigger one without the latter being in motion, whereas the bigger sphere cannot move upon any axis whatever without the smaller one being in motion. For whenever the bigger sphere moves, it necessarily, by means of its movement, sets the smaller one in motion, except in the case in which its motion is on an axis passing through the two centres. From this demonstrative premiss and from the demonstrated fact that vacuum does not exist and from the assumptions regarding eccentricity, it follows necessarily that when the higher sphere is in motion it must move the sphere beneath it with the same motion and around its own centre. Now we do not find that this is so. We find rather that neither of the two spheres, the containing and the contained, is set in motion by the movement of the other nor does it move around the other's centre or poles, but that each of them has its own particular motion. Hence necessity obliges the belief that between every two spheres there are bodies other than those of the spheres. Now if this be so, how many obscure points remain? Where will you suppose the centres of those bodies existing between every two spheres to be? And those bodies should likewise have their own particular motion. Thābit[1] has explained this in a treatise of his and has demonstrated what we have said, namely, that there must be the body of a sphere between every two spheres. All this I did not explain to you when you read under my guidance, for fear of confusing you with regard to that which it was my purpose to make you understand.

As for the inclination and deviation that are spoken of regarding the latitude of Venus and Mercury, I have explained to you by word of mouth and I have shown you that it is impossible to conceive their existence in those bodies. For the rest Ptolemy has said explicitly, as you have seen, that one was unable to do this, stating literally: 'No one should think that these principles and those similar to them may only be put into effect with difficulty, if his reason for doing this be that he regards that which we have set forth as he would regard things obtained by artifice and the subtlety of art and which may only be realized with difficulty. For human matters should not be compared to those that are divine.' This is, as you know, the text of his statement. I have indicated to you the passages from which the true reality of everything I have mentioned to you becomes manifest, except for what I have told you regarding the examination of where the points lie that are the centres of the eccentric circles. For I have never come across anybody who has paid attention to this. However, this shall become clear to you through the knowledge of the measure of the diameter of every sphere and what the distance is between the two centres as compared with half the diameter of the earth, according to what has been demonstrated by al-Qabīsī[2] in the *Epistle Concerning the Distances*. If you examine those distances, the truth of the point to which I have drawn your attention will become clear to you.

Consider now how great these difficulties are. If what Aristotle has stated with regard to natural science is true, there are no epicycles or eccentric circles and everything revolves round the centre of the earth. But in that case how can the various motions of the stars come about? Is it in any way possible that motion should be on the one hand circular, uniform, and perfect, and that on the other hand the things that are observable should be observed in consequence of it, unless this be accounted for by making use of one of the two principles or

1. Better known by the name of Ibn Qurra; died 901. (Ed.)
2. 11th-century mathematician and astronomer; his treatise deals with the distances of the planets from the earth. (Ed.)

of both of them? This consideration is all the stronger because of the fact that if one accepts everything stated by Ptolemy concerning the epicycle of the moon and its deviation towards a point outside the centre of the world and also outside the centre of the eccentric circle, it will be found that what is calculated on the hypothesis of the two principles is not at fault by even a minute. The truth of this is attested by the correctness of the calculations—always made on the basis of these principles—concerning the eclipses and the exact determination of their times as well as of the moment when it begins to be dark and of the length of time of the darkness. Furthermore, how can one conceive the retro-gradation of a star, together with its other motions, without assuming the existence of an epicycle? On the other hand, how can one imagine a rolling motion in the heavens or a motion around a centre that is not immobile? This is the true perplexity.

However, I have already explained to you by word of mouth that all this does not affect the astronomer. For his purpose is not to tell us in which way the spheres truly are, but to posit an astronomical system in which it would be possible for the motions to be circular and uniform and to correspond to what is apprehended through sight, regardless of whether or not things are thus in fact. You know already that in speaking of natural science, Abū Bakr ibn al-Sā'igh expresses a doubt whether Aristotle knew about the eccentricity of the sun and passed over it in silence—treating of what necessarily follows from the sun's inclination, inasmuch as the effect of eccentricity is not distinguishable from that of inclination—or whether he was not aware of eccentricity. Now the truth is that he was not aware of it and had never heard about it, for in his time mathematics had not been brought to perfection. If, however, he had heard about it, he would have violently rejected it; and if it were to his mind established as true, he would have become most perplexed about all his assumptions on the subject. I shall repeat here what I have said before. All that Aristotle states about that which is beneath the sphere of the moon is in accordance with reasoning; these are things that have a known cause, that follow one upon the other, and concerning which it is clear and manifest at what points wisdom and natural providence are effective. However, regarding all that is in the heavens, man grasps nothing but a small measure of what is mathematical; and you know what is in it.

*An exposition and critique of the atomism of the
Mutakallimūn*

141

THE FIRST PREMISS

Its meaning is that they thought that the world as a whole—I mean to say, every body in it—is composed of very small particles that, because of their subtlety, are not subject to division. The individual particle does not possess quantity in any respect. However, when several are aggregated, their aggregate possesses quantity and has thus become a body. If two particles are aggregated together, then according to the statements of some of them, every particle has in that case become a body, so that there are two bodies. All these particles are alike and similar to one another, there being no difference between them in any respect whatever. And, as they say, it is impossible that a body should exist in any respect except it be composed of these particles, which are alike in such a way that they are adjacent to one another. In this way, according to them, generation consists in aggregation, and corruption in separation. They do not, however, call this process corruption, but say that there are the following generations: aggregation and separation, motion and rest. They also say that these particles are not restricted in their existence,[1] as was believed by Epicurus and others who affirmed the existence of such particles; for they say that God, may He be exalted, creates these substances constantly whenever He wishes, and that their annihilation is likewise possible. Further on, I shall let you hear their opinions regarding the annihilation of substance.[2]

THE SECOND PREMISS

The assertion concerning the vacuum. The men concerned with the roots (i.e. the Mutakallimūn), believe likewise that vacuum exists and that it is a certain space[3] or spaces in which there is nothing at all, being accordingly empty of all bodies, devoid of all substance. This premiss is necessary for them because of their belief in the first premiss. For if the world were full of the particles in question, how can a thing in motion move? It would also be impossible to represent to oneself that bodies can penetrate one another. Now there can be no aggregation and no separation of these particles except through their motions. Accordingly they must of necessity resort to the affirmation of vacuum so that it should be possible for these particles to aggregate and to separate and so that it should be possible for a moving thing to move in this vacuum in which there is no body and none of these substances.[4]

THE THIRD PREMISS

This is their saying that time is composed of instants, by which they mean that there are many units of time that, because of the shortness of their duration, are not divisible. This premiss is also necessary for them because of the first premiss. For they undoubtedly had seen Aristotle's demonstrations, by means of which he has demonstrated that distance, time, and locomotion are all three of them equal as far as existence is concerned. I mean to say that their relation to one another is the same and that when one of them is divided the other two are likewise divided and in the same proportion. Accordingly they knew necessarily

1. Perhaps in point of numbers. IbnTibbon's translation of the phrase could be rendered: do not exist from of old. (Trs.)
2. 'Substance' in this context means 'atom'. (Trs.)
3. When used by philosophers, the word *bu'd* often means 'dimension'. (Trs.)
4. Here the word denotes 'atoms'. (Trs.)

that if time were continuous and infinitely divisible, it would follow of necessity that the particles that they had supposed to be indivisible would also be divisible. Similarly if distance were supposed to be continuous, it would follow of necessity that the instant that had been supposed to be indivisible would be divisible—just as Aristotle had made it clear in the *Akroasis*. Therefore they supposed that distance is not continuous, but composed of parts at which divisibility comes to an end, and that likewise the division of time ends with the instants that are not divisible. For example, an hour consists of sixty minutes, a minute of sixty seconds, and a second of sixty thirds. And thus this division of time ends up accordingly with parts constituting, for instance, tenths or something even briefer, which cannot in any respect be separated in their turn into parts and are not subject to division, just as extension is not subject to it. Consequently, time becomes endowed with position and order. In fact they have no knowledge at all of the true reality of time. And this is only appropriate with regard to them; for seeing that the cleverest philosophers were confused by the question of time and that some of them did not understand its notion—so that Galen could say that it is a divine thing, the true reality of which cannot be perceived—this applies all the more to those who pay no attention to the nature of any thing.

CONSEQUENCE OF THE THREE PREMISSES

Hear now what they were compelled to admit as a necessary consequence of these three premisses and what they therefore believed. They said that motion is the passage of an atom belonging to these particles from one atom to another that is contiguous to it. It follows that no movement can be more rapid than another movement. In accordance with this assumption, they said that when you see that two things in motion traverse two different distances in the same time, the cause of this phenomenon does not lie in the greater rapidity of the motion of the body traversing the longer distance; but the cause of this lies in the motion that we call slower being interrupted by a greater number of units of rest than is the case with regard to the motion we call more rapid, which is interrupted by fewer units of rest. And when the example of an arrow shot from a strong bow was alleged as an objection against them, they said that the motions of the arrow were also interrupted by units of rest. In fact your thinking that a certain object is moving in continuous motion is due to an error of the senses, for many of the objects of the perception of the senses elude the latter, as they lay down in the twelfth premiss. In consequence, it was said to them: Have you seen a millstone making a complete revolution? Has not the part that is at its circumference traversed the distance represented by the bigger circle in the same time in which the part near the centre has traversed the distance represented by the smaller circle? Accordingly the motion of the circumference is more rapid than the motion of the inner circle. And there is no opportunity for you to assert that the motion of the latter part is interrupted by a greater number of units of rest as the whole body is one and continuous, I mean the body of the millstone. Their answer to this objection is that the various portions of the millstone become separated from one another in the course of its revolution and that the units of rest that interrupt the motion of all the revolving portions that are near the centre are more numerous than the units of rest that interrupt the motion of the parts that are farther off from the centre. Thereupon it was said to them: How then do we perceive the millstone as one body that cannot be broken up even by hammers? One must accordingly assume that when it turns round, it splits into pieces; and when it comes to rest, it is welded up and becomes as it was before. How is it that one does not perceive its portions as separated from one another? Thereupon, in order to reply to this, they had recourse to the same twelfth premiss, which states that one should not take into account the apprehensions of the senses, but rather the testimony of the intellect.

You should not think that the doctrines I have explained to you are the most abhorrent of the corollaries necessarily following from those three premisses, for the doctrine that necessarily follows from the belief in the existence of a vacuum is even stranger and more abhorrent. Furthermore, the doctrine that I have mentioned to you with regard to motion is not more abhorrent than the assertion going with this view that the diagonal of a square is equal to one of its sides, so that some of them say that the square is a non-existent thing. To sum up: By virtue of the first premiss all geometrical demonstrations become invalid, and they belong to either one or the other of two categories. Some of them are absolutely invalid, as for instance those referring to the properties of incommensurability and commensurability of lines and planes and the existence of rational and irrational lines and all that are included in the tenth book of Euclid and those that resemble them. As for the others, the demonstrations proving them are not cogent, as when we say we want to divide a line into two equal halves. For in the case in which the number of its atoms[1] is odd, the division of the line into two equal parts is impossible according to their assumption. Know, moreover, that the Banū Shākir[2] have composed the famous *Book of Ingenious Devices*, which includes one hundred odd ingenious devices, all of them demonstrated and carried into effect. But if vacuum had existed, not one of them would have been valid, and many of the contrivances to make water flow would not have existed. In spite of this, the lives [of the Mutakallimūn] have been spent in argumentation with a view to establishing the validity of these premisses and others resembling them.

1. It is to be borne in mind throughout this chapter that the same Arabic term is being translated variously as 'atom(s)' or 'substance(s)'. (Trs.)
2. A family of scientists of the 9th century who wrote a book on the construction of pumps and hydraulic machines. (Ed.)

ROBERT GROSSETESTE

The rainbow

142

Of the things that are under the same analogous thing, some are symmetrical therein, as in generated nature, and the mutual difference is as it is in opposite species of which none is the cause of the other; and some, on the contrary, are so constituted in themselves that one is the cause of the other and under the other in causal order, as for example the echo, the rainbow, and the appearance of images in a mirror are as opposite species under one analogous genus which is repercussion, and all are as one in the genus in which through one middle term it is possible to demonstrate one analogous effect. And regarding each separately, by means of the proper middle term the same effect is derived, as we said in the first book about the four proportionals. Since they are interchangeable, what is demonstrated in the demonstration in the fifth book of Euclid is demonstrated also in regard to lines, surfaces, numbers and solids separately in their own sciences. But the echo is the repercussion of sound from an obstacle, just as the appearance of images is the repercussion of visual rays from the surface of a mirror and the rainbow is the repercussion or refraction of rays of the sun in a concave aqueous cloud. For, when light diffusing itself in a straight line comes to an obstacle preventing its advance, it is collected in the place of incidence on the obstacle; and because its nature is to diffuse and generate in a straight line, when it cannot generate itself by advancing directly, it can generate itself only by turning back if the obstacle is an opaque body. Or, if the obstacle is a transparent body, it generates itself off the line and is not direct but penetrates the transparent object at an angle, as the ray of the sun falling on transparent water is reflected back from the surface of the water, as from a mirror, and also penetrates the water, making an angle at the surface itself, and this is properly called a refraction of the ray. Therefore, as light is reflected or refracted at the obstacle, so the rainbow is the reflexion or refraction of the light of the sun in a watery cloud and the appearance of images the reflexion of the visual ray at the mirror. For the visual ray is light passing out from the luminous visual spirit to the obstacle, because vision is not completed solely in the reception of the sensible form without matter, but is completed in the reception just mentioned and in the radiant energy going forth from the eye. But the substance of sound is light incorporated in the most subtle air, and when the sounding body is struck violently parts of it are separated from their natural position in the whole sounding body. But a natural power sends the parts passing away from the natural position back to the natural position, and the strength of its return makes it again progress beyond the natural position, and once again it returns to the natural position, and this may happen several times till the parts are at length quiescent. And so the natural power produces sound by sending back the parts passing out from the natural position when the object has been struck violently and it causes vibrations in them. In the vibrating parts going forth from their natural position there takes place an extension of the longitudinal and compression of the latitudinal diameter, and when they return to the natural position there is in them an alternate constriction and extension of these diameters. And so, when this motion of extension and constriction in the same object, according to the different diameters, reaches the light incorporated in the most subtle air which is in the sounding body, sound results. For every natural body has in itself a celestial luminous nature and luminous fire, and the first incorporation of it is in the most subtle air. Hence, when the sounding body is struck and vibrating, a similar vibration and similar motion must take place in the surrounding contiguous air, and this generation progresses in every direction in straight lines. When, however, the generation comes to an obstacle in which the air cannot generate a similar motion in the above-mentioned way, the expanding parts of the air are beaten or driven back on themselves and the vibration and motion in the air are again cast back and generated in the reverse direction, and the sound returns because of the obstacle, as with the visual rays. For, since it cannot generate itself in rectilinear progression, it regenerates itself by turning back. For the expanding parts of the air colliding with the obstacle must necessarily expand in the reverse direction, and so this repercussion extending to the light which is in the most subtle air is the returning sound, and this is an echo. Therefore, though what in these three cases in substance and truth is the reversal of light, in the rainbow it is the reversal of light because of cloud. An image is the reversal of visual light, that is the reversal of incorporeal light in the way described. [. . .]

Subordinate to this third part of optics is the science of the rainbow. Now a rainbow cannot be produced by means of solar rays passing in a straight line from the sun and falling into the concavity of a cloud, for they would make a continuous illumination in the cloud not in the shape of a bow, but in the shape of the opening on the side towards the sun through which the rays would enter the concavity of the cloud. Nor can a rainbow be produced by the reflection of the rays of the sun from the convexity of mist descending from the cloud, as from a convex mirror, in such a way that the concavity of the cloud may receive the reflected rays and thus a rainbow appear, because if that were so the shape of all rainbows would not be an arc, and it would happen that in proportion as the sun was higher so would the rainbow be bigger and higher, and in proportion as the sun was lower so would the rainbow be smaller, of which the contrary is manifest to the senses. Therefore the rainbow must be produced by the refraction of rays of the sun in the mist of a convex cloud. For I hold that the exterior of a cloud is convex and the interior of it concave, as is clear from the nature of the light and heavy. That which we see of a cloud must be less than a hemisphere, though it may look like a hemisphere, and when mist descends from the concavity of the cloud this mist must be pyramidally convex at the top, descending to the ground, and therefore more condensed near the earth than in the upper part. There will therefore be altogether four transparent media through which a ray of the sun penetrates, namely, the pure air containing the cloud, secondly the cloud itself, thirdly the highest and rarer part of the mist coming from the cloud, fourthly the lower and denser part of the same mist. It must therefore follow from what has been said above about the refraction of rays and the size of the angle of refraction at the junction of two transparent media, that the rays of the sun are refracted first at the junction of air and cloud, and then at the junction of cloud and mist. Because of these refractions the rays run together in the density of the mist and are refracted there again as from a pyramidal cone and spread out, not into the round pyramid, but into a figure like the curved surface of a round pyramid expanded in the opposite direction to the sun. It is for this reason that its shape is an arc and that with us the rainbow is not seen in the south. And, since the apex of the figure mentioned above is near the earth and its expansion is in the opposite direction to the sun, half of this figure or more than half must fall on the surface of the earth, and the remaining half or less than half must fall on the cloud opposite the sun. Therefore, when the sun is near rising or setting the rainbow appears semi-circular and is greater, and when the sun is in other positions the rainbow appears as part of a semi-circle, and in proportion as the sun is higher, so is the part of the rainbow smaller. Because of this, in many places, when the sun has risen to the zenith, no rainbow can appear at noon. What Aristotle said about the variegated bow at sunrise and sunset being of small measure is not to be understood as referring to smallness of size but to smallness of illumination, which comes about because of the passage of rays through a multitude of vapours at this time more than at other times. This Aristotle himself later suggests, saying this to be because of the diminution of that which shines because of the sun's rays in the clouds.

ROGER BACON

Mathematics as the basis of science

143

In which it is proved by reason that every science needs mathematics

What has been demonstrated by means of authority about mathematics as a whole can now be demonstrated in the same way by means of reason. And, firstly, because the other sciences make use of mathematical examples, but examples are given in clarification of the matters with which those sciences are concerned; wherefore if the examples are not understood those matters whose comprehension they were adduced to further are not understood. For, since no change is found in natural things without increase and decrease and these are not found without change, Aristotle could not clearly demonstrate the difference between increase and change by means of any natural example, because to a certain extent they always go together. Because of this he put forward a mathematical example of a quadrangle which increased by the addition of a gnomon[1] and is not changed. This example cannot be comprehended before Proposition 22 of Book VI of the *Elements*, for it is proved in that proposition in the sixth book that a lesser quadrangle is wholly similar to a greater; and therefore the lesser is not altered when from being lesser it becomes greater through the addition of a gnomon. [. . .]

Seventhly, where the same things are not known to us and known to nature, there is born in us a progression from those things better known to us to those better known to nature, or absolutely. And we know more easily the things which are better known to us, and arrive with great difficulty at the things which are better known to nature. And the things which are known to nature are badly and imperfectly understood by us, because our intellect is situated in relation to the things which are known to nature as is the eye of a bat towards sunlight, as Aristotle states in Book II of the *Metaphysics*; chief among these are God and the angels, and the future life, and celestial things, and some created things which are more exalted than others (because the more exalted they are the less they are known to us). And these are called things known to nature, and absolutely. Consequently on the contrary, where the same things are known to us and known to nature, we make considerable progress with regard to things which are known to nature and to all things included in it and can make contact with them in order to understand them properly. But, as Averroes says in Book I of the *Physics* and Book VII of the *Metaphysics* and in his commentary on Book III of *The Heaven and the World*, in mathematics alone are the same things known to us and known to nature, or absolutely. Therefore, as in mathematics we make complete contact with the things which are known to us, so we do with things known to nature, or absolutely; for that reason we reach absolutely the innermost parts of that science. Therefore, since we cannot manage that in the other sciences, it is clear that this one is the better known. Wherefore the source of our knowledge is attainable through it.

Now, eighthly, every doubt is resolved by that which is certain, and every error is removed by genuine truth. But in mathematics we can arrive at complete truth without error, and at certainty without doubt in everything, since in it it is possible for a demonstration to be given by means of the proper and necessary cause, and the demonstration causes the truth to be known. And similarly in mathematics it is possible for a sensible example to be given for everything, and a sensible test in figuring and counting, in order that everything should be clear to the sense; consequently there can be no uncertainty in it. But in the other sciences, once the assistance of mathematics is eliminated, there are so many uncertainties, so many opinions, so many errors of human origin, that they cannot be disentangled, which is obvious, since there is in them

1. A gnomon in this context means two strips meeting at a right angle. (Ed.)

no demonstration arising out of their own capacity through a proper and necessary cause, because there is no necessity about natural things on account of the generation and corruption of their proper causes as well as of their effects. In metaphysics there can be no demonstration except through effect, since spiritual things are discovered by means of their corporeal effects and the creator by means of his creation, as is shown in that science. In morals, as Aristotle teaches, there can be no demonstrations by proper causes; and similarly there can obviously be no very convincing demonstrations in logic nor in grammar, because of the poverty of the material with which those sciences are concerned. And consequently in mathematics alone can there be any convincing demonstrations by means of a necessary cause. And consequently there alone can man from the capacity of the science itself arrive at the truth. Similarly in the other sciences there are uncertainties and opinions and contradictions on our part, so that there is hardly agreement on a single insignificant question or on a single sophism; for in them there are by their very nature no experiments in figuring and counting by which everything could be verified. And consequently only in mathematics is there certainty without doubt.

It therefore appears that, if in the other sciences we are to arrive at certainty without doubt and at truth without error, we must set the foundations of knowledge on mathematics, insofar as disposed through it we can attain to certainty in the other sciences, and to truth through the exclusion of error. And this reasoning can be made clearer through a comparison, and the principle is also set out in Book IX of Euclid. For, as knowledge of the conclusion is related to knowledge of the premisses in such a way that if there is any mistake or uncertainty about these no truth or certainty about the conclusion can be achieved through them, because doubt cannot be resolved by doubt, nor truth by falsehood although it can be syllogized from false premisses, the syllogism being an inference and not a proof, the same is true of the sciences as a whole, so that those in which there are powerful and numerous uncertainties, opinions and errors, at least, I say, on our part, must have these uncertainties and untruths purged by some science well known to us in which we do not hesitate nor make mistakes. For, as conclusions and the principles proper to them form parts of the sciences as a whole, in the same way as one part is related to another part as conclusion to premisses, so is one science related to another science, so that a science which is full of uncertainties and riddled with opinions and obscurities cannot be made certain nor demonstrated nor verified except by another science which is known and verified and certain and plain to us, as a conclusion is through its premisses. But, as has already been expounded, only mathematics is known to us and verified in ultimate certainty and verification. Wherefore all the other sciences have to be known and established by means of it.

And, since it has been demonstrated through the property of the science itself that mathematics is prior to the others and useful and necessary to them, it is now demonstrated by reasons extracted from its actual subject matter. And firstly so because there is born in us a progression from sense to intellect, since if the sense is defective the science arising from that sense is defective, as is said in Book I of the *Posterior Analytics*, since, as the sense progresses, in the same way progresses the human intellect. But quantity is in the highest degree a question of sense, because it is a question of common sense, and sensed by the other senses, and nothing can be sensed without quantity; consequently the intellect can make its greatest progress concerning quantity. Secondly, because the very act of understanding is not achieved without continuous quantity, since Aristotle states in his book on *Memory and Recollection* that our whole understanding is involved with continuity and time. Therefore we understand quantities and bodies by the intuition of the intellect, because their species are to be found in the intellect. The species of incorporeal things are not received by our intellect in this way; alternatively if they do appear in it, as Avicenna says in Book III of the *Metaphysics*, we do not appreciate the fact because of the more powerful preoccupation of our intellect with bodies and quantities. And therefore by way of argumentation and the study of corporeal things and quantities we investigate the notion of incorporeal things, as Aristotle does in Book XI of the *Metaphysics*. Because of this the intellect will make its greatest progress concerning quantity itself, in that quantities and bodies, insofar as they are of this sort, belong properly to the human intellect according to the ordinary condition of understanding. Everything is because of something, and that thing is to a greater degree.

But the final reason for full confirmation can be extracted from the experience of wise men, for all wise men of old worked on mathematics in order to know everything, as we have seen with certain men of our own time, and heard about concerning others who gained knowledge of the whole of science through the mathematics which they knew well. For very famous men have been found, such as Bishop Robert of Lincoln and Brother Adam of Marsh and many others, who knew through the power of mathematics how to explain the causes of everything and how to expound satisfactorily human and divine matters. For the certainty of this is evident in the writings of these men, like those on the impressions, like those on the rainbow and on the comets, and on the generation of heat, and on the investigation of the places of the world, and on celestial bodies, and on other matters of which both theology and philosophy make use. Because of this it is clear that mathematics is entirely necessary and useful to the other sciences.

These reasons are universal, but in the particular it can be shown by recourse to all the branches of philosophy how all things are known through the application of mathematics; and this is in effect to show that the other sciences should not be known by means of the dialectical and sophistical arguments usually advanced, but by mathematical demonstrations which penetrate the truths and the works of the other sciences and regulate them, and without which they cannot be understood nor made clear nor taught nor learnt. Indeed, if anyone should penetrate the particular by applying the power of mathematics to the individual sciences, he would see that nothing remarkable about them can be known without mathematics. But this would in effect be to establish certain methods in all the sciences and to verify by means of mathematics everything which is necessary to the other sciences. But this is not at present a matter under investigation.

144

Concerning the movement of the balance

Now it is clear that a depressed weight is less heavy positionally for this reason, that it makes less of a direct descent on the diameter passing through the centre of revolution towards the centre of the earth; therefore, according to Jordanus,[1] it will be less heavy positionally. The actual truth demands this, as is explicable by a figure; and this kind of figuration resolves objections, and there can be no kind of intellectual satisfaction without a figure. Therefore let a circle be drawn about the centre of revolution, which is O, and the weights will be revolved on the circumference of this circle, and let the diameter AB be drawn parallel to the horizon; and let another diameter bisecting that one be drawn towards the centre of the earth, to be called DC, and let equal arcs be marked on each of the semi-circles on either side of the horizontal diameter, and from either end of it and from the ends of the arcs let lines be drawn in each semi-circle parallel to each other and to the horizontal diameter, to be called FH, GP, TQ, SR, which all cut across the diameter directed towards the centre of the earth. According to Jordanus and his commentator, those parallel lines must cut unequal parts from the diameter directed towards the centre of the earth, so that that parallel line which is nearer to the horizontal diameter will cut off a greater part of the other diameter than the further parallel line, so that TQ will separate a greater part of the diameter DC than will SR, so that the part of diameter DC which falls between AB and TQ is greater than the part of that diameter which falls between TQ and SR, and in the same way the part of the diameter DC which is between AB and GP will be greater than the part which is between GP and FH. And, consequent upon this, if one parallel is taken in one semi-circle and another in the other, which two are equally distant from the diameter parallel to them, these parallel lines must cut off equal parts of the diameter directed towards the centre of the earth, so that TQ and GP cut off equal parts of DC, and so do SR and FH, as is stated by Jordanus' twenty-sixth proposition in the work *On Triangles*. Therefore, if the parts of the diameter directed towards the centre of the earth which are cut off by parallel lines are unequal, in such a way that those parts of the diameter which are cut off by the parallel lines nearest to the horizontal diameter are the greater, then first we must realize that the beam of the balance lies on the horizontal diameter and the pendant is straight along

1. German mathematician; died 1237. (Ed.)

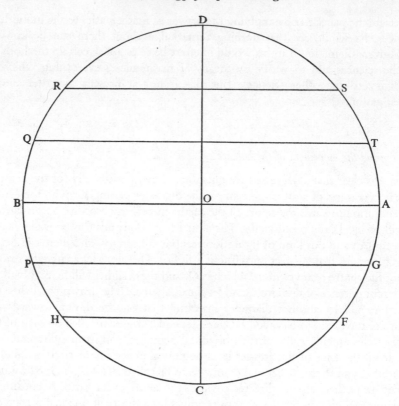

the diameter directed towards the centre in order for the balance to be in a horizontal position and also its arms; then let the balance be moved, and let one part of the balance be lifted to the first parallel in the upper semi-circle, and let the other part be depressed to the end of the first parallel in the lower semi-circle, so that the beam lies in the position of a line *TP*, and the upper part of the balance is at *T* and the other at *P*. If, therefore, *P* descends to *H*, the end of the other parallel line, it will pass on the diameter directed towards the centre that part of it which is between the parallels *GP* and *FH*, which is less than that part of the diameter which is between *TQ* and *AB*, as is clear from what was said above. Therefore if *T* were lowered as far as *A* it would make more of a direct descent on the diameter directed towards the centre of the earth than *P* does when lowered to H; therefore *T* is heavier positionally than *P*. And again *T* descends towards the centre of the earth. But *P* because of the curve of the circle is turned away from the centre, and is at least less directed towards the centre, as is obvious to sense. Therefore it remains that for this reason it is still less heavy. And because this is so, the first objection is resolved; for, although the part of the beam which is *P* is descending nearer the centre of the earth, as is clear if a straight line is drawn from it to *O* in the centre

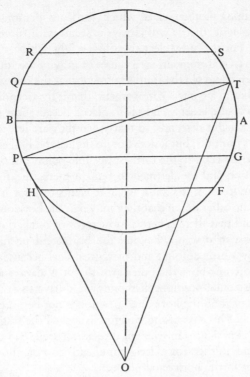

of the earth because that line is shorter than the line which is drawn from *T*, as is obvious to the sense, and is therefore heavier insofar as it has more of a downward position, nevertheless, because *P* does not move towards the centre along a straight line *PO* but instead follows the circulation of the circumference of the circle and that circulation causes it to make less of a direct approach along the diameter directed towards the centre than *T* makes and turns and bends *P* away from the centre so that it is not directed straight towards the centre as *T* is when it descends, *T* must be heavier than *P* while they are in such a position. And therefore if they are left to themselves *T* will descend and *P* will follow its descent in contrary motion, that is by ascending to the horizontal position. And therefore *P* will not always descend, but the two causes of gravity set out here prevail over the one mentioned by the objection. Moreover that gravity is not compulsive, for it is slight and not to be felt, and therefore does not operate here, in the same way as a single feather if placed on one of two equal weights when they are in a horizontal position would not cause a downward movement, and nevertheless strictly speaking the arm bearing the feather is the heavier, because the feather has a certain weight. And therefore the same is true here, because the descending arm when at the end of its descent is not far from a horizontal position and the gravity it acquires is slight and not to be felt, and consequently the added weight is incalculable.

And another thing is then clear which Jordanus demonstrates about this kind of greater weight, that it will always descend with more velocity so that it is never raised. For it is now clear that this weight achieves nothing appreciable, and that the two aforementioned causes of gravity are found in the higher arm. Indeed, when he says in the third argument that the one which follows the descent of the other in a contrary movement, that is by ascending, is less heavy, I entirely agree. For the lesser weight is so placed because it is moving upwards when the greater weight descends, so that when they are left to themselves and are equal that is its position; but it does not go higher and higher on this account, nor does the other descend to the furthest possible point, but they merely reach a horizontal position and are destined to remain there, and for this reason the objection infers more than it should when it wants to infer from the fact that one goes up and the other down that they may pass the horizontal position. And if it were to be said that there is here a movement of oscillation and that therefore the superior arm descends beyond the horizontal position and for that reason passes it by a little amount and by a great deal because this passage is of one kind, and likewise about the other arm that it is always ascending after it has passed the horizontal position, then one would have to say that this descent of the upper arm beyond the horizontal position is not because of the nature of the weight itself, but because of the violent changes of the fragments of the air. For when the air experiences movement it retains it well, and therefore its parts oscillate hither and thither for a long time, and do not allow the weight to settle immediately in the horizontal position.

Mechanically propelled vehicles

145

Machines for navigation can be made without rowers so that the largest ships on rivers or seas will be moved by a single man in charge with greater velocity than if they were full of men. Also cars can be made so that without animals they will move with unbelievable rapidity; such we opine were the scythe-bearing chariots with which the men of old fought. Also flying machines can be constructed so that a man sits in the midst of the machine revolving some engine by which artificial wings are made to beat the air like a flying bird. Also a machine small in size for raising or lowering enormous weights, than which nothing is more useful in emergencies. For by a machine three fingers high and wide and of less size a man could free himself and his friends from all danger of prison and rise and descend. Also a machine can easily be made by which one man can draw a thousand to himself by violence against their wills, and attract other things in like manner. Also machines can be made for walking in the sea and rivers, even to the bottom without danger. For Alexander the Great employed such, that he might see the secrets of the deep, as Ethicus the astronomer tells. These machines were made in antiquity and they have certainly been made in our times, except possibly a flying machine which I have not seen nor do I know of anyone who has, but I know an expert who has thought out the way to make one. And such things can be made almost without limit, for instance, bridges across rivers without piers or other supports, and mechanisms, and unheard of engines.

NICOLE ORESME

The motion of the earth

146

But, subject to correction, it seems to me that one could very well maintain the last opinion, namely that the earth is turned by a diurnal movement, and not the sky. And first, I wish to affirm that the contrary cannot be demonstrated from observation; second, that it cannot be demonstrated by reasoning; and third, I shall show why it cannot be demonstrated. As for the first point, observation shows us that the sun and the moon and several of the stars rise and set from day to day, and that some turn around the arctic pole. And this could not come about except by the movement of the sky, as was shown in the 16th chapter. And therefore the sky is turned by a diurnal movement. Another argument is this: if the earth is thus moved, it makes a complete revolution in a natural day. And therefore we and the houses and the trees are moved towards the east very quickly, and thus it would seem to us that the air would blow very strongly from the east, and would make a rushing noise, just as it does against a bolt shot from a crossbow, and the noise would be even louder; yet the contrary is evident from experience. The third is proposed by Ptolemy: if one were in a ship which was moving rapidly towards the east and if one were to fire an arrow straight upwards, it would not fall in the ship, but very far away, towards the west. And similarly, if the earth is moved very quickly in turning from west to east, supposing that one were to throw a stone directly upwards, it would not fall in the place from which it started but far away towards the west; and the contrary indeed happens. It seems to me that my criticism of these arguments could be used against all others of a like nature brought forward in this connection. And therefore I suppose first that the whole corporeal machine, namely all the mass of all the bodies in the world, is divided into two parts: one is the sky with the fiery sphere and the upper region of the air, and all this part, according to Aristotle in the first book of his *Meteorology*, turns in a diurnal movement. The other part is the remainder, namely the middle and the lower region of the air, the water, the earth and the mixed bodies, and, according to Aristotle, all this part is unaffected by the diurnal movement. Again, I suppose that movement of one body cannot be perceived except in relation to another. And for this reason, if a man is in a ship called *a* which is moved very gently, either quickly or slowly, and if this man can see nothing except another ship called *b* which is moved exactly like *a*, I say that it will seem to this man that neither ship is moving. And if *a* is at rest, and *b* is moved, it seems evident to him that *b* is moved; and if *a* is moved and *b* is at rest, it seems to him as before that *a* is motionless and *b* is moving. And thus, if *a* remained still for an hour and *b* was moved, while on the other hand in the following hour it was *a* which was moved and *b* which remained motionless, this man would not be able to perceive this change or variation, but it would seem to him all the time that *b* was moving; and this is plain from experience. And the cause of this is that these two bodies *a* and *b* are in exactly the same relation to each other when *a* is moved and *b* is at rest as they are, when, *e converso*, *b* changes position and *a* is motionless. And it is apparent from the fourth book of *La Perspective* by Witelo[1] that one only perceives motion insofar as one sees the position of a body changing in relation to that of another. I say, therefore, that if of the two parts of the world above mentioned, that above was today moved with a diurnal movement, and that below were not, and if tomorrow the contrary took place so that the lower part were moved with a diurnal movement, and the other (that is the sky, etc.) remained still, we would not be able to perceive this change at all, but everything would seem the same today and tomorrow in this respect. And it would seem to us at all times that the part in which we are remained motionless and that the other was moved continually, just as it seems to a man who is in a moving ship that the trees in the distance are changing their position. And likewise, if a man were in the sky, which was moved by a diurnal movement, and this man could see the earth clearly, and hills, valleys, rivers, towns and castles distinctly, it would seem to him that the earth was moved with a diurnal movement, just as it seems to us on earth that the sky moves. And similarly, if the earth were moved with a diurnal movement, and the sky were not, it would seem to us that it was at rest and that the sky was moved; and this can easily be imagined by anyone who has good powers of understanding. And in this the refutation of the first argument is clearly apparent, for one would say that the sun and the stars thus seem to rise and set and the sky to turn because of the movement of the earth and of the elements in which we live. The second argument can be answered by the fact that the earth alone is not moved, but with the air and the water as has been said, although the water and the air of the lower part can be moved in other directions by the winds or other agents. A similar case is that of a moving ship with enclosed air, for it would seem to anyone in that air that it was still. As for the third argument, which seems more difficult to refute, concerning the arrow or stone thrown upwards, etc., one could say that the arrow fired up is moved very quickly towards the east with the air through which it passes, and with all the mass of the lower part of the world which is moved with a diurnal movement; and for this reason, the arrow falls back to the place on the earth from which it started. And this seems possible if one considers a similar case; suppose a man to be in a ship travelling very swiftly

1. Polish mathematician, second half of the 13th century, famous for his book on optics. (Ed.)

towards the east, without his perceiving this motion, and suppose him to move his hand quickly down against the mast of the ship, it would seem to him that his hand was not moved except in a straight line; and thus, according to this opinion, the case of the arrow which rises or falls straight up or straight down seems to us similar. Again, within the ship which moves as has been described, there can be movements along, across, up, down, and in all directions, and these seem exactly the same as if the ship were at rest. And for this reason, if a man in this ship were going towards the west less quickly than it is going to the east, it would seem to him that he is approaching the west and yet he is actually going eastwards; and similarly in the case mentioned before, all movements on the earth would be the same as if it were at rest. Again, to elucidate further the reply to the third argument after this imaginary example, I wish to add another natural one which is true according to Aristotle, and which supposes that in the upper region of the air there is a portion of pure fire called *a* which is very light, and for this reason it rises as high as it possibly can to the place called *b* near to the concave surface of the sky. I say that this case resembles that of the arrow above mentioned, and it would happen in this one that the movement of *a* would be composed of straight, and in part, circular, movements, for the region of the air and the sphere of fire through which *a* passed are turned, according to Aristotle, by a circular movement. And indeed, if they were not thus moved, *a* would go straight up along the line *ab*; but since by a circular and diurnal movement *b* is meanwhile moved to the place *c*, it is plain that *a* in rising describes the line *ac*, and that the movement of *a* is composed of straight and circular movement. And thus the movement of the arrow would be similar, as has been stated, and of such a composition or mixture of movements as was said in the third chapter of the first part. I conclude therefore that one could not by any observation show that the sky was moved by a diurnal movement and that the earth was not thus moved.

NICOLAS CUSANUS

147

How the universe, which is only a restricted form of
maximum, is a likeness of the absolute

If we are ready to go on and make an adroit use of what learned ignorance has taught us in the previous chapters, in particular if we simply recall that all things are, or owe their existence to, the Absolute Maximum, we will be able to discover a good deal about the world or universe. For me the universe is only a restricted form of maximum. It is restricted or concrete, because it holds all its being from the Absolute; and because it is a maximum, it reproduces the Absolute Maximum in the greatest way possible. We may affirm, therefore, that all we learned in the First Book of the Absolute Maximum as belonging to

Him without any restriction whatsoever, may be applied in a relative way to the restricted maximum.

Let us take some examples to help the student in his inquiry. God is the absolute Maximum and absolute unity, and, as such, He forestalls and unites things different and distant; e.g. contradictories, between which there is no mean, are identified in Him. In an absolute way He is what all things are; He is the absolute beginning in all, the end of all and the entity of all. Just as the infinite line is all the figures so the Absolute Maximum, in its infinite simplicity and unity, is all things without plurality. The world or universe is also a maximum though a limited maximum; it anticipates in its unity limited opposites like contraries; within its limited existence it is what all things are; and in a restricted sense it is the beginning, end and being of things. It is infinitely contracted to the relatively infinite; and just as the relative maximum line is relatively all figures so the limited maximum—the universe—in its relative simplicity and unity is all things without plurality.

148

Corollaries of movement

The fact that the ignorance which is learning has shown the truth of the foregoing doctrine will perhaps be a surprise to those who had not heard of such teaching before. By it we now know that the universe is a trinity; that there is not a being in the universe which is not a unity composed of potency, act and the movement connecting them and that none of these three is capable of absolute subsistence without the others, with the result that they are necessarily found in all things in the greatest diversity of degrees—in degrees so different that it is impossible to find in the universe two beings perfectly equal in all things. Consequently, once we have taken the different movements of the stars into account, we see that it is impossible for the motor of the world to have the material earth, air, fire or anything else for a fixed, immovable centre. In movement there is no absolute minimum, like a fixed centre, since necessarily the minimum and the maximum are identical.

Therefore the centre and the circumference are identical. Now the world has no circumference. It would certainly have a circumference if it had a centre, in which case it would contain within itself its own beginning and end; and that would mean that there was some other thing which imposed a limit to the world—another being existing in space outside the world. All of these conclusions are false. Since, then, the world cannot be enclosed within a material circumference and centre, it is unintelligible without God as its centre and circumference. It is not infinite, yet it cannot be conceived as finite, since there are no limits within which it is enclosed.

The earth, which cannot be the centre, must in some way be in motion; in fact, its movement even must be such that it could be infinitely less. Just as the earth is not the centre of the world, so the circumference of the world is not the sphere of the fixed stars, despite the fact that by comparison the earth seems nearer the centre and heaven nearer the circumference. The earth, then, is not the centre of the eighth or any other sphere, and the appearance above the horizon of the six stars is no proof that the earth is at the centre of the eighth sphere. If even at some distance from the centre it were revolving on its axis through the poles, in such a way that one part would be facing upwards towards one pole and the other part facing downwards towards the other pole, then, it is evident, that to men as distant from the poles as the horizon only half of the sphere would be visible. Further, the centre itself of the world is no more within than outside the earth; and this earth of ours has no centre nor has any other sphere a centre. Since the centre is a point equidistant from the circumference, and since it is impossible to have a sphere or circle so perfect that a more perfect one could not be given, it clearly follows that a centre could always be found that is truer and more exact than any given centre. Only in God are we able to find a centre which is with perfect precision equidistant from all points, for He alone is infinite equality. God, ever to be blessed, is, therefore, the centre of the world: He it is who is centre of the earth, of all spheres and of all things in the world; and at the same time He is the infinite circumference of all.

149

Conditions of the earth

The ancient philosophers did not reach these truths we have just stated, because they lacked learned ignorance. It is now evident that this earth really moves though to us it seems stationary. In fact, it is only by reference to something fixed that we detect the movement of anything. How would a person know that a ship was in movement, if, from the ship in the middle of the river, the banks were invisible to him and he was ignorant of the fact that water flows? Therein we have the reason why every man, whether he be on earth, in the sun or on another planet, always has the impression that all other things are in movement whilst he himself is in a sort of immovable centre; he will certainly always choose poles which will vary accordingly as his place of existence is the sun, the earth, the moon, Mars, etc. In consequence, there will be a *machina mundi* whose centre, so to speak, is everywhere, whose circumference is nowhere, for God is its circumference and centre and He is everywhere and nowhere. [. . .]

Because the earth is smaller than the sun and is influenced by it is not a reason for calling it baser, for the entire region of the earth, which stretches to the circumference of fire, is great. From shadow and eclipses we know that the earth is smaller than the sun; yet despite that we do not know to what extent the region of the sun is greater or smaller than the region of the earth. It cannot be exactly equal to it, for no star can be equal to another. And the earth is not the smallest star, for eclipses have shown us that it is larger than the moon; and some say that it is larger even than Mercury, larger, perhaps, than all other stars. From size, therefore, no proof can be alleged of its baseness.

Even the influence asserted on it is not a proof of its imperfection; perhaps it, as a star, has a similar influence on the sun and its region, as already stated. We have no knowledge from experience of that influence, since we have no experience beyond that of our existence in the centre where the influences merge. Even if, in fact, we consider the earth as potency, the sun as its formal act or soul, and the moon as their connecting medium, the result would be the mutual relation of the influences of those stars situated within the one region (others like Mercury, Venus, etc., are above this region, according to the ancients and even some moderns); then the correlation of influence is clearly such that one could not exist without another. Likewise, in different degrees this influence, one and threefold, will be found in all. As regards these points, it is evidently impossible for man to discover whether the region of the earth is in degree more perfect or less perfect by comparison with the regions of the other stars such as the sun, moon and the others.

SECTION III
Copernicus to Pascal

Section III: Copernicus to Pascal

INTRODUCTION

The main historical significance of the period discussed in this chapter (in other words the hundred or so years from the mid-sixteenth to the mid-seventeenth century) is that it saw the undermining of the foundations of the Aristotelian view of nature, and abolished most of its scientific presuppositions. This also terminated the medieval scientific approach itself, which was based mainly on Aristotelian doctrine. However, the passages from the scientific literature of the period included here show not only that the thinkers of the time did away with ideas and conceptions that had prevailed for more than 1,500 years and with such powerful factors as medieval theology. They demonstrate too that during this time was provided the basis for the new conceptual structure of the Newtonian era which succeeded it. During these three generations—from the death of Copernicus to the birth of Newton—facts of the utmost importance were discovered. At the same time there appeared the first formation of the theoretical framework within which Newton and his successors in following centuries erected the whole complex edifice of the new science. During the same period too the new philosophy with its two main streams of empiricism and rationalism aided the scientists in their efforts, as it struggled to free itself from the fetters of Scholastic theology and from those of the Aristotelian world-view. It helped to prepare the way for a new view of a world in which the scientific approach would become an inseparable element.

Historians of science usually take the year 1543 as the beginning of the new era, for this was the year in which Copernicus' *On the Revolutions of Celestial Bodies* (*De Revolutionibus*) was published. There are weighty reasons for this. The prime impact of the Copernican revolution was certainly a psychological one. Displacing the earth from the centre of the planetary system (which in the sixteenth century was still identical with the centre of the closed universe) dealt a fatal blow to the anthropocentric view. In the full sense of the term, it opened up new horizons to science, giving it a boundless urge for an understanding of nature based on scientific investigation free from any tutelage. The heliocentric theory provided the psychological impetus to a tendency which spread in ever widening circles during the next hundred years; this was the acknowledgement of scientific endeavour as an autonomous area of human thought. The hostile attitude of the Church felt all through Europe since the end of the seventeenth century did not so much frighten the public as eventually further this tendency of establishing the independence of science from theology.

With the publication of Copernicus' book, the scientific world was faced with a situation entirely unprecedented in its history. The geocentric theory of

Ptolemy and the new ideas of Copernicus were both founded on the same astronomical data, known from the time that the *Almagest* was written. The acceptance of the heliocentric theory depended upon theoretical considerations only, and the decision in favour of one or other had to be based on epistemological or metaphysical grounds. The two introductions prefacing the book—that of its publisher, Osiander, and that of its author—are typical of the two opposing kinds of approach which we find throughout the whole history of science—positivism and realism. Osiander argues that the aim of science is only a methodological one. As we cannot *know* the truth we need to be content with the adaptation of our calculations to known facts; that theory which makes these calculations easier and which enables us to achieve more factual results is the better one. In Osiander's words we can distinguish a note of apologetics for the new theory, one not as yet accepted by the learned who in classical times had adopted the Ptolemaic system. Although Osiander was prepared to support Copernicus' theory for these methodological reasons, it is obvious that he preferred the theory which had held sway for hundreds of years.

On the other hand, Copernicus declares in his introduction that he has found a new truth. His evidence is that his theory discloses intrinsic relations between the different empirical data, so uniting them in one coherent system where no single item may be omitted without destroying the whole. In this context Copernicus emphasizes that, according to this theory, the loops in the paths of the planets reflect the movement of the earth in its orbit round the sun. Thus he is able to demonstrate the necessary connection between the size and the number of these loops and the distances from the sun of Mars, Jupiter and Saturn, while no such necessary connection exists in the Ptolemaic system. Copernicus argues that his theory is the true one because it is structurally more simple—this structural simplicity in a theory being decisive for its truth. This conception of simplicity is outside the sphere of epistemology, becoming metaphysical as soon as one asserts, as Copernicus did, that nature prefers simplicity in all its manifestations. Such a metaphysical view was held by many scientists through the ages, from Aristotle to the physicists of the eighteenth century who formulated the principle of least action.

From a methodological point of view, too, the Copernican theory brought an immense advance to applied astronomy. This was because it enabled astronomers to determine the relative distances of the planets from the sun, thus preparing the ground for Kepler's researches. However, it is worthwhile having a closer look at Copernicus' arguments against the geocentric system. Here we find a certain qualitative anticipation (although still expressed in Aristotelian terms) of the basic principles of classical mechanics—as for example the law of inertia and the notion of universal gravity. These concepts could, of course, be developed fully only after the consolidation of the heliocentric system.

Many decades had to pass before the heliocentric system with all itsi mplications was accepted by the majority of scientists, and they in the late sixteenth century constituted only a very small group in Europe. About forty years after the publication of Copernicus' book, the Dominican Friar, Giordano Bruno, came forward as an enthusiastic advocate of the new theory in the cultural centres of Western Europe; he also boldly developed it further. Following Nicolas of Cusa, and more than a hundred years after him, Bruno once more raised the problem of the infinity of the universe. However, whereas Nicolas proceeded from a theologico-philosophical basis, the impulse which brought Bruno to his abnegation of the closed Aristotelian world was the Copernican system, in which the earth was displaced from the centre of the universe and reduced to the status of one of the planets revolving round it. Having embraced the Copernican theory, Bruno went even further. He regarded the fixed stars as suns dispersed throughout the whole of an infinite universe, and thus our sun as one of an infinite number. Bruno therefore in turn displaced the sun from the central position which Copernicus in his closed universe had still attributed to it. The publicity which Bruno brought to the Copernican theory, and which brought him to the stake, intensified the opposition of the Church. This eventually led to her well-known dispute with Galileo.

Amongst other supporters of the heliocentric theory was William Gilbert, physician to Queen Elizabeth I. His book *De Magnete* (1600) was one of the first scientific works of the new era. Although Gilbert's awkward language and scientific style make his book appear old-fashioned, his approach is eminently scientific and systematic. In his thorough, detailed research into the properties of magnetic materials, we observe the spirit of the new era, with its inclination towards experimentation and its awareness of the dangers of excessive generalization. Gilbert's hypothesis of the earth itself as a huge magnet was the basis of his explanation for the phenomena of the inclination and declination of the magnetic needle. It served as a starting-point for his general hypothesis on the cosmic significance of magnetism. His book had a major impact on the scientists of the seventeenth century; Kepler for instance used the ideas in his attempt to explain the elliptical form he discovered for the planetary orbits.

Gilbert's book may be seen as the first step towards the remarkable scientific activity which appeared in England fifty years later. This activity was also influenced decisively by one of Gilbert's great contemporaries, the philosopher and statesman Francis Bacon. From the middle of the seventeenth century, scientific thought had been exposed to the philosophies of two new main schools—those of the empiricists and of the rationalists. Bacon, first in the line of the English empiricist philosophers of the seventeenth and eighteenth centuries, was not a scientist as such. Above all, he was not interested in the mathematical approach to science. However, more than anybody else he influenced a whole generation of great scientists, including the founders of the Royal Society of London. This he did by determining the path of research on the basis of experimentation and induction, and by urging the scientists to turn away from scholasticism, to look at nature itself, and not to rely on the writings of former ages. Bacon's notion of induction was too simple and inadequate in several respects; his great merit, however, is that he warned the scientist of the danger of the prejudices arising from human nature, habit, and the old-fashioned style of education.

Whereas Bacon and empiricist philosophy exerted their influences on the development of science mainly in England, the rationalist philosophy of René Descartes was the principal authority on the Continental mainland. Not only was Descartes one of the greatest philosophers of all time, he was also a great scientist. His work on geometry, the deductive science *par excellence*, together with his belief in Man's reason as the only secure basis for the understanding of Nature and of Man himself, led Descartes to the conclusion that although observation and experiment are necessary, the main foundation of scientific research is deduction. His rules for philosophical thought and the systematization of science are more definitive and more comprehensive than Bacon's aphoristic remarks. In two fields Descartes exerted an influence greater than that of anyone else in modern scientific thought, although he failed in his attempt to apply all of his theories in practice. Firstly there was his concept that all physical phenomena can be described geometrically, because in the last resort matter is only three-dimensional extension; therefore all its different forms cannot contain anything that is not contained in geometrical laws and in mathematical relations. This conception is also the basis of continuum theory and of its negation of the vacuum from which Descartes had deduced his cosmology. Here he writes that the universe consists of three kinds of substance—matter that radiates light and heat, aether which moves in a circular fashion, and 'dark matter'. The second field in which Descartes' influence on modern science was unique was his extremely mechanistic view of nature. He felt that the physical world and all its phenomena operate like a machine and its parts; it can therefore be explained by mechanical examples and analogies. Here again there exists a close connection with continuum theory, for every mechanical action can be shown to be the result of the direct contact of adjoining bodies or of adjoining parts of bodies. Descartes discussed the laws of motion in greater detail, and was the first to define the law of inertia in its general form. He also clarified the law of conservation of the 'quantity of motion' without, however, distinguishing between collisions of elastic bodies and those of non-elastic ones. Descartes regarded quantity of motion as the basic concept of dynamics.

Descartes' theory of vortices was the first modern attempt at a mechanical explanation of planetary movement. (It is analogous with the hydrodynamics of eddies in water.) Although this theory was refuted by Newton forty years later, it survived for a long time, because, like every mechanical model, it attracted people by its ease of visualization. In biology too, Descartes' views were wholly mechanistic, in spite of his dualism which divided the world into two separate spheres, the 'extensional' one of matter, and the 'thinking' one of mind. He explained all physiological functions by pure mechanism and, seeing all bodies as machines, recognized no difference between the human body and those of all other living beings.

However, notwithstanding his mechanistic approach, his philosophy of deduction, and his tendency for general hypotheses, Descartes did not adopt a dogmatic stand in his epistemology. There is a great measure of plausibility in the deductions from the visible to the invisible, yet it holds no absolute certainty. We may care to compare the scientist to the decoder of a cypher, but even if he succeeded in this task he can never claim that his conclusions are absolutely certain.

As far as we can judge from his astronomical writings Johann Kepler embodied in his scientific personality the whole spectrum of inclinations, including the opposing ones, that are active in a great scientist. We cannot separate these without falsifying the character of his scientific consciousness. Kepler's famous achievements, in particular the three laws concerning the movements of the planets, which form a basis of Newton's law of gravity, are evidence of his greatness as a scientist, his accuracy, his imagination, and his outstanding mathematical ability. He raised astronomy to a level that in the era of the telescope opened up new horizons. After the death of his mentor Tycho Brahe in 1601, he finished tabulating the data of planetary movements. On the basis of his accurate calculations, and in particular of his extensive observations of the movements of Mars, he came to the conclusion which abolished the monopoly of the circle in the orbits of heavenly bodies. He established that each planet moves in an ellipse, the sun being at one of the foci. Kepler's second law, that concerning the areas swept out, is in fact the principle of conservation of angular momentum applied to the motion of a planet. The third law, published ten years after the others, states the relation between the distance of a planet from the sun and the period of its revolution; Kepler regarded it as the culmination of his life's work. If we wish to understand his incredulous excitement at the discovery, we must look at the psychological motivations that influenced his lifelong researches.

Kepler was an enthusiastic Neoplatonist; his belief in the harmonic structure of the universe was fused with mystical Christianity. The basis of his scientific beliefs was the Copernican theory; starting from this, he set out to discover this harmony in the planetary system. First signs of his eminently structural approach can be detected in *Mysterium Cosmographicum*, his first treatise (1596). In this work he tried to relate the distances of the six planetary spheres to the dimensions of the five Platonic bodies enclosed between them. After his discovery of the elliptical orbits of the planets and the variation of their velocities with distance from the sun, he revived the old Pythagorean idea of the musical harmony of the spheres. An example of this was his attempt to discover a relation between the speeds of a planet at parhelion and at aphelion, such a relation being assumed to represent a musical harmony. He saw his third law in fact as definite evidence for his firm belief in the existence of harmonic numerical ratios in the structure of the universe.

Galileo and Kepler, those two great contemporaries, represent two opposite types of man of science. Kepler, the mystic, all his life tried to find a cosmic formula to express the structural harmony of the universe, doing his work in mathematics and in astronomy with the conception of the cosmos as a whole. Galileo, the rationalist, on the other hand avoided all-embracing hypotheses and saw the task of science to be the clarification and systematic and detailed analysis of particular facts whose theoretical and experimental elucidation would widen our knowledge of the physical world. Admittedly Gilbert preceded Galileo in concentrating on the examination of particular phenomena. However, Galileo's selection of phenomena for consideration, his beautifully lucid scientific language, and the mathematical formulations of his experimental results make him the pioneer of specialization, and characteristic of the modern research scientist. Galileo's mathematical definitions of velocity and acceleration and their dependence on time, his laws concerning falling and thrown bodies, and that of the oscillation of the pendulum, form the basis of the development of mechanics in the seventeenth and eighteenth centuries. In the whole history of science, there is no other writer as clear and as brilliant as Galileo. In his popular works he succeeded in explaining his scientific results, and, at the same time, in dealing a fatal blow to Aristotelian physics. His conclusive arguments against Aristotle and the Scholastics, consisting as they did of a detailed physical analysis, went much deeper than the attacks of Bacon and Descartes; they had the greatest influence on the development of physical thought.

Galileo's achievements in astronomy were of no less importance than those in mechanics; his researches here were major factors in the development of scientific thought as an autonomous field of human endeavour. His first observations with the help of a telescope in 1609 provided almost conclusive proof for the Copernican theory. They involved the changes in the appearance of Venus with position in relation to the sun; the four moons of Jupiter, which together represent a miniature model of the heliocentric system; and sunspots, which demonstrate the sun's axial rotation and enable us to draw conclusions regarding the rotation of the planets, including our own. His courageous stand against Aristotle's cosmology, which by implication was equivalent to an attack on the cosmology adopted by the Church, brought an end to concepts held in awe since Hellenistic times. The changes occurring in sunspot patterns prove that there is generation and decay in the heavenly sphere as well as in ours; like Philoponus eleven hundred years earlier, Galileo came to the conclusion that celestial and terrestrial phenomena have the same nature. In this connection he formulated his classic *credo* saying that he preferred the changeable and passing to the unchangeable and static, a programmatic statement of that dynamic approach which would become the prevalent view of the Newtonian era.

Sunspots represent one of the focal notions in Galileo's fight for independence of scientific thought, that fight which led to his famous trial by the Inquisition in 1633. In his letters to the Grand Duchess Christina, he states that all natural phenomena and all that is written in the Bible make but two aspects of one and the same truth. He stresses, however, that the Bible is expressed in a language suited to the imperfect understanding of human beings, whereas the language of nature is unambiguous. Thus the scientist should take natural phenomena as the guide to his considerations. In the course of time this doctrine determined the development of science.

Galileo's writings, in particular his *Dialogo* of 1632 on the 'two chief systems, Ptolemaic and Copernican', and his *Discorsi* of 1638 on 'two new sciences—mechanics and the laws of falling bodies'—provide a mine of fruitful ideas and explanations in all spheres of science. In this Anthology we can give only a few examples of his writings—among them the passage on the speed of light and its measurement, and the passage about the stone thrown up into the air from a ship in motion. Galileo's ideas about the stone falling along the mast of the ship were proved valid in 1640 by the famous experiment of Pierre Gassendi which took place in the Mediterranean near Marseilles. This experiment provided evidence for the law of inertia, and at the same time it demonstrated the classical principle of relativity. Gassendi, together with Mersenne, Descartes, Fermat, and Pascal, was one of that group of great scientists who in France laid the foundations of science before the establishment of the Académie des Sciences in Paris. Gassendi's book on Epicurus' theory of atoms was an important step in the development of the corpuscular concept of matter in modern times; he greatly influenced the scientists of the seventeenth century, including for example Robert Boyle. The book shows too the sympathetic approach of revitalized science to antiquity—to Plato, Epicurus, Archimedes and the other great Greek natural philosophers. This attitude was in no way diminished by its absolute negation of the Aristotelian and Scholastic world-views.

A typical example of the overthrow of Aristotelian ideas by experimental science is given by the researches into the physical properties of air (which until the beginning of the modern era had been regarded as one of the four basic elements), and by those into the existence and properties of vacuum. This work involved, among others, Torricelli in Italy, and Pascal in France. Later, after the German Guericke had invented the air-pump, it spread to England with the systematic research of Boyle and Hooke.

III: COPERNICUS TO PASCAL

NICOLAUS COPERNICUS

Osiander's[1] introduction to De revolutionibus orbium coelestium

150

Since the newness of the hypotheses of this work—which sets the earth in motion and puts an immovable sun at the centre of the universe—has already received a great deal of publicity, I have no doubt that certain of the savants have taken grave offence and think it wrong to raise any disturbance among liberal disciplines which have had the right set-up for a long time now. If, however, they are willing to weigh the matter scrupulously, they will find that the author of this work has done nothing which merits blame. For it is the job of the astronomer to use painstaking and skilled observation in gathering together the history of the celestial movements, and then—since he cannot by any line of reasoning reach the true causes of these movements—to think up or construct whatever causes or hypotheses he pleases such that, by the assumption of these causes, those same movements can be calculated from the principles of geometry for the past and for the future too. This author is markedly outstanding in both of these respects: for it is not necessary that these hypotheses should be true, or even probable; but it is enough if they provide a calculus which fits the observations—unless by some chance there is anyone so ignorant of geometry and optics as to hold the epicycle of Venus as probable and to believe this to be a cause why Venus alternately precedes and follows the sun at an angular distance of up to 40° or more. For who does not see that it necessarily follows from this assumption that the diameter of the planet in its perigee should appear more than four times greater, and the body of the planet more than sixteen times greater, than in its apogee? Nevertheless the experience of all the ages is opposed to that. There are also other things in this discipline which are just as absurd, but it is not necessary to examine them right now. For it is sufficiently clear that this art is absolutely and profoundly ignorant of the causes of the apparent irregular movements. And if it constructs and thinks up causes—and it has certainly thought up a good many—nevertheless it does not think them up in order to persuade anyone of their truth but only in order that they may provide a correct basis for calculation. But since for one and the same movement varying hypotheses are proposed from time to time, as eccentricity or epicycle for the movement of the sun, the astronomer much prefers to take the one which is easiest to grasp. Maybe the philosopher demands probability instead; but neither of them will grasp anything certain or hand it on, unless it has been divinely revealed to him. Therefore

1. Andreas Osiander (1498–1552) supervised the last stages of the printing of *De revolutionibus*. His introduction appeared anonymously: the author's identity was discovered by Kepler. (Ed.)

let us permit these new hypotheses to make a public appearance among old ones which are themselves no more probable, especially since they are wonderful and easy and bring with them a vast storehouse of learned observations. And as far as hypotheses go, let no one expect anything in the way of certainty from astronomy, since astronomy can offer us nothing certain, lest, if anyone take as true that which has been constructed for another use, he should go away from this discipline a bigger fool than when he came to it. Farewell.

Preface and dedication to Pope Paul III

151

[...] But perhaps Your Holiness will not be so much surprised at my giving the results of my nocturnal study to the light—after having taken such care in working them out that I did not hesitate to put in writing my conceptions as to the movement of the Earth—as you will be eager to hear from me what came into my mind that in opposition to the general opinion of mathematicians and almost in opposition to common sense I should dare to imagine some movement of the Earth. And so I am unwilling to hide from Your Holiness that nothing except my knowledge that mathematicians have not agreed with one another in their researches moved me to think out a different scheme of drawing up the movements of the spheres of the world. For in the first place mathematicians are so uncertain about the movements of the sun and moon that they can neither demonstrate nor observe the unchanging magnitude of the revolving year. Then in setting up the solar and lunar movements and those of the other five wandering stars, they do not employ the same principles, assumptions or demonstrations for the revolutions and apparent movements. For some make use of homocentric circles only, others of eccentric circles and epicycles, by means of which, however, they do not fully attain what they seek. For although those who have put their trust in homocentric circles have shown that various different movements can be composed of such circles, nevertheless they have not been able to establish anything for certain that would fully correspond to the phenomena. But even if those who have thought up eccentric circles seem to have been able for the most part to compute the apparent movements numerically by those means, they have in the meanwhile admitted a great deal which seems to contradict the first principles of regularity of movement. Moreover, they have not been able to discover or to infer the chief point of all, i.e. the form of the world and the certain commensurability of its parts. But they are in exactly the same fix as someone taking from different places hands, feet, head and the other limbs—shaped very beautifully but not with reference to one body and without correspondence to one another—so that such parts made up a monster rather than a man. And so, in the process of demonstration which they call 'method', they are found either to have omitted something necessary or to have admitted something foreign which by no means pertains to the matter; and they would by no means have been in this fix, if they had followed sure principles. For if the hypotheses they assumed were not false, everything which followed from the hypotheses would have been verified without fail; and though what I am saying may be obscure right now, nevertheless it will become clearer in the proper place.

Accordingly, when I had meditated upon this lack of certitude in the traditional mathematics concerning the composition of movements of the spheres of the world, I began to be annoyed that the philosophers, who in other respects had made a very careful scrutiny of the least details of the world, had discovered no sure scheme for the movements of the machinery of the world, which has been built for us by the Best and Most Orderly Workman of all. Wherefore I took the trouble to reread all the books by philosophers which I could get hold of, to see if any of them even supposed that the movements of the spheres of the world were different from those laid down by those who taught mathematics in the schools. And as a matter of fact, I found first in Cicero that Hicetas[1] thought that the Earth moved. And afterwards I found in Plutarch that there were some others of the same opinion: I shall copy out his words here, so that they may be known to all:

'Some think that the Earth is at rest; but Philolaus, the Pythagorean, says that it moves around the fire with an obliquely circular motion, like the sun and moon. Heracleides of Pontus[2] and Ecphantus, the Pythagorean, do not give the Earth any movement of locomotion, but rather a limited movement of rising and setting around its centre, like a wheel.'

Therefore I also, having found occasion, began to meditate upon the mobility of the Earth. And although the opinion seemed absurd, nevertheless because I knew that others before me had been granted the liberty of constructing whatever circles they pleased in order to demonstrate astral phenomena, I thought that I too would be readily permitted to test whether or not, by laying down that the Earth had some movement, demonstrations less shaky than those of my predecessors could be found for the revolutions of the celestial spheres.

And so, having laid down the movements which I attribute to the Earth farther on in the work, I finally discovered by the help of long and numerous observations that if the movements of the other wandering stars are correlated with the circular movement of the Earth, and if the movements are computed in accordance with the revolution of each planet, not only do all their phenomena follow from that but also this correlation binds together so closely the order and magnitudes of all the planets and of their spheres or orbital circles and the heavens themselves that nothing can be shifted around in any part of them without disrupting the remaining parts and the universe as a whole.

Accordingly, in composing my work, I adopted the following order: in the

---

1. Hicetas and his pupil Ecphantus were of the later Pythagorean School of the 4th century B.C. (Ed.)
2. Heracleides of Pontus (4th century B.C.) was a pupil of Plato. He put forward the hypothesis of the axial rotation of the earth and of the revolution of Mercury and Venus round the sun. (Ed.)

first book I describe all the locations of the spheres or orbital circles together with the movements which I attribute to the Earth, so that this book contains as it were the general set-up of the universe. But afterwards in the remaining books I correlate all the movements of the other planets and their spheres or orbital circles with the mobility of the Earth, so that it can be gathered from that how far the apparent movements of the remaining planets and their orbital circles can be saved by being correlated with the movements of the Earth. And I have no doubt that talented and learned mathematicians will agree with me, if—as philosophy demands in the first place—they are willing to give not superficial but profound thought and effort to what I bring forward in this work in demonstrating these things. And in order that the unlearned as well as the learned might see that I was not seeking to flee from the judgement of any man, I preferred to dedicate these results of my nocturnal study to Your Holiness rather than to anyone else; because, even in this remote corner of the earth where I live, you are held to be the most eminent both in the dignity of your order and in your love of letters and even of mathematics; hence, by the authority of your judgement you can easily provide a guard against the bites of slanderers, despite the proverb that there is no medicine for the bite of a sycophant. [. . .]

152

Chapter 6. On the immensity of the Heavens in relation to the magnitude of the Earth

It can be understood that this great mass which is the Earth is not comparable with the magnitude of the heavens, from the fact that the boundary circles—for that is the translation of the Greek '*horizons*'—cut the whole celestial sphere into two halves; for that could not take place if the magnitude of the Earth in comparison with the heavens, or its distance from the centre of the world, were considerable. For the circle bisecting a sphere goes through the centre of the sphere, and is the greatest circle which it is possible to circumscribe.

Now let the horizon be the circle *ABCD*, and let the Earth, where our point of view is, be *E*, the centre of the horizon by which the visible stars are separated from those which are not visible. Now with a dioptra or horoscope or level placed at *E*, the beginning of Cancer is seen to rise at point *C*; and at the same moment the beginning of Capricorn appears to set at *A*. Therefore, since *AEC* is in a straight line with the dioptra, it is clear that this line is a diameter of the ecliptic, because the six signs bound a semi-circle, whose centre *E* is the same as that of the horizon. But when a revolution has taken place and the beginning of Capricorn arises at *B*, then the setting of Cancer will be visible at *D*, and *BED* will be a straight line and a diameter of the ecliptic. But it has already been seen that the line *AEC* is a diameter of the same circle; therefore, at their common section, point *E* will be their centre. So in this way the horizon always bisects the ecliptic, which is a great circle of the sphere. But on a sphere, if a circle bisects one of the great circles, then the circle bisecting is a great circle. Therefore the horizon is a great circle; and its centre is the same as that of the ecliptic, as far as appearance goes; although nevertheless the line passing through the centre of the Earth and the line touching to the surface are necessarily different; but on

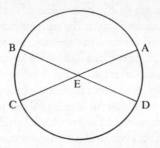

The horizon with the earth at its centre

account of their immensity in comparison with the Earth they are like parallel lines, which on account of the great distance between the termini appear to be one line, when the space contained between them is in no perceptible ratio to their length, as has been shown in optics.

From this argument it is certainly clear enough that the heavens are immense in comparison with the Earth and present the aspect of an infinite magnitude, and that in the judgement of sense-perception the Earth is to the heavens as a point to a body and as a finite to an infinite magnitude. But we see that nothing more than that has been shown, and it does not follow that the Earth must rest at the centre of the world. And we should be even more surprised if such a vast world should wheel completely around during the space of twenty-four hours rather than that its least part, the Earth, should. For saying that the centre is immovable and that those things which are closest to the centre are moved least does not argue that the Earth rests at the centre of the world. That is no different from saying that the heavens revolve but the poles are at rest and those things which are closest to the poles are moved least. In this way Cynosura (the pole star) is seen to move much more slowly than Aquila or Canicula because, being very near to the pole, it describes a smaller circle, since they are all on a single sphere, the movement of which stops at its axis and which does not allow any of its parts to have movements which are equal to one another. And nevertheless the revolution of the whole brings them round in equal times but not over equal spaces.

The argument which maintains that the Earth, as a part of the celestial sphere and as sharing in the same form and movement, moves very little because very near to its centre advances to the following position: therefore the Earth will move, as being a body and not a centre, and will describe in the same time arcs similar to, but smaller than, the arcs of the celestial circle. It is clearer than daylight how false that is; for there would necessarily always be noon at one place and midnight at another, and so the daily risings and settings could not take place, since the movement of the whole and the part would be one and inseparable.

But the ratio between things separated by diversity of nature is so entirely different that those which describe a smaller circle turn more quickly than those which describe a greater circle. In this way Saturn, the highest of the wandering stars, completes its revolution in thirty years, and the moon which is without doubt the closest to the Earth completes its circuit in a month, and finally the Earth itself will be considered to complete a circular movement in the space of a day and a night. So this same problem concerning the daily revolution comes up again. And also the question about the place of the Earth becomes even less certain on account of what was just said. For the demonstration proves nothing except that the heavens are of an indefinite magnitude with respect to the Earth. But it is not at all clear how far this immensity stretches out. On the contrary, since the minimal and indivisible corpuscles, which are called atoms, are not perceptible to sense, they do not, when taken in twos or in some small number, constitute a visible body; but they can be taken in such a large quantity that there will at last be enough to form a visible magnitude. So it is as regards the place of the Earth; for although it is not at the centre of the world, nevertheless the distance is as nothing, particularly in comparison with the sphere of the fixed stars.

153

Chapter 7. *Why the ancients thought the Earth was at rest at the middle of the world as its centre*

Wherefore for other reasons the ancient philosophers have tried to affirm that the Earth is at rest at the middle of the world, and as principal cause they put forward heaviness and lightness. For Earth is the heaviest element; and all things of any weight are borne towards it and strive to move towards the very centre of it.

For since the Earth is a globe towards which from every direction heavy things by their own nature are borne at right angles to its surface, the heavy things would fall on one another at the centre if they were not held back at the surface; since a straight line making right angles with a plane surface where it touches a sphere leads to the centre. And those things which are borne towards the centre seem to follow along in order to be at rest at the centre. All the more then will the Earth be at rest at the centre; and, as being the receptacle for falling bodies, it will remain immovable because of its weight.

They strive similarly to prove this by reason of movement and its nature. For

Aristotle says that the movement of a body which is one and simple is simple, and the simple movements are the rectilinear and the circular. And of rectilinear movements, one is upwards, and the other is downwards. As a consequence, every simple movement is either towards the centre, i.e. downwards, or away from the centre, i.e. upwards, or around the centre, i.e. circular. Now it belongs to earth and water, which are considered heavy, to be borne downwards, i.e. to seek the centre: for air and fire, which are endowed with lightness, move upwards, i.e. away from the centre. It seems fitting to grant rectilinear movement to these four elements and to give the heavenly bodies a circular movement around the centre. So Aristotle. Therefore, said Ptolemy of Alexandria, if the Earth moved, even if only by its daily rotation, the contrary of what was said above would necessarily take place. For this movement which would traverse the total circuit of the Earth in twenty-four hours would necessarily be very headlong and of an unsurpassable velocity. Now things which are suddenly and violently whirled around are seen to be utterly unfitted for reuniting, and the more unified are seen to become dispersed, unless some constant force constrains them to stick together. And a long time ago, he says, the scattered Earth would have passed beyond the heavens, as is certainly ridiculous; and *a fortiori* so would all the living creatures and all the other separate masses which could by no means remain unshaken. Moreover, freely falling bodies would not arrive at the places appointed them, and certainly not along the perpendicular line which they assume so quickly. And we would see clouds and other things floating in the air always borne towards the west.

154

Chapter 8. Answers to the aforesaid reasons and their inadequacy

For these and similar reasons they say that the Earth remains at rest at the middle of the world and that there is no doubt about this. But if someone opines that the Earth revolves, he will also say that the movement is natural and not violent. Now things which are according to nature produce effects contrary to those which are violent. For things to which force or violence is applied get broken up and are unable to subsist for a long time. But things which are caused by nature are in a right condition and are kept in their best organization. Therefore Ptolemy had no reason to fear that the Earth and all things on the Earth would be scattered in a revolution caused by the efficacy of nature, which is greatly different from that of art or from that which can result from the genius of man. But why didn't he feel anxiety about the world instead, whose movement must necessarily be of greater velocity, the greater the heavens are than the Earth? Or have the heavens become so immense, because an unspeakably vehement motion has pulled them away from the centre, and because the heavens would fall if they came to rest anywhere else?

Surely if this reasoning were tenable, the magnitude of the heavens would extend infinitely. For the farther the movement is borne upwards by the vehement force, the faster will the movement be, on account of the ever-increasing circumference which must be traversed every twenty-four hours: and conversely, the immensity of the sky would increase with the increase in movement. In this way, the velocity would make the magnitude increase infinitely, and the magnitude the velocity. [. . .]

But let us leave to the philosophers of nature the dispute as to whether the world is finite or infinite, and let us hold as certain that the Earth is held together between its two poles and terminates in a spherical surface. Why therefore should we hesitate any longer to grant to it the movement which accords naturally with its form, rather than put the whole world in a commotion—the world whose limits we do not and cannot know? And why not admit that the appearance of daily revolution belongs to the heavens but the reality belongs to the Earth? And things are as when Aeneas said in Virgil: 'We sail out of the harbour, and the land and the cities move away.' As a matter of fact, when a ship floats on over a tranquil sea, all the things outside seem to the voyagers to be moving in a movement which is the image of their own, and they think on the contrary that they themselves and all the things with them are at rest. So it can easily happen in the case of the movement of the Earth that the whole world should be believed to be moving in a circle. Then what would we say about the clouds and the other things floating in the air or falling or rising up, except that not only the Earth and the watery element with which it is conjoined are moved in this way but also no small part of the air and whatever other things have a similar kinship with the Earth? whether because the neighbouring air, which is mixed with earthy and watery matter, obeys the same nature as the Earth or because the movement of the air is an acquired one, in which it participates without resistance on account of the contiguity and perpetual rotation of the Earth. [. . .]

In addition, there is the fact that the state of immobility is regarded as more noble and godlike than that of change and instability, which for that reason should belong to the Earth rather than to the world. I add that it seems rather absurd to ascribe movements to the container or to that which provides the place and not rather to that which is contained and has a place, i.e. the Earth. And lastly, since it is clear that the wandering stars are sometimes nearer and sometimes farther away from the Earth, then the movement of one and the same body around the centre—and they mean the centre of the Earth—will be both away from the centre and towards the centre. Therefore it is necessary that movement around the centre should be taken more generally; and it should be enough if each movement is in accord with its own centre. You see therefore that for all these reasons it is more probable that the Earth moves than that it is at rest—especially in the case of the daily revolution, as it is the Earth's very own. And I think that is enough as regards the first part of the question.

155

Chapter 9. Whether many movements can be attributed to the Earth, and concerning the centre of the world

Therefore, since nothing hinders the mobility of the Earth, I think we should now see whether more than one movement belongs to it, so that it can be regarded as one of the wandering stars. For the apparent irregular movement of the planets and their variable distances from the Earth—which cannot be understood as occurring in circles homocentric with the Earth—make it clear that the Earth is not the centre of their circular movements. Therefore, since there are many centres, it is not foolhardy to doubt whether the centre of gravity of the Earth rather than some other is the centre of the world. I myself think that gravity or heaviness is nothing except a certain natural appetency implanted in the parts by the divine providence of the universal Artisan, in order that they should unite with one another in their oneness and wholeness and come together in the form of a globe. It is believable that this effect is present in the sun, moon, and the other bright planets and that through its efficacy they remain in the spherical figure in which they are visible, though they nevertheless accomplish their circular movements in many different ways. Therefore if the Earth too possesses movements different from the one around its centre, then they will necessarily be movements which similarly appear on the outside in the many bodies; and we find the yearly revolution among these movements. For if the annual revolution were changed from being solar to being terrestrial, and immobility were granted to the sun, the risings and settings of the signs and of the fixed stars—whereby they become morning or evening stars—will appear in the same way; and it will be seen that the stoppings, retrogressions, and progressions of the wandering stars are not their own, but are a movement of the Earth and that they borrow the appearances of this movement. Lastly, the sun will be regarded as occupying the centre of the world. And the ratio of order in which these bodies succeed one another and the harmony of the whole world teaches us their truth, if only—as they say—we would look at the thing with both eyes.

156

Chapter 10. On the order of the celestial orbital circles

I know of no one who doubts that the heavens of the fixed stars is the highest up of all visible things. We see that the ancient philosophers wished to take the order of the planets according to the magnitude of their revolutions, for the reason that among things which are moved with equal speed those which are the more distant seem to be borne along more slowly, as Euclid proves in his *Optics*. And so they think that the moon traverses its circle in the shortest period of time, because being next to the Earth, it revolves in the smallest circle. But they think that Saturn, which completes the longest circuit in the longest period of time, is the highest. Beneath Saturn, Jupiter. After Jupiter, Mars.

There are different opinions about Venus and Mercury, in that they do not have the full range of angular elongations from the sun that the others do. Wherefore some place them above the sun, as Timaeus does in Plato; some, beneath the sun, as Ptolemy and a good many moderns. [. . .]

Furthermore, how unconvincing is Ptolemy's argument that the sun must occupy the middle position between those planets which have the full range of angular elongation from the sun and those which do not is clear from the fact that the moon's full range of angular elongation proves its falsity.

But what cause will those who place Venus below the sun, and Mercury next, or separate them in some other order—what cause will they allege why these planets do not also make longitudinal circuits separate and independent of the sun, like the other planets—if indeed the ratio of speed or slowness does not falsify their order? Therefore it will be necessary either for the Earth not to be the centre to which the order of the planets and their orbital circles is referred, or for there to be no sure reason for their order and for it not to be apparent why the highest place is due to Saturn rather than to Jupiter or some other planet. Wherefore I judge that what Martianus Capella[1]—who wrote the *Encyclopedia*— and some other Latins took to be the case is by no means to be despised. For they hold that Venus and Mercury circle around the sun as a centre; and they hold that for this reason Venus and Mercury do not have any farther elongation from the sun than the convexity of their orbital circles permits; for they do not make a circle around the earth as do the others, but have perigee and apogee interchangeable in the sphere of the fixed stars. Now what do they mean except that the centre of their spheres is around the sun? Thus the orbital circle of Mercury will be enclosed within the orbital circle of Venus—which would have to be more than twice as large—and will find adequate room for itself within that amplitude. Therefore if anyone should take this as an occasion to refer Saturn, Jupiter, and Mars also to this same centre, provided he understands the magnitude of those orbital circles to be such as to comprehend and encircle the Earth remaining within them, he would not be in error, as the table of ratios of their movements makes clear. For it is manifest that the planets are always nearer the Earth at the time of their evening rising, i.e. when they are opposite to the sun and the Earth is in the middle between them and the sun. But they are farthest away from the Earth at the time of their evening setting, i.e. when they are occulted in the neighbourhood of the sun, namely, when we have the sun between them and the Earth. All that shows clearly enough that their centre is more directly related to the sun and is the same as that to which Venus and Mercury refer their revolutions. But as they all have one common centre, it is necessary that the space left between the convex orbital circle of Venus and the concave orbital circle of Mars should be viewed as an orbital circle or sphere

1. Roman writer of the beginning of the 5th century A.D. (Ed.)

homocentric with them in respect to both surfaces, and that it should receive the Earth and its satellite the moon and whatever is contained beneath the lunar globe. For we can by no means separate the moon from the Earth, as the moon is incontestably very near to the Earth—especially since we find in this expanse a place for the moon which is proper enough and sufficiently large. Therefore we are not ashamed to maintain that this totality—which the moon embraces— and the centre of the Earth too traverse that great orbital circle among the other wandering stars in an annual revolution around the sun; and that the centre of the world is around the sun. I also say that the sun remains forever immobile and that whatever apparent movement belongs to it can be verified to the mobility of the Earth; that the magnitude of the world is such that, although the distance from the sun to the Earth in relation to whatsoever planetary sphere you please possesses magnitude which is sufficiently manifest in proportion to these dimensions, this distance, as compared with the sphere of the fixed stars, is

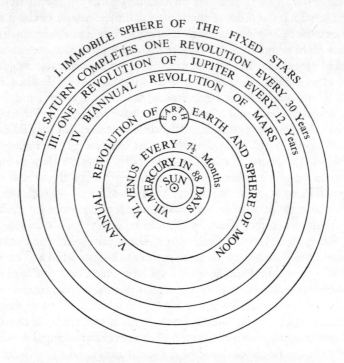

The heliocentric system of Copernicus

imperceptible. I find it much more easy to grant that than to unhinge the understanding by an almost infinite multitude of spheres—as those who keep the earth at the centre of the world are forced to do. But we should rather follow the wisdom of nature, which, as it takes very great care not to have produced anything superfluous or useless, often prefers to endow one thing with many effects. And though all these things are difficult, almost inconceivable, and quite contrary to the opinion of the multitude, nevertheless in what follows we will with God's help make them clearer than day—at least for those who are not ignorant of the art of mathematics.

Therefore if the first law is still safe—for no one will bring forward a better one than that the magnitude of the orbital circles should be measured by the magnitude of time—then the order of the spheres will follow in this way—beginning with the highest: the first and highest of all is the sphere of the fixed stars, which comprehends itself and all things, and is accordingly immovable. In fact it is the place of the universe, i.e. it is that to which the movement and position of all the other stars are referred. For in the deduction of terrestrial movement, we will however give the cause why there are appearances such as to make people believe that even the sphere of the fixed stars somehow moves. Saturn, the first of the wandering stars, follows; it completes its circuit in thirty years. After it comes Jupiter moving in a twelve-year period of revolution. Then Mars, which completes a revolution every two years. The place fourth in order is occupied by the annual revolution in which we said the Earth together with the orbital circle of the moon as an epicycle is comprehended. In the fifth place, Venus, which completes its revolution in seven and a half months. The sixth and final place is occupied by Mercury, which completes its revolution in a period of eighty-eight days. In the centre of all rests the sun. For who would place this lamp of a very beautiful temple in another or better place than this wherefrom it can illuminate everything at the same time? As a matter of fact, not unhappily do some call it the lantern; others, the mind and still others, the pilot of the world. Trismegistus calls it a 'visible god'; Sophocles' *Electra*, 'that which gazes upon all things'. And so the sun, as if resting on a kingly throne, governs the family of stars which wheel around. Moreover, the Earth is by no means cheated of the services of the moon; but, as Aristotle says in the *De Animalibus*, the Earth has the closest kinship with the moon. The Earth moreover is fertilized by the sun and conceives offspring every year.

Therefore in this ordering we find that the world has a wonderful commensurability and that there is a sure bond of harmony for the movement and magnitude of the orbital circles such as cannot be found in any other way. For now the careful observer can note why progression and retrogradation appear greater in Jupiter than in Saturn and smaller than in Mars; and in turn greater in Venus than in Mercury. And why these events of reversal appear more often in Saturn than in Jupiter, and even less often in Mars and Venus than in Mercury. In addition, why when Saturn, Jupiter, and Mars are in opposition (to the mean position of the sun) they are nearer to the Earth than at the time of their occultation and their reappearance. And especially why at the times when Mars is in opposition to the sun, it seems to equal Jupiter in magnitude and to be distinguished from Jupiter only by a reddish colour, but when discovered through careful observation by means of a sextant is found with difficulty among the stars of second magnitude. All these things proceed from the same cause, which resides in the movement of the Earth.

But that there are no such appearances among the fixed stars argues that they are at an immense height away, which makes the circle of annual movement or its image disappear from before our eyes since every visible thing has a certain distance beyond which it is no longer seen, as is shown in optics. For the scintillation of their lights shows that there is a very great distance between Saturn the highest of the planets and the sphere of the fixed stars. It is by this mark in particular that they are distinguished from the planets, as it is proper to have the greatest difference between the moved and the unmoved. How exceedingly fine is the godlike work of the Best and Greatest Artist!

GIORDANO BRUNO

An infinity of worlds

157

Make then your forecasts, my lords Astrologers, with your slavish physicians, by means of those astrolabes with which you seek to discern the fantastic nine moving spheres; in these you finally imprison your own minds, so that you appear to me but as parrots in a cage, while I watch you dancing up and down, turning and hopping within those circles. We know that the Supreme Ruler cannot have a seat so narrow, so miserable a throne, so straight a tribunal, so scanty a court, so small and feeble a simulacrum that a phantasm can bring to birth, a dream shatter, a delusion restore, a chimera disperse, a calamity diminish, a misdeed abolish and a thought renew it again, so that indeed with a puff of air it were brimful and with a single gulp it were emptied. On the contrary we recognize a noble image, a marvellous conception, a supreme figure, an exalted shadow, an infinite representation of the represented infinity, a spectacle worthy of the excellence and supremacy of Him who transcendeth understanding, comprehension or grasp. Thus is the excellence of God magnified and the greatness of his kingdom made manifest; he is glorified not in one, but in countless suns; not in a single earth, a single world, but in a thousand thousand, I say in an infinity of worlds.

Infinite space

158

Speakers:

ELPINO

PHILOTHEO (occasionally called THEOPHILO)

FRACASTORO

BURCHIO

ELPINO: How is it possible that the universe can be infinite?

PHILOTHEO: How is it possible that the universe can be finite?

ELPINO: Do you claim that you can demonstrate this infinitude?

PHILOTHEO: Do you claim that you can demonstrate this finitude?

ELPINO: What is this spreading forth?

PHILOTHEO: What is this limit?

FRACASTORO: To the point, to the point, if you please. Too long you have kept us in suspense.

BURCHIO: Come quickly to argument, Philotheo, for I shall be vastly amused to hear this fable or fantasy.

FRACASTORO: More modestly, Burchio. What wilt thou say if truth doth ultimately convince thee?

BURCHIO: Even if this be true I do not wish to believe it, for this *Infinite* can neither be understood by my head nor brooked by my stomach. Although, to tell the truth, I could yet hope that Philotheo were right, so that if by ill luck I were to fall from this world I should always find myself on firm ground.

ELPINO: Certainly, O Theophilo, if we wish to judge by our senses, yielding suitable primacy to that which is the source of all our knowledge, perchance we shall not find it easier to reach the conclusion you expressed than to take the contrary view. Now be so kind as to begin my enlightenment.

PHILOTHEO: No corporeal sense can perceive the infinite. None of our senses could be expected to furnish this conclusion; for the infinite cannot be the object of sense-perception; therefore he who demandeth to obtain this knowledge through the senses is like unto one who would desire to see with his eyes both substance and essence. And he who would deny the existence of a thing merely because it cannot be apprehended by the senses, nor is visible, would presently be led to the denial of his own substance and being. Wherefore there must be some measure in the demand for evidence from our sense-perception, for this we can accept only in regard to sensible objects, and even there it is not above all suspicion unless it cometh before the court aided by good judgement. It is the part of the intellect to judge, yielding due weight to factors absent and separated by distance of time and by space intervals. And in this matter our sense-perception doth suffice us and doth yield us adequate testimony, since it is unable to gainsay us; moreover it advertiseth and confesseth his own feebleness and inadequacy by the impression it giveth us of a finite horizon, an impression moreover which is ever changing. Since then we have experience that sense-perception deceiveth us concerning the surface of this globe on which we live, much more should we hold suspect the impression it giveth us of a limit to the starry sphere.

ELPINO: Of what use then are the senses to us? Tell me that.

PHILOTHEO: Solely to stimulate our reason, to accuse, to indicate, to testify in part; not to testify completely, still less to judge or to condemn. For our senses, however perfect, are never without some perturbation. Wherefore truth is in but véry small degree derived from the senses as from a frail origin, and doth by no means reside in the senses.

ELPINO : Where then?

PHILOTHEO : In the sensible object as in a mirror. In reason, by process of argument and discussion. In the intellect, either through origin or by conclusion. In the mind, in its proper and vital form.

ELPINO : On, then, and give your reasons.

PHILOTHEO : I will do so. If the world is finite and if nothing lieth beyond, I ask you *where* is the world? *Where* is the universe? Aristotle replieth, it is in itself. The convex surface of the primal heaven is universal space, which being the primal container is by naught contained. For position in space is no other than the surfaces and limit of the containing body, so that he who hath no containing body hath no position in space. What then dost thou mean, O Aristotle, by this phrase, that 'space is within itself'? What will be thy conclusion concerning that which is beyond the world? If thou sayest, there is nothing, then the heaven and the world will certainly not be anywhere.

FRACASTORO : The world will then be nowhere. Everything will be nowhere.

PHILOTHEO : The world is something which is past finding out. If thou sayest (and it certainly appeareth to me that thou seekest to say something in order to escape Vacuum and Nullity), if thou sayest that beyond the world is a divine intellect, so that God doth become the position in space of all things, why then thou thyself wilt be much embarrassed to explain to us how that which is incorporeal [yet] intelligible, and without dimension can be the very position in space occupied by a dimensional body; and if thou sayest that this incorporeal space containeth as it were a form, as the soul containeth the body, then thou dost not reply to the question of that which lieth beyond, nor to the inquiry concerning that which is outside the universe. And if thou wouldst excuse thyself by asserting that where naught is, and nothing existeth, there can be no question of position in space nor of beyond or outside, yet I shall in no wise be satisfied. For these are mere words and excuses, which cannot form part of our thought. For it is wholly impossible that in any sense or fantasy (even though there may be various senses and various fantasies), it is I say impossible that I can with any true meaning assert that there existeth such a surface, boundary or limit, beyond which is neither body, nor empty space, even though God be there. For divinity hath not as aim to fill space, nor therefore doth it by any means appertain to the nature of divinity that it should be the boundary of a body. For aught which can be termed a limiting body must either be the exterior shape or else a containing body. And by no description of this quality canst thou render it compatible with the dignity of divine and universal nature.

BURCHIO : Certainly I think that one must reply to this fellow if a person would stretch out his hand beyond the convex sphere of heaven, the hand would occupy no position in space nor any place, and in consequence would not exist.

PHILOTHEO: I would add that no mind can fail to perceive the contradiction implicit in this saying of the Peripatetic. Aristotle defined position occupied by a body not as the containing body itself, nor as a certain [part of] space, but as a surface of the containing body. Then he affirmeth that the prime, principal and greatest space is that to which such a definition least and by no means conformeth, namely, the convex surface of the first [outermost] heaven. This is the surface of a body of a particular sort, a body which containeth only, and is not contained. Now for the surface to be a position in space, it need not appertain to a contained body but it must appertain to a containing body. And if it be the surface of a containing body and yet be not joined to and continuous with the contained body, then it is a space without position, since the first [outermost] heaven cannot be a space except in virtue of the concave surface thereof, which is in contact with the convex surface of the next heaven. Thus we recognize that this definition is vain, confused and self-destructive, the confusion being caused by that incongruity which maintaineth that naught existeth beyond the firmament.

The non-existence of a centre and circumference in the universe

159

PHILOTHEO: It would be impossible to find another person who in the name of philosophy could invent vainer suppositions and fabricate such foolish and contrary reasons to accommodate such levity as is discernible in his arguments. As for what he saith concerning the spaces occupied by bodies, and of the determinate upper, lower, and intermediate, I would like to know against what opinion he is arguing. For all who posit a body of infinite size, ascribe to it neither centre nor boundary. For he who speaketh of emptiness, the void, or the infinite ether, ascribeth to it neither weight nor lightness, nor motion, nor upper, nor lower, nor intermediate regions; assuming moreover that there are in this space those countless bodies such as our earth and other earths, our sun and other suns, which all revolve within this infinite space, through finite and determined spaces or around their own centres. Thus we on earth say that the earth is in the centre; and all philosophers ancient and modern of whatever sect will proclaim without prejudice to their own principles that here is indeed the centre; just as we say that we are as it were at the centre of that [universally] equidistant circle which is the great horizon and the limit of our own encircling ethereal region, so without doubt those who inhabit the moon believe themselves to be at the centre [of a great horizon] that encircleth this earth, the sun and the other stars, and that is the boundary of the radii of their own horizon. Thus the earth no more than any other world is at the centre; and no points constitute definite determined poles of space for our earth, just as she herself is not a definite and determined pole to any other point of the ether, or of the world space; and the same is true of all other bodies. From various points of view these may all be regarded either as centres, or as points of the circumference, as poles, or zeniths and so forth. Thus the earth is not in the centre of the universe; it is central only to our own surrounding space.

The infinite number of solar systems in the universe

160

ELPINO: There are then innumerable suns, and an infinite number of earths revolve around those suns, just as the seven we can observe revolve around this sun which is close to us.

PHILOTHEO: So it is.

ELPINO: Why then do we not see the other bright bodies which are earths circling around the bright bodies which are suns? For beyond these we can detect no motion whatever; and why do all other mundane bodies (except those known as comets) appear always in the same order and at the same distance?

PHILOTHEO: The reason is that we discern only the largest suns, immense bodies. But we do not discern the earths because, being much smaller, they are invisible to us. Similarly it is not impossible that other earths revolve around our sun and are invisible to us on account either of greater distance or of smaller size, or because they have but little watery surface, or because such watery surface is not turned towards us and opposed to the sun, whereby it would be made visible as a crystal mirror which receiveth luminous rays; whence we perceive that it is not marvellous or contrary to nature that often we hear that the sun hath been partially eclipsed though the moon hath not been interpolated between him and our sight. There may be innumerable watery luminous bodies—that is, earths consisting in part of water—circulating around the sun, besides those visible to us; but the difference in their orbits is indiscernible by us on account of their great distance, wherefore we perceive no difference in the very slow motion discernible of those visible above or beyond Saturn; still less doth there appear any order in the motion of all around the centre, whether we place our earth or our sun as that centre.

ELPINO: How then wouldst thou maintain that all of these bodies, however far from their centre, that is from the sun, can nevertheless participate in the vital heat thereof?

PHILOTHEO: Because the further they are from the sun, the larger is the circle of their orbit around it; and the greater their orbit, the more slowly they accomplish their journey round the sun; the more slowly they move, the more they resist the hot flaming rays of the sun.

ELPINO: You maintain then that though so distant from the sun, these bodies can derive therefrom all the heat that they need. Because, spinning at a greater rate around their own centre and revolving more slowly around the sun, they can derive not only as much heat but more still if it were needed; since by the more rapid spin around her own centre, such part of the convexity of the earth as hath not been sufficiently heated is the more quickly turned to a position to receive heat; while from the slower progress around the fiery central body, she stayeth to receive more firmly the impression therefrom, and thus she will receive fiercer flaming rays.

PHILOTHEO: That is so.

ELPINO: Therefore you consider that if the stars beyond Saturn are really motionless as they appear, then they are those innumerable suns or fires more or less visible to us around which travel their own neighbouring earths which are not discernible by us.

THEOPHILO: Yes, we should have to argue thus, since all earths merit the same amount of heat, and all suns merit the same amount.

ELPINO: Then you believe that all those are suns?

PHILOTHEO: Not so, for I do not know whether all or whether the majority are without motion, or whether some circle around others, since none hath observed them. Moreover they are not easy to observe, for it is not easy to detect the motion and progress of a remote object, since at a great distance change of position cannot easily be detected, as happeneth when we would observe ships in a high sea. But however that may be, the universe being infinite, there must ultimately be other suns. For it is impossible that heat and light from one single body should be diffused throughout immensity, as was supposed by Epicurus if we may credit what others relate of him. Therefore it followeth that there must be innumerable suns, of which many appear to us as small bodies; but that star will appear smaller which is in fact much larger than that which appeareth much greater.

WILLIAM GILBERT

161

The loadstone has parts distinct in their natural power, and poles conspicuous for their property

The stone itself manifests many qualities which, though known afore this, yet, not having been well investigated, are to be briefly indicated in the first place so that students may understand the powers of loadstone and iron, and not be troubled at the outset through ignorance of reasonings and proofs. In the heaven astronomers assign a pair of poles for each moving sphere: so also do we find in the terrestrial globe natural poles pre-eminent in virtue, being the points that remain constant in their position in respect to the diurnal rotation, one tending to the Bears and the seven stars; the other to the opposite quarter of the heaven. In like manner the loadstone has its poles, by nature northern and southern, being definite and determined points set in the stone, the primary boundaries of motions and effects, the limits and governors of the many actions and virtues. However, it must be understood that the strength of the stone does not emanate from a mathematical point, but from the parts themselves, and that while all those parts in the whole belong to the whole, the nearer they are to the poles of the stone the stronger are the forces they acquire and shed into other bodies: these poles are observant of the earth's poles, move towards them, and wait upon them. Magnetick poles can be found in every magnet, in the powerful and mighty (which Antiquity used to call the masculine) as well as in the weak, feeble and feminine; whether its figure is due to art or to chance, whether long, flat, square, three-cornered, polished; whether rough, broken, or unpolished; always the loadstone contains and shows its poles. But since the spherical form, which is also the most perfect, agrees best with the earth, being a globe, and is most suitable for use and experiment, we accordingly wish our principal demonstrations by the stone to be made with a globe-shaped magnet as being more perfect and adapted for the purpose. Take, then, a powerful loadstone, solid, of a just size, uniform, hard, without flaw; make of it a globe

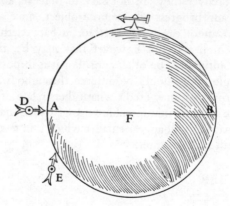

The magnetic terrella

upon the turning tool used for rounding crystals and some other stones, or with other tools as the material and firmness of the stone requires, for sometimes it is difficult to be worked. The stone thus prepared is a true, homogeneous offspring of the earth and of the same shape with it: artificially possessed of the orbicular form which nature granted from the beginning to the common mother earth: and it is a physical corpuscle imbued with many virtues, by means of which many abstruse and neglected truths in philosophy buried in piteous darkness may more readily become known to men. This round stone is called by us a μικρόγη or *terrella*. To find, then, the poles conformable to the earth's, take the round stone in hand, and place upon the stone a needle or wire of iron: the ends of the iron move upon their own centre and suddenly stand still. Mark the stone with ochre or with chalk where the wire lies and sticks: move the middle or centre of the wire to another place, and so on to a third and a fourth, always marking on the stone along the length of the iron where it remains at rest: those lines show the meridian circles, or the circles like meridians on the stone, or terrella, all of which meet as will be manifest at the poles of the stone. By the circles thus continued the poles are made out, the Boreal as well as the southern, and in the middle space betwixt these a great circle may be drawn for an æquator, just as Astronomers describe them in the heavens and on their own globes, or as Geographers do on the terrestrial globe: for that line so drawn on this our terrella is of various uses in our demonstrations and experiments magnetical. Poles are also found in a round stone by a versorium, a piece of iron touched with a loadstone, and placed upon a needle or point firmly fixed on a foot so as to turn freely about in the following way: On the stone AB the versorium is placed in such a way that the versorium may remain in equilibrium: you will mark with chalk the course of the iron when at rest: Move the instrument to another spot, and again make note of the direction and aspect: do the same thing in several places, and from the concurrence of the lines of direction you will find one pole at the point A, the other at B. A versorium placed near the stone also indicates the true pole; when at right angles it eagerly beholds the stone and seeks the pole itself directly, and is turned in a straight line through the axis to the centre of the stone. For instance, the versorium D faces towards A and F, the pole and centre, whereas E does not exactly respect either the pole A or the centre F. A bit of rather fine iron wire, of the length of a barley-corn, is placed on the stone, and is moved over the regions and surface of the stone, until it rises to the perpendicular: for it stands erect at the actual pole, whether boreal or austral; the further from the pole, the more it inclines from the vertical. The poles thus found you shall mark with a sharp file or gimlet.

162

Loadstone seems to attract loadstone when in natural position: but repels it when in a contrary one, and brings it back to order

First of all we must declare, in familiar language, what are the apparent and common virtues of the stone; afterwards numerous subtilities, hitherto abstruse and unknown, hidden in obscurity, are to be laid open, and the causes of these (by the unlocking of nature's secrets) made evident, in their place, by fitting terms and devices. It is trite and commonplace that loadstone draws iron; in the same way too does loadstone attract loadstone. Place the stone which you have seen to have poles clearly distinguished, and marked austral and boreal, in its vessel so as to float; and let the poles be rightly arranged with respect to the plane of the horizon, or, at any rate not much raised or awry: hold in your hand another stone the poles of which are also known; in such a way that its austral pole may be towards the boreal pole of the one that is swimming, and near it, sideways: for the floating stone forthwith follows the other stone (provided it be within its force and dominion) and does not leave off nor forsake it until it adheres; unless by withdrawing your hand, you cautiously avoid contact. In like manner if you set the boreal pole of the one you hold in your hand opposite the austral pole of the swimming stone, they rush together and follow each other in turn. For contrary poles allure contrary. If, however, you apply in the same way the northern to the northern, and the austral to the austral pole, the one stone puts the other to flight, and it turns aside as though a pilot were pulling at the helm and it makes sail in the opposite ward as one that ploughs the sea, and neither stands anywhere, nor halts, if the other is in pursuit. For stone disposeth stone; the one turns the other around, reduces it to range, and brings it back to harmony with itself. When, however, they come together and are conjoined according to the order of nature, they cohære firmly mutually. For instance, if you were to set the boreal pole of that stone which is in your hand before the tropic of Capricorn of a round floating loadstone (for it will be well to mark out on the round stone, that is the terrella, the mathematical circles as we do on a globe itself), or before any point between the æquator and the austral pole; at once the swimming stone revolves, and so arranges itself that its austral pole touches the other's boreal pole, and forms a close union with it. In the same way, again, at the other side of the æquator, with the opposite poles, you

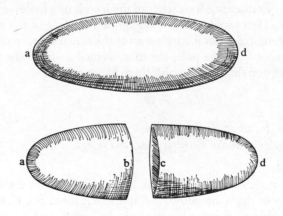

The poles of a magnet

may produce similar results; and thus by this art and subtilty we exhibit attraction, repulsion, and circular motion for attaining a position of agreement and for declining hostile encounters. Moreover 'tis in one and the same stone that we are thus able to demonstrate all these things and also how the same part of one stone may on division become either boreal or austral. Let *AD* be an oblong stone, in which A is the northern, D the southern pole; cut this into two equal parts, then set part A in its vessel on the water, so as to float. And you will then see that A the northern point will turn to the south, as before; in like manner also the point D will move to the north, in the divided stone, as in the whole one. Whereas, of the parts B and C, which were before continuous, and are now divided, the one is southern B, the other northern C. B draws C, desirous to be united, and to be brought back into its pristine continuity: for these which are now two stones were formed out of one: and for this cause C of the one turning itself to B of the other, they mutually attract each other, and when freed from obstacles and relieved of their own weight, as upon the surface of water, they run together and are conjoined. But if you direct the part or point A to C in the other stone, the one repels or turns away from the other: for so were nature perverted, and the form of the stone perturbed, a form that strictly keeps the laws which it imposed upon bodies; hence, when all is not rightly ordered according to nature, comes the flight of one from the other's perverse position and from the discord, for nature does not allow of an unjust and inequitable peace, or compromise: but wages war and exerts force to make bodies acquiesce well and justly. Rightly arranged, therefore, these mutually attract each other; that is, both stones, the stronger as well as the weaker, run together, and with their whole forces tend to unity, a fact that is evident in all magnets, not in the Æthiopian only, as Pliny supposed. [. . .] The pole of a stone more persistently attracts and more rapidly seizes the corresponding part (which they term the adverse part) of another stone; for instance, North pulls South; just so it also summons iron with more vehemence, and the iron cleaves to it more firmly whether it has been previously excited by the magnet, or is untouched. For thus, not without reason hath it been ordained by nature, that the parts nearer to the pole should more firmly attract; but that at the pole itself should be the seat, the throne, as it were, of a consummate and splendid virtue, to which magnetical bodies on being brought are more vehemently attracted, and from which they are with utmost difficulty dislodged. So the poles are the parts which more particularly spurn and thrust away things strange and alien perversely set beside them.

163

That the globe of the earth is magnetick, and a magnet; and how in our hands the magnet stone has all the primary forces of the earth, while the earth by the same powers remains constant in a fixed direction in the universe

Prior to bringing forward the causes of magnetical motions, and laying open the proofs of things hidden for so many ages, and our experiments (the true foundations of terrestrial philosophy), we have to establish and present to the view of the learned our New and unheard of doctrine about the earth; and this, when argued by us on the grounds of its probability, with subsequent experiments and proofs, will be as certainly assured as anything in philosophy ever has been considered and confirmed by clever arguments or mathematical proofs. The terrene mass, which together with the vasty ocean produces the sphærick figure and constitutes our globe, being of a firm and constant substance, is not easily changed, does not wander about, and fluctuates with uncertain motions, like the seas, and flowing waves; but holds all its volume of moisture in certain beds and bounds, and as it were in oft-met veins, that it may be the less diffused and dissipated at random. Yet the solid magnitude of the earth prevails and reigns supreme in the nature of our globe. Water, however, is attached to it, and as an appendage only, and a flux emanating from it; whose force from the beginning is conjoined with the earth through its smallest parts, and is innate in its substance. This moisture the earth as it grows hot throws off freely when it is of the greatest possible service in the generation of things. [. . .] But the loadstone and all magneticks, not the stone only, but every magnetick homogenic substance, would seem to contain the virtue of the earth's core and of its inmost bowels, and to hold within itself and to have conceived that which is the secret and inward principle of its substance; and it possesses the actions peculiar to the globe of attracting, directing, disposing, rotating, stationing itself in the universe, according to the rule of the whole, and it contains and regulates the dominant powers of the globe; which are the chief tokens and proofs of a certain distinguishing combination, and of a nature most thoroughly conjoint. For if among actual bodies one sees something move and breathe, and experience sensations, and be inclined and impelled by reason, will one not, knowing and seeing this, conclude that it is a man or something rather like a man, than that it is a stone or a stick? The loadstone far excels all other bodies known to us in virtues and properties pertaining to the common mother: but those properties have been far too little understood or realized by philosophers: for to its body bodies magnetical rush in from all sides and cleave to it, as we see them do in the case of the earth. It has poles, not mathematical points, but natural termini of force excelling in primary efficiency by the co-operation of the whole: and there are poles in like manner in the earth which our forefathers sought ever in the sky: it has an æquator, a natural dividing line between the two poles, just as the earth has: for of all lines drawn by the mathematicians on the terrestrial globe, the æquator is the natural boundary, and is not, as will hereafter appear, merely a mathematical circle. It, like the earth, acquires Direction and stability towards North and South, as the earth does; also it has a circular motion towards the position of the earth, wherein it adjusts itself to its rule: it follows the ascensions and declinations of the earth's poles, and conforms exactly to the same, and by itself raises its own poles above the horizon naturally according to the law of the particular country and region, or sinks below it. The loadstone derives temporary properties, and acquires its verticity from the earth, and iron is affected by the verticity of the globe even as iron is by a loadstone: Magneticks are conformable to and are regulated by the earth, and are subject to the earth in all their motions. All its movements harmonize with, and strictly wait upon, the geometry and form of the earth, as we shall afterwards prove by most conclusive experiments and diagrams; and the chief parts of the visible earth is also magnetical, and has magnetick motions, although it be disfigured by corruptions and mutations without end. Why then do we not recognize this the chief homogenic substance of the earth, likest of substances to its inner nature and closest allied to its very marrow? [. . .]

164

Why at the pole itself the coition is stronger than in the
other parts intermediate between the æquator and the pole;
and on the proportion of forces of the coition in various
parts of the earth and of the terrella

Observation has already been made that the highest power of alluring exists in the pole, and that it is weaker and more languid in the parts adjacent to the æquator. And as this is apparent in the declination, because that disponent and rotational virtue has an augmentation as one proceeds from the æquator towards the poles: so also the coition of magneticks grows increasingly fresh by the same steps, and in the same proportion. For in the parts more remote from the poles the loadstone does not draw magneticks straight down towards

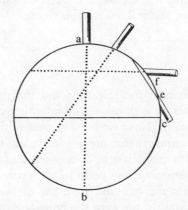

The distribution of magnetic attraction over the terrella

its own viscera; but they tend obliquely and they allure obliquely. For as the smallest chords in a circle differ from the diameter, so much do the forces of attracting differ between themselves in different parts of the terrella. For since attraction is coition towards a body, but magneticks run together by their versatory tendency, it comes about that in the diameter drawn from pole to pole the body appeals directly, but in other places less directly. So the less the magnetick is turned toward the body, the less, and the more feebly, does it approach and adhere. Just as if *AB* were the poles and a bar of iron or a magnetick fragment *C* is allured at the part *E*; yet the end laid hold of does not tend towards the centre of the loadstone, but verges obliquely towards the pole; and a chord drawn from that end obliquely as the attracted body tends is short; therefore it has less vigour and likewise less inclination. But as a greater chord proceeds from a body at *F*, so its action is stronger; at *G* still longer; longest at *A*, the pole (for the diameter is the longest way) to which all the parts from all sides bring assistance, in which is constituted, as it were, the citadel and tribunal of the whole province, not from any worth of its own, but because a force resides in it contributed from all the other parts, just as all the soldiers bring help to their own commander. Wherefore also a slightly longer stone attracts more than a spherical one, since the length from pole to pole is extended, even if the stones are both from the same mine and of the same weight and size. The way from pole to pole is longer in a longer stone, and the forces brought together from other parts are not so scattered as in a round magnet and terrella, and in a narrow one they agree more and are better united, and a united stronger force excels and is pre-eminent. A much weaker office, however, does a plane or oblong stone perform, when the length is extended according to the leading of the parallels, and the pole stops neither on the apex nor in the circle and orbe, but is spread over the flat. Wherefore also it invites a friend wretchedly, and feebly retains him, so that it is esteemed as one of an abject and contemptible class, according to its less apt and less suitable figure.

165

Of the globe of Earth as a loadstone

Hitherto we have spoken of the loadstone and magnetic bodies, how they conspire together and act on each other, and how they conform themselves to the terrella and to the earth. Now we have to treat of the globe of earth itself separately. All the experiments that are made on the terrella, to show how magnetic bodies conform themselves to it, may—at least the principal and most striking of them—be shown on the body of the earth; to the earth, too, all magnetized bodies are associate. And first, on the terrella the equinoctial circle, the meridians, parallels, the axis, the poles, are natural limits: similarly on the earth these exist as natural and not merely mathematical limits. As on the periphery of a terrella a loadstone or the magnetick needle takes direction to the pole, so on the earth there are revolutions special, manifest, and constant, from both sides of the equator: iron is endowed with verticity by being stretched towards the pole of the earth as towards the pole of a terrella; again, by being laid down and suffered to grow cool lying towards the earth's pole, after its prior verticity has been destroyed by fire, it acquires now verticity conformed to the position earthward. And iron rods that have for a long time lain in the poleward direction acquire verticity simply by regarding the earth; just as the same rods, if they be pointed towards the pole of a loadstone, though not touching it, receive polar force. There is no magnetic body that draws nigh in any way to a loadstone which does not in like manner obey the earth. As a loadstone is more powerful at one end and at one side of the equator, so the same thing is shown with a small terrella on a large one. According to the difference in amount and mode of friction in magnetizing a piece of iron at a terrella, it will be powerful or weak in performing its functions. In movements towards the body of the earth, just as on a terrella, variation is produced by unlikeness and inequality of prominences and by imperfections of the surface; and all variation of the versorium or the mariner's compass all over the earth and everywhere at sea—a thing that has so bewildered men's minds—is found and recognized through the same causes. The dip of the magnetic needle (that wonderful turning of magnetic bodies to the body of the terrella by formal progression) is seen also in the earth most clearly. And that one experiment reveals plainly the grand magnetic nature of the earth, innate in all the parts thereof and diffused throughout. The magnetic energy, therefore, exists in the earth just as in the terrella, which is a part of the earth and homogenic in nature with it, but by art made spherical so it might correspond to the spherical body of the earth and be in agreement with the earth's globe for the capital experiments.

FRANCIS BACON

The unsatisfactory state of science

166

APHORISMS: BOOK I

1. Man, the minister and interpreter of Nature, does and understands so much as he may have discerned concerning the order of Nature by observing or by meditating on facts: he knows no more, he can do no more.

2. Neither the bare hand nor the intellect left to itself have much power; results are produced by instruments and helps; which are needed as much for the understanding as for the hand. And as instruments of use to the hand either rouse or regulate its movements, so instruments for the mind either prompt the intellect or defend it from error.

3. Human knowledge and power coincide, because ignorance of the cause hinders the production of the effect. For Nature is not conquered save by obedience: and what in contemplation stands as a cause, the same in operation stands as a rule.

4. With a view to results man can do nothing but apply and remove natural bodies: everything else Nature performs within.

5. The Mechanician, the Mathematician, the Physician, the Alchemist, and the Magician are wont to interfere with Nature (as far as regards results): but all, as things now stand, with slight endeavours and small success.

6. It is madness, and a downright contradiction, to think that those things which never have been done as yet, can be done except by means never as yet tried.

7. The productions of mind and hand seem numerous enough in books and manufactures. But all that variety consists in a wonderful subtilty, and in subdivisions of a few well-known things; but not in a number of axioms.

8. Even the results which have been already discovered are due to Chance and Experience rather than to the Sciences: for those Sciences which we now have are nothing more than certain refinements on things previously discovered; not methods of discovery or plans for obtaining new results.

9. The cause, however, and root of almost all evils in the Sciences is this one; that while we falsely admire and exalt the powers of the human mind, we do not seek its true aids.

10. The subtilty of Nature far surpasses the subtilty of sense and intellect; so that men's fair meditations, speculations and reasonings are a kind of insanity, only there is no one standing by to notice it.

11. Just as the Sciences which now prevail are useless for the discovery of results, so the logic also which now prevails is useless for the discovery of Sciences.

12. The Logic which is in vogue is rather potent for the confirming and fixing of errors (which are based on vulgar opinions) than for the investigation of Truth: so that it is more harmful than useful.

13. The Syllogism is not applied at all to the principles of Science—is applied in vain to the medial axioms, since it is no match for the subtilty of Nature. And so it constrains Assent, not Things.

14. Syllogism is composed of propositions, propositions of words, words are the symbols of conceptions. And so if the conceptions themselves (which are the foundation of the whole thing) be confused and rashly abstracted from things, there is no solidity in what is built upon them. And so our only hope is in a true Induction.

18. What things have been hitherto discovered in the Sciences are of such a kind as to be subjects of common conceptions; but in order that we may penetrate to the inner and more remote secrets of Nature, it is necessary that conceptions as well as axioms should be abstracted from things by a more certain and a better made way, and that a thoroughly improved and more certain method of applying the Intellect should be brought into use.

19. There are and can be but two ways of investigating and discovering Truth. The one leaps from the senses and particulars to the most general axioms, and from these as first principles, and their unshaken truth, judges on and discovers medial axioms: and this way is in vogue. The other raises axioms from the

senses and particulars, by ascending steadily, step by step, so that at last the most general may be reached; and this way is the true one, but untried.

The four kinds of 'phantom'

167

38. The phantoms and false conceptions which have hitherto preoccupied man's intellect, and are deeply rooted in it, not merely so besiege men's minds that truth can with difficulty approach; but also, where truth has a passage allowed her, they will again occur and be troublesome in the Instauration of the Sciences, unless men be forewarned and defend themselves against them as far as possible.

39. There are four kinds of phantoms which lay siege to human minds. To them (for instruction's sake) we give names; and call the first kind *Phantoms of the Tribe*; the second, *Phantoms of the Cave*; the third, *Phantoms of the Market-Place*; and the fourth, *Phantoms of the Theatre*.

40. The raising of conceptions and axioms by means of true *Induction* is certainly the proper remedy for driving and clearing out phantoms: yet there is great use in pointing out the phantoms. For right teaching about phantoms stands towards the Interpretation of Nature, as the teaching about confutation of Sophisms does to the common Logic.

41. The *Phantoms of the Tribe* are founded in Human Nature itself, and in the very tribe or race of man. For it is a false assertion, that 'human sense is the measure of things'; whereas on the contrary, all perceptions both of sense and of mind are measured by the standard of man, not by the standard of the Universe; and the human intellect is like an uneven mirror catching the rays of things, which mingles its own nature with the nature of things, and distorts and corrupts it.

42. The *Phantoms of the Cave* are the phantoms of the individual. For each man has (besides the generic aberrations of human nature) some individual cave or den which breaks and corrupts the light of nature; either by reason of the peculiar and singular nature of each; or by reason of education and conversation of man with man; or by reason of the reading of books, and the authority of those whom each man studies and admires; or by reason of differences of impressions as they occur in a mind preoccupied and predisposed, or even and sedate; or the like. So that evidently the human spirit (according as it is placed in each individual) is a various thing, and altogether disturbed and, as it were, the creature of circumstance. Whence Heraclitus hath well said that 'men seek knowledge in lesser worlds, not in the greater and common world'.

43. There are also phantoms arising as it were from the intercourse and society of men with one another. These we call *Phantoms of the Market-Place*, on account of the commerce and consort of men. For men associate by means of discourse; but words are imposed at the will of the vulgar: and so a bad and foolish imposition of words besieges the intellect in strange ways. Neither do the

definitions or explanations wherewith learned men have been accustomed to fortify and clear themselves in some cases, in any way set the matter to rights. But words clearly put a force on the intellect, disturb everything, and lead men on to empty and innumerable controversies and fictions.

44. Lastly there are phantoms which have entered into the minds of men from the different dogmas of Philosophical systems, and even from perverted laws of Demonstration. These we name *Phantoms of the Theatre*: because we count that each Philosophy received or invented is like a Play brought out and acted, creating each its own fictitious and scenic world. Nor do we only speak of those philosophies and sects which are now flourishing, nor even of the ancient ones, since plenty more such Plays might be composed and got up; for the causes of errors utterly diverse may be nevertheless almost common. Nor again do we only understand this of the universal philosophies, but even of very many other principles and axioms of the sciences, which have gotten strength through tradition, credence, and neglect.

The new concept of induction

168

70. Experience is far the best Demonstration: provided it adheres to the experiment itself. For if it pass on to other instances which are thought similar, it is fallacious, unless that passage be made duly and in order. But the method of experiment which men now make use of is blind and stupid: and so, while they wander and stray without any certain way, but only take counsel from the occurrence of circumstances, they are carried about to many points, but advance little; and are sometimes glad, sometimes distracted; and ever are finding something farther to seek after. For so it mostly happens that men make their experiments lightly and as it were in play, by varying little by little experiments already known; and if the thing does not succeed, they grow disgusted, and leave off their attempt. Even if they gird themselves more seriously, constantly, and laboriously for experiments; yet they spend their toil in turning over some one; as Gilbert did with the magnet, the Chemists with Gold. This, however, men do with as unskilful as futile a purpose. For no one happily explores the nature of a thing in the thing itself; but the inquiry should be extended to what is more general. But if men even build up some science and dogmas from experiments, yet they almost always turn aside with a hasty and untimely eagerness to practical application; not only for the sake of the use and fruit of that same application to practice, but also that in some new work they may snatch for themselves as it were a pledge that they will not employ themselves uselessly in the rest; and they even make the most of themselves to others, in order that they may gain a higher reputation in those subjects on which they are engaged. Whence it comes that, like Atalanta, they go aside to take up the golden apple, so meanwhile interrupting their course and letting victory slip

out of their hands. But in the true course of experiment, and the carrying it on to new effects, the Divine Wisdom and Order are entirely to be taken as our examples. Now God on the first day of Creation, created only Light; and gave a whole day for that work, and on that day created no material object. Similarly, in experience of every kind, first the discovery of causes and true Axioms is to be made; and light-bringing not fruit-bringing experiments to be sought for. And Axioms rightly discovered and established supply practical uses not scantily but in crowds; and draw after themselves bands and troops of effects. [. . .]

169

100. But not only is a greater abundance of experiments to be sought for and procured, and of another kind too from those hitherto gathered; but also an entirely different Method, Arrangement and Process are to be introduced for the continuation and promotion of experience. For vague experience, which follows only itself (as had been said above) is a mere groping, which stupifies men rather than informs them. But when experience shall proceed under a fixed Law, in due series and sequence, then we may have better Hopes for the sciences.

105. Moreover, in forming an Axiom, a form of *Induction*, different to that hitherto in use, must be thought out, and this too, not only for testing and discovering principles (as they call them), but also for lesser, then intermediate, and finally for all Axioms. For that Induction which proceeds by simple enumeration is a puerile thing, and concludes uncertainly, and is exposed to danger from any contradictory instance, and for the most part pronounces from fewer instances than it ought, and of these only from such as are at hand. But the Induction which will be useful for the discovery and demonstrations of Sciences and Arts, ought to separate Nature by due rejections and exclusions; and then, after a sufficient number of negatives, to conclude upon affirmatives; a thing which hitherto has not been done, nor indeed attempted, save only by Plato, who for the formation of Definitions and Ideas certainly uses this form of Induction up to a certain point. But for the good and legitimate appointing of this Induction or Demonstration, very many things are to be made use of which have as yet entered into the head of no man; so that more labour must be spent upon it than has hitherto been spent upon the syllogism. And the assistance of this Induction must be had not only to discover Axioms, but to limit conceptions also. And on this Induction depends our greatest hope.

170

APHORISMS: BOOK II

2. The unhappy case of Human Knowledge, as it is now, is even manifested by what is ordinarily asserted. It is rightly laid down that 'true knowledge is knowledge by causes'. Also the establishment of four Causes is not bad:

Material, Formal, Efficient, Final. Of these, however, the Final Cause is so far from profiting us, that it even corrupts Knowledge, except in Morals. The discovery of Form is held to be hopeless. And the Efficient and Material Causes (such as are now sought for and received, i.e. those remote from, and without *Latent Process* towards Form) are slovenly and superficial, and of scarcely any avail for true and active Knowledge. Nor have we forgotten how we above noticed and corrected the error of the human kind in assigning to Forms the first qualities of Essence. For although in Nature nothing really exists except individual bodies, which produce individual pure acts according to Law; yet in science, that Law itself and the investigation, discovery, and unfolding thereof are the foundation both of knowledge and of practice. This *Law*, then, *and its Paragraphs* we mean when we speak of Forms; especially as this word has grown into common use, and is of familiar occurrence.

16. And so there must be made an entire solution and the separation of Nature; not by fire, indeed, but by the mind, as by a Divine Fire. And so the first duty of true *Induction* (with a view to true discovery of Forms) is the *rejection* or *exclusion* of single Natures, which are not found in any Instance in which the given Nature is present; or which are found in any Instance from which the given Nature is absent; or are found to increase in any Instance, while the given Nature decreases, or to decrease, while the given Nature increases. Then after *Rejection* and *Exclusion* made in proper ways, there will remain (as a Residuum)—volatile opinions having passed off like smoke—Form, affirmative, solid, true, and well limited. And this is a short matter to tell; though one arrives at it after many windings.

JOHANN KEPLER

The harmonic structure of the universe

171

The theory that the universe has a spherical boundary was fully treated by Aristotle, who based his arguments mainly on the excellence of the spherical figure. As a result Copernicus's outer sphere of the fixed stars, though it lacks motion, retains the Aristotelian shape, but now it receives within its bosom the sun as its centre. The other stellar orbits are also circular, as is proved by the rotating motion of the stars. This makes it sufficiently clear that God utilized circular shape in his construction of the universe. There are, however, three types of quantity discernible in the world, namely the shape, number and magnitude of bodies, and of these it is only shape that we find to be circular. For magnitude has nothing to do with the fact that one figure concentrically inserted in another, as sphere in sphere or circle in circle, will coincide at all points or at no point. And the spherical, too, since it is one and unique in shape, which is its type of quantity, cannot be the subject of any number other than a trinity. If God, therefore, had thought of circular shape alone in his making of the world, there would only be the sun at the centre which he intended as an image of the Father, the sphere of the fixed stars, or the waters mentioned in Genesis, at the circumference, as an image of the Son, and the air of heaven, that expanse and firmament that fills all things, as an image of the Holy Spirit. Apart from this, I say, there would be nothing else whatever in the fabric of the universe. But of course there is also number, that of the countless fixed stars and the more limited list of moving ones. And there are diverse magnitudes, too, in the heavens. We are therefore forced to seek the causes of these not in the circular but in the rectilinear. For if we do not, we shall be driven to admit that God acted arbitrarily in the universe, even though perfectly good rational procedures were open to him. And this is a conclusion I will not accept on anyone's authority, even if I were to limit my observation to the fixed stars, the collocation of which is admittedly the most confused, for they rather resemble seed cast at random by a sower. [. . .]

If we are to make a selection among bodies, getting rid of the confused throng of irregular ones, let us retain only those which are universally plane, equilateral and equiangular. We shall then be left with those five regular bodies to which the Greeks gave the following names: the cube or six-faced body, the pyramid or four-faced, the dodecahedron or twelve-faced, the icosahedron or twenty-faced, the octahedron or eight-faced. The proof that there cannot be more than these five you will find in Euclid, Book Thirteen, in the scholion following Proposition 18. [. . .]

We have, then, circularity of orbit because of planetary motion and we have the bodies themselves because of number and magnitude. There is nothing to add but Plato's saying that 'God always geometrises'. In the fabric of mobile things He inserted these bodies between the orbits, and the orbits between the bodies in such a way that none of the bodies was left without being clothed within and without by a mobile orbit. From Propositions 13 to 17 of Euclid's Thirteenth Book we can learn how naturally suitable these five bodies are for such inscription and circumscription. And if these five bodies are inserted in order into the circles that come between them and enclose them the number of circles will come to six.

So, then, if we find that the structure of the universe was ever described by positing six mobile orbits around an immobile sun, such an astronomical theory was indeed correct. Now we find that Copernicus does have six orbits, and each pair of them in such proportion to each other that the five bodies can be accurately fitted between them. That can be taken as the summing-up of what I am going to say. Copernicus must therefore be followed until someone either produces hypotheses better suited to these investigations, or else allows that there may creep by sheer chance into calculation and thought an idea which is actually based on the very principles of nature. Copernicus established his points by arguing merely from appearances, effects and consequences. He arrived at them like a man walking with the aid of a stick, to quote a phrase of his own, often repeated to Rheticus.[1] It was more a lucky conjecture than a soundly based one, as he himself realized. What, I now ask, could be more wonderful, what could be more convincing in thought and word, than to see his points fully verified by deductions *a priori*, from causes, from the idea of creation?

Perhaps someone will take and deride my philosophic reasonings, in mere mockery, without offering any reasoning of his own, just because I, an upstart at the end of the ages, produce these tenets where the great philosophic luminaries of antiquity said nothing. Then I shall have to summon for him Pythagoras, a leading author and teacher from the most ancient age, of whom there is frequent mention in the schools. When, over two thousand years ago, he observed the importance of the five bodies, he took the same line of argument

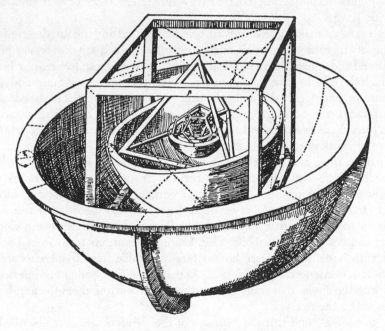

The spheres of the six planets and the five Platonic bodies

that I do now. It was not unworthy of the Creator's care, he thought, to think of these mathematical bodies, and God therefore accommodated non-mathematical objects to them, considering them as they are in their physical reality, each endowed with its own accidental qualities. For God, said Pythagoras, fitted the earth to the cube because both are stable, though this is not an essential property

1. Rheticus (1514–1576) was professor of mathematics in Wittenberg and a pupil of Copernicus. In 1540 he wrote a first account of Copernicus' theory. (Ed.)

of the cube. And to heaven he allotted the dodecahedron because both revolve; to fire he gave the pyramid for such is the shape of leaping flame; he allotted the other two bodies to air and water because of similar mutual affinity. Pythagoras, however, had no Copernicus to point out to him at the start the contents of the universe. Given such guidance, no doubt he would have found out the reason why things were so, and as a result this harmony of the heavens would now be as familiar to people as the five bodies themselves, as readily accepted as the idea of a moving sun and a fixed earth, which held sway for so long.

But now at long last we must test whether among the orbits of Copernicus the proportions of the regular bodies hold. For a start let us examine the matter somewhat roughly. The greatest difference of distance, according to Copernicus, is between Jupiter and Mars, as you will see in the explanation of Copernicus's hypotheses which I give along with Table I, and also in chapters 14 and 15 below. For the distance from the sun to Mars is less than a third of the distance to Jupiter. Let us seek then a mathematical body which produces the largest interval between its circumscribed circle and its inscribed one, if the reader will pardon a usage that speaks of the hollow in place of the solid. The body required is the tetrahedron or pyramid. Between Jupiter and Mars, then, there is the pyramid. After these the greatest difference in distance from the sun is produced by Jupiter and Saturn. That of Jupiter is something over half that of Saturn. A similar difference occurs in the circles inside and outside the cube. Therefore the orbit of Saturn encircles a cube and that cube encloses the orbit of Jupiter.

The relationship of distance is almost equal for Venus and Mercury. Quite a similar relationship exists between the two circles of the octahedron. The orbit of Venus therefore encloses that body and the orbit of Mercury is enclosed by it.

The remaining two relationships, between Venus and Earth and between Earth and Mars are very small and almost equal, the interior being three-quarters or two-thirds of the exterior. In the icosahedron and the dodecahedron the intervals of the two circles are the same, being the smallest among the regular bodies. So it is highly likely that Mars encloses Earth by means of one of those two bodies, and that the Earth is spaced from Venus by the other.

That is why, if I am asked why there are only six planetary orbits, my reply will be as follows: there ought not to be more than five inter-relationships, for that is the total number of the regular mathematical bodies, and six terms will fully provide that number of relationships.

In defence of the Copernican system

172

What was the ancients' argument from place?
They considered that the region of the elements was around the centre of the world; the heavens at the surface. Therefore to the bodies made up of elements belongs rectilinear movement, which has a beginning and an end and which,

disbursed according to the contrary principles of heaviness and lightness, brings any of those bodies back into its own place; and hence in proportion to different nearness to the natural place, or mark, there are different speeds, and finally pure rest. But the celestial bodies move everlastingly in the circular expanse of the world: and that argues that they are neither heavy nor light, and that they are not moved for the sake of rest or for the sake of occupying a place—for they are always circling in their place—but that accordingly they are moved only in order to be moved; and so their movement must be regular, and the form of their movement must be other than rectilinear, namely suited to an eternity of movement, that is, returning into itself.

What answer do you make to this third argument?
Not every irregularity of movements comes from heaviness and lightness, the properties of the elements; but some comes from the change of the distance too, as is clear in the case of the lever and the balance; and this cause produces intensification and remission of movements, as has been explained so far. We must, however, remark that there is nevertheless some kinship between the principles of heaviness and lightness in the elements and the natural inertia of the planetary globe with respect to movement, but no irregularity of movement is explained by this kinship.

But as regards the figure of the movement, the argument concludes nothing more than we can grant, namely that the movement bends back into itself. And not only the circular but also the elliptical are of such a kind; and so the assumptions are not denied. For in truth bodies which revolve around their axes are moved only in order that by their everlasting motion they may obey some necessity of their own globe—some bodies indeed in order that they may carry the planets around themselves in everlasting circles.

State the fourth argument of the ancients which was taken from the circular figure
They philosophized that of all movements which return into themselves, the circular is the most simple and the most perfect and that something of straightness is mixed in with all the others, such as the oval and similar figures: accordingly this circular movement is most akin to the very simple nature of the bodies, to the motors, which are divine minds—for its beauty and perfection is somehow of the mind—and finally to the heavens, which have a spherical figure.

How must this be refuted?
To this I make answer as follows: Firstly, if the celestial movements were the work of mind, as the ancients believed, then the conclusion that the routes of the planets are perfectly circular would be plausible. For then the form of movement conceived by the mind would be to the virtue a rule and mark to which the movement would be referred. But the celestial movements are not the work of mind but of nature, that is, of the natural power of the bodies, or else a work of the soul acting uniformly in accordance with those bodily powers:

and that is not proved by anything more validly than by the observation of the astronomers, who, after rightfully removing the deceptions of sight, find that the elliptical figure of revolution is left in the real and very true movement of the planet; and the ellipse bears witness to the natural bodily power and to the emanation and magnitude of its form. [. . .]

Then what causes do you bring forward as to why, although all the routes of the primary planets are arranged around the sun, nevertheless the angles—in which as if from the centre of the sun, the different parts of the route of one planet are viewed—are not completed by the planet in proportional times?

Two causes concur, the one optical, the other physical, and each of almost equal effect. The first cause is that the route of the planet is not described around the sun at an equal distance everywhere; but one part of it is near the sun, and the opposite part is so much the farther away from the sun. But of equal things, the near are viewed at a greater angle, and the far away, at a smaller; and of those which are viewed at an equal angle, the near are smaller, and the far away are greater.

The other cause is that the planet is really slower at its greater distance from the sun, and faster at its lesser.

Therefore if the two causes are made into one, it is quite clear that of two arcs which are equal to sight, the greater time belongs to the arc which is greater in itself, and a much greater time on account of the real slowness of the planet in that farther arc.

But could not one cause suffice, so that, because generally the orbit of the planet draws as far away from the sun on one side as it draws near on the other, we might make such a great distance that all this apparent irregularity might be explained merely by this unequal distance of the parts of the orbit?

Observations do not allow us to make the inequality of the distances as great as the inequality of the time wherein the planet makes equal angles at the sun; but they bear witness that the inequality of the distances is sufficient to explain merely half of this irregularity: therefore the remainder comes from the real acceleration and slowing up of the planet.

What are the laws and the instances of this speed and slowness?

There is a genuine instance in the lever. For there, when the arms are in equilibrium, the ratio of the weights hanging from each arm is the inverse of the ratio to the arms. For a greater weight hung from the shorter arm makes a moment equal to the moment of the lesser weight which is hung from the longer arm. And so, as the short arm is to the long, so the weight on the longer arm is to the weight on the shorter arm. And if in our mind we remove the other arm, and if instead of the weight on it we conceive at the fulcrum an equal power to lift up the remaining arm with its weight; then it is apparent that this power at the fulcrum does not have so much might over a weight which is distant as it does over the same weight when near. So too astronomy bears witness concerning the planet that the sun does not have as much power to move it and to make it revolve when the planet is farther away from the sun in a straight line, as it does when the interval is decreased. And, in brief, if on the orbit of the planet you take arcs which are of equal length the ratio between the distances of each arc from the sun is the same as the ratio of the times which the planet spends in those arcs. [. . .]

The motions of the planets and universal harmony

173

First of all, my readers should know that the ancient astronomical hypotheses of Ptolemy, in the fashion in which they have been unfolded in the *Theoricae* of Peurbach[1] and by the other writers of epitomes, are to be completely removed from this discussion and cast out of the mind. For they do not convey the true layout of the bodies of the world and the polity of the movements. [. . .]

To arrive at the movements between which the consonances have been set up, once more I impress upon the reader that in the *Commentaries on Mars* I have demonstrated from the sure observations of Brahe that daily arcs, which are equal in one and the same eccentric circle, are not traversed with equal speed; but that these differing delays in equal parts of the eccentric observe the ratio of their distances from the sun, the source of movement; and conversely, that if equal times are assumed, namely, one natural day in both cases, the corresponding true diurnal arcs of one eccentric orbit have to one another the ratio which is the inverse of the ratio of the two distances from the sun. Moreover, I demonstrated at the same time that the planetary orbit is elliptical and the sun, the source of movement, is at one of the foci of this ellipse; and so, when the planet has completed a quarter of its total circuit from its aphelion, then it is exactly at its mean distance from the sun, midway between its greatest distance at the aphelion and its least at the perihelion. But from these two axioms it results that the diurnal mean movement of the planet in its eccentric is the same as the true diurnal arc of its eccentric at those moments wherein the planet is at the end of the quadrant of the eccentric measured from the aphelion, although that true quadrant appears still smaller than the just quadrant. [. . .]

174

So far we have dealt with the different delays or arcs of one and the same planet. Now we must also deal with the comparison of the movements of two planets. Here take note of the definitions of the terms which will be necessary for us. We give the name of nearest apsides of two planets to the perihelion of the

1. Georg Peurbach (1423–1461), Austrian astronomer and mathematician. (Ed.)

upper and the aphelion of the lower, notwithstanding that they tend not towards the same region of the world but towards distinct and perhaps contrary regions. By extreme movements understand the slowest and the fastest of the whole planetary circuit; by converging or converse extreme movements, those which are at the nearest apsides of two planets—namely, at the perihelion of the upper planet and the aphelion of the lower; by diverging or diverse, those at the opposite apsides—namely, the aphelion of the upper and the perihelion of the lower. Therefore again, a certain part of my *Mysterium Cosmographicum*, which was suspended twenty-two years ago, because it was not yet clear, is to be completed and herein inserted. For after finding the true intervals of the orbits by the observations of Tycho Brahe and continuous labour and much time, at last, at last the right ratio of the periodic times to the orbits

> though it was late, looked to the unskilled man,
> yet looked to him, and, after much time, came;

and, if you want the exact time, was conceived mentally on the 8th of March in this year One Thousand Six Hundred and Eighteen but unfelicitously submitted to calculation and rejected as false; finally, summoned back on the 15th of May, with a fresh assault undertaken, outfought the darkness of my mind by the great proof afforded by my labour of seventeen years on Brahe's observations and meditation upon it uniting in one concord, in such fashion that I first believed I was dreaming and was presupposing the object of my search among the principles. But it is absolutely certain and exact that the ratio which exists between the periodic times of any two planets is precisely the ratio of the $\frac{3}{2}$th power of the mean distances, i.e. of the orbits themselves; provided, however, that the arithmetic mean between both diameters of the elliptic orbit be slightly less than the longer diameter. And so if any one take the period, say, of the Earth, which is one year, and the period of Saturn, which is thirty years, and extract the cube roots of this ratio and then square the ensuing ratio by squaring the cube roots, he will have as his numerical products the most just ratio of the distances of the Earth and Saturn from the sun. For the cube root of 1 is 1, and the square of it is 1; and the cube root of 30 is greater than 3, and therefore the square of it is greater than 9. And Saturn, at its mean distance from the sun, is slightly higher than nine times the mean distance of the Earth from the sun.

175

Accordingly, since we see that the universal harmonies of all six planets cannot take place by chance, especially in the case of the extreme movements, all of which we see concur in the universal harmonies—except two, which concur in harmonies closest to the universal—and since much less can it happen by chance that all the pitches of the system of the octave (as set up in Book III) by means of harmonic divisions are designated by the extreme planetary movements, but least of all that the very subtle business of the distinction of the celestial consonances into two modes, the major and minor, should be the outcome of chance, without the special attention of the Artisan: accordingly it follows I say, that He, the Artisan of the celestial movements Himself, should have conjoined to the five regular solids the harmonic ratios arising from the regular plane figures, and out of both classes should have formed one most perfect archetype of the heavens: in order that in this archetype, as through the five regular solids the shapes of the spheres shine through on which the six planets are carried, so too through the consonances, which are generated from the plane figures, and deduced from them in Book III, the measures of the eccentricities in the single planets might be determined so as to proportion the movements of the planetary bodies; and in order that there should be one tempering together of the ratios and the consonances, and that the greater ratios of the spheres should yield somewhat to the lesser ratios of the eccentricities necessary for procuring the consonances, and conversely those in especial of the harmonic ratios which had a greater kinship with each solid figure should be adjusted to the planets—insofar as that could be effected by means of consonances. And in order that, finally, in that way both the ratios of the spheres and the eccentricities of the single planets might be born of the archetype simultaneously, while from the amplitude of the spheres and the bulk of the bodies the periodic times of the single planets might result.

While I struggle to bring forth this process into the light of human intellect by means of the elementary form customary with geometers, may the Author of the heavens be favourable, the Father of intellects, the Bestower of mortal senses, Himself immortal and superblessed, and may He prevent the darkness of our mind from bringing forth in this work anything unworthy of His Majesty, and may He effect that we, the imitators of God by the help of the Holy Ghost, should rival the perfection of His works in sanctity of life, for which He chose His church throughout the Earth and, by the blood of His Son, cleansed it from sins, and that we should keep at a distance all the discords of enmity, all contentions, rivalries, anger, quarrels, dissensions, sects, envy, provocations, and irritations arising through mocking speech and the other works of the flesh; and that along with myself, all who possess the spirit of Christ will not only desire but will also strive by deeds to express and make sure their calling, by spurning all crooked morals of all kinds which have been veiled and painted over with the cloak of zeal or of the love of truth or of singular erudition or modesty over against contentious teachers, or with any other showy garment. Holy Father, keep us safe in the concord of our love for one another, that we may be one, just as Thou art one with Thy Son, Our Lord, and with the Holy Ghost, and just as through the sweetest bonds of harmonies Thou hast made all Thy works one; and that from the bringing of Thy people into concord the body of Thy Church may be built up in the Earth, as Thou didst erect the heavens themselves out of harmonies.

176

The things which have been said up to now will become clearer from the history of my discoveries. Since I had fallen into this speculation twenty-four years ago, I first inquired whether the single planetary spheres are equal distances apart from one another (for the spheres are apart in Copernicus, and do not touch one another), that is to say, I recognized nothing more beautiful than the ratio of equality. But this ratio is without head or tail: for this material equality furnished no definite number of mobile bodies, no definite magnitude for the intervals. Accordingly, I meditated upon the similarity of the intervals to the spheres, i.e. upon the proportionality. But the same complaint followed. For although to be sure, intervals which were altogether unequal were produced between the spheres, yet they were not unequally equal, as Copernicus wishes, and neither the magnitude of the ratio nor the number of the spheres was given. I passed on to the regular plane figures: intervals were formed from them by the ascription of circles. I came to the five regular solids: here both the number of the bodies and approximately the true magnitude of the intervals was disclosed, in such fashion that I summoned to the perfection of astronomy the discrepancies remaining over and above. Astronomy was perfect these twenty years; and behold! there was still a discrepancy between the intervals and the regular solids, and the reasons for the distribution of unequal eccentricities among the planets were not disclosed. That is to say, in this house the world, I was asking not only why stones of a more elegant form but also what form would fit the stones, in my ignorance that the Sculptor had fashioned them in the very articulate image of an animated body. So, gradually, especially during these last three years, I came to the consonances and abandoned the regular solids in respect to minutiae, both because the consonances stood on the side of the form which the finishing touch would give, and the regular solids, on that of the material—which in the world is the number of bodies and the rough-hewn amplitude of the intervals—and also because the consonances gave the eccentricities, which the regular solids did not even promise—that is to say, the consonances made the nose, eyes, and remaining limbs a part of the statue, for which the regular solids had prescribed merely the outward magnitude of the rough-hewn mass.

GALILEO GALILEI

Nature and the Scriptures: two aspects of one truth

177

As the Holy Scriptures and nature equally have their origin in the divine word, the one as given by the Holy Spirit and the other as the most observant executor of the orders of God; and as furthermore it is the custom of Scripture, in order to accommodate itself to the common understanding, to say many things which apparently deviate from the absolute truth as far as the literal significance of the word is concerned; and as on the other hand nature is inexorable and immutable and never transcends the limits of the laws imposed on her, because she does not care whether her hidden reasons and modes of operation are or are not susceptible to the capacity of men, it seems that not one of the natural phenomena which sense experience reveals to us or which are proved to us by the necessary demonstrations can be held in any doubt whatsoever, or rejected, because of some passages in the Scriptures which in their literal wording seem to be at variance with it. And it seems furthermore that not everything that is said in the Scriptures is bound to the same precision as every phenomenon of nature, and that God reveals Himself in no smaller measure in the phenomena of nature than in the sacred words of Scripture. And possibly that is what Tertullian had in mind when he said: 'We assert that God is recognized first in nature, and after that in His teachings; in nature, i.e. through His creations; in His teachings, i.e. through the sermons.'

The existence and significance of sun spots

178

SALVIATI: But you, Simplicio, what have you thought of to reply to the opposition of these importunate spots which have come to disturb the heavens, and worse still, the Peripatetic philosophy? It must be that you, as its intrepid defender, have found a reply and a solution which you should not deprive us of.

SIMPLICIO: I have heard different opinions on this matter. Some say 'They are stars which, like Venus and Mercury, go about the sun in their proper orbits, and in passing under it present themselves to us as dark; and because there are many of them, they frequently happen to collect together, and then again to separate.' Others believe them to be figments of the air; still others, illusions of the lenses; and still others, other things. But I am most inclined to believe—yes, I think it certain—that they are a collection of various different opaque objects, coming together almost accidentally; and therefore we often see that in one spot there can be counted ten or more such tiny bodies of irregular shape that look like snowflakes, or tufts of wool, or flying moths. They change places with each other, now separating and now congregating, but mostly right under the sun, about which, as their centre, they move. But it is not therefore necessary to say that they are generated or decay. Rather, they are sometimes hidden behind the body of the sun; at other times, though far from it, they cannot be seen because of their proximity to its immeasurable light. For in the sun's eccentric sphere there is established a sort of onion composed of various folds, one within another, each being studded with certain little spots, and moving; and although their movements seem at first to be inconstant and irregular, nonetheless it is said to be ultimately observed that after a certain time the same spots are sure to return. This seems to me to be the most appropriate expedient that has so far been found to account for such phenomena, and at the same time to maintain the incorruptibility and ingenerability of the heavens. And if this is not enough, there are more brilliant intellects who will find better answers.

SALVIATI: If what we are discussing were a point of law or of the humanities, in which neither true nor false exists, one might trust in subtlety of mind and readiness of tongue and in the greater experience of the writers, and expect him who excelled in those things to make his reasoning most plausible, and one might judge it to be the best. But in natural sciences, whose conclusions are true and necessary and have nothing to do with human will, one must take care not to place oneself in the defence of error; for here a thousand Demostheneses and a thousand Aristotles would be left in the lurch by every mediocre wit who happened to hit upon the truth for himself. Therefore, Simplicio, give up this idea and this hope of yours that there may be men so much more learned, erudite, and well-read than the rest of us as to be able to make that which is false become true in defiance of nature. And since among all opinions that have thus far been produced regarding the essence of sunspots, this one you have just explained appears to you to be the correct one, it follows that all the rest are false. Now to free you also from that one—which is an utterly delusive chimera—I shall, disregarding the many improbabilities in it, convey to you but two observed facts against it.

One is that many of these spots are seen to originate in the middle of the solar disc, and likewise many dissolve and vanish far from the edge of the sun, a necessary argument that they must be generated and dissolved. For without generation and corruption, they could appear there only by way of local motion, and they all ought to enter and leave by the very edge.

The other observation, for those not in the rankest ignorance of perspective, is that from the changes of shape observed in the spots, and from their apparent changes in velocity, one must infer that the spots are in contact with the sun's body, and that, touching its surface, they are moved either with it or upon it, and in no sense revolve in circles distant from it. Their motion proves this by appearing to be very slow around the edge of the solar disc, and quite fast towards its centre; the shapes of the spots prove the same by appearing very narrow around the sun's edge in comparison with how they look in the vicinity of the centre. For around the centre they are seen in their majesty and as they really are; but around the edge, because of the curvature of the spherical surface, they show themselves foreshortened. These diminutions of both motion and shape, for anyone who knows how to observe them and calculate diligently, correspond exactly to what ought to appear if the spots are contiguous to the sun, and hopelessly contradict their moving in distant circles, or even at small intervals from the solar body.

The superiority of change to immutability

179

SAGREDO: I cannot without great astonishment—I might say without great insult to my intelligence—hear it attributed as a prime perfection and nobility of the natural and integral bodies of the universe that they are invariant, immutable, inalterable, etc., while on the other hand it is called a great imperfection to be alterable, generable, mutable, etc. For my part I consider the earth very noble and admirable precisely because of the diverse alterations, changes, generations, etc., that occur in it incessantly. If, not being subject to any changes, it were a vast desert of sand or a mountain of jasper, or if at the time of the flood the waters which covered it had frozen, and it had remained an enormous globe of ice where nothing was ever born or ever altered or changed, I should deem it a useless lump in the universe, devoid of activity and, in a word, superfluous and essentially non-existent. This is exactly the difference between a living animal and a dead one; and I say the same of the moon, of Jupiter, and of all other world globes.

The deeper I go in considering the vanities of popular reasoning, the lighter and more foolish I find them. What greater stupidity can be imagined than that of calling jewels, silver and gold 'precious', and earth and soil 'base'? People who do this ought to remember that if there were as great a scarcity of soil as of jewels or precious metals, there would not be a prince who would not spend a bushel of diamonds and rubies and a cartload of gold just to have enough earth to plant a jasmine in a little pot, or to sow an orange seed and watch it sprout, grow, and produce its handsome leaves, its fragrant flowers, and fine fruit. It is scarcity and plenty that make the vulgar take things to be precious or worthless; they call a diamond very beautiful because it is like pure water, and then would not exchange one for ten barrels of water. Those who so greatly exalt incorruptibility, inalterability, etc., are reduced to talking this way, I believe, by their great desire to go on living, and by the terror they have of death. They do not reflect that if men were immortal, they themselves would never have come into the world. Such men really deserve to encounter a Medusa's head which would transmute them into statues of jasper or of diamond, and thus make them more perfect than they are.

The concept of inertia

180

SALVIATI: You say, then, that since when the ship stands still the rock falls to the foot of the mast, and when the ship is in motion it falls apart from there, then conversely, from the falling of the rock at the foot it is inferred that the ship stands still, and from its falling away it may be deduced that the ship is moving. And since what happens on the ship must likewise happen on the land, from the falling of the rock at the foot of the tower one necessarily infers the immobility of the terrestrial globe. Is that your argument?

SIMPLICIO: That is exactly it, briefly stated, which makes it easy to understand.

SALVIATI: Now tell me: If the stone dropped from the top of the mast when the ship was sailing rapidly fell in exactly the same place on the ship to which it fell when the ship was standing still, what use could you make of this falling with regard to determining whether the vessel stood still or moved?

SIMPLICIO: Absolutely none; just as by the beating of the pulse, for instance, you cannot know whether a person is asleep or awake, since the pulse beats in the same manner in sleeping as in waking.

SALVIATI: Very good. Now, have you ever made this experiment of the ship?

SIMPLICIO: I have never made it, but I certainly believe that the authorities who adduced it had carefully observed it. Besides, the cause of the difference is so exactly known that there is no room for doubt.

SALVIATI: You yourself are sufficient evidence that those authorities may have offered it without having performed it, for you take it as certain without having done it, and commit yourself to the good faith of their dictum. Similarly it not only may be, but must be that they did the same thing too—I mean, put faith in their predecessors, right on back without ever arriving at anyone who had performed it. For anyone who does will find that the experiment shows exactly the opposite of what is written; that is, it will show that the stone always falls in the same place on the ship, whether the ship is standing still or moving with any speed you please. Therefore, the same cause holding good on the earth as on the ship, nothing can be inferred about the earth's motion or rest from the stone falling always perpendicularly to the foot of the tower.

SIMPLICIO: If you had referred me to any other agency than experiment, I think that our dispute would not soon come to an end; for this appears to me to be a thing so remote from human reason that there is no place in it for credulity or probability.

SALVIATI: For me there is, just the same.

SIMPLICIO: So you have not made a hundred tests, or even one? And yet you so freely declare it to be certain? I shall retain my incredulity, and my own confidence that the experiment has been made by the most important authors who make use of it, and that it shows what they say it does.

SALVIATI: Without experiment, I am sure that the effect will happen as I tell you, because it must happen that way; and I might add that you yourself also know that it cannot happen otherwise, no matter how you may pretend not to know it—or give that impression. But I am so handy at picking people's brains that I shall make you confess this in spite of yourself [. . .]

Now tell me: Suppose you have a plane surface as smooth as a mirror and made of some hard material like steel. This is not parallel to the horizon, but somewhat inclined, and upon it you have placed a ball which is perfectly spherical and of some hard and heavy material like bronze. What do you believe this will do when released? Do you not think, as I do, that it will remain still?

SIMPLICIO: If that surface is tilted?

SALVIATI: Yes, that is what was assumed.

SIMPLICIO: I do not believe that it would stay still at all; rather, I am sure that it would spontaneously roll down.

SALVIATI: Pay careful attention to what you are saying, Simplicio, for I am certain that it would stay wherever you placed it.

SIMPLICIO: Well, Salviati, so long as you make use of assumptions of this sort I shall cease to be suprised that you deduce such false conclusions.

SALVIATI: Then you are quite sure that it would spontaneously move downward?

SIMPLICIO: What doubt is there about this?

SALVIATI: And you take this for granted not because I have taught it to you—indeed, I have tried to persuade you to the contrary—but all by yourself, by means of your own common sense.

SIMPLICIO: Oh, now I see your trick; you spoke as you did in order to get me out on a limb, as the common people say, and not because you really believed what you said.

SALVIATI: That was it. Now how long would the ball continue to roll, and how fast? Remember that I said a perfectly round ball and a highly polished surface, in order to remove all external and accidental impediments. Similarly, I want you to take away any impediment of the air caused by its resistance to separation, and all other accidental obstacles, if there are any.

SIMPLICIO: I completely understand you, and to your question I reply that the ball would continue to move indefinitely, as far as the slope of the surface extended, and with a continually accelerated motion. For such is the nature of heavy bodies, which *vires acquirunt eundo*; and the greater the slope, the greater would be the velocity.

SALVIATI: But if one wanted the ball to move upward on this same surface, do you think it would go?

SIMPLICIO: Not spontaneously, no; but drawn or thrown forcibly, it would.

SALVIATI: And if it were thrust along with some impetus impressed forcibly upon it, what would its motion be, and how great?

SIMPLICIO: The motion would constantly slow down and be retarded, being contrary to nature, and would be of longer or shorter duration according to the greater or lesser impulse and the lesser or greater slope upward.

SALVIATI: Very well; up to this point you have explained to me the events of motion upon two different planes. On the downward inclined plane, the heavy moving body spontaneously descends and continually accelerates, and to keep it at rest requires the use of force. On the upward slope, force is needed to thrust it along or even to hold it still, and motion which is impressed upon it continually diminishes until it is entirely annihilated. You say also that a difference in the two instances arises from the greater or lesser upward or downward slope of the plane, so that from a greater slope downward there follows a greater speed, while on the contrary upon the upward slope a given movable body thrown with a given force moves farther according as the slope is less.

Now tell me what would happen to the same movable body placed upon a surface with no slope upward or downward.

SIMPLICIO: Here I must think a moment about my reply. There being no downward slope, there can be no natural tendency towards motion; and there being no upward slope, there can be no resistance to being moved, so there would be an indifference between the propensity and the resistance to motion. Therefore it seems to me that it ought naturally to remain stable. But I forgot; it was not so very long ago that Sagredo gave me to understand that that is what would happen.

SALVIATI: I believe it would do so if one set the ball down firmly. But what would happen if it were given an impetus in any direction?

SIMPLICIO: It must follow that it would move in that direction.

SALVIATI: But with what sort of movement? One continually accelerated, as on the downward plane, or increasingly retarded as on the upward one?

SIMPLICIO: I cannot see any cause for acceleration or deceleration, there being no slope upward or downward.

SALVIATI: Exactly so. But if there is no cause for the ball's retardation, there ought to be still less for its coming to rest; so how far would you have the ball continue to move?

SIMPLICIO: As far as the extension of the surface continued without rising or falling.

SALVIATI: Then if such a space were unbounded, the motion on it would likewise be boundless? That is, perpetual?

SIMPLICIO: It seems so to me, if the movable body were of durable material.

SALVIATI: That is of course assumed, since we said that all external and accidental impediments were to be removed, and any fragility on the part of the moving body would in this case be one of the accidental impediments.

Now tell me, what do you consider to be the cause of the ball moving spontaneously on the downward inclined plane, but only by force on the one tilted upward?

SIMPLICIO: That the tendency of heavy bodies is to move towards the centre of the earth, and to move upward from its circumference only with force; now the downward surface is that which gets closer to the centre, while the upward one gets farther away.

SALVIATI: Then in order for a surface to be neither downward nor upward, all its parts must be equally distant from the centre. Are there any such surfaces in the world?

SIMPLICIO: Plenty of them; such would be the surface of our terrestrial globe if it were smooth, and not rough and mountainous as it is. But there is that of the water, when it is placid and tranquil.

SALVIATI: Then a ship, when it moves over a calm sea, is one of these movables which courses over a surface that is tilted neither up nor down, and if all external and accidental obstacles were removed, it would thus be disposed to move incessantly and uniformly from an impulse once received?

SIMPLICIO: It seems that it ought to be.

SALVIATI: Now as to that stone which is on top of the mast; does it not move, carried by the ship, both of them going along the circumference of a circle about its centre? And consequently is there not in it an ineradicable motion, all external impediments being removed? And is not this motion as fast as that of the ship?

SIMPLICIO: All this is true, but what next?

SALVIATI: Go on and draw the final consequence by yourself, if by yourself you have known all the premisses.

SIMPLICIO: By the final conclusion you mean that the stone, moving with an indelibly impressed motion, is not going to leave the ship, but will follow it, and finally will fall at the same place where it fell when the ship remained motionless.

A critique of Aristotle's 'natural motion'

181

SALVIATI: How many questions we should have to be diverted into, if we wished to settle all the difficulties that are linked together, one in consequence of another! You call that principle external, preternatural, and constrained which moves heavy projectiles upward, but perhaps it is no less internal and natural than that which moves them downward. It may perhaps be called external and constrained while the movable body is joined to its mover; but once separated, what external thing remains as the mover of an arrow or a ball? It must be admitted that the force which takes this on high is no less internal than that which moves it down. Thus I consider the upward motion of heavy bodies due to received impetus to be just as natural as their downward motion dependent upon gravity.

SIMPLICIO: This I shall never admit, because the latter has a natural and perpetual internal principle while the former has a finite and constrained external one.

SALVIATI: If you flinch from conceding to me that the principles of motion of heavy bodies downward and upward are equally internal and natural, what would you do if I were to tell you that they may also be one and the same?

SIMPLICIO: I leave it to you to judge.

SALVIATI: Rather, I want you to be the judge. Tell me, do you believe that contradictory internal principles can reside in the same natural body?

SIMPLICIO: Absolutely not.

SALVIATI: What would you consider to be the natural intrinsic tendencies of earth, lead and gold, and in brief of all very heavy materials? That is, towards what motion do you believe that their internal principle draws them?

SIMPLICIO: Motion towards the centre of heavy things; that is, to the centre of the universe and of the earth, whither they would be conducted if not impeded.

SALVIATI: So that if the terrestrial globe were pierced by a hole which passed through its centre, a cannon ball dropped through this and moved by its natural and intrinsic principle would be taken to the centre, and all this motion would be spontaneously made and by an intrinsic principle. Is that right?

SIMPLICIO: I take that to be certain.

SALVIATI: But having arrived at the centre is it your belief that it would pass on beyond, or that it would immediately stop its motion there?

SIMPLICIO: I think it would keep on going a long way.

SALVIATI: Now wouldn't this motion beyond the centre be upward, and according to what you have said preternatural and constrained? But upon what other principle will you make it depend, other than the very one which has brought the ball to the centre and which you have already called intrinsic and natural? Let me see you find an external thrower who shall overtake it once more to throw it upward.

And what is said thus about motion through the centre is also to be seen up here by us. For the internal impetus of a heavy body falling along an inclined plane which is bent at the bottom and deflected upward will carry the body upward also, without interrupting its motion at all. A ball of lead hanging from a thread and moved from the perpendicular descends spontaneously, drawn by its internal tendency, without pausing to rest it goes past the lowest point and without any supervening mover it moves upward. I know you will not deny that the principle which moves heavy bodies downward is as natural and internal to these as the principle which moves light ones upward is to those. Hence I ask you to consider a ball of wood which, descending through the air from a great height and therefore moved by a natural principle, meets with deep water and continues its descent; without any external mover it submerges for a long stretch; and yet the downward motion through water is preternatural to it. Still, it depends upon a principle which is internal and not external to the ball. Thus you see how a movable body may be moved with contrary motions by the same internal principle.

The telescope gives support of the Copernican theory

SALVIATI: [. . .] Next in Venus, which at its evening conjunction when it is beneath the sun ought to look almost forty times as large as in its morning conjunction, and is seen as not even doubled, it happens in addition to the effects of irradiation that it is sickle-shaped, and its horns, besides being very thin, receive the sun's light obliquely and therefore very weakly. So that because it is small and feeble, it makes its irradiation less ample and lively than when it shows itself to us with its entire hemisphere lighted. But the telescope plainly shows us its horns to be as bounded and distinct as those of the moon, and they are seen to belong to a very large circle, in a ratio almost forty times as great as the same disc when it is beyond the sun, towards the end of its morning appearances.

SAGREDO: O Nicholas Copernicus, what a pleasure it would have been for you to see this part of your system confirmed by so clear an experiment!

SALVIATI: Yes, but how much less would his sublime intellect be celebrated among the learned! For as I said before, we may see that with reason as his guide he resolutely continued to affirm what sensible experience seemed to contradict. I cannot get over my amazement that he was constantly willing to persist in saying that Venus might go around the sun and be more than six times as far from us at one time as at another, and still look always equal, when it should have appeared forty times larger. [. . .]

It remains for us to remove what would seem to be a great objection to the motion of the earth. This is that though all the planets turn about the sun, the earth alone is not solitary like the others, but goes together in the company of the moon and the whole elemental sphere around the sun in one year, while at the same time the moon moves around the earth every month. Here one must once more exclaim over and exalt the admirable perspicacity of Copernicus, and simultaneously regret his misfortune at not being alive in our day. For now Jupiter removes this apparent anomaly of the earth and moon moving conjointly. We see Jupiter, like another earth, going around the sun in twelve years accompanied not by one but by four moons, together with everything that may be contained within the orbits of its four satellites.

SAGREDO: And what is the reason for your calling the four Jovian planets 'moons'?

SALVIATI: That is what they would appear to be to anyone who saw them from Jupiter. For they are dark in themselves, and receive their light from the sun; this is obvious from their being eclipsed when they enter into the cone of Jupiter's shadow. And since only that hemisphere of theirs is illuminated which faces the sun, they always look entirely illuminated to us who are outside their orbits and closer to the sun; but to anyone on Jupiter they would look completely lighted only when they were at the highest points of their circles. In the lowest part—that is, when between Jupiter and the sun—they would appear horned from Jupiter. In a word, they would make for Jovians the same changes of shape which the moon makes for us Terrestrials.

Now you see how admirably these [. . .] notes harmonize with the Copernican system, when at first they seemed so discordant with it. From this, Simplicio will be much better able to see with what great probability one may conclude that not the earth, but the sun, is the centre of rotation of the planets. And since this amounts to placing the earth among the world bodies which indubitably move about the sun (above Mercury and Venus but beneath Saturn, Jupiter and Mars), why will it not likewise be probable, or perhaps even necessary, to admit that it also goes around?

The velocity of light

183

SAGREDO: But of what kind and how great must we consider this speed of light to be? Is it instantaneous or momentary or does it like other motions require time? Can we not decide this by experiment?

SIMPLICIO: Everyday experience shows that the propagation of light is instantaneous; for when we see a piece of artillery fired, at great distance, the flash reaches our eyes without lapse of time; but the sound reaches the ear only after a noticeable interval.

SAGREDO: Well, Simplicio, the only thing I am able to infer from this familiar bit of experience is that sound, in reaching our ear, travels more slowly than light; it does not inform me whether the coming of light is instantaneous or whether, although extremely rapid, it still occupies time. An observation of this kind tells us nothing more than one in which it is claimed that 'As soon as the sun reaches the horizon its light reaches our eyes'; but who will assure me that these rays had not reached this limit earlier than they reached our vision?

SALVIATI: The small conclusiveness of these and other similar observations once led me to devise a method by which one might accurately ascertain whether illumination, i.e. the propagation of light, is really instantaneous. The fact that the speed of sound is as high as it is, assures us that the motion of light cannot fail to be extraordinarily swift. The experiment which I devised was as follows:

Let each of two persons take a light contained in a lantern, or other receptacle, such that by the interposition of the hand, the one can shut off or admit the light to the vision of the other. Next let them stand opposite each other at a distance of a few cubits and practice until they acquire such skill in uncovering and occulting their lights that the instant one sees the light of his companion he will uncover his own. After a few trials the response will be so prompt that without sensible error the uncovering of one light is immediately followed by the uncovering of the other, so that as soon as one exposes his light he will instantly see that of the other. Having acquired skill at this short distance let the two experimenters, equipped as before, take up positions separated by a distance of two or three miles and let them perform the same experiment at night, noting carefully whether the exposures and occultations occur in the same manner as at short distances; if they do, we may safely conclude that the propagation of light is instantaneous; but if time is required at a distance of three miles which, considering the going of one light and the coming of the other, really amounts to six, then the delay ought to be easily observable. If the experiment is to be made at still greater distances, say eight or ten miles, telescopes may be employed, each observer adjusting one for himself at the place where he is to make the experiment at night; then although the lights are not large and are therefore invisible to the naked eye at so great a distance, they can readily be covered and uncovered since by aid of telescopes, once adjusted and fixed, they will become easily visible.

SAGREDO : This experiment strikes me as a clever and reliable invention. But tell us what you conclude from the results.

SALVIATI : In fact I have tried the experiment only at a short distance, less than a mile, from which I have not been able to ascertain with certainty whether the appearance of the opposite light was instantaneous or not.

All bodies fall with equal velocity: refutation of Aristotle

184

SIMPLICIO : So far as I remember, Aristotle inveighs against the ancient view that a vacuum is a necessary prerequisite for motion and that the latter could not occur without the former. In opposition to this view Aristotle shows that it is precisely the phenomenon of motion, as we shall see, which renders untenable the idea of a vacuum. His method is to divide the argument into two parts. He first supposes bodies of different weights to move in the same medium; then supposes, one and the same body to move in different media. In the first case, he supposes bodies of different weight to move in one and the same medium with different speeds which stand to one another in the same ratio asthe weights; so that, for example, a body which is ten times as heavy as another will move ten times as rapidly as the other. In the second case he assumes that the speeds of one and the same body moving in different media are in inverse ratio to the densities of these media; thus, for instance, if the density of water were ten times that of air, the speed in air would be ten times greater than in water. From this second supposition, he shows that, since the tenuity of a vacuum differs infinitely from that of any medium filled with matter however rare, any body which moves in a plenum through a certain space in a certain time ought to move through a vacuum instantaneously; but instantaneous motion is an impossibility; it is therefore impossible that a vacuum should be produced by motion.

SALVIATI : The argument is, as you see, *ad hominem*, that is, it is directed against those who thought the vacuum a prerequisite for motion. Now if I admit the argument to be conclusive and concede also that motion cannot take place in a vacuum, the assumption of a vacuum considered absolutely, and not with reference to motion, is not thereby invalidated. But to tell you what the ancients might possibly have replied and in order to better understand just how conclusive Aristotle's demonstration is, we may, in my opinion, deny both of his assumptions. And as to the first, I greatly doubt that Aristotle ever tested by experiment whether it be true that two stones, one weighing ten times as much as the other, if allowed to fall, at the same instant, from a height of, say, 100 cubits, would so differ in speed that when the heavier had reached the ground, the other would not have fallen more than 10 cubits.

SIMPLICIO: His language would seem to indicate that he had tried the experiment, because he says: *We see the heavier*; now the word *see* shows that he had made the experiment.

SAGREDO: But I, Simplicio, who have made the test can assure you that a cannon ball weighing one or two hundred pounds, or even more, will not reach the ground by as much as a span ahead of a musket ball weighing only half a pound, provided both are dropped from a height of 200 cubits.

SALVIATI: But, even without further experiment, it is possible to prove clearly, by means of a short and conclusive argument, that a heavier body does not move more rapidly than a lighter one provided both bodies are of the same material and in short such as those mentioned by Aristotle. But tell me, Simplicio, whether you admit that each falling body acquires a definite speed fixed by nature, a velocity which cannot be increased or diminished except by the use of force or resistance.

SIMPLICIO: There can be no doubt but that one and the same body moving in a single medium has a fixed velocity which is determined by nature and which cannot be increased except by the addition of momentum or diminished except by some resistance which retards it.

SALVIATI: If then we take two bodies whose natural speeds are different, it is clear that on uniting the two, the more rapid one will be partly retarded by the slower, and the slower will be somewhat hastened by the swifter. Do you not agree with me in this opinion?

SIMPLICIO: You are unquestionably right.

SALVIATI: But if this is true, and if a large stone moves with a speed of, say, eight while a smaller moves with a speed of four, then when they are united, the system will move with a speed less than eight; but the two stones when tied together make a stone larger than that which before moved with a speed of eight. Hence the heavier body moves with less speed than the lighter; an effect which is contrary to your supposition. Thus you see how, from your assumption that the heavier body moves more rapidly than the lighter one, I infer that the heavier body moves more slowly.

SIMPLICIO: I am all at sea because it appears to me that the smaller stone when added to the larger increases its weight and by adding weight I do not see how it can fail to increase its speed or, at least, not to diminish it.

SALVIATI: Here again you are in error, Simplicio, because it is not true that the smaller stone adds weight to the larger.

SIMPLICIO: This is, indeed, quite beyond my comprehension.

SALVIATI: It will not be beyond you when I have once shown you the mistake under which you are labouring. Note that it is necessary to distinguish between heavy bodies in motion and the same bodies at rest. A large stone placed in a balance not only acquires additional weight by having another stone placed upon it, but even by the addition of a handful of hemp its weight is augmented six to ten ounces according to the quantity of hemp. But if you tie the hemp to the stone and allow them to fall freely, from some height, do you believe that the hemp will press down upon the stone and thus accelerate its motion or do you think the motion will be retarded by a partial upward pressure? One always feels the pressure upon his shoulders when he prevents the motion of a load resting upon him; but if one descends just as rapidly as the load would fall how can it gravitate or press upon him? Do you not see that this would be the same as trying to strike a man with a lance when he is running away from you with a speed which is equal to, or even greater than, that with which you are following him? You must therefore conclude that, during free and natural fall, the small stone does not press upon the larger and consequently does not increase its weight as it does when at rest.

SIMPLICIO: But what if we should place the larger stone upon the smaller?

SALVIATI: Its weight would be increased if the larger stone moved more rapidly; but we have already concluded that when the small stone moves more slowly it retards to some extent the speed of the larger, so that the combination of the two, which is a heavier body than the larger of the two stones, would move less rapidly, a conclusion which is contrary to your hypothesis. We infer therefore that large and small bodies move with the same speed provided they are of the same specific gravity.

SIMPLICIO: Your discussion is really admirable; yet I do not find it easy to believe that a bird-shot falls as swiftly as a cannon ball.

SALVIATI: Why not say a grain of sand as rapidly as a grindstone? But, Simplicio, I trust you will not follow the example of many others who divert the discussion from its main intent and fasten upon some statement of mine which lacks a hair's-breadth of the truth and, under this hair, hide the fault of another which is as big as a ship's cable. Aristotle says that 'an iron ball of one hundred pounds falling from a height of 100 cubits reaches the ground before a one-pound ball has fallen a single cubit'. I say that they arrive at the same time. You find, on making the experiment, that the larger outstrips the smaller by two finger-breadths, that is, when the larger has reached the ground, the other is short of it by two finger-breadths; now you would not hide behind these two fingers the 99 cubits of Aristotle, nor would you mention my small error and at the same time pass over in silence his very large one. Aristotle declares that bodies of different weights, in the same medium, travel (insofar as their motion depends upon gravity) with speeds which are proportional to their weights; this he illustrates by use of bodies in which it is possible to perceive the pure and unadulterated effect of gravity, eliminating other considerations, for example, figure as being of small importance, influences which are greatly dependent upon the medium which modifies the single effect of gravity alone. Thus we observe that gold, the densest of all substances, when beaten out into a very thin leaf, goes floating through the air; the same thing happens with stone when ground into a very fine powder. But if you wish to maintain the general proposition you will have to show that the same ratio of speeds is preserved in the case of all heavy bodies, and that a stone of 20 pounds moves ten times as rapidly as one of two; but I claim that this is false and that, if they fall from a height of 50 or 100 cubits, they will reach the earth at the same moment.

SIMPLICIO: Perhaps the result would be different if the fall took place not from a few cubits but from some thousands of cubits.

SALVIATI: If this were what Aristotle meant you would burden him with another error which would amount to a falsehood; because, since there is no such sheer height available on earth, it is clear that Aristotle could not have made the experiment; yet he wishes to give us the impression of his having performed it when he speaks of such an effect as one which we see.

SIMPLICIO: In fact, Aristotle does not employ this principle, but uses the other one which is not, I believe, subject to these same difficulties.

SALVIATI: But the one is as false as the other; and I am surprised that you yourself do not see the fallacy and that you do not perceive that if it were true that, in media of different densities and different resistances, such as water and air, one and the same body moved in air more rapidly than in water, in proportion as the density of water is greater than that of air, then it would follow that any body which falls through air ought also to fall through water. But this conclusion is false inasmuch as many bodies which descend in air not only do not descend in water, but actually rise.

Experiments with the pendulum

185

SALVIATI: The facts set forth by me up to this point and, in particular, the one which shows that difference of weight, even when very great, is without effect in changing the speed of falling bodies, so that as far as weight is concerned they all fall with equal speed: this idea is, I say, so new and at first glance so remote from fact, that if we do not have the means of making it just as clear as sunlight, it had better not be mentioned; but having once allowed it to pass my lips I must neglect no experiment or argument to establish it.

SAGREDO: Not only this but also many other of your views are so far removed from the commonly accepted opinions and doctrines that if you were to publish them you would stir up a large number of antagonists; for human nature is such that men do not look with favour upon discoveries—either of truth or fallacy—in their own field, when made by others than themselves. They call him an innovator of doctrine, an unpleasant title, by which they hope to cut those knots which they cannot untie, and by subterranean mines they seek to destroy structures which patient artisans have built with customary tools. But as for ourselves who have no such thoughts, the experiments and arguments which you have thus far adduced are fully satisfactory; however, if you have any experiments which are more direct or any arguments which are more convincing we will hear them with pleasure.

SALVIATI: The experiment made to ascertain whether two bodies, differing greatly in weight will fall from a given height with the same speed offers some difficulty; because, if the height is considerable, the retarding effect of the medium, which must be penetrated and thrust aside by the falling body, will be greater in the case of the small momentum of the very light body than in the case of the great force of the heavy body; so that, in a long distance, the light body will be left behind; if the height be small, one may well doubt whether there is any difference; and if there be a difference it will be inappreciable.

It occurred to me therefore to repeat many times the fall through a small height in such a way that I might accumulate all those small intervals of time that elapse between the arrival of the heavy and light bodies respectively at their common terminus, so that this sum makes an interval of time which is not only observable, but easily observable. In order to employ the slowest speeds possible and thus reduce the change which the resisting medium produces upon the simple effect of gravity it occurred to me to allow the bodies to fall along a plane slightly inclined to the horizontal. For in such a plane, just as well as in a vertical plane, one may discover how bodies of different weight behave: and besides this, I also wished to rid myself of the resistance which might arise from contact of the moving body with the aforesaid inclined plane. Accordingly I took two balls, one of lead and one of cork, the former more than a hundred times heavier than the latter, and suspended them by means of two equal fine threads, each four or five cubits long. Pulling each ball aside from the perpendicular, I let them go at the same instant, and they, falling along the circumferences of circles having these equal strings for semi-diameters, passed beyond the perpendicular and returned along the same path. This free vibration repeated a hundred times showed clearly that the heavy body maintains so nearly the period of the light body that neither in a hundred swings nor even in a thousand will the former anticipate the latter by as much as a single moment, so perfectly do they keep step. We can also observe the effect of the medium which, by the resistance which it offers to motion, diminishes the vibration of the cork more than that of the lead, but without altering the frequency of either; even when the arc traversed by the cork did not exceed five or six degrees while that of the lead was fifty or sixty, the swings were performed in equal times.

SIMPLICIO: If this be so, why is not the speed of the lead greater than that of the cork, seeing that the former traverses sixty degrees in the same interval in which the latter covers scarcely six?

SALVIATI: But what would you say, Simplicio, if both covered their paths in the same time when the cork, drawn aside through thirty degrees, traverses an arc of sixty, while the lead pulled aside only two degrees traverses an arc of four? Would not then the cork be proportionately swifter? And yet such is the experimental fact. But observe this: having pulled aside the pendulum of lead, say through an arc of fifty degrees, and set it free, it swings beyond the perpendicular almost fifty degrees, thus describing an arc of nearly one hundred degrees; on the return swing it describes a little smaller arc; and after a large number of such vibrations it finally comes to rest. Each vibration, whether of ninety, fifty, twenty, ten, or four degrees occupies the same time: accordingly the speed of the moving body keeps on diminishing since in equal intervals of time, it traverses arcs which grow smaller and smaller.

Precisely the same things happen with the pendulum of cork, suspended by a string of equal length, except that a smaller number of vibrations is required to bring it to rest, since on account of its lightness it is less able to overcome the resistance of the air; nevertheless the vibrations, whether large or small, are all performed in time-intervals which are not only equal among themselves, but also equal to the period of the lead pendulum. Hence it is true that, if while the lead is traversing an arc of fifty degrees the cork covers one of only ten, the cork moves more slowly than the lead; but on the other hand it is also true that the cork may cover an arc of fifty while the lead passes over one of only ten or six; thus, at different times, we have now the cork, now the lead, moving more rapidly. But if these same bodies traverse equal arcs in equal times we may rest assured that their speeds are equal.

Fundamental concepts of the new kinematics

186

My purpose is to set forth a very new science dealing with a very ancient subject. There is, in nature, perhaps nothing older than motion, concerning which the books written by philosophers are neither few nor small; nevertheless I have discovered by experiment some properties of it which are worth knowing and which have not hitherto been either observed or demonstrated. Some superficial observations have been made, as, for instance, that the free motion of a heavy falling body is continuously accelerated; but to just what extent this acceleration occurs has not yet been announced; for so far as I know, no one has yet pointed out that the distances traversed, during equal intervals of time, by a body falling from rest, stand to one another in the same ratio as the odd numbers beginning with unity.

It has been observed that missiles and projectiles describe a curved path of some sort; however, no one has pointed out the fact that this path is a parabola. But this and other facts, not few in number or less worth knowing, I have succeeded in proving; and what I consider more important, there have been opened up to this vast and most excellent science, of which my work is merely the beginning, ways and means by which other minds more acute than mine will explore its remote corners.

This discussion is divided into three parts; the first part deals with motion which is steady or uniform; the second treats of motion as we find it accelerated in nature; the third deals with the so-called violent motions and with projectiles.

The acceleration of freely falling bodies

187

SALVIATI: The properties belonging to uniform motion have been discussed in the preceding section; but accelerated motion remains to be considered.

And first of all it seems desirable to find and explain a definition best fitting natural phenomena. For anyone may invent an arbitrary type of motion and discuss its properties; thus, for instance, some have imagined helices and conchoids as described by certain motions which are not met with in nature, and have very commendably established the properties which these curves possess in virtue of their definitions; but we have decided to consider the phenomena of bodies falling with an acceleration such as actually occurs in nature and to make this definition of accelerated motion exhibit the essential features of observed accelerated motions. And this, at last, after repeated efforts we trust we have succeeded in doing. In this belief we are confirmed mainly by the consideration that experimental results are seen to agree with and exactly correspond with those properties which have been, one after another, demonstrated by us. Finally, in the investigation of naturally accelerated motion we were led, by hand as it were, in following the habit and custom of nature herself, in all her various other processes, to employ only those means which are most common, simple and easy.

For I think no one believes that swimming or flying can be accomplished in a manner simpler or easier than that instinctively employed by fishes and birds.

When, therefore, I observe a stone initially at rest falling from an elevated position and continually acquiring new increments of speed, why should I not believe that such increases take place in a manner which is exceedingly simple and rather obvious to everybody? If now we examine the matter carefully we find no addition or increment more simple than that which repeats itself always in the same manner. This we readily understand when we consider the intimate relationship between time and motion; for just as uniformity of motion is defined by and conceived through equal times and equal spaces (thus we call a motion uniform when equal distances are traversed during equal time-intervals), so also we may, in a similar manner, through equal time-intervals, conceive additions of speed as taking place without complication; thus we may picture to our mind a motion as uniformly and continuously accelerated when, during any equal intervals of time whatever, equal increments of speed are given to it. Thus if any equal intervals of time whatever have elapsed, counting from the time at which the moving body left its position of rest and began to descend, the amount of speed acquired during the first two time-intervals will be double that acquired during the first time-interval alone; so the amount added during three of these time-intervals will be treble; and that in four, quadruple that of the first time-interval. To put the matter more clearly, if a body were to continue its motion with the same speed which it had acquired during the first time-interval and were to retain this same uniform speed, then its motion would be twice as slow as that which it would have if its velocity had been acquired during *two* time-intervals.

And thus, it seems, we shall not be far wrong if we put the increment of speed as proportional to the increment of time; hence the definition of motion which we are about to discuss may be stated as follows: A motion is said to be uniformly accelerated, when starting from rest, it acquires, during equal time-intervals, equal increments of speed.

SAGREDO: Although I can offer no rational objection to this or indeed to any other definition, devised by any author whomsoever, since all definitions are arbitrary, I may nevertheless without offence be allowed to doubt whether such a definition as the above, established in an abstract manner, corresponds to and describes that kind of accelerated motion which we meet in nature in the case of freely falling bodies. And since the Author apparently maintains that the motion described in his definition is that of freely falling bodies, I would like to clear my mind of certain difficulties in order that I may later apply myself more earnestly to the propositions and their demonstrations.

SALVIATI: It is well that you and Simplicio raise these difficulties. They are, I imagine, the same which occurred to me when I first saw this treatise, and which were removed either by discussion with the Author himself, or by turning the matter over in my own mind.

SAGREDO: When I think of a heavy body falling from rest, that is, starting with zero speed and gaining speed in proportion to the time from the beginning of the motion; such a motion as would, for instance, in eight beats of the pulse acquire eight degrees of speed; having at the end of the fourth beat acquired four degrees; at the end of the second, two; at the end of the first, one; and since time is divisible without limit, it follows from all these considerations that if the earlier speed of a body is less than its present speed in a constant ratio, then there is no degree of speed however small (or, one may say, no degree of slowness however great) with which we may not find this body travelling after starting from infinite slowness, i.e. from rest. So that if that speed which it had at the end of the fourth beat was such that, if kept uniform, the body would traverse two miles in an hour, and if keeping the speed which it had at the end of the second beat, it would traverse one mile an hour, we must infer that, as the instant of starting is more and more nearly approached, the body moves so slowly that, if it kept on moving at this rate, it would not traverse a mile in an hour, or in a day, or in a year or in a thousand years; indeed, it would not traverse a span in an even greater time; a phenomenon which baffles the imagination, while our senses show us that a heavy falling body suddenly acquires great speed.

SALVIATI: This is one of the difficulties which I also at the beginning experienced, but which I shortly afterwards removed; and the removal was effected by the very experiment which creates the difficulty for you. You say the experiment appears to show that immediately after a heavy body starts from rest it acquires a very considerable speed: and I say that the same experiment makes clear the fact that the initial motions of a falling body, no matter how heavy, are very slow and gentle. Place a heavy body upon a yielding material, and leave it there without any pressure except that owing to its own weight; it is clear that if one lifts this body a cubit or two and allows it to fall upon the same material, it will, with this impulse, exert a new and greater pressure than that caused by its mere weight; and this effect is brought about by the [weight of the] falling body together with the velocity acquired during the fall, an effect which will be greater and greater according to the height of the fall, that is, according to the height of the fall, velocity of the falling body becomes greater. From the quality and intensity of the blow we are thus enabled to accurately estimate the speed of a falling body. But tell me, gentlemen, is it not true that if a block be allowed to fall upon a stake from a height of 4 cubits and drives it into the earth, say, four finger-breadths, that coming from a height of 2 cubits it will drive the stake a much less distance, and from the height of one cubit a still less distance; and finally if the block be lifted only one finger-breadth how much more will it accomplish than if merely laid on top of the stake without percussion? Certainly very little. If it be lifted only the thickness of a leaf, the effect will be altogether imperceptible. And since the effect of the blow depends upon the velocity of this striking body, can anyone doubt the motion is very slow and the speed more than small whenever the effect [of the blow] is imperceptible? See now the power of the truth; the same experiment which at first glance seemed to show one thing, when more carefully examined, assures us of the contrary.

But without depending upon the above experiment, which is doubtless very conclusive, it seems to me that it ought not to be difficult to establish such a fact by reasoning alone. Imagine a heavy stone held in the air at rest; the support is removed and the stone set free; then since it is heavier than the air it begins to fall, and not with uniform motion but slowly at the beginning and with a continuously accelerated motion. Now since velocity can be increased and diminished without limit, what reason is there to believe that such a moving body starting with infinite slowness, that is, from rest, immediately acquires a speed of ten degrees rather than one of four, or of two, or of one, or of a half, or of a hundredth; or, indeed, of any of the infinite number of small values [of speed]? Pray listen. I hardly think you will refuse to grant that the gain of speed of the stone falling from rest follows the same sequence as the diminution and loss of this same speed when, by some impelling force, the stone is thrown to its former elevation: but even if you do not grant this, I do not see how you can doubt that the ascending stone, diminishing in speed, must before coming to rest pass through every possible degree of slowness.

SIMPLICIO: But if the number of degrees of greater and greater slowness is limitless, they will never be all exhausted, therefore such an ascending heavy body will never rest, but will continue to move without limit always at a slower rate; but this is not the observed fact.

SALVIATI: This would happen, Simplicio, if the moving body were to maintain its speed for any length of time at each degree of velocity; but it merely passes each point without delaying more than an instant: and since each time-interval however small may be divided into an infinite number of instants, these will always be sufficient [in number] to correspond to the infinite degrees of diminished velocity.

That such a heavy rising body does not remain for any length of time at any given degree of velocity is evident from the following: because if, some time-interval having been assigned, the body moves with the same speed in the last as in the first instant of that time-interval, it could from this second degree of elevation be in like manner raised through an equal height, just as it was transferred from the first elevation to the second, and by the same reasoning would pass from the second to the third and would finally continue in uniform motion forever.

SAGREDO: From these considerations it appears to me that we may obtain a proper solution of the problem discussed by philosophers, namely, what causes the acceleration in the natural motion of heavy bodies? [. . .]

SALVIATI: The present does not seem to be the proper time to investigate the cause of the acceleration of natural motion concerning which various opinions have been expressed by various philosophers, some explaining it by attraction to the centre, others by repulsion between the very small parts of the body, while still others attribute it to a certain stress in the surrounding medium which closes in behind the falling body and drives it from one of its positions to another. Now, all these fantasies, and others too, ought to be examined; but it is not really worth while. At present it is the purpose of our Author merely to investigate and to demonstrate some of the properties of accelerated motion (whatever the cause of this acceleration may be)—meaning thereby a motion, such that the momentum of its velocity goes on increasing after departure from rest, in simple proportionality to the time, which is the same as saying that in equal time-intervals the body receives equal increments of velocity; and if we find the properties [of accelerated motion] which will be demonstrated later are realized in freely falling and accelerated bodies, we may conclude that the assumed definition includes such a motion of falling bodies and that their speed goes on increasing as the time and the duration of the motion.

Experiments with the inclined plane

188

SALVIATI: The request which you, as a man of science, make, is a very reasonable one; for this is the custom—and properly so—in those sciences where mathematical demonstrations are applied to natural phenomena, as is seen in the case of perspective, astronomy, mechanics, music, and others where the principles, once established by well-chosen experiments, become the foundations of the entire superstructure. I hope therefore it will not appear to be a waste of time if we discuss at considerable length this first and most fundamental question upon which hinge numerous consequences of which we have in this book only a small number, placed there by the Author, who has done so much to open a pathway hitherto closed to minds of speculative turn. So far as experiments go they have not been neglected by the Author; and often, in his company, I have attempted in the following manner to assure myself that the acceleration actually experienced by falling bodies is that above described.

A piece of wooden moulding or scantling, about 12 cubits long, half a cubit wide, and three finger-breadths thick, was taken; on its edge was cut a channel a little more than one finger in breadth; having made this groove very straight, smooth, and polished, and having lined it with parchment, also as smooth and polished as possible, we rolled along it a hard, smooth, and very round bronze ball. Having placed this board in a sloping position, by lifting one end some one or two cubits above the other, we rolled the ball, as I was just saying, along the channel, noting, in a manner presently to be described, the time required to make the descent. We repeated this experiment more than once in order to measure the time with an accuracy such that the deviation between two observations never exceeded one-tenth of a pulse-beat. Having performed this operation and having assured ourselves of its reliability, we now rolled the ball only one-quarter the length of the channel; and having measured the time of its descent, we found it precisely one-half of the former. Next we tried other distances, comparing the time for the whole length with that for the half, or with that for two-thirds, or three-fourths, or indeed for any fraction; in such experiments, repeated a hundred times, we always found that the spaces traversed were to each other as the squares of the times, and this was true for all inclinations of the plane, i.e. of the channel, along which we rolled the ball. We also observed that the times of descent, for various inclinations of the plane, bore to one another precisely that ratio which, as we shall see later, the Author had predicted and demonstrated for them.

For the measurement of time, we employed a large vessel of water placed in an elevated position; to the bottom of this vessel was soldered a pipe of small diameter giving a thin jet of water, which we collected in a small glass during the time of each descent, whether for the whole length of the channel or for a part of its length; the water thus collected was weighed, after each descent, on a very accurate balance; the differences and ratios of these weights gave us the differences and ratios of the times, and this with such accuracy that although the operation was repeated many, many times, there was no appreciable discrepancy in the results.

SIMPLICIO : I would like to have been present at these experiments; but feeling confidence in the care with which you performed them, and in the fidelity with which you relate them, I am satisfied and accept them as true and valid.

Inertial motion on a horizontal plane

189

Furthermore, we may remark that any velocity once imparted to a moving body will be rigidly maintained as long as the external causes of acceleration or retardation are removed, a condition which is found only on horizontal planes; for in the case of planes which slope downwards there is already present a cause of acceleration, while on planes sloping upward there is a retardation; from this it follows that motion along a horizontal plane is perpetual; for, if the velocity be uniform, it cannot be diminished or slackened, much less destroyed. Further, although any velocity which a body may have acquired through natural fall is permanently maintained so far as its own nature is concerned, yet it must be remembered that if, after descent along a plane inclined downwards, the body is deflected to a plane inclined upward, there is already existing in this latter plane a cause of retardation; for in any such plane the same body is subject to a natural acceleration downwards. Accordingly we have here the superposition of two different states, namely, the velocity acquired during the preceding fall which if acting alone would carry the body at a uniform rate to infinity, and the velocity which results from a natural acceleration downwards common to all bodies.

RENÉ DESCARTES

190

Rules for the direction of the mind

Rule II

*Only those objects should engage our attention to the
sure and indubitable knowledge of which our mental
powers seem to be adequate*

All science is certain, evident knowledge, and he who doubts many things is
not more learned that he who has never thought about these things; on the
contrary, he seems even less learned if he has conceived some false opinion
about them. And therefore it is better never to study than to turn our attention
to such difficult topics than, being unable to distinguish the true from the
false, we are forced to accept doubtful conclusions as certain, since in such
cases there is not as much hope of increasing your knowledge as there is danger
of diminishing it. And so, in accordance with this rule, we reject all knowledge
which is merely probable, and judge that only those things should be believed
which are perfectly known, and about which we can have no doubts. Learned
men may perhaps persuade themselves, nevertheless, that hardly any such
knowledge exists, because, due to a common fault of human nature, they have
failed to consider such knowledge for the very reason that it is so simple and
obvious to everyone. Nevertheless, I advise them that there is much more such
knowledge than they think, and knowledge such as will suffice to provide
certain demonstrations of innumerable propositions which they have hitherto
been unable to discuss except as merely probable. And because they have
believed it unworthy of a learned man to confess there is something he does not
know, they have been accustomed so to elaborate their pretended reasons that
they then have gradually persuaded themselves and so offered these reasons to
others as genuine.

And truly, if we make good use of this rule, there will be very few things
left to the investigation of which it will be permissible to direct our efforts.
For there is hardly any question in the sciences concerning which talented
men do not often differ among themselves. But whenever the judgements of two
persons concerning the same thing are opposed, it is certain that at least one
of them is wrong, and there is not even one of them who seems to have knowledge. For if one person's argument were certain and evident, he could propose
it in such a way to the other one that even the latter's mind would eventually be
convinced. Therefore, we do not seem able to acquire perfect knowledge of
anything about which there are merely plausible opinions of this sort, because
we have no right, without vanity, to hope to achieve more ourselves than others

have accomplished. And thus, if we argue correctly, only arithmetic and geometry remain of the sciences already discovered, to which we are reduced by the observation of this rule. [. . .]

A little while before, we said that of the disciplines known to others, only arithmetic and geometry are free from the taint of any falsity or uncertainty. In order to inquire more carefully why this should be so, we should note that there are two paths available to us leading to the knowledge of reality, namely, experience and deduction. It should also be noted that our experiences with things often lead to error, but 'deduction', or the pure logical inference from one thing to another, can never be performed improperly by an intellect which is in the least degree rational, although it may escape our attention if we do not happen to notice it. And it seems to me that there is little benefit in those chains of reasoning by which logicians believe they can regulate human reasoning, though I do not deny that they are very suitable for other purposes. For no error which can happen to human beings—I say to human beings, not to animals—arises from incorrect inference, but only from the fact that certain poorly understood experiences are falsified, or certain opinions are accepted rashly and without justification.

From this we can easily understand why arithmetic and geometry are much more certain than the other disciplines, because only these are concerned with an object so pure and simple that nothing need be assumed which experience has rendered uncertain, and because they consist wholly of rationally deduced consequences. These disciplines are therefore the easiest and most obvious of all, and they have the kind of characteristic we require, since, except by oversight, it hardly seems that in this case it is human to err. Nevertheless, it ought not to be surprising that many people prefer to turn their attention to other arts or to philosophy: this happens because each person takes the liberty of prophesying with greater confidence in an obscure matter than in one that is evident, and it is much easier to make some conjecture concerning any subject that happens to arise rather than to reach the real truth about one question, however trivial it may be.

Now from all of this it is to be concluded, not that arithmetic and geometry are the only subjects to be studied, but only that in seeking the correct path to truth we should be concerned with nothing about which we cannot have a certainty equal to that of the demonstrations of arithmetic and geometry.

191

Rule VII

If we wish our science to be complete, those matters which promote the end we have in view must one and all be scrutinized by a movement of thought which is continuous and nowhere interrupted; they must also be included in an enumeration which is both adequate and methodical

[. . .] Furthermore, we must note that by adequate enumeration or induction is only meant that method by which we may attain surer conclusions than by any other type of proof, with the exception of simple intuition. But when the knowledge of some matter cannot be reduced to this, we must cast aside all syllogistic fetters and employ induction, the only method left us, but one in which all confidence should be reposed. For whenever single facts have been immediately deduced the one from the other, they have been already reduced, if the inference was evident, to a true intuition. But if we infer any single thing from various and disconnected facts, often our intellectual capacity is not so great as to be able to embrace them all in a single intuition; in which case our mind should be content with the certitude attaching to this operation. It is in precisely similar fashion that though we cannot with one single gaze distinguish all the links of a lengthy chain, yet if we have seen the connection of each with its neighbour, we shall be entitled to say that we have seen how the first is connected with the last. [. . .]

Further, while now the enumeration ought to be complete, now distinct, there are times when it need have neither of these characters; it was for this reason that I said only that it should be adequate. For if I want to prove by enumeration how many genera there are of corporeal things, or of those that in any way fall under the senses, I shall not assert that they are just so many and no more, unless I previously have become aware that I have included them all in my enumeration, and have distinguished them each separately from all the others. But if in the same way I wish to prove that the rational soul is not corporeal, I do not need a complete enumeration; it will be sufficient to include all bodies in certain collections in such a way as to be able to demonstrate that the rational soul has nothing to do with any of these. If, finally, I wish to show by enumeration that the area of a circle is greater than the area of all other figures whose perimeter is equal, there is no need for me to call in review all other figures; it is enough to demonstrate this of certain others in particular, in order to get thence by induction the same conclusion about all the others.

I added also that the enumeration ought to be methodical. This is both because we have no more serviceable remedy for the defects already instanced, than to scan all things in an orderly manner; and also because if often happens that if each single matter which concerns the quest in hand were to be investigated separately, no man's life would be long enough for the purpose, whether because they are far too many, or because it would chance that the same things had to be repeated too often. But if all these facts are arranged in the best order, they will for the most part be reduced to determinate classes, out of which it will be sufficient to take one example for exact inspection, or some one feature in a single case, or certain things rather than others, or at least we shall never have to waste our time in traversing the same ground twice. The advantage of this course is so great that often many particulars can, owing to a well-devised arrangement, be gone over in a short space of time and with little trouble, though at first view the matter looked immense.

But this order which we employ in our enumerations can for the most part be varied and depends upon each man's judgement. For this reason, if we would elaborate it in our thought with greater penetration, we must remember what was said in our fifth proposition. There are also many of the trivial things of man's devising, in the discovery of which the whole method lies in the disposal of this order. Thus if you wish to construct a perfect anagram by the transposition of the letters of a name, there is no need to pass from the easy to the difficult, nor to distinguish absolute from relative. Here there is no place for these operations; it will be sufficient to adopt an order to be followed in the transpositions of the letters which we are to examine, such that the same arrangements are never handled twice over. The total number of transpositions should, e.g. be split up into definite classes, so that it may immediately appear in which there is the best hope of finding what is sought. In this way the task is often not tedious but merely child's play. [. . .]

192

Rule IX

We ought to give the whole of our attention to the most insignificant and most easily mastered facts, and remain a long time in contemplation of them until we are accustomed to behold the truth clearly and distinctly

Having set forth the two operations of our intellect, intuition and deduction, upon which alone we have said we can depend for the acquisition of knowledge, we continue to explain, in this and the following rule, how we can render ourselves more able to accomplish this, and at the same time to improve the two principal faculties of the mind, that is, perspicacity in intuiting every object distinctly and sagacity in deducing one thing from another skilfully.

How the intuition of the mind should be used can be learned by a comparison with the perception of the eyes. For whoever tries to observe many objects at the same time with a single glance sees none of them distinctly; and likewise, whoever is accustomed to consider many things at the same time by a single act of understanding is confused in his mind. But those artisans who work on minute tasks, and are accustomed to direct their glance attentively to a single point, acquire by experience the ability to distinguish any number of minute and tiny objects perfectly. And in the same way those who never divide their attention among many objects at the same time, but always direct it wholly upon the simplest and easiest matters, are the ones who become perspicacious.

It is, moreover, a common failing among mortals to consider difficult matters most attractive; and many consider that they know nothing when they discover a very obvious and simple cause of something, and at the same time admire the sublime and profound reasoning of philosophers, even though this reasoning depends upon principles which have never been sufficiently examined by anyone. How foolish they are who prefer darkness to light! And we should notice that those who truly know recognize each truth with equal ease whether they have derived it from a simple source or from a difficult one, for they comprehend each truth by a similar but unique and distinct act, as soon as they have reached it. The difference lies wholly in the chain of reasoning, which certainly ought to be longer if it leads to a truth further removed from the first and most absolute principles.

Everybody should therefore become accustomed to consider such simple things and so few at a time that they think they never know anything which they do not grasp as distinctly as that which they know most distinctly of all. Some persons, no doubt, are much better equipped at birth to do this than others, but anyone can make himself much more skilful by study and practice. And there is one thing, it seems to me, which should be pointed out above all, namely, that each person should become firmly persuaded that even the most obscure sciences should be deduced, not from impressive and difficult matters, but only from the easy and more obvious ones.

Thus, for example, if I should wish to inquire whether any natural force can travel instantaneously to a distant point and pass through all intermediate points, I will not immediately consider magnetism, nor the influence of the stars, nor even the velocity of light, in order to find out whether these forces are instantaneous, for this would be more difficult to investigate than the original question; but I will rather think about the motion of bodies in space, because nothing in this category can be more apparent. And I will notice that a stone cannot travel instantaneously from one place to another, because it is a body; but that a power, such as that which moves the stone cannot be communicated otherwise than instantaneously if it goes by itself from one object to another. For example, if I move one end of a very long rod, I easily conceive that the power which moves that end of the rod also necessarily moves all the other parts at the same instant, because then it is communicated by itself, and does not exist in some body, such as the stone, by which it is carried.

In the same way, if I wish to learn how contrary effects can be produced simultaneously by one and the same simple cause, I will not borrow examples from the doctors, whose drugs expel certain humours and retain others; nor will I talk nonsense about the moon, saying that it heats by its light and cools by an occult quality; but rather I will consider a balance, in which the same weight causes one pan to rise at one and the same instant that it depresses the other, and other similar cases.

193

That the perceptions of the senses do not teach us what is
really in things, but merely that whereby they are useful or
hurtful to man's composite nature

It will be sufficient for us to observe that the perceptions of the senses are related simply to the intimate union which exists between body and mind, and that while by their means we are made aware of what in external bodies can profit or hurt this union, they do not present them to us as they are in themselves unless occasionally and accidentally. For after this observation we shall without difficulty set aside all the prejudices of the senses and in this regard rely upon our understanding alone, by reflecting carefully on the ideas implanted therein by nature.

194

That the nature of body consists not in weight, nor in
hardness, nor colour and so on, but in extension alone

In this way we shall ascertain that the nature of matter or of body in its universal aspect, does not consist in its being hard, or heavy, or coloured, or one that affects our senses in some other way, but solely in the fact that it is a substance extended in length, breadth and depth. For as regards hardness we do not know anything of it by sense, excepting that the portions of the hard bodies resist the motion of our hands when they come in contact with them; but if, whenever we moved our hands in some direction, all the bodies in that part retreated with the same velocity as our hands approached them, we should never feel hardness; and yet we have no reason to believe that the bodies which recede in this way would on this account lose what makes them bodies. It follows from this that the nature of body does not consist in hardness. The same reason shows us that weight, colour, and all the other qualities of the kind that is perceived in corporeal matter, may be taken from it, it remaining meanwhile entire: it thus follows that the nature of body depends on none of these.

195

*That it is contrary to reason to say that there is a vacuum or
space in which there is absolutely nothing*

As regards a vacuum in the philosophic sense of the word, i.e. a space in which
there is no substance, it is evident that such cannot exist, because the extension
of space or internal place, is not different from that of body. For, from the mere
fact that a body is extended in length, breadth or depth, we have reason to
conclude that it is a substance, because it is absolutely inconceivable that nothing
should possess extension, we ought to conclude also that the same is true of the
space which is supposed to be void, i.e. that since there is in it extension, there is
necessarily also substance.

196

*That a vacuum, in the ordinary sense, does not exclude all
body*

And when we take this word vacuum in its ordinary sense, we do not mean a
place or space in which there is absolutely nothing, but only a place in which
there are none of those things which we expected to find there. Thus because
a pitcher is made to hold water, we say that it is empty when it contains nothing
but air; or if there are no fish in a fish-pond, we say that there is nothing in it,
even though it be full of water; similarly we say a vessel is empty, when, in
place of the merchandise which it was designed to carry, it is loaded only with
sand, so that it may resist the impetuous violence of the wind; and finally we
say in the same way that a space is empty when it contains nothing sensible,
even though it contain created matter and self-existent substance; for we are not
wont to consider things excepting those with which our senses succeed in
presenting us. And if, in place of keeping in mind what we should comprehend
by these words—vacuum and nothing—we afterwards suppose that in the space
which is termed vacuum there is not only nothing sensible, but nothing at all,
we shall fall into the same error as if, because a pitcher is usually termed empty
since it contains nothing but air, we were therefore to judge that the air contained in it is not a substantive thing.

197

That this confirms what was said of rarefaction

After we have thus remarked that the nature of material substance consists only
in its being an extended thing, or that its extension is not different from what
has been attributed to space however empty, it is easy to discover that it is
impossible that any one of these parts should in any way occupy more space at
one time than another, and thus that it may be rarefied otherwise than in the

manner explained above; or again it is easy to perceive that there cannot be more matter or corporeal substance in a vessel when it is filled with gold or lead, or any other body that is heavy and hard, than when it only contains air and appears to be empty; for the quantity of the parts of matter does not depend on their weight or hardness, but only on the extension which is always equal in the same vessel.

198

That from this may be demonstrated the non-existence of atoms

We also know that there cannot be any atoms or parts of matter which are indivisible of their own nature as certain philosophers have imagined. For however small the parts are supposed to be, yet because they are necessarily extended we are always able in thought to divide any one of them into two or more parts; and thus we know that they are divisible. For there is nothing which we can divide in thought, which we do not thereby recognize to be divisible; and therefore if we judged it to be indivisible, our judgement would be contrary to the knowledge we have of the matter. And even should we suppose that God had reduced some portion of matter to a smallness so extreme that it could not be divided into smaller, it would not for all that be properly termed indivisible. For though God had rendered the particle so small that it was beyond the power of any creature to divide it, He could not deprive Himself of His power of division, because it is absolutely impossible that He should lessen His own omnipotence as was said before. And therefore, absolutely speaking, its divisibility remains to the smallest extended particle because from its nature it is such.

199

That the philosophy of Democritus is not less different from ours than from the vulgar

But Democritus also imagined that there were certain corpuscles that had various figures, sizes and motions, from the heaping together and mutual concourse of which all sensible bodies took their origin; and nevertheless his philosophy is by common consent universally rejected. To this I reply that it never was rejected by anyone because in it he considered bodies smaller than those that can be perceived by the senses, and attributed to them various sizes, figures and motions, for no one can doubt that there are in reality many such, as has been already shown. But this philosophy was rejected in the first place, because it presupposed certain indivisible corpuscles, which hypothesis I also completely reject; in the second place it was rejected because Democritus imagined a void about them, which I demonstrate to be an impossibility; in the third place because he attributed to them gravity, the existence of which I deny in any body insofar as it is considered by itself, because this is a quality depending on the relationship in respect of situation and motion which bodies bear to one another; and finally because he had not explained in detail how all things arose from the concourse of the corpuscles alone, or, if he explained it in regard to certain cases, his reasoning was not in all cases by any means coherent or such as was capable of proving to us that all nature can be explained in the same way. If we are to judge of his opinions from what has been preserved regarding his opinions, this at least is the verdict we must give on his philosophy. I leave it to others to judge as to whether what I have written in philosophy possesses sufficient coherence in itself and whether it is fertile enough in yielding us conclusions.

200

That God is the first cause of motion and that he always
preserves it in the universe in the same quantity

After considering the nature of motion, we must treat of its cause; in fact, of two sorts of cause. First, the universal and primary cause—the general cause of all the motions in the universe; secondly the particular cause that makes any given piece of matter assume a motion that it had not before.

As regards the general cause, it seems clear to me that it can be none other than God himself. He created matter along with motion and rest in the beginning; and now, merely by his ordinary co-operation, he preserves just the quantity of motion and rest in the material world that he put there in the beginning. Motion, indeed, is only a state of the moving body; but it has a certain definite quantity, and it is readily conceived that this quantity may be constant in the universe as a whole, while varying in any given part. We must reckon the quantity of motion in two pieces of matter as equal if one moves twice as fast as the other, and this in turn is twice as big as the first; again, if the motion of one piece of matter is retarded, we must assume an equal acceleration of some other body of the same size. Further, we conceive it as belonging to God's perfection, not only that he should in himself be unchangeable, but also that his operation should occur in a supremely constant and unchangeable manner. Therefore, apart from the changes of which we are assured by manifest experience or by divine revelation, and about which we can see, or believe by faith, that they take place without any change in the Creator, we must not assume any others in the works of God, lest they should afford an argument for his being inconstant. Consequently it is most reasonable to hold that, from the mere fact that God gave pieces of matter various movements at their first creation, and that he now preserves all this matter in being in the same way as he first created it, he must likewise always preserve in it the same quantity of motion.

201

That the sky is liquid

[. . .] we think that the sky, as well as the sun and the fixed stars, is made of liquid matter. This view is now commonly accepted by all astronomers, because they realize that it is almost impossible to explain the phenomena satisfactorily in any other way.

202

That it carries with it any bodies contained within it

But it seems to me that several are mistaken, for, in attributing to the sky the properties of a liquid, they think of it as a completely empty void, not only offering no resistance to the movement of other bodies, but also having no power to move them and carry them with it. For, apart from the fact that the existence of such a void in nature is impossible, all liquids have this in common: that the reason why they offer no resistance to the movements of other bodies is not that they consist of less material substance, but that they are equally or more disturbed, and that their small parts can easily be made to move in all directions; and in cases when they are caused to move all together in one direction, this means that they are forced to carry with them any bodies which they contain and surround on all sides, and which are in no way prevented from following them by any outside cause, although these bodies may be completely at rest, hard and solid, as follows obviously from what was said above about the nature of liquid bodies.

203

That the earth rests in its own sky, but that it is still carried along
by it

[. . .] since we can see that the earth is not supported by pillars, nor suspended in the air by ropes, but that it is completely surrounded by a very liquid sky, we think it is at rest, and that it has no inclination towards movement, because we do not notice it moving; but let us not think that this prevents it being carried away in the path of the sky and following its motion, without, however, moving of its own accord; just as a vessel which is neither propelled by the wind nor by oars, nor held in position by anchors, remains at rest in the middle of the sea, although perhaps the ebb or flow of this great mass of water carries it imperceptibly with it.

204

That all planets are carried round the sun by the sky
which contains them

Having by these arguments removed any doubts one might have about the movement of the earth, let us assume that the material of the heaven where the planets are circulates ceaselessly, like a whirlpool with the sun as its centre, and that the parts which are near the sun move more quickly than those which are a certain distance from it, and that all the planets (among whose number we include from now on the earth) always remain suspended between the same parts of this heavenly matter; for only thus, and without using any other tools, shall we find a simple explanation of all the things we notice about them. For, in the same way as in the windings of rivers where the water is turned back on itself, and turning round thus makes circles, if some bits of straw or other very light bodies are floating about in this water, we can see that it carries them

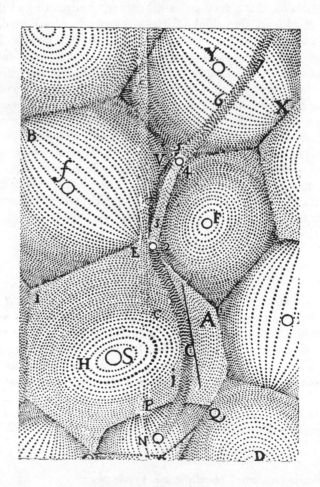

The vortices of the aether

along and makes them move around with it; and even among these straws one notices that there are often some which also turn round on their own axis; and that those nearest to the centre of the whirlpool which contains them complete their circle sooner than those which are further away; and finally that, although these whirlpools of water always seem to turn in a circle, they hardly ever describe completely perfect circles, sometimes stretching longer and sometimes wider, so that all points of the circumference they describe are not equidistant from the centre; so it may easily be imagined that these same things happen to the planets; and that is all that is necessary to explain the phenomena relating to them.

205

What sensation is, and how it operates

[. . .] For up to this point I have described the earth, and all the visible world, as if it were simply a machine in which there was nothing to consider but the figure and movements of its parts, and yet our senses cause other things to be presented to us, such as colours, smells, sounds, and other such things, of which, if I did not speak, it might be thought that I had omitted the main part of the explanation of the objects of nature.

We must know, therefore, that although the mind of man informs the whole body, it yet has its principal seat in the brain, and it is there that it not only understands and imagines, but also perceives; and this by means of the nerves which are extended like filaments from the brain to all the other members, with which they are so connected that we can hardly touch any part of the human body without causing the extremities of some of the nerves spread over it to be moved; and this motion passes to the other extremities of those nerves which are collected in the brain round the seat of the soul, as I have just explained quite fully enough in the fourth chapter of the *Dioptrics*. But the movements which are thus excited in the brain by the nerves, affect in diverse ways the soul or mind, which is intimately connected with the brain, according to the diversity of the motions themselves. And the diverse affections of our mind, or thoughts that immediately arise from these motions, are called perceptions of the senses, or, in common language, sensations.

206

How we may arrive at a knowledge of the figures,
magnitudes and motions of the insensible particles of
bodies

But since I assign determinate figures, magnitudes and motions to the insensible particles of bodies, as if I had seen them, whereas I admit that they do not fall under the senses, someone will perhaps demand how I have come to my knowledge of them. To this I reply that I first considered generally the most simple and best understood principles implanted in our understanding by nature, and examined the principal differences that could be found between the magnitudes, figures and situations of bodies insensible on account of their smallness alone, and what sensible effects could be produced by the various ways in which they impinge on one another. And finally, when I found like effects in the bodies perceived by our senses, I considered that they might have been produced from a similar concourse of such bodies, especially as no other mode of explaining them could be suggested. And for this end the example of certain bodies made by art was of service to me, for I can see no difference between these and natural bodies, excepting that the effects of machines depend for the most part on the operation of certain instruments, which, since men necessarily make them, must always be large enough to be capable of being easily perceived by the senses. The effects of natural causes, on the other hand, almost always depend on certain organs minute enough to escape every sense. And it is certain that there are no rules in mechanics which do not hold good in physics, of which mechanics forms a part or species so that all that is artificial is also natural; for it is not less natural for a clock, made of the requisite number of wheels, to indicate the hours, than for a tree which has sprung from this or that seed, to produce a particular fruit. Accordingly, just as those who apply themselves to the consideration of automata, when they know the use of a certain machine and see some of its parts, easily infer from these the manner in which others which they have not seen are made, so from considering the sensible effects and parts of natural bodies, I have endeavoured to discover the nature of the imperceptible causes and insensible parts contained in them.

207

That touching the things which our senses do not perceive,
it is sufficient to explain what the possibilities are about
the nature of their existence, though perhaps they are not
what we describe them to be

But here it may be said that although I have shown how all natural things can be formed, we have no right to conclude on this account that they were produced by these causes. For just as there may be two clocks made by the same workman, which though they indicate the time equally well and are externally in all respects similar, yet in nowise resemble one another in the composition of their wheels, so doubtless there is an infinity of different ways in which all things that we see could be formed by the great Artificer without it being possible for the mind of man to be aware of which of these means he has chosen to employ. This I most freely admit; and I believe that I have done all that is required of me if the causes I have assigned are such that they correspond to all the phenomena manifested by nature without inquiring whether it is by their means or by others that they are produced. And it will be sufficient for the usages of life to know such causes, for medicine and mechanics and in general all these arts to which the knowledge of physics subserves, have for their end only those effects which are sensible, and which are accordingly to be reckoned among the phenomena of nature. [. . .]

208

That nevertheless there is a moral certainty that everything
is such as it has been shown to be

But nevertheless, that I may not injure the truth, we must consider two kinds of certainty and first of all what has moral certainty; that is, a certainty which suffices for the conduct of life, though if we regard the absolute power of God, what is morally certain may be uncertain. So those who have never visited Rome do not doubt its being a city in Italy, although it may very well be that all those from whom they have heard about it have deceived them. Again, if, for instance, anyone wishing to read a letter written in Latin characters that are not placed in their proper order, takes it into his head to read B wherever he finds A and C where he finds B, thus substituting for each letter the one following it in the alphabet, and if he in this way finds that there are certain Latin words composed of these, he will not doubt that the true meaning of the writing is contained in these words, though he may discover this by conjecture, and although it is possible that the writer did not arrange the letters in this order of succession, but in some other, and thus concealed another meaning in it: for this is so unlikely to occur especially when the cipher contains many words that it seems incredible. But they who observe how many things regarding the magnet, fire, and the fabric of the whole world, are here deduced from a very small number of principles, although they considered that I had taken up these principles at random and without good grounds, they will yet acknowledge that it could hardly happen that so much would be coherent if they were false.

PIERRE GASSENDI

Atoms and the void: Epicurus revived

209

It is now advisable to look at the opinions of those who think it most likely that the first and universal matter, from which all things came, was atoms. To quote the words of Aneponymus,[1]

> There is no opinion so false that it has no truth mixed with it; but that truth is unfortunately obscured by the very mixture.

1. Aneponymus was the pseudonym of William of Conches (1080–1145) whose *Dragmaticon* was printed in 1567 as *Dialogus de substantiis physicis*. (Ed.)

He goes on,

> In that they say the cosmos is made out of atoms, the Epicureans are right; but it is
> nonsense to say, as they do, that these atoms were uncreated, and rushed hither and
> thither through the great void, finally being united in four great substances.

This is most suggestive: clearly there is nothing to stop one arguing the
hypothesis that the matter of the cosmos and the objects in it is atoms, provided
that everything false in this view is cut out. To put it more plainly, we must
do away with the notion that atoms are eternal and uncreated and infinite in
number, even granted any sort of shapes you like. Then it may be conceded
that atoms are the primal matter, which God in the beginning created finite,
which he shaped into this visible cosmos, which he ordered and permitted to
pass through its various changes, and out of which, finally, all bodies that are
in nature are made. Worked out in this fashion, there is nothing exceptionable
in the hypothesis, as there is in that of Aristotle, and of others who similarly
make the primal matter eternal and uncreated. The theory has this advantage
too, that it satisfactorily explains how aggregation and dissolution into the
original forms take place; how something comes to be a solid body, and large
or small, tenuous or dense, soft or hard, delicate or gross, and so forth. Attributes
of this sort are not so readily deducible from the other hypotheses, by which
matter is declared to be, for example, infinitely divisible; or uncompounded,
as they more or less say; or shaped, but not given enough variety of shapes;
or endowed with primary and secondary qualities, but qualities either insufficient
for or incongruous with the variety of natural objects.

Next, we must do away with the notion that atoms have a self-generated
motive power or impetus, a movement of such a sort that from all eternity they
have been making, and still do make, random fluctuations. We can then concede
that atoms are in motion, energized by that power of moving and acting which
God endowed them with at their creation, and through which He works, insofar
as His maintenance of all things requires His direction of them. At a stroke this
corrects the idea of self-generated power, and so indeed must be corrected the
hypotheses that attribute motion and action to matter, such as the Platonic, that
makes matter fluctuate blindly from all eternity. The motion of matter, then,
has to this extent been regularized by God. (It is apparently this idea of the
eternity of movement that makes Aristotle couple Plato with Leucippus, the
author of the atomic hypothesis, inasmuch as both of them say, 'There has always
been movement.') This corrected idea of ours has the advantage that it shows
clearly the origin and real root of all motion and action in causes normally
labelled secondary; other hypotheses cannot make this so clear, especially as
regards that essence which they would have as the origin of all motion and
action, because in one place they want whatever being it is to come from
matter, and in another, they say that matter is inert and quite without any motive
or active power. It is worth noticing that Plato, though not in so many words

using the idea of atoms, still described this essence as being of a fineness which only the intellect, not the senses, can perceive; and about these very small particles he says,

It is entirely reasonable that God should provide for their number, motion and other properties, insofar as nature, bowing to necessity, will endure it.

Accordingly we can assume that in the beginning God created a multitude of atoms, as great as was necessary for the formation of this entire cosmos. Not that God was obliged to create atoms separately, which He then would have united in larger and larger aggregations, till the cosmos was formed; but in creating a whole mass of matter, which was reducible to particles, and hence composed of the smallest bodies and ultimate particles, He must be assumed to have created these particles in the process. It can also be assumed that the atoms each took from their Creator, in however small a measure, their own weights, sizes and shapes in all their inconceivable variety, and likewise the impetus appropriate to each for motion, effort and change, as for being loosened, rising up, leaping forward, striking, repelling, bouncing back, seizing, enfolding, holding and retaining, etc., as He saw fit to give them for their assigned ends and purposes.

Again, we may suppose that when God at first bade land and sea bring forth life, and produce plants and animals, He made what one might call a seedbed containing all things generative: that is to say, He composed from selected atoms the seminal principles of all things, out of which afterwards the propagation of species would occur by generation. These seminal principles were scattered throughout the whole reproductive universe, not evenly, but as was appropriate in each area. And though they themselves may dissolve into their component atoms, the atoms may come together to form the seminal principles again, being as it were of the same nature, and sharing ways of combining and developing. Hence began that continual generation and corruption which still goes on and will always do so, at least while the mass of atoms continues unexhausted, supplying eternally both the matter from which bodies are formed and the motive force or cause by which they are formed. It is essential to grasp this point, in order to see that atoms are so constituted that all things are formed, not from other things, nor by reciprocal action, but from their own seminal principles; so that atoms are the first principles or primal matter of all things. [. . .] Moving on in this discussion, let me now examine whether the authors of the atomic hypothesis posited not merely atoms, but the void itself, as the original of all things. This is a common accusation, foisting on them the absurdity of creating things not just from matter, but from nothingness itself. Following that, I shall reply to the main objections to the atomist position, which, as they are brought up in their various ways, have been carefully collected by Lactantius.

The examination first: in the works of Servius, we find that Epicurus declared,

'There are two principles, Body and Void.' Plutarch, attacking Colotes, repeats the point: 'The principles of Epicurus are Infinity and the Void.' Nor is this confined to Epicurus, for Leucippus and (according to Clemens Alexandrinus) Metrodorus both say, 'There are two final principles, namely Body and Void'. Stobaeus confirms this of Metrodorus, and Cicero confirms it of Leucippus, adding the name of Democritus. Aristotle reports it of the first two, where he says in the *Metaphysics*, 'They allege that the basic principles are Body and Void,' and of Democritus specifically, where in the *Physics* he proves that the basic principles are of opposite natures, or where he says that Democritus 'believes in the Solid and the Void'. But taking this further, one cannot say that either Epicurus or the others felt that things are composed of these two principles, atoms and void. What has misled most people is that they posited these two uncreated and incorruptible elements, and declared that the universe, or the frame of nature, was constituted from these divided principles. But they did not on that account believe that these two principles actually are present as original component parts in all things which are subject to growth and decay. Granted that in some sense they may be called principles or elements of the universe, they are not thereby both principles or constituent elements of all generative matter. Only atoms enter into that; the void provides only place, and the means whereby one thing is separated from another. Being bodiless itself, it is incapable of begetting things which have body. Even if bodies are mingled with the void, that does not make it part of them, any more than the air in our nostrils, bones, blood or lungs is part of us. So little is it part of them, that it does not even stick to them, nor can it accompany them, but must constantly encounter changing matter as things change their places.

Lactantius has the following objection:

> Possibly where small objects are concerned, atoms may suffice to constitute them, but when it is a question of a whole cosmos—and Epicurus has said that the cosmos is made from atoms—the numbers required reach lunatic proportions.

Having raised the question of an infinity of worlds, he continues,

> What colossal force must have been in these atoms, to make masses so inconceivably large out of particles so small?

But really, if you imagine that the whole earth may be a part of some great cosmos, it will strike you at once that of course the entire earth is itself composed of masses in exactly this relation to the whole, like Mount Atlas or the Caucasus. By the same reasoning, the mountain is composed of many agglomerated masses of earth or rock; the rock is composed of stones, the stones of smaller stones, and the small stones from particles, like grains of sand. So there is nothing against the idea that the whole cosmos is composed of particles no bigger than grains of sand.

This being accepted, it has been clearly established long ago by the convincing argument of Archimedes that even if these grains of sand were so small that one poppy seed would make ten thousand of them, a number expressing the quantity of grains needed to make up the cosmos, as its extent was then popularly known, would need only fifty-two zeros (i.e. 10^{52}); and not more than sixty-four would be needed to express that number that would make up that incredibly vast cosmos posited by Aristarchus and Copernicus (10^{64}). To come down to the atomic scale: imagine one of these particles to be composed of 10^6 particles of the ultimate size; then multiply by 10^{64}, and the number of these ultimate particles will still need no more than seventy zeros (10^{70}). And if you think that even so we have not reached the limit of divisibility, split each of these again into 10^6 parts, and the resulting number will need only seventy-six zeros (10^{76}). Divide further if you like, and you will see how easily you can find figures to express even then the number of atoms needed to make up the cosmos.

Testing the law of inertia

210

Our friend [François] Luillier hinted in his last letter that you have shown some misgiving towards the things I had written to him about experiments made by me concerning the motion of projectiles. I am not surprised at this, as I have observed that nothing seems more unlikely, and I have known other outstanding men to whom the thing seemed no less improbable so far as they themselves had tested it. The noble Governor [of Provence, Louis Emmanuel de Valois, Comte d'Alais] recently gave me the opportunity of once again investigating the truth of the matter when, setting out for Marseilles, he wanted me included in the company. Assuredly, as he is extremely learned and devotes whatever time he has over from matters of state to liberal studies, he conducted on the journey several discussions on the subject of motion. I myself recounted on the way both my own observations and those which Galileo collected to add to the theorem: 'If the object on which we stand is moved all our movements and those of things movable by us happen and appear just as though it were stationary.'

Since not so much the learned Governor as some other members of the company thought that the matter was beyond all belief, they were amazed when, told to make the experiment both in the coach and on horseback, they realized that things thrown upwards returned to the hand just the same whether they themselves were stationary or being carried along at top speed; that things dropped from the hand fell in the same line as that of the horse or carriage; that things thrown out in front struck the ground neither nearer nor further away; that things pushed out backwards left no more quickly from the rear; that things aimed to the right or to the left were kept to a straight line on the level. Later, in order that they should be completely convinced, they had to be taken to sea to observe, on a swiftly moving and on a stationary ship, whether a stone thrown upwards from the foot of the mast along the length of the mast would always keep the same distance from that mast, both going up and coming down, and would fall back again at the foot of the mast, or on exactly the same spot; whether a stone dropped from a hand stationary at the top of the mast and without any force, would fall at that same foot; whether, when thrown with equal force from the stern to the bow and from the bow to the stern, the stone would appear to be borne by an equal impetus and to travel an equal distance; whether, if it were impelled crossways, sideways and for whatever reason precisely, a not unequal motion would be demonstrated. But the matter was so successful that they were no longer in doubt, since they observed these and other things on board a trireme sailing so fast that it covered four miles in the space of quarter of an hour on the open and calm sea. As a miracle the most impressive was that a stone dropped without force or thrown from the tip of the mast did not either leave the mast in the direction of the stern nor was it touched by it from the direction of the bow, but continued to fall at the same distance away, or along a line parallel to it.

EVANGELISTA TORRICELLI

The mechanical cause of the 'horror vacui'

211

My most Illustrious and Revered Patron,

A few weeks ago I sent certain demonstrations of mine concerning the area of the Cycloid to Signor Antonio Nardi, asking him, when he had looked at them, kindly to send them direct to you, Sir, or to Signor Magiotti.

I mentioned to you before that various physical experiments were being made with regard to the vacuum: not merely in order to produce a vacuum, but with a view to making an instrument that should show the changes in the air, which at certain times is heavy and gross and at others light and fine. Many people have said it is impossible to make a vacuum, others that it can be made, but only with difficulty and by going against Nature; I do not know, indeed, whether anyone has said that it can be obtained without difficulty, and without going against Nature. My argument was as follows: if one were to find an obvious cause to which to attribute the resistance which one encounters in trying to make a vacuum, then it would be vain to seek to attribute to the vacuum itself these effects which clearly derive from other causes. Furthermore, by making a few simple calculations, I found that the cause adduced by me (i.e. the weight of the air) might alone be expected to create more resistance than in fact we encounter when seeking to create a vacuum. I say this lest some Philosopher, seeing himself bound to admit that the weight of the air causes the resistance which we encounter when making a vacuum, rather than concede the importance of the part played by the ¦weight of the air, should persist in asserting that Nature too concurs in resisting the creation of a vacuum.

We live submerged at the bottom of a sea of elemental air, the weight of which has been demonstrated beyond doubt: we know in fact that the heaviest air near the earth's surface weighs about the four-hundredth part of the weight of water. The authors of the *Crepusculi*[1] have affirmed, moreover, that the visible and vaporous air rises above us to a height of approximately fifty or fifty-four miles. I am not entirely convinced of this, since I could show that, if this were the case, the creation of a vacuum ought to give rise to far more resistance than in fact it does. But they might have recourse to the argument that the weight noted by Galileo refers to the lowest air, inhabited by men and beasts, and not to the air above the mountain tops, where it is extremely pure, and much lighter than the four-hundredth part of the weight of water.

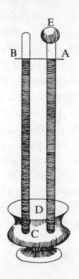

Torricelli's barometer

We made a number of glass vessels, similar to these marked *A* and *B* in the diagram above, and with necks four feet long. These we filled with mercury, stopped their mouths with a finger, and turned them upside down in a basin containing mercury (*C*). When we did this we could see the mercury leave the vessel, without anything particular happening inside the vessel as it did so; and, moreover, the neck *AD* of the vessel in every case remained full of mercury to a

---

1. *Liber de crepusculis* was written in the 11th century by Ibn Muadh, but was usually attributed to Alhazen. Its author attempts to estimate the height of the humid atmosphere from the apparent altitude of the sun at sunrise and sunset. (Ed.)

height of two feet six inches. To show that the body of the vessel was completely empty we filled the basin underneath it with water as far as *D*, and then raised it very gradually until its mouth reached the level of the water: when this happened we saw the mercury run down its neck, and the water rush up with fearful force to fill it, right up to the mark *E*.

We conducted our argument whilst the vessel *AE* remained empty, and the mercury was supported, in spite of its great weight, in the neck of the vessel *AC*. Up to now it has been thought that the force which sustained the mercury, against its natural inclination to fall, was contained inside the vessel *AE*, and belonged to the vacuum, or to the highly rarefied substance, there enclosed. But I claim that this force is external, and comes from outside the vessel. The air weighing on the surface of the mercury in the basin rises to a height of fifty miles. What wonder, then, if, inside the glass *CE*, where it encounters neither attraction nor resistance, since there is nothing there, the mercury rises to the height at which it balances the weight of the air pushing on it from outside. Moreover, water, in a similar but much taller vessel, will rise to a height of nearly thirty-six feet, i.e. as much higher than mercury as mercury is heavier than water, in order to reach equilibrium with the same pressure felt by both.

My argument was confirmed by another experiment, using simultaneously the vessel *A* and the tube *B*. The fact that the mercury stopped at the same level *AB* in both vessels indicates almost without a doubt that the virtue sustaining it was not inside them. For the vessel *AE* ought, in the latter case, to have had more force than the tube *B*, owing to the greater quantity of rarefied matter contained by it, which should have given it a far more vigorous power of attraction than that offered by the tiny space *B*. I have also tried out this principle in the explanation of all the sorts of resistance encountered in the various effects attributed to the vacuum. Up to now I have not met with any difficulties in this. I know that many objections will occur to you with regard to this argument, but hope, too, that the more you think about it the less you will object to it.

BLAISE PASCAL

212

*The mass of the air has weight, and with this weight
presses upon all the bodies it surrounds*

It is no longer open to discussion that the air has weight. It is common knowledge that a balloon is heavier when inflated than when empty, which is proof enough. For if the air were light, the more the balloon was inflated, the lighter the whole would be, since there would be more air in it. But since, on the contrary, when more air is put in, the whole becomes heavier, it follows that each part has a weight of its own, and consequently that the air has weight.

Whoever wishes for more elaborate proofs can find them in the writings of those who have devoted special treatises to the subject.

If it be objected that air is light when pure, but that the air that surrounds us is not pure, being mixed with vapour and impurities which alone give it weight, my answer is brief: I am not acquainted with 'pure' air, and believe that it might be very difficult to find it. But throughout this treatise I am referring solely to the air such as we breathe, regardless of its component elements. Whether it be compound or simple, that is the body which I call the air, and which I declare to have weight. This cannot be denied, and I require nothing more for my further proof.

This principle being laid down, I will now proceed to draw from it certain consequences.

1. Since every part of the air has weight, it follows that the whole mass of the air, that is to say, the whole sphere of the air, has weight, and as the sphere of the air is not infinite in extent, but limited, neither is the weight of the whole mass of the air infinite.

2. The mass of the water of the sea presses with its weight that part of the earth which is beneath it; if it surrounded the whole earth instead of only a part, its weight would press upon the whole surface of the earth. In the same way, since the mass of the air covers the whole face of the earth, its weight presses upon the earth at every point.

3. Just as the bottom of a bucket containing water is pressed more heavily by the weight of the water when it is full than when it is half empty, and the more heavily the deeper the water is, similarly the high places of the earth, such as the summits of mountains, are less heavily pressed than the lowlands are by the weight of the mass of the air. This is because there is more air above the lowlands than above the mountain tops; for all the air along a mountain side presses upon the lowlands but not upon the summit, being above the one but below the other.

4. Bodies immersed in water are pressed on all sides by the weight of the water above them, as we have shown in the *Treatise on the Equilibrium of Liquids*. In the same way bodies in the air are pressed on all sides by the weight of the air above them.

5. Animals in water do not feel its weight, neither do we the weight of the air and for the same reason. Just as it would be a mistake to infer that, because we do not feel the weight of the water when immersed in it, water has no weight; so it would be a mistake to infer that air has no weight because we do not feel its pressure. We have shown the reason for this in the *Treatise on the Equilibrium of Liquids*.

6. If there were collected a great bulk of wool, say twenty or thirty fathoms high, this mass would be compressed by its own weight; the bottom layers would be far more compressed than the middle or top layers, because they are pressed by a greater quantity of wool. Similarly the mass of the air, which is a compressible and heavy body like wool, is compressed by its own weight, and

the air at the bottom, in the lowlands, is far more compressed than the higher layers on the mountain tops, because it bears a greater load of air.

7. In the case of that bulk of wool, if a handful of it were taken from the bottom layer, compressed as it is, and lifted, in the same state of compression, to the middle of the mass, it would expand of its own accord; for it would then be nearer the top and subjected there to the pressure of a smaller quantity of wool. Similarly if a body of air, as found here below in its natural state of compression, were by some device transferred to a mountain top, it would necessarily expand and come to the condition of the air around it on the mountain; for then it would bear a lesser weight of air than it did below. Hence if a balloon, only half inflated—not fully so, as they generally are—were carried up a mountain, it would necessarily be more inflated at the mountain top, and would expand in the degree to which it was less burdened. The difference will be visible, provided the quantity of air along the mountain slopes, from the pressure of which it is now relieved, has a weight great enough to cause a sensible effect.

There is so necessary a bond between these consequences and their principle that if the principle is true the consequences will be true also. Since, therefore, it is acknowledged that the air, reaching from the earth to the periphery of its sphere, has weight, all the conclusions we have inferred from this fact are equally correct.

But, however certain these conclusions may be deemed, it appears to me that all who accept them would nevertheless be eager to see this last consequence confirmed by experiment, because it involves all the others and indeed directly verifies the principle itself. There is no doubt that if a balloon such as we have described were seen to expand as it was lifted up, the conclusion could not be avoided that the expansion was due to a pressure, which was greater below than above. Nothing else could cause that expansion, the more so as the mountains are colder than the lowlands. The compression of the air in the balloon could have no other cause than the weight of the mass of the air, since this air was taken in its actual condition at low altitudes and was uncompressed, the balloon being even limp and only half inflated. This would be proof positive that air has weight; that the mass of the air is heavy; that its weight presses all the bodies it contains; that its pressure is greater on the lowlands than on the highlands; that it compresses itself by its own weight, and is more highly compressed below than above. And, since in physical science experience is far more convincing than argument, I do not doubt that everyone will wish to see this reasoning confirmed by experiment. Moreover, should the experiment be performed, I should enjoy this advantage: that if no expansion of the balloon were observed even on the highest mountains, my conclusions, nevertheless, would not be invalidated; for I might then claim that the mountains were still not high enough to cause a perceptible difference. Whereas if a considerable and very marked change occurred, say of one-eighth or one-ninth in volume, the proof, to me, would be absolutely convincing, and there could remain no doubt as to the truth of all that I had asserted.

But I delay too long. It is time to say, in a word, that the trial has been made and with the following successful result.

An experiment made at two high places, the one about 500 fathoms higher than the other
If one takes a balloon half-filled with air, shrunken and flabby, and carries it by a thread to the top of a mountain 500 fathoms high, it will expand of its own accord as it rises, until at the top it will be fully inflated as if more air had been blown into it. As it is brought down it will gradually shrink by the same degrees, until at the foot of the mountain it has resumed its former condition.

This experiment proves all that I have said of the mass of the air, with wholly convincing force; but it must be fully confirmed, since the whole of my discourse rests on this foundation. Meanwhile it remains to be pointed out only that the mass of the air weighs more or less at different times, according as it is more charged with vapour or more contracted by cold.

Let it then be set down, (1) that the mass of the air has weight; (2) that its weight is limited; (3) that it is heavier at some times than at others; (4) that its weight is greater in some places than in others, as in lowlands; (5) that by its weight it presses all the bodies it surrounds, the more strongly when its weight is greater.

213

Conclusion

I have recorded in the preceding Treatise all the general effects which have been heretofore ascribed to nature's effort to avoid a vacuum, and have shown that it is utterly wrong to attribute them to that imaginary cause. I have demonstrated, on the contrary, by absolutely convincing arguments and experiments that the weight of the mass of the air is their real and only cause. Consequently, it is now certain that nature nowhere produces any effects in order to avoid a vacuum.

It is not difficult to demonstrate, furthermore, that nature does not abhor a vacuum at all. This manner of speaking is improper, since created nature, which is the nature under consideration, is not animated, and can have no passions. Such language is in fact metaphorical, and means nothing more than that nature makes the same efforts to avoid a vacuum as if she abhorred it. Those who use this phrase mean that it is the same thing to say that nature abhors a vacuum as to say that nature makes great efforts to prevent a vacuum. Now, since I have shown that nature does nothing at all to avoid a vacuum, the conclusion is that nature does not abhor it. To carry out the metaphor: just as we say of a man that a thing is indifferent to him when his actions never betray any movement of desire for, or of aversion to, that thing, so should we say of nature that it is supremely indifferent to a vacuum, since it never does anything either to seek or to avoid it. (I am here still using the word 'vacuum' to mean a space empty of all bodies which our senses can apprehend.) [. . .]

So positively was this believed that philosophers have made it one of the most general principles of their science and the foundation of the treatises on the vacuum. It is and has been didactically asserted every day in all the schoolrooms in the world, ever since books were written. Everyone has firmly believed it, and it has remained uncontradicted down to our own time.

This fact perhaps may open the eyes of those who dare not doubt an opinion which has always been universally entertained; for simple workmen have been able to prove in this instance that all the great men we call philosophers were wrong. Galileo declares in his dialogues that Italian plumbers taught him that water rises in pumps only to a certain height: whereupon he himself confirmed the statement as others did also, afterward, first in Italy and later in France, by using quicksilver, which is easier to handle but provides merely several other ways of making the same demonstration.

Before men gained that knowledge, there was no incentive to prove that the weight of the air was the cause of water rising in pumps; since, the weight of the air being limited, it could not produce an unlimited effect.

But all these experiments were insufficient to show that the air does produce those effects: they had rid us of one error but left us in another. They taught us, to be sure, that water rises only to a certain height, but they did not teach us that it rises higher in low-lying places. On the contrary, the belief was held that it always rises to the same height, in every place on the earth. And since the weight of the air never entered anybody's head, it was vaguely thought that the nature of the pump was such that it lifted water to a limited height and no further. Indeed Galileo took that to be the natural height of a pump, and called it *la altessa limitatissima*. How indeed could it have been imagined that that height was different in different places? Certainly, it would seem improbable. Yet that last error again put out of the question the proof that the weight of the air causes these effects; since, because this weight would be greater at the foot than at the top of a mountain, its effects, obviously, would be proportionately greater there.

That is why I decided that the proof could be obtained only by experimenting in two places, one some four or five hundred fathoms above the other. I chose for my purpose the Puy-de-Dôme mountain in Auvergne, for the reasons that I have set forth in a little paper which I printed as early as the year 1648, immediately after the experiment had proved successful.

This experiment revealed the fact that water rises in pumps to very different heights, according to the variation of altitudes and weathers, but is always in proportion to the weight of the air. It perfected our knowledge of these effects, and put an end to all doubting; it showed their real cause, which was not abhorrence of a vacuum, and shed on the subject all the light that could be wished for.

Try now to explain otherwise than by the weight of the air why suction pumps do not raise water so high by one-quarter on the top of Puy-de-Dôme in Auvergne as at Dieppe; why the same siphon lifts water and draws it over at Dieppe and not at Paris; why two polished bodies in close contact are easier to separate on a steeple than on the street level; why a completely sealed bellows is easier to open on a house-top than in the yard below; why, when the air is more heavily charged with vapours, the piston of a syringe is harder to withdraw; and lastly why all these effects are invariably proportional to the weight of the air, as effects are to their cause.

Does nature abhor a vacuum more in the highlands than in the lowlands? In damp weather more than in fine? Is not its abhorrence the same on a steeple, in an attic, and in the yard? Let all the disciples of Aristotle collect the profoundest writings of their master and of his commentators in order to account for these things by abhorrence of a vacuum if they can. If they cannot, let them learn that experiment is the true master that one must follow in Physics; that the experiment made on mountains has overthrown the universal belief in nature's abhorrence of a vacuum, and given the world the knowledge, never more to be lost, that nature has no abhorrence of a vacuum, nor does anything to avoid it; and that the weight of the mass of the air is the true cause of all the effects hitherto ascribed to that imaginary cause.

SECTION IV

The Royal Society to Laplace

Section IV: The Royal Society to Laplace

INTRODUCTION

DURING the century and a half from the foundation of the Royal Society to the early 1800s, modern science began to crystallize and to establish itself in the shape of what is usually called the classical era, which came to an end at the beginning of this century. From about the middle of the seventeenth century regular meetings began to take place in London and Oxford of groups of professional and amateur scientists whose discussions were to lead to a turning-point in the history of the relations between science and society. In 1662 King Charles II gave official sanction to this activity by granting a charter which recognized and gave moral and material support to the academic body henceforth to be called the Royal Society of London. The second half of that century saw the first beginnings of other academies of science in Europe. In Italy, one such was founded in Florence in 1657, mainly by pupils of Galileo; it was, however, forced to close ten years later, in part because of pressures from the Church. There were similar developments in France, the French Academy of Sciences being founded in Paris in 1667 by Louis XIV, in Prussia in 1700, this organization having its seat in Berlin, and in Russia in St Petersburg (1724).

Royal and state recognition of science as an activity useful to society provided such research with considerable momentum and vastly enhanced its prestige, giving its work a formal organizational basis. In this period too useful international scientific communications developed and strengthened, largely by means of the various scientific periodicals published by the academies.

Thomas Sprat's book *The History of the Royal Society*, published five years after the group's charter, is a most interesting document reflecting as it does the new scientific spirit and its approach to experimental activity. The first part contains a brief survey of the earlier history of science; the second opens with the words of the Royal Charter empowering the Society to carry out various activities useful for the advancement of science. This part of the book also contains excerpts or summaries of several investigations carried out by Fellows of the Society, as well as the outstanding proposal of Robert Hooke for systematic research in meteorology. The third part is essentially an apologia of experimental research, written in true Baconian spirit. The profound influence exerted by Francis Bacon on this generation of scientists is noticeable throughout the whole book, for example in the answer to the arguments regarding the uncertainties of experiments.

One of the principal founding Fellows of the Royal Society was Robert Boyle. His article, *The Spring of the Air* (1660), was published in the form of a letter to his nephew. The investigations of Boyle and Hooke, together with the earlier ones of Torricelli and Pascal, made a decisive contribution to the under-standing of the physical properties of air. Boyle gave two possible explanations for the elasticity of air—the elasticity of its primary particles, and the vortex-like state of motion of the aether in which the particles were conceived to be supported. From the time of Descartes to the end of the nineteenth century, the aether was an essential component in the theory of many diverse physical phenomena. In the course of time these explanations became so complicated that scientists were forced to abandon the hypothesis altogether. In his article Boyle mentioned too the use he made of the air pump built in 1658, a few years after that of Guericke in Magdeburg. This pump also played a part in his discovery of the pressure-volume behaviour of gases, expressed in the law now commonly named after him.

In the Society's *Transactions* for the year 1675 there appeared a short paper by the Danish astronomer Olaf Rømer. His studies of the eclipses of one of Jupiter's satellites provided convincing evidence for the finite speed of propaga-tion of light. The fact that an event is detected only at the moment when the radiation announcing it has reached the observer created a revolution in Man's consciousness of time, a revolution which came to a climax in Einstein's state-ments concerning the relativity of the simultaneity of events. Christiaan Huygens, the great Dutch scientist (whose work on 'centrifugal' force, the pendulum, and the collision of elastic bodies helped to prepare the ground for Newtonian physics), referred to Rømer's discovery in his important treatise on light. Huygens' hypothesis of the wave-like propagation of light was based partly on a model of collisions between the elastic particles of which the medium of this propagation was supposed to consist. Huygens' approach is typical of the mechanistic view of his generation. Although the propagation of light, like that of sound, is in effect the propagation of a certain physical state, it can be explained easily by a mechanical model; Huygens stressed the necessity for such a mechanical conception also for non-mechanical phenomena. His hypo-thesis of elementary waves was the first effective step in the development of the wave theory of light so successfully extended by Thomas Young and Augustin Fresnel in the early nineteenth century, and a few decades later by Maxwell's electromagnetic theory. Newton, however, rejected Huygens' hypothesis, and because of his overriding authority this long constituted a serious handicap to the further progress of the wave theory.

Within the frame of an anthology such as this, it is difficult to provide an adequate account of Newton's work or to do full justice to his historical signifi-cance. Even today, in the second half of the twentieth century, one can still without exaggeration call him the greatest mathematical physicist and one of the foremost experimentalists in the history of physics. Newton's personality was a highly complex one, involving many contradictory features, and histor-ians of science have tended to emphasize one aspect or another according to the philosophy prevailing in their time. This Anthology contains passages from two of Newton's chief works, *Philosophiae Naturalis Principia Mathematica* (1687) and *Opticks* (1704). His short preface to the first edition of the *Principia* clearly demonstrates his well-defined views on the task of the scientist in the explanation of the phenomena of nature. Newton's work itself is the most clear and convincing example of his conception that mechanics can be developed and presented mathematically as a consistent system capable of application to all dynamical phenomena. It must be remembered too that Newton's success was all the more astonishing in that in spite of being an inventor of calculus, he composed the mathematical part of his book by drawing exclusively on that classical method of synthetic geometry which had been in use since the Hellenistic period of ancient Greece. The formal structure of the *Principia* shows Newton's endeavour to transform science into a mathematical discipline. In close analogy to geometrical description, his study commenced with definitions of physical quantities, including statements on the nature of space, time and motion. From there he progressed to axioms, such as his 'Laws of Motion', and to derivation of theorems which are proved mathematically and which evolve from one another with a gradual increase of complexity. The validity of Newton's three 'Laws of Motion', the backbone of his theory of force, depends on certain metaphysical assumptions concerning the absolute character of space and time. In his famous experiment of the rotating vessel, Newton attempted to establish practically the existence of absolute space. However, the questions which were subsequently raised about empty space as such as the cause of physical actions were never satisfactorily answered until Mach's criticism and Einstein's theory led to a new conception of it. Newton himself had no doubts about absolute space, for he regarded it as the sensorium of God insofar as it constituted a confluence between the supreme metaphysical being and physical events. Moreover the tremendous success of his concept of force as the cause of acceleration provided for him a convincing confirmation of his presuppositions.

Newton's conception culminated in his theory of universal gravitation. Here he proved mathematically the identity of the cause which accelerates freely falling bodies near the surface of the earth with that which keeps the moon in its path around it, as well as the planets and comets in theirs around the sun. His theory also explained the tidal movements of the oceans as being caused in the first instance by the gravitational attraction of the moon. Newton's laws of dynamics can logically be linked with the three kinematical laws of Kepler, thus clearly demonstrating that Newton himself complied with the rules of scientific reasoning recommended by him to his fellow scientists. In his essay *On the System of the World*, a supplement to the third part of the *Principia*, he anticipates the idea of an artificial satellite (one which became reality in 1957) by explaining how a missile projected with sufficient initial speed will not fall to ground but will follow an orbit around the earth.

Without expressly mentioning Descartes, Newton attacked his hypothesis of the aether vortices and proved mathematically that it would lead to results contrary to Kepler's third law. Newton is careful not to draw excessive conclusions from his mathematico-physical methods. The gist of his case is that he succeeded in representing many important phenomena as consequences of his mathematical laws of force, based on mechanical principles. But he does not claim to have discovered the 'real' nature of all these forces, in particular that of gravity. He does not feel in other words that he has succeeded in explaining this force by a mechanical model, as Descartes and other mechanistic scientists asserted of theirs. In the *General Scholium* at the end of his book there is the often quoted sentence '*hypotheses non fingo*'. However, the term 'hypothesis' is not without ambiguity.

Descartes' hypothesis of the aether had to be rejected because it could not conform with observed phenomena. In other parts of the *Principia* however, and in particular in the queries at the end of his *Opticks* (published about twenty years later), there are passages which reveal that throughout his long career Newton never ceased to ponder over the 'real', or mechanical, nature of the force of gravity. He even attempted to explain it by the aether hypothesis and with the assumption that density gradients in this tenuous stuff could produce gravitation as well as the refraction of light. More than that, he expressed opinions that the possible explanation of other physical and even physiological phenomena (such as transmission of sensory information through the nerves) could lie in the aether hypothesis. To summarize—Newton indeed framed hypotheses, like any other scientist, but he refrained from publishing them as fully fledged theories because he was not able to verify them directly or indirectly by correlation with known facts or by discovering new facts that could be predicted by them.

Newton's constant doubts and ponderings provide a sharp contrast to the introduction to the second edition of the *Principia*, written by his pupil Robert Cotes. Cotes' interpretation of Newton's ideas is typical of the image which the Newtonian school created of its founder's views, similar to the dogmatism which the followers of Marx ascribed to *his* doctrine. The positivistic trend of the eighteenth and nineteenth centuries also had its share in the creation of a distorted and simplified picture of Newton's scientific personality. The passages quoted from the last part of the *Principia* indicate clearly that Newton did not believe that the marvellous regularity of the solar system could be explained by his law of gravitation alone; he attributed it to the wisdom of the Creator. Seventy years later Kant in his cosmogony attempted to prove this to be the product of an evolutionary process based on mechanical phenomena which in principle can be derived from Newton's laws.

Newton's classical experiments with the prism, and his explanations for the colours produced by the refraction of light and for the optical phenomena of laminar layers, superseded the former theories. This work thus helped to pave the way for the wave theory of light of the nineteenth century. Despite his belief in the corpuscular nature of light, Newton carefully avoided committing himself in this respect; his felicitous expression 'fits of easy reflection and trans-mission', alluding to the periodical phenomena, did not prevent progress towards acceptance of the wave conception of light.

Although he 'stood on the shoulders of giants', Newton was without doubt the father of modern physics as a science striving for comprehensive explanations of physical phenomena with their unification in a single system of thought. Most physical research in the eighteenth century involved an elaboration of subjects connected with Newton's work, especially that in his *Principia*. These investigations established what we have come to call today 'classical mechanics'. Among the few critical reactions to Newton's metaphysical assumptions were the arguments of his great contemporary and rival Leibniz against his views of absolute space and time, as developed in his controversy with Clarke, Newton's champion. The ideas of modern physics have vindicated Leibniz's arguments as against those of Clarke, and thus in effect against Newton's views in the matter.

Generally speaking, the work of the eighteenth-century physicists did not lead to basic innovations but provided considerable formal improvements for Newton's theory. This activity was concerned mainly with the establishment of general principles from which his theory could be derived in a more elegant and fruitful way, with the application of calculus to such developments, and with the extension of the laws of dynamics to the movement of rigid bodies. The vectors 'force' and 'quantity of movement', the main dynamical concepts of the *Principia*, were supplemented by 'work' and '*vis viva*' (kinetic energy) which, being scalars, lend themselves more easily to mathematical treatment. D'Alembert, in his essay on dynamics, showed the futility of the argument between Leibniz, who regarded '*vis viva*' as the basic concept, and the Cartesians, who gave preference to 'quantity of movement'. Both expressions are nothing more than different aspects of the same fundamental idea in dynamics. The important researches of d'Alembert, Johann Bernoulli, Euler, and Lagrange led to the systematization of Newtonian mechanics; this eventually culminated in the principle of least action and the law of conservation of energy.

Bernoulli and d'Alembert attempted to reduce the principles of dynamics to the laws of statics which had already developed in the preceding century. To this end they made use of the concept of 'virtual velocities'—velocities which can be reconciled with the conditions of a system in a state of equilibrium. Some passages quoted in this Anthology give a hint of these lines of thought, but detailed description here would take us too far afield.

The principle of least action turned out to be the most comprehensive of all the principles of dynamics. In its most simple form it states that of all the possible paths which a body can follow in the interval between two given times, the actual one is characterized by being the one for which the physical quantity called the action of the body has its minimum value. Action is the product of the quantity of movement of the body and the distance it covers during the interval of time in question. In its final formulation the principle is named after Hamilton, who gave it the mathematical expression now in use; even today, in an age of relativity and quantum mechanics, it continues to be of primary importance. For the historian of ideas the principle of least action is of special interest, because its birth can be connected with the teleological conception which was the cornerstone of Aristotle's philosophy of nature. In 1712, Leibniz expressed the view that in the investigation of nature the search after mechanical causes (what Aristotle called 'efficient causes') alone will not do; one has rather to go further and to penetrate to the final causes, the true manifestations of the wisdom of God. The ideas of Leibniz were extended by the French scientist Maupertuis who maintained that he had actually expressed this final cause in his principle of least action—the fact that the actual path of a body is unique by reason of its simplicity indicates the forethought of the Creator. Although this teleological interpretation was not accepted by physicists, the principle of least action retains its importance because of its universal applicability in all disciplines of physical science.

Another subject of supreme importance for science began to rise in prominence in the eighteenth century; this is that of probability. Mathematical probability had its first beginnings in the seventeenth century, when it was applied to games of chance and when statistical data relating to mass phenomena (such as mortality) were collected. Jakob Bernoulli's book *Ars conjectandi* gave to this aspect of probability an analytical expression which became famous under the name of the 'law of large numbers'. Another member of the family, Daniel Bernoulli, was the first to introduce statistical ideas to the consideration of a physical phenomenon, when he explained the pressure of a gas in a closed vessel by the random motion of its particles and their impinging on the walls of the vessel.

At the end of the eighteenth century Laplace developed the first scientific theory of probability, one based on logical principles. In an essay brilliant in its clarity and style, he considers the philosophical aspect of probability. He opens with his deterministic *credo*, the same as that of the whole Newtonian period. Mathematical astronomy (to which Laplace contributed so much by his theory of perturbation) provides clear evidence for the truth of the basic tenets of determinism. This is that knowledge of the positions and velocities of all bodies in the universe at a certain moment, together with the forces acting on them, will allow one to determine its state at any other moment. A supreme being (the 'Laplacian intellect') in possession of all these data for a given instant of time would thus be able to describe every future event by evolution from them. Similarly earlier events may be known, because the present is a necessary effect of each one. Only lack of such knowledge can prevent us from grasping the concatenation of events in all their details. Laplace saw in the calculus of probability a substitute for the law of causality, thus enabling us to acquire statistical knowledge at least of repetitive events or of mass phenomena. The conception remained in force until the advent of quantum mechanics. This regards probability as a primary concept, and therefore has abolished the classical notion of causality.

Another important sector in the physical sciences commenced its development proper in the eighteenth century. This was cosmology, the science concerned with the structure and evolution of the universe. The early workers were content to produce hypotheses about the origin of the solar system. In the twentieth century, however, the subjects received a fresh impetus when the various theories about the development of the galaxies, and about the cosmology of the universe as a whole, began to be discussed. The father of modern cosmology is Immanuel Kant, whose book, *The History of Nature and Theory of the Heavens*, was published in 1755. According to him, the solar system developed from a primordial state of chaos. He assumed this chaotic state to consist of a vast distribution of dust particles within the volume delimited by the orbit of Saturn (the most distant planet known in his time). In the regions of higher density arising from chance fluctuations the sun, and later the planets, were formed by a gradual aggregation caused by the gravitational attractions of the particles. There are several weaknesses in Kant's explanation. It requires a sense of axial rotation of the planets opposite to that of their revolution around the sun, in contradiction to observation. Kant also sinned against the law of conservation of angular momentum by assuming that all the dust particles eventually start to rotate in one sense. Today, after the principle of conservation of parity has been discarded in whole areas of atomic and sub-atomic physics, we may tend perhaps to be more lenient towards this transgression in the macroscopic sphere. Kant's cosmology is of great historical interest, in particular his apologia for his purely mechanistic explanation of cosmic evolution. This, he said, would not lead to atheism because it strengthened the belief in the existence of a Creator who was able to imbue matter with all the properties enabling it to transform itself from a chaotic state into a state of order. Laplace (who did not know of Kant's work) refrained in his cosmology from making the mistakes of the other. The gist of *his* hypothesis was that the primordial state of the solar system was that of a slowly rotating tenuous sphere of gas. The subsequent contraction of this sphere, its increasing angular velocity, and the interplay of gravitational and centrifugal forces eventually lead to the formation of rings (similar to those of Saturn) and their breaking up into planets. Laplace's theory served as a model for many subsequent hypotheses.

Euler's greatest contribution to physics like that of Lagrange was the derivation of the differential equations of mechanics which later could be shown to be the consequence of Hamilton's principle. He was also a great popularizer of science, as proved by his *Letters to a German Princess*. In his explanation of light he made use of the well-known analogies between the propagation of sound in air and that of light in the aether. The results are beautiful, but in many respects too simplistic. In Euler's times, however, there flourished hypotheses about all sorts of subtle fluid and imponderable substances. An example is the electric fluid, which was a central concept in the first theories of electricity of the days of Franklin and of some French and British scientists. Some of these hypotheses are reflected in Euler's explanation of the connection between electricity and aether density.

The excerpts from the scientific literature of the eighteenth century include in addition a passage from the work of the Croat Jesuit Roger Boscovich, published in 1763. Boscovich's view on the nature of matter differed from that generally accepted since the days of Gassendi. He did not believe in material particles of finite extension but regarded atoms as pointlike centres of force in the space continuum, immaterial seats of forces of attraction and of repulsion varying in intensity with distance. Boscovich's ideas in fact represent a continuum theory of matter in which the atoms are singularities, the source of all physical action in the continuum. His work is of historical significance because of the great influence it exerted on Faraday, who in turn developed the field concept which to a large extent governed the physical science of the nineteenth century.

IV: THE ROYAL SOCIETY TO LAPLACE

THOMAS SPRAT

The foundation of the Royal Society

214

Charles *the second, by the Grace of God, of* England, Scotland, France, *and* Ireland
King, Defender of the Faith, &c. *To all unto whom these presents shall come, Greeting.
Having long resolv'd within our self to promote the welfare of Arts and Sciences, as well
as that of our Territories and Dominions, out of our Princely affection to all kind of
Learning, and more particular favour to Philosophical Studies. Especially those which
indeavour by solid Experiments either to reform or improve Philosophy. To the intent
therefore that these kinds of study, which are no where yet sufficiently cultivated, may
flourish in our Dominions; and that the Learned world may acknowledge us to be, not
only the Defender of the Faith, but the Patron and Encourager of all sorts of useful
knowledge.*

*Know ye, that we out of our special Grace, certain knowledge, and meer motion, have
given and granted, and do by these presents give and grant for us, our Heirs, and
Successors, That there shall be for ever a Society, consisting of a President, Council, and
Fellows, which shall be called by the name of the President, Council, and Fellows of the
Royal Society of* London, *for and improving of Natural knowledge, of which Society
we do by these presents declare our self to be Founder and Patron. And we do hereby
make and constitute the said Society by the name,* &c. *to be a Body corporate, to be
continued under the same name in a perpetual succession;* [. . .]

*Furthermore, Libertie is granted to the said Society, lawfully to make and hold
meetings of themselves, for the searching out and discovery of Natural Things, and
Transaction of other Businesses relating to the said Society, when and as often as shall
be requisite, in any Colledge, Hall, or other Convenient place in* London, *or within 10
miles thereof.* [. . .]

*Moreover, on the behalf of the Society, it is granted unto the President and Council,
that they may assemble and meet together in any Colledge, Hall, or other convenient
place in* London, *or within ten miles thereof (due and lawful summons of all the
Members of the Council to extraordinary meetings being always premised) and that they
being so met together, have full power and authority from time to time, to make,
constitute, and establish such Laws, Statutes, Orders, and Constitutions, which shall
appear to them to be good, useful, honest, and necessary, according to their judgments and
discretions, for the Government, Regulation and Directions of the Royal Society, and
every Member thereof: And to do all things concerning the Government, Estate, Goods,
Lands, Revenues, as also the Businesses and Affairs of the said Society: All which
Laws, Statutes, Orders,* &c. *so made, His Majesty so wills and commands, that they*

be from time to time inviolably observed, according to the tenor and effect of them: Provided that they be reasonable and not repugnant or contrary to the Laws, Customs, &c. of his Kingdom of England. [. . .]

And for the greater advantage and success of the Society in their Philosophical Studies and Indeavours, full Power and Authority is granted unto them, to require, take, and receive, from time to time, dead bodies of Persons executed, and the same to anatomize, to all intents and purposes, and in as ample manner and form as the Colledge of Physitians, and Company of Chirurgions of London (*by what names soever the said two Corporations are or may be called*) *have had and made use of, or may have and use the said Bodies.*

And for the improvement of such Experiments, Arts, and Sciences as the Society may be imploy'd in, full Power and Authority is granted unto them from time to time by Letters under the hand of the President in the presence of the Council, to hold Correspondence and Intelligence with any Strangers, whether private Persons, or Collegiate Societies or Corporations, without any Interruption or Molestation whatsoever: Provided that this Indulgence or Grant be extended to no further use than the particular Benefit and Interest of the Society, in Matters Philosophical, Mathematical, and Mechanical.

Full Power and Authority is also granted on the behalf of the Society to the Council, to erect and build one or more Colledges within London, *or ten miles thereof, of what form or quality soever, for Habitation, Assembling, or Meeting of the President, Council and Fellows, about any affairs and businesses of the Society.* [. . .]

215

An objection answered concerning the uncertainty of experiments

As I am now passing away from their *Experiments*, and *Observations*, which have been their proper, and principal work: there comes before me an *Objection*, which is the more to be regarded, because it is rais'd by the *Experiments* themselves. For it is their common complaint, that there is a great *nicety*, and *contingency*, in the making of many *Experiments*: that their success is very often various, and inconstant, not only in the hands of *different* but even of the same *Triers*. From hence they suggest their fears, that this continuance of *Experimenters*, of which we talk so much, will not prove so advantageous, though they shall be all equally cautious in *observing*, and faithful in recording their *Discoveries*: because it is probable, that the *Trials* of Future Ages will not agree with those of the present, but frequently thwart, and contradict them.

The *Objection* is strong, and material; and I am so far from diminishing the weight of it, that I am rather willing to add more to it. I confess many *Experiments* are obnoxious to failing; either by reason of some *circumstances*, which are scarce discernible, till the work be over: or from the diversity of *Materials*, whereof some may be *genuine*, some *sophisticated*, some *simple*, some *mix'd*, some *fresh*, some may have lost their *virtue*. And this is chiefly remarkable, in *Chymical Operations*, wherein if the dissolvents be ill prepar'd, if the *Spirits* be too much, or too little purify'd, if there be the least alteration, in the degree of *Fire*, the quantity of *Matter*, or by the negligence of those that attend it, the whole course will be overthrown, or chang'd from its first purpose.

But what is now to be concluded from hence? shall this *instability*, and *Casualty* of *Experiments*, deter us from labouring in them at all? or should it not rather excite us to be more curious and watchful in their *process*? It is to be allow'd that such *undertakings* are wonderfully hazardous and difficult; why else does the *Royal Society* indeavour to preserve them from degenerating, by so many *forewarnings*, and *Rules*, and a *Method* so severe? It is granted, that their *event* is often uncertain, and not answerable to our expectations. But that only ought to admonish us, of the undispensable necessity of a jealous, and exact *Inquiry*. If the uncertainty proceeded from a constant irregularity of *Nature*, we had reason then to despair: but seeing it for the most part arises only from some defect or change in our progress, we should thence learn, first to correct our own miscarriages, before we cease to hope for the *success*.

Let then the *Experiment* be often renew'd. If the same kinds and proportions of *Ingredients* be us'd, and the same circumstances be punctually observ'd, the *effect* without all question will be the same. If some little variation of any of these, has made any alteration, a judicious, and well practis'd *Trier* will soon be able to discern the *cause* of it; and to rectifie it, upon the next repetition. If the difference of *time*, or *place*, or *matter*, or *Instruments*, will not suffer the product to be just the same in all points: yet something else will result, that may prove perhaps as beneficial. If we cannot always arrive at the main end of our *Labours*, some less unsought *Curiosities* will arise. If we cannot obtain that which shall be useful for practice, there may something appear that may instruct.

It is stranger that we are not able to inculcate into the minds of many men, the necessity of that *distinction* of my Lord *Bacons*, that there ought to be *Experiments* of *Light*, as well as of *Fruit*. It is their usual word, *What solid good will come from thence?* They are indeed to be commended for being so severe *Exactors* of *goodness*. And it were to be wish'd, that they would not only exercise this vigour, about *Experiments*, but on their own *lives*, and *actions*: that they would still question with themselves, in all that they do; what *solid good* will come from thence? But they are to know, that in so large, and so various an *Art* as this of *Experiments*, there are many degrees of usefulness: some may serve for real, and plain *benefit*, without much *delight*: some for *teaching* without apparent *profit*: some for *light* now, and for *use* hereafter; some only for *ornament*, and *curiosity*. If they will persist in contemning all *Experiments*, except those which bring with them immediate *gain*, and a present *harvest*: they may as well cavil at the Providence of God, that he has not made all the seasons of the year, to be times of *mowing*, *reaping*, and *vintage*.

ROBERT HOOKE

216

A method for making a history of the weather

For the better making a History of the Weather, I conceive it requisite to observe,

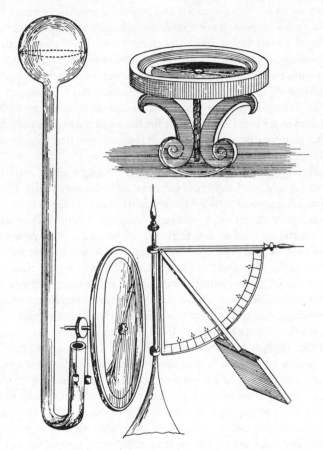

Hooke's meteorological recording instruments

1. The Strength and Quarter of the Winds, and to register the Changes as often as they happen: both which may be very conveniently shewn, by a small addition to an ordinary Weather-clock.

2. The Degrees of Heat and Cold in the Air; which will be best observed by a sealed *Thermometer*, graduated according to the Degrees of *Expansion*, which bear a known proportion to the whole bulk of Liquor, the beginning of which

278

gradation should be that dimension which the Liquor hath, when encompassed with Water, just beginning to freeze, and the degrees of *Expansion*, either greater or less, should be set or marked above it or below it.

3. The Degrees of Dryness and Moisture in the Air; which may be most conveniently observed by a *Hygroscope*, made with the single beard of a wild Oat perfectly ripe, set upright and headed with an *Index*, after the way described by *Emmanuel Magnan*; the conversions and degrees of which, may be measured by divisions made on the rim of a Circle, in the *Center* of which, the *Index* is turned round: The beginning or Standard of which Degree of *Rotation*, should be that, to which the *Index* points, when the beard, being throughly wet, or covered, with Water, is quite unwreathed, and becomes straight. But because of the smallness of this part of the Oat, the cod of a wild *Vetch* may be used instead of it, which will be a much larger *Index*, and will be altogether as sensible of the changes of the Air.

4. The degrees of Pressure in the Air: which may be several wayes observed, but best of all with an Instrument with Quicksilver, contrived so, as either by means of water or an *Index*, it may sensibly exhibit the minute variations of that Action.

5. The constitution and face of the Sky or Heavens; and this is best done by the eye; here should be observed, whether the Sky be clear or clouded; and if clouded, after what manner; whether with high Exhalations or great white clouds, or dark thick ones. Whether those Clouds afford Fogs or Mists, or Sleet, or Rain, or Snow, &c. Whether the under side of those Clouds be flat or waved and irregular, as I have often seen before thunder. Which way they drive, whether all one way, or some one way, some another; and whether any of these be the same with the Wind that blows below; the Colour and face of the Sky at the rising and setting of the Sun and Moon; what Haloes or Rings may happen to encompass those Luminaries, their bigness form and number.

6. What Effects are produc'd upon other bodies: As what Aches and Distempers in the bodies of men: what Diseases are most rife, as Colds, Fevours, Agues, &c. What putrefactions or other changes are produc'd in other Bodies; As the sweating of Marble, the burning blew of a Candle, the blasting of Trees and Corn; the unusual sprouting growth, or decay of any Plants or Vegetables: the putrefaction of bodies not usual; the plenty or scarcity of Insects; of several Fruits, Grains, Flowers, Roots, Cattel, Fishes, Birds, any thing notable of that kind. What conveniences or inconveniences may happen in the year, in any kind, as by floods, droughts, violent showers, &c. What nights produce dews and hoar-frosts, and what not?

7. What Thunders and Lightnings happen, and what Effects they produce; as souring Beer or Ale, turning Milk, killing Silk worms, &c.

8. Any thing extraordinary in the Tides; as double Tides later or earlier, greater or less Tides than ordinary, Rising or drying of Springs; Comets or

unusual Apparitions, new Stars, *Ignes fatui* or shining Exhalations, or the like.

These should all or most of them be diligently observed and registered by some one, that is always conversant in or neer the same place.

Now that these and some other, hereafter to be mentioned, may be registred so as to be most convenient for the making of comparisons, requisite for the raising *Axioms*, whereby the Cause or Laws of Weather may be found out; it will be desirable to order them so, that the Scheme of a whole Moneth, may at one view be presented to the Eye: And this may conveniently be done on the pages of a Book in folio, allowing fifteen dayes for one side, and fifteen for the other. Let each of those pages be divided into nine Columes, and distinguished by perpendicular lines; let each of the first six Columes be half an inch wide, and the three last equally share the remaining of the side.

Let each Colume have the title of what it is to contain, in the first at least, written at the top of it: As, let the first Colume towards the left hand, contain the dayes of the Moneth, or place of the Sun, and the remarkable hours of each day. The second, the Place, Latitude, Distance, Ages and Phaces of the Moon. The third the Quarters and strength of Winds. The fourth the Heat and Cold of the season. The fifth the Dryness and Moisture of it. The sixth the Degrees of pressure. The seventh the faces and appearances of the Sky. The eighth the Effects of the Weather upon other bodies, Thunders, Lightnings, or any thing extraordinary. The ninth general Deductions, Corollaries or Syllogisms, arising from the comparing the several *Phaenomena* together.

That the Columes may be large enough to contain what they are designed for, it will be necessary, that the particulars be expressed with some Characters, as brief and compendious as is possible. The two first by the Figures and Characters of the Signs commonly us'd in Almanacks. The Winds may be exprest by the Letters, by which they are exprest in small Sea-Cards: and the degrees of strength by 1, 2, 3, 4, &c. according as they are marked in the contrivance of the Weather-cock. The degrees of Heat and Cold may be exprest by the Numbers appropriate to the Divisions of the *Thermometer*. The Dryness and Moisture, by the Divisions in the rim of the *Hydroscope*. The pressure by Figures denoting the height of the *Mercurial Cylinder*. But for the faces of the Sky, they are so many, that many of them want proper names; and therefore it will be convenient to agree upon some determinate ones, by which the most usual may be in brief exprest. As let *Cleer* signifie a very cleer Sky without any Clouds or Exhalations: *Checker'd* a cleer Sky, with many great white round Clouds, such as are very usual in Summer. *Hazy*, a Sky that looks whitish, by reason of the thickness of the higher parts of the Air, by some Exhalation not formed into Clouds. *Thick*, a Sky more whitened by a greater company of Vapours: these do usually make the *Luminaries* look bearded or hairy, and are oftentimes the cause of the appearance of Rings and Haloes about the *Sun* as well as the *Moon*. *Overcast*, when the Vapours so whiten and thicken the Air, that the *Sun* cannot break through; and of this there are very many degrees, which may be exprest by a *little, much, more, very much overcast*, &c. Let *Hairy* signifie a Sky that hath many small, thin and high Exhalations, which resemble locks of hair, or flakes of Hemp or Flax: whose varieties may be exprest by *straight* or *curv'd*, &c. according to the resemblance they bear. Let *Water'd* signifie a Sky that has many high thin and small Clouds, looking almost like water'd Tabby, called in some places a Mackeril Sky. Let a Sky be called *Waved*, when those Clouds appear much bigger and lower, but much after the same manner. *Cloudy*, when the Sky has many thick dark Clouds. *Lowring*, when the Sky is not very much overcast, but hath also underneath many thick dark Clouds which threaten rain. The significa- tion of *gloomy, foggy, misty, sleeting, driving, rainy, snowy*, reaches or racks *variable*, &c. are well known, they being very commonly used. There may be also several faces of the Sky compounded of two or more of these, which may be intelligibly enough exprest by two or more of these names. It is likewise desirable, that the particulars of the eighth and ninth Columes may be entered in as little room, and as few words as are sufficient to signifie them intelligibly and plainly.

It were to be wisht that there were divers in several parts of the World, but especially in distant parts of this Kingdom, that would undertake this work, and that such would agree upon a common way somewhat after this manner, that as neer as could be, the same method and words might be made use of. The benefit of which way is easily enough conceivable.

As for the Method of using and digesting those so collected Observations; That will be more advantageously considered when the *Supellex* is provided; A Workman being then best able to fit and prepare his Tools, for his work, when he sees what materials he has to work upon.

ROBERT BOYLE

217

The spring of the air

For the more easy understanding of the experiments triable by our engine, I thought it not superfluous nor unreasonable in the recital of this first of them, to insinuate that notion, by which it seems likely, that most, if not all of them, will prove explicable. Your Lordship will easily suppose, that the notion I speak of is, that there is a spring, or elastical power in the air we live in. By which ἐλατήρ or spring, of the air, that which I mean is this; that our air either consists of or at least abounds with, parts of such a nature, that in case they be bent or compressed by the weight of the incumbent part of the atmosphere, or by any other body, they do endeavour, as much as in them lieth, to free them- selves from that pressure, by bearing against the contiguous bodies that keep them bent; and, as soon as those bodies are removed, or reduced to give them way, by presently unbending and stretching out themselves, either quite, or so far forth as the contiguous bodies that resist them will permit, and thereby expanding the whole parcel of air, these elastical bodies compose.

This notion may perhaps be somewhat further explained, by conceiving the air near the earth to be such a heap of little bodies, lying one upon another, as may be resembled to a fleece of wool. For this (to omit other likenesses betwixt them) consists of many slender and flexible hairs; each of which may indeed, like a little spring, be easily bent or rolled up; but will also, like a spring, be still endeavouring to stretch itself out again. For though both these hairs, and the aëreal corpuscles to which we liken them, do easily yield to external pressures; yet, each of them (by virtue of its structure) is endowed with a power or principle of self-dilatation; by virtue whereof, though the hairs may by a man's hand be bent and crouded closer together, and into a narrower room than suits best with the nature of the body; yet, whilst the compression lasts, there is in the fleece they compose an endeavour outwards, whereby it continually thrusts against the hand that opposes its expansion. And upon the removal of the external pressure, by opening the hand more or less, the compressed wool doth, as it were, spontaneously expand or display itself towards the recovery of its former more loose and free condition, till the fleece hath either regained its former dimensions, or at least approached them as near as the compressing hand (perchance not quite opened) will permit. This power of self-dilatation is somewhat more conspicuous in a dry spunge compressed, than in a fleece of wool. But yet we rather chose to employ the latter on this occasion, because it is not, like a spunge, an entire body, but a number of slender and flexible bodies, loosely complicated, as the air itself seems to be.

There is yet another way to explicate the spring of the air; namely, by supposing with that most ingenious gentleman, Monsieur *Des Cartes*, that the air is nothing but a congeries or heap of small and (for the most part) of flexible particles, of several sizes, and of all kinds of figures, which are raised by heat (especially that of the sun) into that fluid and subtle ethereal body that surrounds the earth; and by the restless agitation of that celestial matter, wherein those particles swim, are so whirled around, that each corpuscle endeavours to beat off all others from coming within the little sphere requisite to its motion about its own centre; and in case any, by intruding into that sphere, shall oppose its free rotation, to expel or drive it away: so that, according to this doctrine, it imports very little, whether the particles of the air have the structure requisite to springs, or be of any other form (how irregular soever) since their elastical power is not made to depend upon their shape or structure, but upon the vehement agitation, and (as it were) brandishing motion, which they receive from the fluid aether, that swiftly flows between them, and whirling about each of them (independently from the rest) not only keeps those slender aëreal bodies separated and stretched out (at least, as far as the neighbouring ones will permit) which otherwise, by reason of their flexibleness and weight, would flag or curl; but also makes them hit against, and knock away each other, and consequently require more room than that, which, if they were compressed, they would take up.

By these two differing ways, my Lord, may the springs of the air be explicated. But though the former of them be that, which by reason of its seeming somewhat more easy, I shall for the most part make use of in the following discourse; yet am I not willing to declare peremptorily for either of them against the other. And indeed, though I have in another treatise endeavoured to make it probable, that the returning of elastical bodies (if I may so call them) forcibly bent, to their former position, may be mechanically explicated; yet I must confess, that to determine whether the motion of restitution in bodies proceed from this, that the parts of a body of a peculiar structure are put into motion by the bending of the spring, or from the endeavour of some subtle ambient body, whose passage may be opposed or obstructed, or else its pressure unequally resisted by reason of the new shape or magnitude, which the bending of a spring may give the pores of it: to determine this, I say, seems to me a matter of more difficulty, than at first sight one would easily imagine it. Wherefore I shall decline meddling with a subject, which is much more hard to be explicated than necessary to be so by him, whose business it is not, in this letter, to assign the adequate cause of the spring of the air, but only to manifest, that the air hath a spring, and to relate some of its effects. [. . .]

Taking it then for granted, that the air is not devoid of weight, it will not be uneasy to conceive, that that part of the atmosphere, wherein we live, being the lower part of it, the corpuscles, that compose it, are very much compressed by the weight of all those of the like nature, that are directly over them; that is, of all the particles of air, that being piled up upon them, reach to the top of the atmosphere. And though the height of this atmosphere, according to the famous *Kepler*, and some others, scarce exceeds eight common miles; yet other eminent and later astronomers would promote the confines of the atmosphere to exceed six or seven times that number of miles. And the diligent and learned *Ricciolo* makes it probable, that the atmosphere may, at least in divers places, be at least fifty miles high. So that, according to a moderate estimate of the thickness of the atmosphere, we may well suppose, that a column of air, of many miles in height, leaning upon some springy corpuscles of air here below, may have weight enough to bend their little springs, and keep them bent; as (to resume our former comparison), if there were fleeces of wool piled up to a mountainous height one upon another, the hairs, that compose the lowermost locks, which support the rest, would, by the weight of all the wool above them, be as well strongly compressed, as if a man should squeeze them together in his hands, or employ any such other moderate force to compress them. So that we need not wonder, that upon the taking off the incumbent air from any parcel of the atmosphere here below, the corpuscles, whereof that undermost air consists, should display themselves, and take up more room than before. [. . .] And, as for the easy yielding of the air to the bodies that move in it, if we consider, that the corpuscles, whereof it consists, though of a springy nature, are yet so very small, as to make up (which it is manifest they do) a fluid body, it will not be difficult to conceive, that in the air, as in other bodies that are fluid, the little bodies it consists of, are in an almost restless motion, whereby they become (as we have more fully discoursed in another treatise) very much disposed to yield to other bodies, or easy to be displaced by them; and that the same corpuscles are likewise so variously moved, as they are entire corpuscles, that if some strive to push a body placed among them towards the right hand (for instance), others, whose motion hath an opposite determination, as strongly thrust the same body towards the left; whereby neither of them proves to be able to move it out of its place, the pressure on all hands being reduced as it were to an equilibrium: so that the corpuscles of the air must be as well sometimes considered under the notion of little springs, which remaining bent, are in their entire bulk transported from place to place; as under the notion of springs displaying themselves, whose parts fly abroad, whilst, as to their entire bulk, they scarce change place: as the two ends of a bow, shot off, fly from one another; whereas the bow itself may be held fast in the archer's hand. And that it is the equal pressure of the air on all sides upon the bodies that are in it, which causeth the easy cession of its parts, may be argued from hence; that if by the help of our engine the air be but in great part, though not totally, drawn away from one side of a body without being drawn away from the other, he that shall think to move that body to and fro, as easily as before, will find himself much mistaken.

OLAF RØMER

218

A demonstration showing the motion of light

Philosophers have been labouring for many years to decide by some Experience, whether the action of Light be conveyed in an instance to distant places, or whether it requireth time. M. *Rømer* of the *R. Academy* of the Sciences hath devised a way, taken from the Observations of the first Satellit of *Jupiter*, by which he demonstrates, that for the distance of about 3000 leagues, such as is very near the bigness of the Diameter of the Earth, Light needs not one second of time.

Let *A* be the *Sun*, *B Jupiter*, *C* the first Satellit of *Jupiter*, which enters into the shadow of *Jupiter*, to come out of it at *D*; and let *EFGHKL* be the *Earth* placed at divers distances from *Jupiter*.

Now, suppose the *Earth*, being in *L* towards the second Quadrature of *Jupiter*, hath seen the first satellit at the time of its emersion or issuing out of the shadow in *D*; and that about $42\frac{1}{2}$ hours after, (*vid.* after one revolution of this Satellit,) the *Earth* being in *K*, do see it returned in *D*; it is manifest, that if the Light require time to traverse the interval LK, the Satellit will be seen returned later in D, than it would have been if the Earth had remained in L, so that the revolution of this Satellit being thus observed by the Emersions, will be retarded by so much time, as the Light shall have taken in passing from L to K, and that, on the contrary, in the other Quadrature FG, where the *Earth* by approaching goes to meet the Light, the revolutions of the Immersions will appear to be shortened by so much, as those of the Emersions had appeared to be lengthened. And because in 42½ hours, which this Satellit very near takes to make one revolution, the distance between the *Earth* and *Jupiter* in both the Quadratures varies at least 210 Diameters of the *Earth*, it follows, that if for the account of every Diameter of the *Earth* there were required a second of time, the Light would take 3½ minutes for each of the intervals GF, KL; which would cause near half a quarter of an hour between two revolutions of the first Satellit, one observed in FG, and the other in KL, whereas there is not observed any sensible difference.

Yet doth it not follow hence, that Light demands no time. For, after M. *Rømer* had examin'd the thing more nearly, he found, that what was not sensible

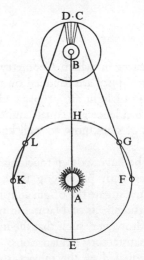

The path of light from Jupiter's satellite to the earth

in two revolutions, became very considerable in many being taken together, and that, for example, forty revolutions observed on the side F, might be sensibly shorter, than forty others observed in any place of the Zodiack where *Jupiter* may be met with; this is in proportion to 22 minutes for the whole interval of HE, which is the double of the interval that is from hence to the Sun.

The necessity of this new Equation of the retardment of Light, is established by all the observations that have been made in the R. *Academy* and in the

Observatory, for the space of eight years, and it hath been lately confirmed by the Emersion of the first Satellit observed at *Paris* the *9th* of *November* last at 5 a Clock, 35′. 45″. at Night, 10 minutes later than it was to be expected, by deducing it from those that had been observed in the Month of *August*, when the *Earth* was much nearer to *Jupiter*: Which M. *Rømer* had predicted to the said Academy from the beginning of *September*.

But to remove all doubt, that this inequality is caused by the retardment of the Light, he demonstrates, that it cannot come from any excentricity, or any other cause of those that are commonly alledged to explicate the irregularities of the *Moon* and the other Planets; though he be well aware, that the first Satellit or *Jupiter* was excentrick, and that, besides, his revolutions were advanced or retarded according as *Jupiter* did approach to or recede from the Sun, as also that the revolutions of the *primum mobile* were unequal; yet saith he, these three last causes of inequality do not hinder the first from being manifest.

CHRISTIAAN HUYGENS

The wave theory of light

219

As happens in all the sciences in which Geometry is applied to matter, the demonstrations concerning Optics are founded on truths drawn from experience. Such are that the rays of light are propagated in straight lines; that the angles of reflexion and of incidence are equal; and that in refraction the ray is bent according to the law of sines, now so well known, and which is no less certain than the preceding laws.

The majority of those who have written touching the various parts of Optics have contented themselves with presuming these truths. But some, more inquiring, have desired to investigate the origin and the causes, considering these to be in themselves wonderful effects of Nature. In which they advanced some ingenious things, but not, however, such that the most intelligent folk do not wish for better and more satisfactory explanations. Wherefore I here desire to propound what I have meditated on the subject, so as to contribute as much as I can to the explanation of this department of Natural Science, which, not without reason, is reputed to be one of its most difficult parts. I recognize myself to be much indebted to those who were the first to begin to dissipate the strange obscurity in which these things were enveloped, and to give us hope that they might be explained by intelligible reasoning. But, on the other hand I am astonished also that even here these have often been willing to offer, as assured and demonstrative, reasonings which were far from conclusive. For I do not find that any one has yet given a probable explanation of the first and most notable phenomena of light, namely why it is not propagated except in straight lines, and how visible rays, coming from an infinitude of diverse places, cross one another without hindering one another in any way.

I shall therefore essay in this book, to give, in accordance with the principles accepted in the Philosophy of the present day, some clearer and more probable reasons, firstly of these properties of light propagated rectilinearly; secondly of light which is reflected on meeting other bodies. Then I shall explain the phenomena of those rays which are said to suffer refraction on passing through transparent bodies of different sorts; and in this part I shall also explain the effects of the refraction of the air by the different densities of the Atmosphere. [. . .]

It is inconceivable to doubt that light consists in the motion of some sort of matter. For whether one considers its production, one sees that here upon the Earth it is chiefly engendered by fire and flame which contain without doubt bodies that are in rapid motion, since they dissolve and melt many other bodies, even the most solid; or whether one considers its effects, one sees that when light is collected, as by concave mirrors, it has the property of burning as a fire does, that is to say it disunites the particles of bodies. This is assuredly the mark of motion, at least in the true Philosophy, in which one conceives the causes of all natural effects in terms of mechanical motions. This, in my opinion, we must necessarily do, or else renounce all hopes of ever comprehending anything in Physics.

And as, according to this Philosophy, one holds as certain that the sensation of sight is excited only by the impression of some movement of a kind of matter which acts on the nerves at the back of our eyes, there is here yet one reason more for believing that light consists in a movement of the matter which exists between us and the luminous body.

Further, when one considers the extreme speed with which light spreads on every side, and how, when it comes from different regions, even from those directly opposite, the rays traverse one another without hindrance, one may well understand that when we see a luminous object, it cannot be by any transport of matter coming to us from this object, in the way in which a shot or an arrow traverses the air; for assuredly that would too greatly impugn these two properties of light, especially the second of them. It is then in some other way that light spreads; and that which can lead us to comprehend it is the knowledge which we have of the spreading of Sound in the air.

We know that by means of the air, which is an invisible and impalpable body, Sound spreads around the spot where it has been produced, by a movement which is passed on successively from one part of the air to another; and that the spreading of this movement, taking place equally rapidly on all sides, ought to form spherical surfaces ever enlarging and which strike our ears. Now there is no doubt at all that light also comes from the luminous body to our eyes by some movement impressed on the matter which is between the two; since, as we have already seen, it cannot be by the transport of a body which passes from one to the other. If, in addition, light takes time for its passage—which we are now going to examine—it will follow that this movement, impressed on the intervening matter, is successive; and consequently it spreads, as Sound does, by spherical surfaces and waves: for I call them waves from their resemblance to. those which are seen to be formed in water when a stone is thrown into it, and which present a successive spreading as circles, though these arise from another cause, and are only in a flat surface. [. . .]

But that which I employed only as a hypothesis, has recently received great seemingness as an established truth by the ingenious proof of Mr Rømer[1] [. . .] It is founded as is the preceding argument upon celestial observations, and proves not only that Light takes time for its passage, but also demonstrates how much time it takes. and that its velocity is even at least six times greater than that which I have just stated. [. . .]

As regards the different modes in which I have said the movements of Sound and of Light are communicated, one may sufficiently comprehend how this occurs in the case of Sound if one considers that the air is of such a nature that it can be compressed and reduced to a much smaller space than that which it ordinarily occupies. And in proportion as it is compressed the more does it exert an effort to regain its volume; for this property along with its penetrability, which remains notwithstanding its compression, seems to prove that it is made up of small bodies which float about and which are agitated very rapidly in the ethereal matter composed of much smaller parts. So that the cause of the spreading of Sound is the effort which these little bodies make in collisions with one another, to regain freedom when they are a little more squeezed together in the circuit of these waves than elsewhere.

But the extreme velocity of Light, and other properties which it has, cannot admit of such a propagation of motion, and I am about to show here the way in which I conceive it must occur. For this, it is needful to explain the property which hard bodies must possess to transmit movement from one to another.

When one takes a number of spheres of equal size, made of some very hard substance, and arranges them in a straight line, so that they touch one another, one finds, on striking with a similar sphere against the first of these spheres, that the motion passes as in an instant to the last of them, which separates itself from the row, without one's being able to perceive that the others have been stirred. And even that one which was used to strike remains motionless with them. Whence one sees that the movement passes with an extreme velocity which is the greater, the greater the hardness of the substance of the spheres.

But it is still certain that this progression of motion is not instantaneous, but successive, and therefore must take time. For if the movement, or the disposition to movement, if you will have it so, did not pass successively through all these spheres, they would all acquire the movement at the same time, and hence would all advance together; which does not happen. For the last one leaves the whole

1. This proof is given in extract 218, page 284. (Ed.)

row and acquires the speed of the one which was pushed. Moreover there are experiments which demonstrate that all the bodies which we reckon of the hardest kind, such as quenched steel, glass, and agate, act as springs and bend somehow, not only when extended as rods but also when they are in the form of spheres or of other shapes. That is to say they yield a little in themselves at the place where they are struck, and immediately regain their former figure. For I have found that on striking with a ball of glass or of agate against a large and quite thick piece of the same substance which had a flat surface, slightly soiled with breath or in some other way, there remained round marks, of smaller or larger size according as the blow had been weak or strong. This makes it evident that these substances yield where they meet, and spring back: and for this time must be required.

Now in applying this kind of movement to that which produces Light there is nothing to hinder us from estimating the particles of the ether to be of a substance as nearly approaching to perfect hardness and possessing a springiness as prompt as we choose. It is not necessary to examine here the causes of this hardness, or of that springiness, the consideration of which would lead us too far from our subject. [. . .]

But though we shall ignore the true cause of springiness we still see that there are many bodies which possess this property; and thus there is nothing strange in supposing that it exists also in little invisible bodies like the particles of the Ether. Also if one wishes to seek for any other way in which the movement of Light is successively communicated, one will find none which agrees better, with uniform progression, as seems to be necessary, than the property of springiness; because if this movement should grow slower in proportion as it is shared over a greater quantity of matter, in moving away from the source of the light, it could not conserve this great velocity over great distances. But by supposing springiness in the ethereal matter, its particles will have the property of equally rapid restitution whether they are pushed strongly or feebly; and thus the propagation of Light will always go on with an equal velocity.

And it must be known that although the particles of the ether are not ranged thus in straight lines, as in our row of spheres, but confusedly, so that one of them touches several others, this does not hinder them from transmitting their movement and from spreading it always forward. As to this it is to be remarked that there is a law of motion serving for this propagation, and verifiable by experiment. It is that when a sphere, such as *A* here, touches several other similar spheres *CCC*, if it is struck by another sphere *B* in such a way as to exert an impulse against all the spheres *CCC* which touch it, it transmits to them the whole of its movement, and remains after that motionless like the sphere *B*. And without supposing that the ethereal particles are of spherical form (for I see indeed no need to suppose them so) one may well understand that this property of communicating an impulse does not fail to contribute to the aforesaid propagation of movement.

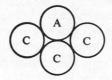

Collision of elastic spheres

Equality of size seems to be more necessary, because otherwise there ought to be some reflexion of movement backwards when it passes from a smaller particle to a larger one, according to the Laws of Percussion which I published some years ago.

However, one will see hereafter that we have to suppose such an equality not so much as a necessity for the propagation of light as for rendering that propagation easier and more powerful; for it is not beyond the limits of probability that the particles of the ether have been made equal for a purpose so important as that of light, at least in that vast space which is beyond the region of atmosphere and which seems to serve only to transmit the light of the Sun and the Stars.

I have then shown in what manner one may conceive Light to spread successively, by spherical waves, and how it is possible that this spreading is accomplished with as great a velocity as that which experiments and celestial observations demand. Whence it may be further remarked that although the particles are supposed to be in continual movement (for there are many reasons for this) the successive propagation of the waves cannot be hindered by this; because the propagation consists nowise in the transport of those particles but merely in a small agitation which they cannot help communicating to those surrounding, notwithstanding any movement which may act on them causing them to be changing positions amongst themselves.

But we must consider still more particularly the origin of these waves, and the manner in which they spread. And, first, it follows from what has been said on the production of Light, that each little region of a luminous body, such as the Sun, a candle, or a burning coal, generates its own waves of which that region is the centre. Thus in the flame of a candle, having distinguished the points A, B, C, concentric circles described about each of these points represent the waves which come from them. And one must imagine the same about every point of the surface and of the part within the flame.

But as the percussions at the centres of these waves possess no regular succession, it must not be supposed that the waves themselves follow one another at

Light waves emitted from the flame of a candle

equal distances: and if the distances marked in the figure appear to be such, it is rather to mark the progression of one and the same wave at equal intervals of time than to represent several of them issuing from one and the same centre.

After all, this prodigious quantity of waves which traverse one another without confusion and without effacing one another must not be deemed inconceivable; it being certain that one and the same particle of matter can serve for many waves coming from different sides or even from contrary directions, not only if it is struck by blows which follow one another closely but even for those which act on it at the same instant. It can do so because the spreading of the movement is successive. This may be proved by the row of equal spheres of hard matter, spoken of above. If against this row there are pushed from two opposite sides at the same time two similar spheres *A* and *D*, one will see each of them rebound with the same velocity which it had in striking, yet the whole row will remain in its place, although the movement has passed along its whole length twice over. And if these contrary movements happen to meet one another at the middle sphere, *B*, or at some other such as *C*, that sphere will yield and act as a spring at both sides, and so will serve at the same instant to transmit these two movements.

Transmission of an elastic impact

But what may at first appear full strange and even incredible is that the undulations produced by such small movements and corpuscles, should spread to such immense distances; as for example from the Sun or from the Stars to us.

For the force of these waves must grow feeble in proportion as they move away from their origin, so that the action of each one in particular will without doubt become incapable of making itself felt to our sight. But one will cease to be astonished by considering how at a great distance from the luminous body an infinitude of waves, though they have issued from different points of this body, unite together in such a way that they sensibly compose one single wave only, which, consequently, ought to have enough force to make itself felt. Thus this infinite number of waves which originate at the same instant from all points of a fixed star, big it may be as the Sun, make practically only one single wave which may well have force enough to produce an impression on our eyes. Moreover from each luminous point there may come many thousands of waves in the smallest imaginable time, by the frequent percussion of the corpuscles which strike the Ether at these points: which further contributes to rendering their action more sensible.

There is the further consideration in the emanation of these waves, that each particle of matter in which a wave spreads, ought not to communicate its

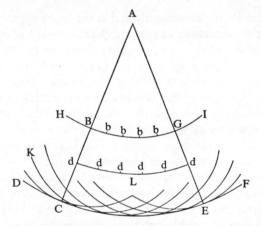

Propagation of light waves

motion only to the next particle which is in the straight line drawn from the luminous point, but that it also imparts some of it necessarily to all the others which touch it and which oppose themselves to its movement. So it arises that around each particle there is made a wave of which that particle is the centre. Thus if *DCF* is a wave emanating from the luminous point *A*, which is its centre, the particle *B*, one of those comprised within the sphere *DCF*, will have made its particular or partial wave *KCL*, which will touch the wave *DCF* at *C* at the same moment that the principal wave emanating from the point *A* has arrived at *DCF*; and it is clear that it will be only the region *C* of the wave *KCL* which will touch the wave *DCF*, to wit, that which is in the straight line drawn through *AB*. Similarly the other particles of the sphere *DCF*, such as *bb*, *dd* etc., will each make its own wave. But each of these waves can be infinitely feeble only as compared with the wave *DCF*, to the composition of which all the others contribute by the part of their surface which is most distant from the centre *A*.

One sees, in addition, that the wave *DCF* is determined by the distance attained in a certain space of time by the movement which started from the point *A*; there being no movement beyond this wave, though there will be in the space which it encloses, namely in parts of the particular waves, those parts which do not touch the sphere *DCF*. And all this ought not to seem fraught with too much minuteness or subtlety, since we shall see in the sequel that all the properties of Light, and everything pertaining to its reflexion and its refraction, can be explained in principle by this means. This is a matter which has been quite unknown to those who hitherto have begun to consider the waves of light, amongst whom are Mr Hooke in his *Micrographia*, and Father Pardies, who, in a treatise of which he let me see a portion, and which he was unable to complete as he died shortly afterward, had undertaken to prove by these waves the effects of reflexion and refraction. But the chief foundation, which consists in the remark I have just made, was lacking in his demonstrations; and for the rest he had opinions very different from mine, as may be will appear some day if his writing has been preserved.

To come to the properties of Light. We remark first that each portion of a wave ought to spread in such a way that its extremities lie always between the same straight lines drawn from the luminous point. Thus the portion *BG* of the wave, having the luminous point *A* as its centre, will spread into the arc *CE* bounded by the straight lines *ABC*, *AGE*. For although the particular waves produced by the particles comprised within the space *CAE* spread also outside this space, they yet do not concur at the same instant to compose a wave which terminates the movement, as they do precisely at the circumference *CE*, which is their common tangent.

And hence one sees the reason why light, at least if its rays are not reflected or broken, spreads only by straight lines, so that it illuminates no object except when the path from its source to that object is open along such lines. For if, for example, there were an opening *BG*, limited by opaque bodies *BH*, *GI*, the wave of light which issues from the point *A* will always be terminated by the straight lines *AC*, *AE*, as has just been shown; the parts of the partial waves which spread outside the space *ACE* being too feeble to produce light there.

Now, however small we make the opening *BG*, there is always the same reason causing the light there to pass between straight lines; since this opening is always large enough to contain a great number of particles of the ethereal matter, which are of an inconceivable smallness; so that it appears that each little portion of the wave necessarily advances following the straight line which comes from the luminous point. Thus then we may take the rays of light as if they were straight lines.

It appears, moreover, by what has been remarked touching the feebleness of the particular waves, that it is not needful that all the particles of the Ether should be equal amongst themselves, though equality is more apt for the propagation of the movement. For it is true that inequality will cause a particle by pushing against another larger one to strive to recoil with a part of its movement; but it will thereby merely generate backwards towards the luminous point some partial waves incapable of causing light, and not a wave compounded of many as *CE* was.

Another property of waves of light, and one of the most marvellous, is that when some of them come from different or even from opposing sides, they produce their effect across one another without any hindrance. When also it comes about that a number of spectators may view different objects at the same time through the same opening, and that two persons can at the same time see one another's eyes. Now according to the explanation which has been given of the action of light, how the waves do not destroy nor interrupt one another when they cross one another, these effects which I have just mentioned are easily conceived. But in my judgement they are not at all easy to explain according to the views of Mr Des Cartes, who makes Light to consist in a continuous pressure merely tending to movement. For this pressure not being able to act from two opposite sides at the same time, against bodies which have no inclination to approach one another, it is impossible so to understand what I have been saying about two persons mutually seeing one another's eyes, or how two torches can illuminate one another.

ISAAC NEWTON

PHILOSOPHIAE NATURALIS PRINCIPIA MATHEMATICA

Newton's Preface to the First Edition

220

Since the ancients (as we are told by Pappus) esteemed the science of mechanics of greatest importance in the investigation of natural things, and the moderns, rejecting substantial forms and occult qualities, have endeavored to subject the phenomena of nature to the laws of mathematics, I have in this treatise cultivated mathematics as far as it relates to philosophy. The ancients considered mechanics in a twofold respect; as rational, which proceeds accurately by demonstration, and practical. To practical mechanics all the manual arts belong, from which mechanics took its name. But as artificers do not work with perfect accuracy, it comes to pass that mechanics is so distinguished from geometry that what is perfectly accurate is called geometrical; what is less so, is called mechanical. However, the errors are not in the art, but in the artificers. He that works with less accuracy is an imperfect mechanic; and if any could work with perfect accuracy, he would be the most perfect mechanic of all, for the description of right lines and circles, upon which geometry is founded, belongs to mechanics. Geometry does not teach us to draw these lines, but requires them to be drawn, for it requires that the learner should first be taught to describe these accurately before he enters upon geometry, then it shows how by these operations problems may be solved. To describe right lines and circles are problems, but not geometrical problems. The solution of these problems is required from mechanics, and by geometry the use of them, when so solved, is shown; and it is the glory of geometry that from those few principles, brought from without, it is able to produce so many things. Therefore geometry is founded in mechanical practice, and is nothing but that part of universal mechanics which accurately proposes and demonstrates the art of measuring. But since the manual arts are chiefly employed in the moving of bodies, it happens that geometry is commonly referred to their magnitude, and mechanics to their motion. In this sense rational mechanics will be the science of motions resulting from any forces whatsoever, and of the forces required to produce any motions, accurately proposed and demonstrated. This part of mechanics, as far as it extended to the five powers[1] which relate to manual arts, was cultivated by the ancients, who considered gravity (it not being a manual power) no otherwise than in moving weights by those powers. But I consider philosophy rather than arts and write not concerning manual but natural powers, and consider chiefly those things which relate to gravity, levity, elastic force, the resistance of fluids, and the like forces, whether attractive or impulsive; and therefore I offer this work as the mathematical principles of philosophy, for the whole burden of philosophy seems to consist in this—from the phenomena of motions to investigate the forces of nature, and then from these forces to demonstrate the other phenomena; and to this end the general propositions in the first and second Books are directed. In the third Book I give an example of this in the explication of the System of the World; for by the propositions mathematically demonstrated in the former Books, in the third I derive from the celestial phenomena the forces of gravity with which bodies tend to the sun and the several planets. Then from these forces, by other propositions which are also mathematical, I deduce the motions of the planets, the comets, the moon, and the sea. I wish we could derive the rest of the phenomena of Nature by the same kind of reasoning from mechanical principles, for I am induced by many reasons to suspect that they may all depend upon certain forces by which the particles of bodies, by some causes hitherto unknown, are either mutually impelled towards one another, and cohere in regular figures, or are repelled and recede from one another. These forces being unknown, philosophers have hitherto attempted the search of Nature in vain; but I hope the principles here laid down will afford some light either to this or some truer method of philosophy. [. . .]

1. Lever, pulley, wedge, screw, and winch. (Ed.)

Cotes' Preface to the Second Edition

221

Those who have treated of natural philosophy may be reduced to about three classes. Of these some have attributed to the several species of things, specific and occult qualities, according to which the phenomena of particular bodies are supposed to proceed in some unknown manner. The sum of the doctrine of the Schools derived from *Aristotle* and the *Peripatetics* is founded on this principle. They affirm that the several effects of bodies arise from the particular natures of those bodies. But whence it is that bodies derive those natures they don't tell us; and therefore they tell us nothing. And being entirely employed in giving names to things, and not in searching into things themselves, they have invented, we may say, a philosophical way of speaking, but they have not made known to us true philosophy.

Others have endeavoured to apply their labours to greater advantage by rejecting that useless medley of words. They assume that all matter is homogeneous, and that the variety of forms which is seen in bodies arises from some very plain and simple relations of the component particles. And by going on from simple things to those which are more compounded they certainly proceed right, if they attribute to those primary relations no other relations than those which Nature has given. But when they take a liberty of imagining at pleasure unknown figures and magnitudes, and uncertain situations and motions of the parts, and moreover of supposing occult fluids, freely pervading the pores of bodies, endued with an all-performing subtilty, and agitated with occult motions, they run out into dreams and chimeras, and neglect the true constitution of things, which certainly is not to be derived from fallacious conjectures, when we can scarce reach it by the most certain observations. Those who assume hypotheses as first principles of their speculations, although they afterwards proceed with the greatest accuracy from those principles, may indeed form an ingenious romance, but a romance it will still be.

There is left then the third class, which possess experimental philosophy. These indeed derive the causes of all things from the most simple principles possible; but then they assume nothing as a principle, that is not proved by phenomena. They frame no hypotheses, nor receive them into philosophy otherwise than as questions whose truth may be disputed. They proceed therefore in a twofold method, synthetical and analytical. From some select phenomena they deduce by analysis the forces of Nature and the more simple laws of forces; and from thence by synthesis show the constitution of the rest. This is that incomparably best way of philosophizing, which our renowned author most justly embraced in preference to the rest, and thought alone worthy to be cultivated and adorned by his excellent labours. Of this he has given us a most illustrious example, by the explication of the System of the World, most happily deduced from the Theory of Gravity. That the attribute of gravity was found in all bodies, others suspected, or imagined before him, but he was the only and the first philosopher that could demonstrate it from appearances, and make it a solid foundation to the most noble speculations. [...] Since, then, all bodies, whether upon earth or in the heavens, are heavy, so far as we can make any experiments or observations concerning them, we must certainly allow that gravity is found in all bodies universally. And in like manner as we ought not to suppose that any bodies can be otherwise than extended, movable, or impenetrable, so we ought not to conceive that any bodies can be otherwise than heavy. The extension, mobility, and impenetrability of bodies become known to us only by experiments; and in the very same manner their gravity becomes known to us. All bodies upon which we can make any observations, are extended, movable, and impenetrable; and thence we conclude all bodies, and those concerning which we have no observations, are extended and movable and impenetrable. So all bodies on which we can make observations, we find to be heavy; and thence we conclude all bodies, and those we have no observations of, to be heavy also. If anyone should say that the bodies of the fixed stars are not heavy because their gravity is not yet observed, they may say for the same reason that they are neither extended nor movable nor impenetrable, because these properties of the fixed stars are not yet observed. In short, either gravity must have a place among the primary qualities of all bodies, or extension, mobility, and impenetrability must not. And if the nature of things is not rightly explained by the gravity of bodies, it will not be rightly explained by their extension, mobility, and impenetrability.

Some I know disapprove this conclusion, and mutter something about occult qualities. They continually are cavilling with us, that gravity is an occult property, and occult causes are to be quite banished from philosophy. But to this the answer is easy: that those are indeed occult causes whose existence is occult, and imagined but not proved; but not those whose real existence is clearly demonstrated by observations. Therefore gravity can by no means be called an occult cause of the celestial motions, because it is plain from the phenomena that such a power does really exist. Those rather have recourse to occult causes, who set imaginary vortices of a matter entirely fictitious and imperceptible by our senses, to direct those motions.

But shall gravity be therefore called an occult cause, and thrown out of philosophy, because the cause of gravity is occult and not yet discovered? Those who affirm this, should be careful not to fall into an absurdity that may overturn the foundations of all philosophy. For causes usually proceed in a continued chain from those that are more compounded to those that are more simple; when we arrived at the most simple cause we can go no farther. Therefore no mechanical account or explanation of the most simple cause is to be expected or given; for if it could be given, the cause were not the most simple. These most simple causes will you then call occult, and reject them? Then you must reject those that immediately depend upon them, and those which depend upon these last, till philosophy is quite cleared and disencumbered of all causes. [. . .]

Some there are who dislike this celestial physics because it contradicts the opinions of *Descartes*, and seems hardly to be reconciled with them. Let these enjoy their own opinion, but let them act fairly, and not deny the same liberty to us which they demand for themselves. Since the *Newtonian* Philosophy appears true to us, let us have the liberty to embrace and retain it, and to follow causes proved by phenomena, rather than causes only imagined and not yet proved. The business of true philosophy is to derive the natures of things from causes truly existent, and to inquire after those laws on which the Great Creator actually chose to found this most beautiful Frame of the World, not those by which he might have done the same, had he so pleased. It is reasonable enough to suppose that from several causes, somewhat differing from one another, the same effect may arise; but the true cause will be that from which it truly and actually does arise; the others have no place in true philosophy. The same motion of the hour-hand in a clock may be occasioned either by a weight hung, or a spring shut up within. But if a certain clock should be really moved with a weight, we should laugh at a man who would suppose it moved by a spring, and from that principle, suddenly taken up without further examination, should go about to explain the motion of the index; for certainly the way he ought to have taken would have been actually to look into the inward parts of the machine, that he might find the true principle of the proposed motion. The like judgement ought to be made of those philosophers who will have the heavens to be filled with a most subtile matter which is continually carried round in vortices. For if they could explain the phenomena ever so accurately by their hypotheses, we could not yet say that they have discovered true philosophy and the true causes of the celestial motions, unless they could either demonstrate that those causes do actually exist, or at least that no others do exist. Therefore if it be made clear that the attraction of all bodies is a property actually existing *in rerum natura*, and if it be also shown how the motions of the celestial bodies may be solved by that property, it would be very impertinent for anyone to object that these motions ought to be accounted for by vortices; even though we should allow such an explication of those motions to be possible. But we allow no such thing; for the phenomena can by no means be accounted for by vortices, as our author has abundantly proved from the clearest reasons. So that men must be strangely fond of chimeras, who can spend their time so idly as in patching up a ridiculous figment and setting it off with new comments of their own.

If the bodies of the planets and comets are carried round the sun in vortices, the bodies so carried, and the parts of the vortices next surrounding them, must be carried with the same velocity and the same direction, and have the same density, and the same inertia, answering to the bulk of the matter. But it is certain, the planets and comets, when in the very same parts of the heavens, are carried with various velocities and various directions. Therefore it necessarily follows that those parts of the celestial fluid, which are at the same distances from the sun, must revolve at the same time with different velocities in different directions; for one kind of velocity and direction is required for the motion of the planets, and another for that of the comets. But since this cannot be accounted for, we must either say that all celestial bodies are not carried about by vortices, or else that their motions are derived, not from one and the same vortex, but from several distinct ones, which fill and pervade the spaces round about the sun.

But if several vortices are contained in the same space, and are supposed to penetrate one another, and to revolve with different motions, then because these motions must agree with those of the bodies carried about by them, which are perfectly regular, and performed in conic sections which are sometimes very eccentric, and sometimes nearly circles, one may very reasonably ask how it comes to pass that these vortices remain entire, and have suffered no manner of perturbations in so many ages from the actions of the conflicting matter. Certainly if these fictitious motions are more compounded and harder to be accounted for than the true motions of the planets and comets, it seems to no purpose to admit them into philosophy, since every cause ought to be more simple than its effect. Allowing men to indulge their own fancies, suppose any man should affirm that the planets and comets are surrounded with atmospheres like our earth, which hypothesis seems more reasonable than that of vortices; let him then affirm that these atmospheres by their own nature move about the sun and describe conic sections, which motion is much more easily conceived than that of the vortices penetrating one another; lastly that the planets and comets are carried about the sun by these atmospheres of theirs: and then applaud his own sagacity in discovering the causes of the celestial motions. He that rejects this fable must also reject the other; for two drops of water are not more like than this hypothesis of atmospheres, and that of vortices. [. . .]

Bodies in going on through a fluid communicate their motion to the ambient fluid by little and little, and by that communication lose their own motion, and by losing it are retarded. Therefore the retardation is proportional to the motion communicated, and the communicated motion, when the velocity of the moving body is given, is as the density of the fluid; and therefore the retardation or resistance will be as the same density of the fluid; nor can it be taken away, unless the fluid, coming about to the hinder parts of the body, restore the motion lost. Now this cannot be done unless the impression of the fluid on the hinder parts of the body be equal to the impression of the fore parts of the body on the fluid; that is, unless the relative velocity with which the fluid pushes the body behind is equal to the velocity with which the body pushes the fluid; that is, unless the absolute velocity of the recurring fluid be twice as great as the absolute velocity with which the fluid is driven forwards by the body, which is impossible. Therefore the resistance of fluids arising from their inertia can by no means be taken away. So that we must conclude that the celestial fluid has no inertia, because it has no resisting force; that is has no force to communicate motion with, because it has no inertia; that it has no force to produce any change in one or more bodies, because it has no force wherewith to communicate motion; that it has no manner of efficacy, because it has no faculty wherewith to produce any change of any kind. Therefore certainly this hypothesis may be justly called ridiculous and unworthy a philosopher, since it is altogether without foundation and does not in the least serve to explain the nature of things. Those who would have the heavens filled with a fluid matter, but suppose it void of any inertia, do indeed in words deny a vacuum, but allow it in fact. For since a fluid matter of that kind can noways be distinguished from empty space, the dispute is now about the names and not the natures of things. [. . .]

Absolute space, time and motion

222

1. Absolute, true and mathematical time, of itself, and from its own nature, flows equably without relation to anything external, and by another name is called duration: relative, apparent, and common time, is some sensible and external (whether accurate or unequable) measure of duration by the means of motion, which is commonly used instead of true time; such as an hour, a day, a month, a year.

II. Absolute space, in its own nature, without relation to anything external, remains always similar and immovable. Relative space is some movable dimension or measure of the absolute spaces; which our senses determine by its position to bodies; and which is commonly taken for immovable space; such is the dimension of a subterraneous, an aerial, or celestial space, determined by its position in respect of the earth. Absolute and relative space are the same in figure and magnitude; but they do not remain always numerically the same. For if the earth, for instance, moves, a space of our air, which relatively and in respect of the earth remains always the same, will at one time be one part of the absolute space into which the air passes; at another time it will be another part of the same, and so, absolutely understood, it will be continually changed.

III. Place is a part of space which a body takes up, and is according to the space, either absolute or relative. I say, a part of space; not the situation, nor the external surface of the body. For the places of equal solids are always equal; but their surfaces, by reason of their dissimilar figures, are often unequal. Positions properly have no quantity, nor are they so much the places themselves, as the properties of places. The motion of the whole is the same with the sum of the motions of the parts; that is, the translation of the whole, out of its place, is the same thing with the sum of the translations of the parts out of their places; and therefore the place of the whole is the same as the sum of the places of the parts, and for that reason, it is internal, and in the whole body.

IV. Absolute motion is the translation of a body from one absolute place into another; and relative motion, the translation from one relative place into another. Thus in a ship under sail, the relative place of a body is that part of the ship which the body possesses; or that part of the cavity which the body fills, and which therefore moves together with the ship: and relative rest is the continuance of the body in the same part of the ship, or of its cavity. But real, absolute rest, is the continuance of the body in the same part of that immovable space, in which the ship itself, its cavity, and all that it contains, is moved. Wherefore, if the earth is really at rest, the body, which relatively rests in the ship, will really and absolutely move with the same velocity which the ship has on the earth. But if the earth also moves, the true and absolute motion of the body will arise, partly from the true motion of the earth, in immovable space, partly from the relative motion of the ship on the earth; and if the body moves also relatively in the ship, its true motion will arise, partly from the true motion of the earth, in immovable space, and partly from the relative motions as well of the ship on the earth, as of the body in the ship; and from these relative motions will arise the relative motion of the body on the earth. As if that part of the earth, where the ship is, was truly moved towards the east, with a velocity of 10010 parts; while the ship itself, with a fresh gale, and full sails, is carried towards the west, with a velocity expressed by 10 of those parts; but a sailor walks in the ship towards the east, with 1 part of the said velocity; then the sailor will be moved truly in immovable space towards the east, with a velocity of 10001 parts, and relatively on the earth towards the west, with a velocity of 9 of those parts.

The experiment of the rotating vessel

223

The effects which distinguish absolute from relative motion are the forces of receding from the axis of circular motion. For there are no such forces in a circular motion purely relative, but in a true and absolute circular motion, they are greater or less, according to the quantity of the motion. If a vessel, hung by a long cord, is so often turned about that the cord is strongly twisted, then filled with water, and held at rest together with the water; thereupon, by the sudden action of another force, it is whirled about the contrary way, and while the cord is untwisting itself, the vessel continues for some time in this motion; the surface of the water will at first be plain, as before the vessel began to move; but after that, the vessel, by gradually communicating its motion to the water, will make it begin sensibly to revolve, and recede by little and little from the middle, and ascend to the sides of the vessel, forming itself into a concave figure (as I have experienced), and the swifter the motion becomes, the higher will the water rise, till at last, performing its revolutions in the same times with the vessel, it becomes relatively at rest in it. This ascent of the water shows its endeavour to recede from the axis of its motion; and the true and absolute circular motion of the water, which is here directly contrary to the relative, becomes known, and may be measured by this endeavour. At first, when the relative motion of the water in the vessel was greatest, it produced no endeavour to recede from this axis; the water showed no tendency to the circumference, nor any ascent towards the sides of the vessel, but remained of a plain surface, and therefore its true circular motion had not yet begun. But afterwards, when the relative motion of the water had decreased, the ascent thereof towards the sides of the vessel proved its endeavour to recede from the axis; and this endeavour showed the real circular motion of the water continually increasing, till it had acquired its greatest quantity, when the water rested relatively in the vessel. And therefore this endeavour does not depend upon any translation of the water in respect of the ambient bodies, nor can true circular motion be defined by such translation.

224

Axioms, or laws of motion

LAW I

Every body continues in its state of rest, or of uniform motion in a right line, unless it is compelled to change that state by forces impressed upon it.

Projectiles continue in their motions, so far as they are not retarded by the resistance of the air, or impelled downwards by the force of gravity. A top, whose parts by their cohesion are continually drawn aside from rectilinear motions, does not cease its rotation, otherwise than as it is retarded by the air. The greater bodies of the planets and comets, meeting with less resistance in freer spaces, preserve their motions both progressive and circular for a much longer time.

LAW II

The change of motion is proportional to the motive force impressed; and is made in the direction of the right line in which that force is impressed.

If any force generates a motion, a double force will generate double the motion, a triple force triple the motion, whether that force be impressed altogether and at once, or gradually and successively. And this motion (being always directed the same way with the generating force), if the body moved before, is added to or subtracted from the former motion, according as they directly conspire with or are directly contrary to each other; or obliquely joined, when they are oblique, so as to produce a new motion compounded from the determination of both.

LAW III

To every action there is always opposed an equal reaction; or, the mutual actions of two bodies upon each other are always equal, and directed to contrary parts.

Whatever draws or presses another is as much drawn or pressed by that other. If you press a stone with your finger, the finger is also pressed by the stone. If a horse draws a stone tied to a rope, the horse (if I may so say) will be equally drawn back towards the stone; for the distended rope, by the same endeavour to relax or unbend itself, will draw the horse as much towards the stone as it does the stone towards the horse, and will obstruct the progress of the one as much as it advances that of the other. If a body impinge upon another, and by its force change the motion of the other, that body also (because of the equality of the mutual pressure) will undergo an equal change, in its own motion, towards the contrary part. The changes made by these actions are equal, not in the velocities but in the motions of bodies; that is to say, if the bodies are not hindered by any other impediments. For, because the motions are equally changed, the changes of the velocities made towards contrary parts are inversely proportional to the bodies. This law takes place also in attractions, as will be proved in the next Scholium.

225

Rules of reasoning in philosophy

RULE I
We are to admit no more causes of natural things than such as are both true and sufficient to explain their appearances.

To this purpose the philosophers say that Nature does nothing in vain, and more is in vain when less will serve; for Nature is pleased with simplicity, and affects not the pomp of superfluous causes.

RULE II
Therefore to the same natural effects we must, as far as possible, assign the same causes.

As to respiration in a man, and in a beast; the descent of stones in Europe and in America; the light of our culinary fire and of the sun; the reflection of light in the earth, and in the planets.

RULE III
The qualities of bodies, which admit neither intensification nor remission of degrees, and which are found to belong to all bodies within the reach of our experiments, are to be esteemed the universal qualities of all bodies whatsoever.

For since the qualities of bodies are only known to us by experiments, we are to hold for universal all such as universally agree with experiments; and such as are not liable to diminution can never be quite taken away. We are certainly not to relinquish the evidence of experiments for the sake of dreams and vain fictions of our own devising; nor are we to recede from the analogy of Nature, which is wont to be simple, and always consonant to itself. We no other way know the extension of bodies than by our senses, nor do these reach it in all bodies; but because we perceive extension in all that are sensible, therefore we ascribe it universally to all others also. That abundance of bodies are hard, we learn by experience; and because the hardness of the whole arises from the hardness of the parts, we therefore justly infer the hardness of the undivided particles not only of the bodies we feel but of all others. That all bodies are impenetrable, we gather not from reason, but from sensation. The bodies which we handle we find impenetrable, and thence conclude impenetrability to be an universal property of all bodies whatsoever. That all bodies are movable, and endowed with certain powers (which we call the inertia) of persevering in their motion, or in their rest, we only infer from the like properties observed in the bodies which we have seen. The extension, hardness, impenetrability, mobility, and inertia of the whole, result from the extension, hardness, impenetrability, mobility, and inertia of the parts, and hence we conclude the least particles of all bodies to be also all extended, and hard and impenetrable, and movable, and endowed with their proper inertia. And this is the foundation of all philosophy. Moreover, that the divided but contiguous particles of bodies may be separated from one another, is matter of observation; and, in the particles that remain undivided, our minds are able to distinguish yet lesser parts, as is mathematically demonstrated. But whether the parts so distinguished, and not yet divided, may, by the powers of Nature, be actually divided and separated from one another, we cannot certainly determine. Yet, had we the proof of but one experiment that any undivided particle, in breaking a hard and solid body, suffered a division, we might by virtue of this rule conclude that the undivided as well as the divided particles may be divided and actually separated to infinity.

Lastly, if it universally appears, by experiments and astronomical observations, that all bodies about the earth gravitate towards the earth, and that in proportion to the quantity of matter which they severally contain; that the moon likewise, according to the quantity of its matter, gravitates towards the earth; that, on the other hand, our sea gravitates towards the moon; and all the planets one towards another; and the comets in like manner towards the sun; we must, in consequence of this rule, universally allow that all bodies whatsoever are endowed with a principle of mutual gravitation. For the argument from the appearances concludes with more force for the universal gravitation of all bodies than for their impenetrability; of which, among those in the celestial regions, we have no experiments, nor any manner of observation. Not that I affirm gravity to be essential to bodies; by their *vis insita* I mean nothing but their inertia. This is immutable. Their gravity is diminished as they recede from the earth.

RULE IV

In experimental philosophy we are to look upon propositions inferred by general induction from phenomena as accurately or very nearly true, notwithstanding any contrary hypotheses that may be imagined, till such time as other phenomena occur, by which they may either be made more accurate, or liable to exceptions.

This rule we must follow, that the argument of induction may not be evaded by hypotheses.

226

General Scholium

The hypothesis of vortices is pressed with many difficulties. That every planet by a radius drawn to the sun may describe areas proportional to the times of description, the periodic times of the several parts of the vortices should observe the square of their distances from the sun; but that the periodic times of the planets may obtain the $\frac{3}{2}$th power of their distances from the sun, the periodic times of the parts of the vortex ought to be as the $\frac{3}{2}$th power of their distances. That the smaller vortices may maintain their lesser revolutions about Saturn, Jupiter, and other planets, and swim quietly and undisturbed in the greater vortex of the sun, the periodic times of the parts of the sun's vortex should be equal; but the rotation of the sun and planets about their axes, which ought to correspond with the motions of their vortices, recede far from all these proportions. The motions of the comets are exceedingly regular, are governed by the same laws with the motions of the planets, and can by no means be accounted for by the hypothesis of vortices; for comets are carried with very eccentric motions through all parts of the heavens indifferently, with a freedom that is incompatible with the notion of a vortex.

Bodies projected in our air suffer no resistance but from the air. Withdraw the air, as is done in *Mr Boyle's* vacuum, and the resistance ceases; for in this void a bit of fine down and a piece of solid gold descend with equal velocity. And the same argument must apply to the celestial spaces above the earth's atmosphere; in these spaces, where there is no air to resist their motions, all bodies will move with the greatest freedom; and the planets and comets will constantly pursue their revolutions in orbits given in kind and position, according to the laws above explained; but though these bodies may, indeed, continue in their orbits by the mere laws of gravity, yet they could by no means have at first derived the regular position of the orbits themselves from those laws.

The six primary planets are revolved about the sun in circles concentric with the sun, and with motions directed towards the same parts, and almost in the same plane. Ten moons are revolved about the earth, Jupiter, and Saturn, in circles concentric with them, with the same direction of motion, and nearly in the planes of the orbits of those planets; but it is not to be conceived that mere mechanical causes could give birth to so many regular motions, since the comets range over all parts of the heavens in very eccentric orbits; for by that kind of motion they pass easily through the orbs of the planets, and with great rapidity; and in their aphelions, where they move the slowest, and are detained the longest, they recede to the greatest distances from each other, and hence suffer the least disturbance from their mutual attractions. This most beautiful system of the sun, planets and comets, could only proceed from the counsel and dominion of an intelligent and powerful Being. And if the fixed stars are the centres of other like systems, these, being formed by the like wise counsel, must be all subject to the dominion of One; especially since the light of the fixed stars is of the same nature with the light of the sun, and from every system light passes into all the other systems, and lest the systems of the fixed stars should, by their gravity, fall on each other, he hath placed those systems at immense distances from one another. [. . .] Hitherto we have explained the phenomena of the heavens and of our sea by the power of gravity, but have not yet assigned the cause of this power. This is certain, that it must proceed from a cause that penetrates to the very centres of the sun and planets, without suffering the least diminution of its force; that operates not according to the quantity of the surfaces of the particles upon which it acts (as mechanical causes used to do), but according to the quantity of the solid matter which they contain, and propagates its virtue on all sides to immense distances, decreasing always as the inverse square of the distances. Gravitation towards the sun is made up out of the gravitations towards the several particles of which the body of the sun is composed; and in receding from the sun decreases accurately as the inverse square of the distances as far as the orbit of Saturn, as evidently appears from the quiescence of the aphelion of the planets; nay, and even to the remotest aphelion of the comets, if those aphelions are also quiescent. But hitherto I have not been able to discover the cause of those properties of gravity from phenomena, and I frame no hypotheses; for whatever is not deduced from the phenomena is to be called an hypothesis; and hypotheses, whether metaphysical or physical, whether of occult qualities or mechanical, have no place in experimental philosophy. In this philosophy particular propositions are inferred from the phenomena, and afterwards rendered general by induction. Thus it was that the impenetrability, the mobility, and the impulsive force of bodies, and the laws of motion and of gravitation, were discovered. And to us it is enough that gravity does really exist, and act according to the laws which we have explained, and abundantly serves to account for all the motions of the celestial bodies, and of our sea.

And now we might add something concerning a certain most subtle spirit which pervades and lies hid in all gross bodies; by the force and action of which spirit the particles of bodies attract one another at near distances, and cohere, if contiguous; and electric bodies operate to greater distances, as well repelling as attracting the neighbouring corpuscles; and light is emitted, reflected, refracted, inflected, and heats bodies; and all sensation is excited, and the members of animal bodies move at the command of the will, namely, by the vibrations of this spirit, mutually propagated along the solid filaments of the nerves, from the outward organs of sense to the brain, and from the brain into the muscles. But these are things that cannot be explained in few words, nor are we furnished with that sufficiency of experiments which is required to an accurate determination and demonstration of the laws by which this electric and elastic spirit operates.

227

The system of the world

1. THE MATTER OF THE HEAVENS IS FLUID

It was the ancient opinion of not a few, in the earliest ages of philosophy, that the fixed stars stood immovable in the highest parts of the world; that under the fixed stars the planets were carried about the sun; that the earth, as one of the planets, described an annual course about the sun, while by a diurnal motion it was in the meantime revolved about its own axis; and that the sun, as the common fire which served to warm the whole, was fixed in the centre of the universe.

This was the philosophy taught of old by *Philolaus*, *Aristarchus* of *Samos*, *Plato* in his riper years, and the whole sect of the *Pythagoreans*; and this was the judgement of *Anaximander*, more ancient still; and of that wise king of the *Romans*, *Numa Pompilius*, who, as a symbol of the figure of the world with the sun in the centre, erected a round temple in honour of *Vesta*, and ordained perpetual fire to be kept in the middle of it.

The *Egyptians* were early observers of the heavens; and from them, probably, this philosophy was spread abroad among other nations; for from them it was, and the nations about them, that the *Greeks*, a people more addicted to the study of philology than of Nature, derived their first, as well as soundest, notions of philosophy; and in the Vestal ceremonies we may yet trace the ancient spirit of the *Egyptians*; for it was their way to deliver their mysteries, that is, their philosophy of things above the common way of thinking, under the veil of religious rites and hieroglyphic symbols.

It is not to be denied that *Anaxagoras*, *Democritus*, and others, did now and then start up, who would have it that the earth possessed the centre of the world, and that the stars were revolved towards the west about the earth quiescent in the centre, some at a swifter, others at a slower rate.

However, it was agreed on both sides that the actions of the celestial bodies were performed in space altogether free and void of resistance. The whim of solid orbs was of a later date, introduced by *Eudoxus*, *Calippus*, and *Aristotle*; when the ancient philosophy began to decline, and to give place to the new prevailing fictions of the *Greeks*.

But, above all things, the phenomena of comets can by no means tolerate the idea of solid orbits. The *Chaldeans*, the most learned astronomers of their time, looked upon comets (which of ancient times before had been numbered among the celestial bodies) as a particular sort of planets, which, describing eccentric orbits, presented themselves to view only by turns, once in a revolution, when they descended into the lower parts of their orbits.

And as it was the unavoidable consequence of the hypothesis of solid orbits, while it prevailed, that the comets should be thrust into spaces below the moon;

so, when later observations of astronomers restored the comets to their ancient places in the higher heavens, these celestial spaces were necessarily cleared of the incumbrance of solid orbits.

2. THE PRINCIPLE OF CIRCULAR MOTION IN FREE SPACES

After this time, we do not know in what manner the ancients explained the question, how the planets came to be retained within certain bounds in these free spaces, and to be drawn off from the rectilinear courses, which, left to themselves, they should have pursued, into regular revolutions in curvilinear orbits. Probably it was to give some sort of satisfaction to this difficulty that solid orbs had been introduced.

The later philosophers pretend to account for it either by the action of certain vortices, as *Kepler* and *Descartes*; or by some other principle of impulse or attraction, as *Borelli*,[1] *Hooke*, and others of our nation; for, from the laws of motion, it is most certain that these effects must proceed from the action of some force or other.

But our purpose is only to trace out the quantity and properties of this force from the phenomena, and to apply what we discover in some simple cases as principles, by which, in a mathematical way, we may estimate the effects thereof in more involved cases; for it would be endless and impossible to bring every particular to direct and immediate observation.

We said, *in a mathematical way*, to avoid all questions about the nature or quality of this force, which we would not be understood to determine by any hypothesis; and therefore call it by the general name of a centripetal force, as it is a force which is directed towards some centre; and as it regards more particularly a body in that centre, we call it circumsolar, circumterrestrial, circumjovial; and so in respect of other central bodies.

3. THE ACTION OF CENTRIPETAL FORCES

That by means of centripetal forces the planets may be retained in certain orbits, we may easily understand, if we consider the motions of projectiles; for a stone that is projected is by the pressure of its own weight forced out of the rectilinear path, which by the initial projection alone it should have pursued, and made to describe a curved line in the air; and through that crooked way is at last brought down to the ground; and the greater the velocity is with which it is projected, the farther it goes before it falls to the earth. We may therefore suppose the velocity to be so increased, that it would describe an arc of 1, 2, 5, 10, 100, 1000 miles before it arrived at the earth, till at last, exceeding the limits of the earth, it should pass into space without touching it.

Let *AFB* represent the surface of the earth, *C* its centre, *VD*, *VE*, *VF* the curved lines which a body would describe, if projected in an horizontal direction from the top of an high mountain successively with more and more velocity; and, because the celestial motions are scarcely retarded by the little or no resistance

1. Giovanni Borelli (1608–1679), Italian physiologist and physicist. (Ed).

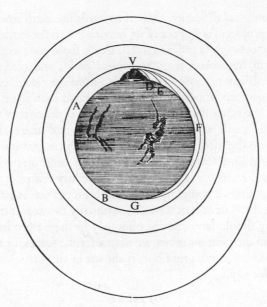

The orbits of a projectile thrown off the earth with different initial velocities

of the spaces in which they are performed, to keep up the parity of cases, let us suppose either that there is no air about the earth, or at least that it is endowed with little or no power of resisting; and for the same reason that the body projected with a less velocity describes the lesser arc *VD*, and with a greater velocity the greater arc *VE*, and, augmenting the velocity, it goes farther and farther to *F* and *G*, if the velocity was still more and more augmented, it would reach at last quite beyond the circumference of the earth, and return to the mountain from which it was projected.

And since the areas which by this motion it describes by a radius drawn to the centre of the earth are (by Prop. I, Book I, *Princip. Math.*) proportional to the times in which they are described, its velocity, when it returns to the mountain, will be no less than it was at first; and, retaining the same velocity, it will describe the same curve over and over, by the same law.

But if we now imagine bodies to be projected in the directions of lines parallel to the horizon from greater heights, as of 5, 10, 100, 1000, or more miles, or rather as many semidiameters of the earth, those bodies, according to their different velocity, and the different force of gravity in different heights, will describe arcs either concentric with the earth, or variously eccentric, and go on revolving through the heavens in those orbits just as the planets do in their orbits.

4. THE CERTAINTY OF THE PROOF

As when a stone is projected obliquely, that is, any way but in the perpendicular direction, the continual deflection thereof towards the earth from the right line in which it was projected is a proof of its gravitation to the earth, no less certain than its direct descent when suffered to fall freely from rest; so the deviation of bodies moving in free spaces from rectilinear paths, and continual deflection therefrom towards any place, is a sure indication of the existence of some force which from all quarters impels those bodies towards that place.

And as, from the supposed existence of gravity, it necessarily follows that all bodies about the earth must press downwards, and therefore must either descend directly to the earth, if they are let fall from rest, or at least continually deviate from right lines towards the earth, if they are projected obliquely; so, from the supposed existence of a force directed to any centre, it will follow, by the like necessity, that all bodies upon which this force acts must either descend directly to that centre, or at least deviate continually towards it from right lines, if otherwise they should have moved obliquely in these right lines.

And how from the motions given we may infer the forces, or from the forces given we may determine the motions, is shown in the first two Books of our *Principles of Philosophy*.

OPTICKS

228

PROPOSITION I, THEOREM I
Lights which differ in colour, differ also in degrees of refrangibility

229

PROPOSITION II, THEOREM II
The light of the sun consists of rays differently refrangible

The PROOF by Experiments

Exper. 3. In a very dark Chamber, at a round Hole, about one third Part of an Inch broad, made in the Shut of a Window, I placed a Glass Prism, whereby the Beam of the Sun's Light, which came in at that Hole, might be refracted upwards toward the opposite Wall of the Chamber, and there form a colour'd Image of the Sun. The Axis of the Prism (that is, the Line passing through the middle of the Prism from one end of it to the other end parallel to the edge of the Refracting Angle) was in this and the following Experiments perpendicular to the incident Rays. About this Axis I turned the Prism slowly, and saw the refracted Light on the Wall, or coloured Image of the Sun, first to descend, and then to ascend. Between the Descent and Ascent, when the Image seemed Stationary, I stopp'd the Prism, and fix'd it in that Posture, that it should be moved no more. For in that Posture the Refractions of the Light at the two Sides of the refracting Angle, that is, at the Entrance of the Rays into the Prism, and at their going out of it, were equal to one another. [. . .] And in this Posture, as the most convenient, it is to be understood that all the Prisms are placed in the following Experiments, unless where some other Posture is described. The Prism therefore being placed in this Posture, I let the refracted Light fall perpendicularly upon a Sheet of white Paper at the opposite Wall of the Chamber, and observed the Figure and Dimensions of the Solar Image formed on the Paper by that Light. This Image was Oblong and not Oval, but terminated with two Rectilinear and Parallel Sides, and two Semicircular Ends. On its Sides it was bounded pretty distinctly, but on its Ends very confusedly and indistinctly, the Light there decaying and vanishing by degrees. The Breadth of this Image answered to the Sun's Diameter, and was about two Inches and the eighth Part of an Inch, including the Penumbra. [. . .] For let *EG* represent the Window-shut, *F* the hole made therein through which a beam of the Sun's Light was

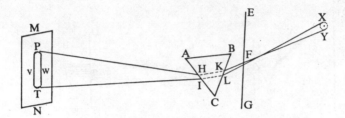

Spectrum created by the refraction of a sunbeam in a prism

transmitted into the darkened Chamber, and *ABC* a Triangular Imaginary Plane whereby the Prism is feigned to be cut transversely through the middle of the Light. Or if you please, let *ABC* represent the Prism itself, looking directly towards the Spectator's Eye with its nearer end: And let *XY* be the Sun, *MN* the Paper upon which the Solar Image or Spectrum is cast, and *PT* the Image itself whose sides towards *v* and *w* are Rectilinear and Parallel, and ends towards *P* and *T* Semicircular. [. . .]

And therefore seeing by Experience it is found that the Image is not round, but about five times longer than broad, the Rays which going to the upper end *P* of the Image suffer the greatest Refraction, must be more refrangible than those which go to the lower end *T*, unless the Inequality of Refraction be casual.

This Image or Spectrum *PT* was coloured, being red at its least refracted end *T*, and violet at its most refracted end *P*, and yellow green and blue in the intermediate Spaces. Which agrees with the first Proposition, that Lights which differ in colour, do also differ in Refrangibility. The length of the Image in the foregoing Experiments, I measured from the faintest and outmost red at one end, to the faintest and outmost blue at the other end, excepting only a little Penumbra, whose breadth scarce exceeded a quarter of an Inch, as was said above.

Exper. 6. In the middle of two thin Boards I made round holes a third part of an Inch in diameter, and in the Window-shut a much broader hole being made to let into my darkened Chamber a large Beam of the Sun's Light; I placed a Prism behind the Shut in that beam to refract it towards the opposite Wall, and close behind the Prism I fixed one of the Boards, in such a manner that the middle of the refracted Light which came through the hole in the first Board, and fell upon the opposite Wall, might pass through the hole in this other Board, and the rest being intercepted by the Board might paint upon it the coloured Spectrum of the Sun. And close behind this Board I fixed another Prism to

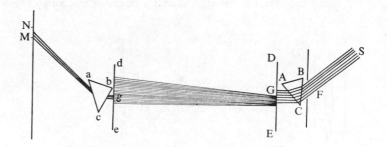

Refraction of light passing through two prisms

refract the Light which came through the hole. Then I returned speedily to the first Prism, and by turning it slowly to and fro about its Axis, I caused the Image which fell upon the second Board to move up and down upon that Board, that all its parts might successfully pass through the hole in that Board and fall upon the Prism behind it. And in the mean time, I noted the places on the opposite Wall to which that Light after its Refraction in the second Prism did pass; and by the difference of the places I found that the Light which being most refracted in the first Prism did go to the blue end of the Image, was again more refracted in the second Prism than the Light which went to the red end of that Image, which proves as well the first Proposition as the second. And this happened whether the Axis of the two Prisms were parallel, or inclined to one another, and to the Horizon in any given Angles.

The hypothesis of 'fits'

231

PROPOSITION XII
Every Ray of Light in its passage through any refracting Surface is put into a certain transient Constitution or State, which in the progress of the Ray returns at equal

Intervals, and disposes the Ray at every return to be easily transmitted through the next refracting Surface, and between the returns to be easily reflected by it. [. . .]

What kind of action or disposition this is; Whether it consists in a circulating or a vibrating motion of the Ray, or of the Medium, or something else, I do not here enquire. Those that are averse from assenting to any new Discoveries, but such as they can explain by an Hypothesis, may for the present suppose, that as Stones by falling upon Water put the Water into an undulating Motion, and all Bodies by percussion excite vibrations in the Air; so the Rays of Light, by impinging any refracting or reflecting Surface, excite vibrations in the refracting or reflecting Medium or Substance, and by exciting them agitate the solid parts of the refracting or reflecting Body, and by agitating them cause the Body to grow warm or hot; that the vibrations thus excited are propagated in the refracting or reflecting Medium or Substance, much after the manner that vibrations are propagated in the Air for causing Sound, and move faster than the Rays so as to overtake them; and that when any Ray is in that part of the vibration which conspires with its Motion, it easily breaks through a refracting Surface, but when it is in the contrary part of the vibration which impedes its Motion, it is easily reflected; and, by consequence, that every Ray is successfully disposed to be easily reflected, or easily transmitted, by every vibration which overtakes it. But whether this Hypothesis be true or false I do not here consider. I content my self with the bare Discovery, that the Rays of Light are by some cause or other alternately disposed to be reflected or refracted for many vicissitudes.

232

DEFINITION

The returns of the disposition of any Ray to be reflected I will call its Fits *of easy* Reflection, *and those of its disposition to be transmitted its* Fits *of easy Transmission, and the space it passes between every return and the next return, the* Interval *of its* Fits.

PROPOSITION XIII

The reason why the Surfaces of all thick transparent Bodies reflect part of the Light incident on them, and refract the rest, is, that some Rays at their Incidence are in Fits of easy Reflexion, and others in Fits of easy Transmission.

This may be gather'd from the 24th Observation, where the Light reflected by thin Plates of Air and Glass, which to the naked Eye appear'd evenly white all over the Plate, did through a Prism appear waved with many Successions of Light and Darkness made by alternate Fits of easy Reflexion and easy Transmission, the Prism severing and distinguishing the Waves of which the white reflected Light was composed, as was explain'd above.

And hence Light is in Fits of easy Reflexion and easy Transmission, before its Incidence on transparent Bodies. And probably it is put into such fits at its first emission from luminous Bodies, and continues in them during all its progress. For these Fits are of a lasting nature, as will appear by the next part of this Book.

In this Proposition I suppose the transparent Bodies to be thick; because if the thickness of the Body be much less than the Interval of the Fits of easy Reflexion and transmission of the Rays, the Body loseth its reflecting power. For if the Rays, which at their entering into the Body are put into Fits of easy Transmission, arrive at the farthest Surface of the Body before they be out of those Fits, they must be transmitted. And this is the reason why Bubbles of Water lose their reflecting power when they grow very thin; and why all opake Bodies, when reduced into very small parts, become transparent.

The properties and actions of the aether

233

Query 18. If in two large tall cylindrical Vessels of Glass inverted, two little Thermometers be suspended so as not to touch the Vessels, and the Air be drawn out of one of these Vessels, and these Vessels thus prepared be carried out of a cold place into a warm one; the Thermometer *in vacuo* will grow warm as much, and almost as soon as the Thermometer which is not *in vacuo*. And when the Vessels are carried back into the cold place, the Thermometer *in vacuo* will grow cold almost as soon as the other Thermometer. Is not the Heat of the warm Room convey'd through the *Vacuum* by the Vibrations of a much subtiler Medium than Air, which after the Air was drawn out remained in the *Vacuum*? And is not this Medium the same with that Medium by which Light is refracted and reflected, and by whose Vibrations Light communicates Heat to Bodies, and is put into Fits of easy Reflexion and easy Transmission? And do not the Vibrations of this Medium in hot Bodies contribute to the intenseness and duration of their Heat? And do not hot Bodies communicate their Heat to contiguous cold ones, by the Vibrations of this Medium propagated from them into the cold ones? And is not this Medium exceedingly more rare and subtile than the Air, and exceedingly more elastick and active? And doth it not readily pervade all Bodies? And is it not (by its elastick force) expanded through all the Heavens?

234

Query 19. Doth not the Refraction of Light proceed from the different density of this Æthereal Medium in different places, the Light receding always from the denser parts of the Medium? And is not the density thereof greater in free and open Spaces void of Air and other grosser Bodies, than within the Pores of Water, Glass, Crystal, Gems, and other compact Bodies? For when Light passes through Glass or Crystal, and falling very obliquely upon the farther Surface thereof is totally reflected, the total Reflexion ought to proceed rather from the density and vigour of the Medium without and beyond the Glass, than from the rarity and weakness thereof.

235

Query 20. Doth not this Æthereal Medium in passing out of Water, Glass, Crystal, and other compact and dense Bodies into empty Spaces, grow denser and denser by degrees, and by that means refract the Rays of Light not in a point, but by bending them gradually in curve Lines? And doth not the gradual condensation of this Medium extend to some distance from the Bodies, and thereby cause the Inflexions of the Rays of Light, which pass by the edges of dense Bodies, at some distance from the Bodies?

236

Query 21. Is not this Medium much rarer within the dense Bodies of the Sun, Stars, Planets and Comets, than in the empty celestial Spaces between them? And in passing from them to great distances, doth it not grow denser perpetually, and thereby cause the gravity of those great Bodies towards one another, and of their parts towards the Bodies; every Body endeavouring to go from the denser parts of the Medium towards the rarer? For if this Medium be rarer within the Sun's Body than at its Surface, and rarer there than at the hundredth part of an Inch from its Body, and rarer there than at the fiftieth part of an Inch from its Body, and rarer there than at the Orb of *Saturn*; I see no reason why the Increase of density should stop any where, and not rather be continued through all distances from the Sun to *Saturn*, and beyond. And though this Increase of density may at great distances be exceeding slow, yet if the elastick force of this Medium be exceeding great, it may suffice to impel Bodies from the denser parts of the Medium towards the rarer, with all that power which we call Gravity. And that the elastick force of this Medium is exceeding great, may be gather'd from the swiftness of its Vibrations. Sounds move about 1140 *English* Feet in a second Minute of Time, and in seven or eight Minutes of Time they move about one hundred *English* Miles. Light moves from the Sun to us in about seven or eight Minutes of Time, which distance is about 70,000,000 *English* Miles, supposing the horizontal Parallax of the Sun to be about 12″. And the Vibrations or Pulses of this Medium, that they may cause the alternate Fits of easy Transmission and easy Reflexion, must be swifter than Light, and by consequence above 700,000 times swifter than Sounds. And therefore the elastick force of this Medium, in proportion to its density, must be above 700,000 × 700,000 (that is, above 490,000,000,000) times greater than the elastick force of the Air is in proportion to its density. For the Velocities of the Pulses of elastic Mediums are in a subduplicate *Ratio* of the Elasticities and the Rarities of the Mediums taken together.

As Attraction is stronger in small Magnets than in great ones in proportion to their Bulk, and Gravity is greater in the Surfaces of small Planets than in those of great ones in proportion of their bulk, and small Bodies are agitated much more by electric attraction than great ones; so the smallness of the Rays of Light may contribute very much to the power of the Agent by which they are refracted. And so if any one should suppose that *Æther* (like our Air) may contain Particles which endeavour to recede from one another (for I do not know what this *Æther* is) and that its Particles are exceedingly smaller than those of Air, or even than those of Light: The exceeding smallness of its Particles may contribute to the greatness of the force by which those Particles may recede from one another, and thereby make that Medium exceedingly more rare and elastick than Air, and by consequence exceedingly less able to resist the motions of Projectiles, and exceedingly more able to press upon gross Bodies, by endeavouring to expand it self.

237

Query 22. May not Planets and Comets, and all gross Bodies, perform their Motions more freely, and with less resistance in this Æthereal Medium than in any Fluid, which fills all Space adequately without leaving any Pores, and by consequence is much denser than Quick-silver or Gold? And may not its resistance be so small, as to be inconsiderable? For instance; If this *Æther* (for so I will call it) should be supposed 700,000 times more elastick than our Air, and above 700,000 times more rare; its resistance would be above 600,000,000 times less than that of Water. And so small a resistance would scarce make any sensible alteration in the Motions of the Planets in ten thousand Years. If any one would ask how a Medium can be so rare, let him tell me how the Air, in the upper parts of the Atmosphere, can be above an hundred thousand times rarer than Gold. Let him also tell me, how an electrick Body, can by Friction emit an Exhalation so rare and subtile, and yet so potent, as by its Emission to cause no sensible Diminution of the weight of the electrick Body, and to be expanded through a Sphere, whose Diameter is above two Feet, and yet to be able to agitate and carry up Leaf Copper, or Leaf Gold, at the distance of above a Foot from the electrick Body? And how the Effluvia of a Magnet can be so rare and subtile, as to pass through a Plate of Glass without any Resistance or Diminution of their Force, and yet so potent as to turn a magnetick Needle beyond the Glass?

238

Query 28. Are not all Hypotheses erroneous, in which Light is supposed to consist in Pression or Motion, propagated through a fluid Medium? For in all these Hypotheses the Phenomena of Light have been hitherto explain'd by supposing that they arise from new Modifications of the Rays; which is an erroneous Supposition.

If Light consisted only in Pression propagated without actual Motion, it would not be able to agitate and heat the Bodies which refract and reflect it. If it consisted in Motion propagated to all distances in an instant, it would require an infinite force every moment, in every shining Particle, to generate that Motion. And if it consisted in Pression or Motion, propagated either in an instant or in time, it would bend into the Shadow. For Pression or Motion cannot be propagated in a Fluid in right Lines, beyond an Obstacle which stops part of the Motion, but will bend and spread every way into the quiescent Medium which lies beyond the Obstacle. Gravity tends downwards, but the Pressure of Water arising from Gravity tends every way with equal Force, and is propagated as readily, and with as much force sideways as downwards, and through crooked passages as through strait ones. The Waves on the Surface of stagnating Water, passing by the sides of a broad Obstacle which stops part of them, bend afterwards and dilate themselves gradually into the quiet Water behind the Obstacle. The Waves, Pulses or Vibrations of the Air, wherein Sounds consist, bend manifestly, though not so much as the Waves of Water. For a Bell or a Cannon may be heard beyond a Hill which intercepts the sight of the sounding Body, and Sounds are propagated as readily through crooked Pipes as through straight ones. But Light is never known to follow crooked Passages nor to bend into the Shadow. For the fix'd Stars by the Interposition of any of the Planets cease to be seen. And so do the Parts of the Sun by the Interposition of the Moon, *Mercury* or *Venus*. The Rays which pass very near to the edges of any Body, are bent a little by the action of the Body, as we shew'd above; but this bending is not towards but from the Shadow, and is perform'd only in the passage of the Ray by the Body, and at a very small distance from it. So soon as the Ray is past the Body, it goes right on. [. . .]

And it is as difficult to explain by these Hypotheses, how Rays can be alternately in Fits of easy Reflexion and easy Transmission; unless perhaps one might suppose that there are in all Space two Æthereal vibrating Mediums, and that the Vibrations of one of them constitute Light, and the Vibrations of the other are swifter, and as often as they overtake the Vibrations of the first, put them into those Fits. But how two *Æthers* can be diffused through all Space, one of which acts upon the other, and by consequence is re-acted upon, without retarding, shattering, dispersing and confounding one anothers' Motions, is inconceivable. And against filling the Heavens with fluid Mediums, unless they be exceeding rare, a great Objection arises from the regular and very lasting Motions of the Planets and Comets in all manner of Courses through the Heavens. For thence it is manifest, that the Heavens are void of all sensible Resistance, and by consequence of all sensible Matter. [. . .]

239

Query 31. Have not the small Particles of Bodies certain Powers, Virtues, or Forces, by which they act at a distance, not only upon the Rays of Light for reflecting, refracting, and inflecting them, but also upon one another for producing a great Part of the Phenomena of Nature? For it's well known, that Bodies act one upon another by the Attractions of Gravity, Magnetism, and Electricity; and these Instances shew the Tenor and Course of Nature, and make it not improbable but that there may be more attractive Powers than these. For Nature is very consonant and conformable to her self. How these Attractions may be perform'd, I do not here consider. What I call Attraction may be perform'd by impulse, or by some other means unknown to me. I use that Word here to signify only in general any Force by which Bodies tend towards one another, whatsoever be the Cause. For we must learn from the Phenomena of Nature what Bodies attract one another, and what are the Laws and Properties of the Attraction, before we enquire the Cause by which the Attraction is perform'd. The Attractions of Gravity, Magnetism, and Electricity, reach to very sensible distances, and so have been observed by vulgar Eyes, and there may be others which reach to so small distances as hitherto escape Observation; and perhaps electrical Attraction may reach to such small distances, even without being excited by Friction. [. . .]

GOTTFRIED WILHELM LEIBNIZ

Conservation of force versus the conservation of quantity
of motion

240

The most famous proposition of the Cartesians is that the same quantity of motion is conserved in things. They have given no demonstration of this, however, for no one can fail to see the weakness of their argument derived from the constancy of God. For although the constancy of God may be supreme, and he may change nothing except in accordance with the laws of the series already laid down, we must still ask what it is, after all, that he has decreed should be conserved in the series—whether the quantity of motion or something different, such as the quantity of force. I have proved that it is rather this latter which is conserved, that this is distinct from the quantity of motion, and that it often happens that the quantity of motion changes while the quantity of force remains permanent. The arguments by which I have shown this and defended it against objections may be read elsewhere. But, since the matter is of great importance, I shall give the heart of my conception in a brief example. Assume two bodies: A with a mass of 4 and a velocity of 1, and B with a mass of 1 and a velocity of 0, that is, at rest. Now imagine that the entire force of A is transferred to B, that is, that A is reduced to rest and B alone moves in its place. We ask what velocity B must assume. According to the Cartesians, the answer is that B should have a velocity of 4, since the original quantity of motion and the present quantity would then be equal, since mass 4 multiplied by velocity 1 is equal to mass 1 multiplied by velocity 4. Thus the increase in velocity is proportional to the decrease of the quantity of the body. But in my opinion the answer should be that *B*, whose mass is 1, will receive the velocity 2, in order to have only as much quantity of power as *A*, whose mass is 4 and whose velocity is 1. I shall explain my reason for this as briefly as possible, lest I appear to have proposed it without any reason. I say, then, that *B* will have only as much force as *A* had previously or that the present and the former force are equal, a thing which is worth proving. To go deeper, namely, and explain the true method of computation—which is the duty of any really universal mathematics, though it has not yet been carried out—it is clear, first of all, that force is doubled, tripled, or quadrupled when its simple quantity is repeated twice, three times, or four times, respectively. So two bodies of equal mass and velocity will have twice as much force as one of them. It does not follow, however, that one body with twice the velocity must have only twice the force of a body with simple velocity, for even though the degree of velocity may be doubled, the subject of this velocity is not itself duplicated, as it is when a body twice as great, or two bodies of the same velocity, are taken in place of one, so that they completely repeat the one in magnitude as well as motion. Similarly 2 pounds elevated to the height of 1 foot are exactly double in essence and power to one elevated the same distance, and two elastic bodies stretched equally are double one of them. But when the two bodies possessing this power are not fully homogeneous and cannot be compared with each other in this way, or reduced to a common measure of matter and force, an indirect comparison must be attempted by comparing their homogeneous effects or causes. Every cause whatever has a force equal to its total effect, or to the effect which it produces in using up its own force. Therefore, since the two bodies mentioned above, *A* with mass 4 and velocity 1, and *B* with mass 1 and velocity 2, are not exactly comparable, and no one quantity possessing force can be designated whose simple repetition will produce both, we must examine their effects. Let us assume, namely, that these two bodies are heavy and that *A* can change its direction and rise; then by virtue of its velocity of 1 it will rise to the height of 1 foot, while *B*, by virtue of its velocity of 2, will rise 4 feet, as Galileo and others have demonstrated. In each case the effect will entirely consume the force and so be equal to the cause which produced it. But these two effects are equal to each other in force or power, namely, the elevation of body *A*, 4 pounds, to 1 foot, and the elevation of body *B*, 1 pound, to 4 feet. Therefore the causes, too, are equal; that is, the body *A* of 4 pounds with velocity of 1 is equal in force or power to the body *B* of 1 pound with velocity 2, as was asserted. But if someone denies that the same power is needed to raise 4 pounds 1 foot and 1 pound 4 feet, or that these two effects are equivalent (though, unless I am mistaken, almost everyone will admit this), he can be convinced by the same principle. If we take a balance with unequal arms, 4 pounds can be raised exactly 1 foot by the descent of 1 pound for 4 feet, and no further work can be done; thus the effect exactly exhausts the power of the cause and is equal to it in force. Let me summarize, therefore. If the whole power of body *A*, 4, with a velocity of 1, is transferred to *B*, 1, *B* must receive a velocity of 2; or, what amounts to the same thing, if *B* is first at rest and *A* in motion, but *A* is then at rest and *B* has been placed in motion, other things remaining equal, the velocity of *B* must be double, since the mass of *A* was quadruple. If, as is popularly held, *B* should receive four times the velocity of *A* because it has one-fourth of its mass, we should have perpetual motion or an effect more powerful than its cause. For, when *A* was moving, it could raise 4 pounds only 1 foot, or 1 pound 4 feet; but later, when *B* moved, it would be able to lift 1 pound 16 feet, for altitudes are as the square of the velocities by force of which bodies are lifted, and four times the velocity will raise a body sixteen times the altitude. With the aid of *B* not only could we thus once more raise *A* for 1 foot, after its descent had given it its original velocity, but we could do many other things besides and thus exhibit perpetual motion, since the original force is restored but there is still more left. Moreover, even though the assumption that the whole force of *A* is transmitted to *B* cannot actually be realized, this does not affect the matter, since we are here concerned with the true calculation, or with the question of how much force *B* would necessarily take on according to this hypothesis. Even if a part of the force is retained and only part transmitted, the same absurdities would still arise, for, if the quantity of motion is to be conserved, the quantity of forces can obviously not always be conserved, since the quantity of motion is known to be the product of mass and velocity, while the quantity of force is, as we have shown, the product of mass and the altitude to which it can be raised by force of its power, altitudes being proportional to the square of the velocities of ascent. Meanwhile this rule can be set up: The same quantity of force as well as of motion is conserved when bodies tend in the same directions both before and after their collision, as well as when the colliding bodies are equal.

'Vis viva' and 'vis morta'

241

Hence force is also of two kinds: the one elementary, which I also call *dead* force, because motion does not yet exist in it but only a solicitation to motion, such as that of the ball in the tube or a stone in a sling even while it is still held by the string; the other is ordinary force combined with actual motion, which I call *living* force [*vis viva*]. An example of dead force is centrifugal force, and likewise the force of gravity or centripetal force; also the force with which a stretched elastic body begins to restore itself. But in impact, whether this arises from a heavy body which has been falling for some time, or from a bow which has been restoring itself for some time, or from some similar cause, the force is living and arises from an infinite number of continuous impressions of dead force. This is what Galileo meant when, in an enigmatic way, he called the force of impact infinite as compared with the simple impulsion of gravity.

Relative space and relative time

242

As for my own opinion, I have said more than once that I hold space to be something merely relative, as time is; that I hold it to be an order of co-existences as time is an order of successions. For space denotes, in terms of possibility, an order of things which exist at the same time, considered as existing together, without inquiring into their particular manner of existing. And when many things are seen together, one perceives that order of things among themselves. [...] The case is the same with respect to time. Supposing anyone should ask why God did not create everything a year sooner, and the same person should infer from thence that God has done something concerning which 'tis not possible there should be a reason why he did it so and not otherwise; the answer is that his inference would be right if time was anything distinct from things existing in time. For it would be impossible there should be any reason why things should be applied to such particular instants rather than to others, their succession continuing the same. But then the same argument proves that instants, considered without the things, are nothing at all and that they consist only in the successive order of things, which order remaining the same, one of the two states, viz., that of a supposed anticipation, would not at all differ, nor could be discerned from the other which now is.

The laws of motion and final causes

243

The supreme wisdom of God has made him choose especially those *laws of motion* which are best adjusted and most fitted to abstract or metaphysical reasons. There is conserved the same quantity of total and absolute force or of action, also the same quantity of relative force or of reaction, and, finally, the same quantity of directive force. Furthermore, action is always equal to reaction, and the entire effect is always equal to its full cause. It is surprising that no reason can be given for the laws of motion which have been discovered in our own time, and part of which I myself have discovered, by a consideration of *efficient causes* or of matter alone. For I have found that we must have recourse to *final causes* and that these laws do not depend upon the *principle of necessity*, as do the truths of logic, arithmetic, and geometry, but upon the *principle of fitness*, that is to say, upon the choice of wisdom. This is one of the most effective and obvious proofs of the existence of God for those who can probe into these matters thoroughly.

PIERRE LOUIS MOREAU DE MAUPERTUIS

The principle of the least quantity of action

244

If we are willing to acknowledge the fact, it will be agreed that the most compelling reason for admitting of the existence only of elastic bodies has been the inability to discover the laws of the communication of motion of hard bodies.

Descartes accepted the fact of the existence of these bodies, and believed that he had discovered their laws of motion. He had started from a relatively convincing principle: *that the quantity of motion is always conserved as the same in nature.* The laws that he deduced from it were false, because the principle was invalid.

The philosophers who came after him were struck by another form of *conservation*: that of what is called *living force*, which is *the product of each mass by the square of its velocity*. These philosophers did not base their laws of motion on this conservation; they deduced this conservation from the laws of motion, seeing it to be a consequence of these laws. However, since the conservation of living force took place only in the collision of elastic bodies, the view that there were no bodies at all other than these in nature became strengthened.

But *the conservation of the quantity of motion is true only in certain cases. The conservation of living force takes place only for certain bodies.* Thus, neither can be taken as a universal principle, or even as a generalized result of the laws of motion.

If we examine the principles on which the authors who gave us these laws based their reasoning and the paths that they followed, it is astounding to see how successfully they arrived at their conclusions, and we can not help thinking that they relied less on these principles than on trial and error. Those of them who reasoned the most correctly recognized that the principle that they were using to explain the communication of motion of elastic bodies could not be applied to the communication of motion of *hard* bodies. Finally, none of the principles that have been used until now, either for the laws of motion of hard bodies or for the laws of motion of elastic bodies, can be extended to the laws of rest.

Following upon so many great men who have applied themselves to this matter, I hardly dare to say that I have discovered the universal principle on which all these laws are based, which is equally applicable to hard bodies and elastic bodies, and which governs the motion and rest of all bodily substances.

It is what I call the principle *of the least quantity of action*, a principle that is so wise and so worthy of the Supreme Being, and to which nature appears to be so constantly subject that she observes it not only in all her changes but still tends to observe it in her permanence. *In the collision of bodies, the motion is distributed in such a way that the quantity of action that is presupposed by the change having taken place is the smallest possible. When at rest, bodies which are being held in equilibrium must be so situated that if any slight motion were imparted to them the quantity of action would be the least.*

This principle is consonant with the conception that we hold of the Supreme Being not only in that He must always act in the wisest manner, but also in that He must always hold everything beneath his domination.

The principle put forward by Descartes seemed to remove the world from the realm of the Divinity; it ruled that whatever the changes that took place in nature, *the same quantity of motion was always conserved*, whereas experiments and reasoning more cogent than his showed the contrary to be true. The principle of the conservation of living force, again, would appear to place the world in some kind of state of independence: whatever the changes that occur in nature, the absolute quantity of this force would always be conserved and could always reproduce the same effects. But for this to be so it would be necessary for there to be only elastic bodies in nature; it would be necessary to exclude hard bodies, that is, to exclude what may possibly be the only kind that exist.

My principle, which is more consonant with the conceptions that we should have of things, leaves the world in continual need of the power of the Creator and is an inevitable consequence of the wisest use of this power. [. . .]

I must now explain what I mean by the quantity of action. When a body is moved from one point to another, a certain action is required in order to effect this. This action depends upon the velocity that the body possesses and the distance that it covers, but it is neither the velocity nor the distance taken separately. The quantity of action is greater, the greater the velocity of the body and the longer the distance that it covers. It is proportional to the sum of the distances, each multiplied by the velocity at which the body covers them. [. . .]

It can not be doubted that all things are controlled by a Supreme Being who, while He has impressed upon matter forces that denote His power, has destined it to carry out effects that mark His wisdom; and the harmony of these two attributes is so perfect that all the effects of nature could without doubt be deduced from each of them taken separately. Blind and inevitable Mechanics follow the designs of the most enlightened and liberal intelligence; and if our minds were sufficiently all-embracing, they would also see the causes of the physical effects, either by calculating the properties of these bodies or by attempting to discover what it would be more fitting to cause them to perform.

The first of these methods is the more within our scope, but it does not take us very far. The second sometimes misleads us, because our comprehension of the purpose of nature is totally inadequate, and because we may be mistaken about the *quantity*, which we should regard as its *consumption* in the production of its effects.

In order to combine wideness of scope with safety in our researches, we must employ both of these methods. Let us calculate the motions of bodies by all means, but let us also consult the designs of the intelligence that causes them to move.

It seems that the ancient philosophers made the initial attempts at this kind of mathematics. They sought metaphysical relationships in the properties of numbers and bodies, and when they said that God's business was geometry, they doubtless meant it only of the branch of learning which compares the workings of His power with the designs of His wisdom.

JAKOB BERNOULLI

The law of large numbers

245

[Previously] it was shown how, from the numbers of cases for proof of something either to exist or not exist and either to indicate it or not indicate it (or even to indicate its converse), its provability and its thence derived probability can be deduced and estimated by means of calculation. This brings us to the point where all that is necessary for the correct calculation of the probability of something is first to determine the numbers of these cases precisely and then to define how much more readily the one set of cases can occur than the other. But it seems to me that it is precisely in this that the real difficulty arises, since it is possible to do so only in very rare instances, and even then almost exclusively in games of chance, which, in order that the players should have equal chances of winning, were devised by their original inventors in such a way that the numbers of cases of either success or failure are certain and known and that all these cases can occur with equal facility. In by far the greater majority of all other matters, whether governed by the laws of nature or by human judgement, this is by no means so. Thus, for instance, with dice the numbers of cases are known, because for each individual die there are obviously as many cases as it has sides; all these cases are of equal likelihood since, owing to the identical shape of the sides and to the uniformly distributed weight of the die, there is no reason why any one side of the die should be more likely to be thrown than another, which would be so if the sides of the die were of different shapes or if one part of the die were made of a heavier material than another part. In the same way, the numbers of cases are known for drawing either a white or a black pebble out of an urn, and it is known that all the pebbles have an equal likelihood of being drawn, because it is known indisputably how many pebbles of each colour there are in the urn and there can be no reason why this or that pebble should be more likely to be drawn than any other. Yet who among mortals could determine the number, for instance, of the diseases (i.e. the equivalent number of cases) that can attack the innumerable parts of the human body at any age and cause death, define how much more likely one disease is than another (e.g. plague than dropsy, dropsy than fever) to destroy man, and thence form a conjecture as to life and death of future generations? Or who could count up the innumerable changes that are undergone daily by the air and from this to conjecture what its composition will be a month or even a year from now? Or again, who could study the nature of the human mind or the wonderful structure of our bodies in sufficient detail to venture, in games that depend wholly or partly upon the acumen of the former or the agility of the latter, to determine the numbers of possibilities for this or that player to win or lose? Since these and all similar matters are dependent upon completely obscure causes which, moreover, in the infinite variety of their combinations will forever elude our diligent attempts at comprehension, it would be quite senseless to set about studying something in this manner.

But there is in fact another course open to us here, by which we can obtain what we are seeking and can determine what we are unable to ascertain *a priori* at least *a posteriori*, i.e. from the result that has been observed in a large number of similar examples. To do so, it has to be assumed that, given identical circumstances, any individual event has the same number of possible cases of either occurring or not occurring henceforth as it has previously been observed to have had of occurring or not occurring. If, for instance, it has been observed by means of trials that, of three hundred men of the same age and physical condition as that of Titius at present, two hundred of them died before a furher ten years had elapsed while the remainder lived longer, it is reasonably safe to assume that there are twice as many possibilities that Titius too will have to pay nature's final debt within the next ten years than possibilities that he will live beyond this time. Just as, if someone has been observing the weather for many years past and has noted how often it has been fine or rainy, or again, if someone has watched two players very often and has seen how often the one or the other has won, from these same observations he can determine the ratio that probably exists between the numbers of possible cases, given circumstances identical with those gone before, for the same events likewise either to occur or not occur henceforth.

But this empirical method of observing the numbers of cases by means of trials is neither new nor uncommon, for the celebrated author of *L'Art de Penser*, a man of great acumen and talent,[1] has described a very similar method in Chapter 12 *et seq.* of the last part of his book, and everyone continually observes the same thing in daily life. And it must be obvious to everyone that, in order to decide about a certain event or other by this method, it is not sufficient to have used only one or two trials, but that a large number of trials is necessary. Indeed, even quite stupid men, guided by some natural instinct or other and completely of their own accord, without any previous training (which is truly remarkable), have at times ascertained that, the more observations of this kind that have been made, the less danger there will be of deviating from the truth. But although this is obvious to everyone from the very nature of things, the proof of it, as based on scientific principles, is by no means generally known, and it is therefore incumbent on me to give it here. I should, however, feel that I had not done

1. Antoine Arnauld (1612–1694), whose book on logic appeared anonymously in 1664. (Ed.)

enough if I were to stop short at the proof of this one point that is known to everyone. Rather, there is also something else which should be taken into consideration, something to which possibly nobody has even given thought before now. Namely, it also remains to investigate whether, when the number of observations is increased in this way, the probability of finding the true ratio between the numbers of possibilities for an event either to occur or not occur respectively also increases continuously, and this to the extent that this probability eventually exceeds any given degree of certainty, or whether the problem has, as it were, its asymptote, i.e. whether there is some fixed degree of certainty which can never be exceeded, however far the number of observations is increased: for instance, that we can never be sure beyond a $\frac{1}{2}$, $\frac{2}{3}$ or $\frac{3}{4}$ degree of certainty that we have determined the true ratio of the possible cases.

To give an example of what I mean, I shall assume that, unknown to you, three thousand white and two thousand black pebbles have been placed in an urn and that you, wishing to calculate their number by means of trials, draw out one pebble after another (but in such a way that each time you replace the pebble that you have just drawn out before you draw out another, so that the number of pebbles in the urn never decreases) and observe how often a white pebble is drawn and how often a black one. The question arises as to whether you would be able to do this often enough for it to be ten, a hundred or a thousand, etc., times more probable (i.e. eventually, morally certain) that the ratio of the number of times that a white pebble is drawn to the number of times that a black pebble is drawn will be the same ratio of one and a half which exists between the numbers of pebbles (or possibilities), rather than some other ratio different from this. If this is not so, I acknowledge that our attempt to determine the numbers of possibilities by means of trials will come to nothing. If, however, it is so, and in this manner we can eventually attain moral certainty (and I shall show that this is in fact so in the next chapter), we can determine the numbers of possibilities *a posteriori* almost as accurately as if they had been known to us *a priori*. And for the purposes of normal daily life, in which the morally certain is regarded as absolutely certain, in accordance with Axiom 9 of Chapter II this is quite sufficient for our method of conjecture to hold good for any topic pertaining to chance no less scientifically than in games of chance. For if instead of the urn we substitute the air, for instance, or the human body, which encompass an enormous number of different kinds of changes and diseases in just the same way as the urn contains pebbles, we shall assuredly, in exactly the same manner by means of observations, be able to determine how much more likely this or that event can occur in these matters.

In order that this should not be misinterpreted, however, it must be emphasized that I do not mean that the ratio between the numbers of cases that we set about determining by means of trials will be found with absolute accuracy (if this were so, the exact opposite would be true, and the probability that the true ratio has been found would decrease with increasing number of observations), but that the ratio will be found to a certain degree of approximation only, i.e. contained between two limits which can, however, be set as close to each other as may be wished. If, in the example of the urn and pebbles given above, we assume two ratios, e.g. 301/200 and 299/200, or 3001/2000 and 2999/2000, etc., one of which is slightly greater and the other slightly smaller than one and a half, it is found that it becomes more and more probable, with any given degree of probability, that the ratio that is found by means of frequently repeated trials will lie within these limits of the ratio of one and a half than outside them.

This, then, is the problem which I have undertaken to publish in the present work, after having pursued it for the last twenty years; both its novelty and its extraordinarily wide applicability, coupled with its equally considerable difficulty, add to the significance and weight of all the other chapters in the book. Before moving on to its solution, however, I shall pause briefly to refute the objections which certain learned persons have raised against these maxims.

Firstly, they argue that the ratio between the pebbles is of a different nature from that between the diseases or the changes undergone by the air, since the number of the former is fixed while that of the latter is undefined and uncertain. My reply to this is that, as far as our knowledge is concerned, both are equally uncertain and undefined. It is, however, as difficult for us to comprehend that something is so in itself and by its very nature as it is for us to comprehend that God has simultaneously created and yet not created something; for everything that God has created has, by this very act, been defined by Him.

Secondly, they argue that the number of pebbles is finite, while that of the diseases, etc., is infinite. My reply to this is that the latter number is stupendously large rather than infinite; but even if we assume that it is infinite, it is known that even between two infinite numbers there can exist a defined ratio which can be expressed in finite numbers either precisely or at least to as close an approximation as anyone could wish. Thus, for instance, there always exists between the circumference of a circle and its diameter a defined ratio that can only be expressed precisely by an infinitely large number of decimal places of Ludolph's number, but which has been confined by Archimedes, Metius and Ludolph[1] himself to limits that are more than sufficiently narrow for its practical use. There is, therefore, no reason why the ratio between two infinite numbers which can, however, itself be expressed in finite numbers to a very close approximation should not likewise be determined by means of a finite number of trials.

Thirdly, they argue that the number of diseases does not remain constant, but that new ones are discovered every day. My reply to this is that I do not deny that the number of diseases can increase in the course of time, and it is certainly true that anyone trying to use present-day observations to reach conclusions regarding antediluvian times would deviate far from the truth. All that this means, however, is that new observations have to be made from time to

---

1. Adrian Metius (1571–1635) and Ludolph van Ceulen (1540–1610) calculated π in several approximations. (Ed.)

time; just as, indeed, they would have to be made in the case of the pebbles, if it were supposed that the number of them in the urn had altered.

DANIEL BERNOULLI

Gas pressure and the corpuscular hypothesis

246

1. We shall now consider elastic fluids, and can formulate a definition of them which encompasses all the properties that have so far been established for them while leaving room for the inclusion of further properties that have not yet

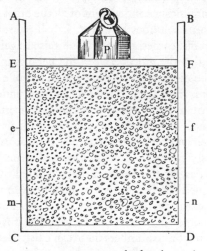

Gas pressure in a cylindrical vessel

been studied in sufficient detail. The principal properties of elastic fluids, however, consist in the following: (i) they have weight, (ii) they spread in all directions unless they are contained; and (iii) they can be continuously compressed more and more by increasing forces of compression. Air satisfies this definition, and it is to air that our present considerations will refer in particular.

2. Imagine, therefore, a vertically positioned cylindrical vessel *ACDB*, and on it a movable lid *EF* on which a weight *P* presses down. The volume *ECDF* contains very small corpuscles that are propelled in different directions in very rapid motion; these corpuscles, which strike against the lid *EF* and support it by their continuously repeated impact, constitute an elastic fluid which expands when the weight *P* is removed or reduced and, when it is increased, contracts and moves down towards the horizontal base *CD* just as if it possessed no elastic properties at all. For whether the corpuscles are at rest or in motion their gravity does not change, so that the base supports sometimes the weight and sometimes the elasticity of the fluid. We shall take air as a fluid of this kind which matches

the principal properties of elastic fluids, and explain some properties that have already been detected in it and also illustrate others that have not yet been investigated in sufficient detail.

3. We shall regard the corpuscles enclosed within the volume of the cylinder as though they were infinite in number, and with them occupying the space *ECDF* we shall say that they form natural air, a standard to which everything is related. Thus, the weight *P* which holds down the lid *EF* in position is no different from the pressure of the atmosphere lying above it, which we shall therefore designate henceforth by *P*.

It should, however, be noted that this pressure is by no means equal to the absolute weight of the vertical cylinder of air lying above the lid *EF* in the atmosphere, as has been inadvisedly stated by some authors; rather, it is equal to the fourth member of the proportion obtaining between the earth's surface, the size of the lid *EF* and the weight of the entire atmosphere on the earth's surface. [. . .]

6. The elasticity of air, however, is increased not only by its contraction but also by increase in heat, and since it is known that heat spreads out in all directions with increasing internal motion of the particles, it follows that increased elasticity of the air, which does not change the volume, indicates intensified motion in the particles of air, which is in good agreement with our hypothesis. For it is obvious that the weight *P* that is required to keep the air contained in the volume *ECDF* will be greater, the greater the velocity with which the particles of air move. Indeed, it is not difficult to see that the weight *P* will be proportional to the square of this velocity, because an increase in velocity increases both the number of impacts and their intensity equally together; separately, each is proportional to the weight *P*.

JOHANN BERNOULLI

The principle of virtual velocities

247

DEFINITION I

1. *Virtual velocities* are those which two or more forces held in equilibrium acquire when a slight motion is imparted to them or if these forces are already in motion. The *virtual velocity* is the component velocity gained or lost by each body, of a velocity that has already been acquired, in an infinitely small period of time and depending on its direction. [. . .]

HYPOTHESIS I
2. *Two acting causes are in equilibrium, or have equal moments, when their absolute forces are in inverse ratio to their virtual velocities, whether the interacting forces are in motion or at rest.*

This is a standard principle of Statics and Mechanics, and I shall not stop to prove it here. I shall rather use it to demonstrate how motion is produced under the effect of a pressure which acts continuously and without any opposition other than that from the inertia of the moving body.

3. Let us assume two bodies *A* and *B* to be at rest, with a compressed spring *C* between them; the spring, in starting to unwind, exerts an equal force on both sides to move the bodies *A* and *B* away from each other. It can be seen that each

Explanation of the virtual velocity

of these bodies will, by its inertia, oppose the motion of the spring with a resistance proportional to its mass. By virtue of the hypothesis taken from Mechanics, it must therefore be true that, since the two opposing forces of the spring are equal, the force of inertia in *A* is to the force of inertia in *B*, or the mass *A* is to the mass *B*, in inverse ratio to what the virtual velocity of body *B* is to the virtual velocity of body *A*. And as the phenomenon is continuous, while the spring, in expanding, accelerates the velocity of these bodies, it is obvious that their accelerations are always in inverse ratio to the masses *A* and *B*, which constitutes a constant ratio. And consequently, the velocities that are acquired respectively in the same period of time, which are nothing other than the sums of the virtual velocities successively produced by the force of the spring, are also in the same ratio. That is to say, the velocity of *B* is to the velocity of *A* as *A* is to *B*. Hence it follows that, when the spring *C* is fully unwound or checked by some obstacle that prevents it from unwinding completely, the two bodies *A* and *B* will continue to move at the last velocities that were acquired under the successive impression of the spring.

JEAN LE ROND D'ALEMBERT

The laws of dynamics based on the laws of statics

248

Motion and its general properties are the first and foremost object of the study of mechanics; this science presupposes the existence of motion and we shall also take this to be accepted by all physicists. Philosophers, on the other hand, are much divided concerning the nature of motion. Nothing would seem more natural, I admit, than to conceive motion as the successive placing of the moving body at different parts of infinite space, which we think of as the place of the bodies: but this idea presupposes a space, the parts of which are penetrable and immobile: yet all are aware that the Cartesians (a sect which does not number very many nowadays) do not recognize that space is different from the bodies; they regard extension and matter as one and the same thing. One must admit that if we start from such a principle, motion will be a most difficult thing to understand and that a Cartesian should rather deny its existence than to attempt to define its nature. Yet, however absurd the opinions of these philosophers seem to us, and however little clarity and precision can be found in those metaphysical principles on which they endeavour to lean, we shall not try to refute them here; we shall be content to remark that in order to have a clear idea of motion, one cannot avoid distinguishing, even if only mentally, between two different sorts of extension: the one will be regarded as impenetrable; this is the one which we generally call bodies; the other, considered simply as extension, without examining whether it is penetrable or not, shall be the measure of the distance between one body and the other; the parts of this second sort of space, considered fixed and immobile, can serve to distinguish between moving bodies and those at rest. It will be thus legitimate to conceive of infinite space as the real or imaginary place of bodies, and to regard motion as the displacement of the moving body from one place to the other. [. . .]

It is thus clearly seen that a body cannot move itself. It cannot be moved from its rest without the action of an external cause. But will it continue to move of itself or will it require repeated action of the cause to be kept moving? Whatever side we take on this issue one thing is incontestable: having presupposed the existence of motion without any other special hypothesis, the simplest law according to which a body could possibly move will be the law of uniformity; and, indeed, this is the law which it must follow, as we shall see in more details in the first chapter of this treatise. Motion is thus uniform by its nature. I admit that the proofs that have been advanced so far for this principle were perhaps far from convincing; I will show in my work the difficulties that occur on opposing them, and also the way I have chosen in order to avoid resolving them. It seems to me that this law of uniformity, essential for motion considered as such, furnishes one of the best reasons to support the measurement of time by uniform motion. Moreover, I believe that I had to enter into details on this topic, although basically this discussion might appear strange in Mechanics.

Having established that there is the force of inertia, i.e. the property of bodies to persevere in their respective state of motion or state of rest, it is clear that a motion that needs a cause to come into existence will not be able to accelerate or retard without an external cause. If so, which are the causes that can create motion or produce changes in the motions of bodies? So far we know of two sorts: the ones which manifest themselves simultaneously with the effect they produce, or rather which they occasion; these are the causes that have their source in the perceivable, mutual interactions of the bodies, as a result of their impenetrability. These are reducible to those of impact and some other effects deriving from it. All the other causes can be recognized only by their effects and we do not know anything about their nature. Such is the cause that makes heavy bodies fall towards the centre of the earth, the cause that retains the planets in their orbits, etc.

We shall soon see how to determine the effects of impact and other causes which can be related to it. Treating here only the second kind of causes, it is clear that when it is a question of the effect produced by these, the effects must always be given independently of the knowledge of the cause, as these cannot be deduced from it. Thus it is that without knowing the cause of gravity, we learn from experience that the spaces covered by a falling body relate to each other as the squares of the times. In general, for non-uniform motions, the causes of which are unknown, it is evident that the effect produced by the cause, be it an instantaneous effect or one that takes finite time, must always be given by an equation connecting times and distances. This effect being known, and the principle of inertia being presupposed, geometry and arithmetic will suffice for the investigation of the properties of this sort of motion. Why should we then have recourse to that nowadays universally accepted principle that the accelerating or retarding force is proportional to the increment of velocity, a principle that is supported only by the indefinite and obscure axiom that the effect is proportional to the cause? We shall not examine whether this principle is a necessary truth but only admit that the proofs advanced for it so far do not seem to us to be flawless. Neither shall we adopt it, with some geometers, as a purely contingent truth; this would destroy the certitude of Mechanics and reduce it to no more than an experimental science. We shall be content with the remark that this principle, true or dubious, clear or obscure, is absolutely unnecessary for Mechanics and thus it must be excluded.

Hitherto we have talked only about the change in the speed of a moving body, due to causes capable of altering its motion; we have not yet investigated what happens if the cause of the motion tends to deflect the body from its original direction. All that the principle of the force of inertia can teach us in this case is that the body has to describe a straight line and move on it uniformly; this does not give us either its speed or its direction. Therefore we must have recourse to a second principle, the one called the principle of the composition of motions, and by it to determine the single motion of a body which tends at the same time to move in different directions with given speeds. A new demonstration of this principle will be found in this work; in it I have proposed to avoid all those difficulties of which this proof commonly suffers without using a great many complicated propositions, for this principle, being one of the first in mechanics, should rest on simple and easy proofs.

Just as the motion of a body undergoing change of direction can be regarded as composed of its previous and newly acquired motions, so the previous motion of the body could be regarded as a composition of a newly received motion and of another which it has lost. It follows from this that the laws of a motion which has been changed by any sort of obstacle depend solely on the laws of the motion which has been destroyed by this same obstacle. Evidently it is enough to decompose the motion of the body before the encounter with the obstacle into two other motions: one that the obstacle would not interfere with and the other which would be annihilated by the encounter. Thus one cannot only demonstrate the laws of the motion changed by insurmountable obstacles, the only ones found so far by this method, but one can also determine in which cases this obstacle will annihilate the motion. As to the laws of motions changed by obstacles which are not insurmountable as such, it is clear for the same reason that in general for the determination of these laws no more is needed than an exact knowledge of the laws of equilibrium.

What is then the general law of the equilibrium of bodies? All geometers agree that two bodies with opposite directions of motion are in equilibrium if their masses are in the inverse ratio to the speeds with which they tend to move; but possibly it is difficult to find a rigorous demonstration for this law that would leave nothing obscure; that is why most of the geometers have preferred to treat this as an axiom instead of attempting to prove it. However, on closer scrutiny, it will be evident that there is only one case where equilibrium manifests itself clearly and distinctly: this is the case when the masses of the two bodies are equal and their velocities equal and of opposite directions. It seems to me that the only way one could follow to prove this principle in the other cases is to reduce these, when possible, to this first, simple and self-evident case. This is the course I have tried to follow; the reader will have to judge whether I have succeeded.

The principle of equilibrium together with the principle of the force of inertia and that of composite motion leads to the solution of all problems connected with the motion of a body insofar as it can be changed through the influence of an impenetrable, mobile obstacle, that is generally by any other body, to which it must communicate motion in order to retain at least part of its own. From a combination of these principles one can easily deduce the laws of motion of bodies that collide in any manner, or that exercise a pull on each other by means of some other body interposed between the two, to which both are connected. [. . .]

When speaking of the force of bodies in motion, one either does not attach any clear idea to the word one utters, or one means only a very general property of moving bodies of being able to overcome the obstacles which get in their way or which resist their motion. This force cannot be directly estimated either from the distance the body has covered, or from the time that it took to cover that distance, or from the simple and abstract consideration of its mass and velocity alone; forces can be estimated only from the obstacles a body encounters and from the resistance these obstacles offer it. The more considerable the obstacle which the body succeeds in overcoming, or which it is capable of resisting, the larger can its force be considered to be, provided one does not wish to represent by that word an imaginary being residing in the body, but uses it only as an abbreviation to express a fact. This is somewhat like saying that one body has twice the speed of another instead of saying that in the same time it covers twice the distance, without claiming by the use of the word 'speed' that it represents a being inherent in the body.

This being agreed upon, it is clear that there are three sorts of obstacles that a moving body can encounter: either invincible obstacles that completely destroy the motion, however large it might have been; or obstacles that offer precisely that resistance which is necessary to destroy the motion of the body, and to accomplish this in one moment: this is the case of equilibrium; or, finally, obstacles that destroy the motion slowly: this is the case of retarded motion. As the invincible obstacles annihilate equally any motion, these cannot help us to estimate the force. One could therefore look for its measure only in the case of equilibrium or of retarded motion. There is general agreement that equilibrium is established between two bodies if the products of their masses by their virtual velocities, i.e. the velocities with which they tend to move, are equal for both bodies. Thus, for the case of equilibrium, the product of mass and velocity, or, what is the same, the quantity of motion, can represent the force. Also, it is generally accepted that in retarded motion the number of obstacles which have been overcome is proportional to the square of the velocity; in this way a body which has compressed a spring when moving with a given velocity, when moving with twice that velocity could compress either at once or in succession not two, but four such springs, nine if its velocity trebled, etc. From this the partisans of the living force conclude that the force of a body in motion is in general proportional to the product of its mass by the square of its velocity. Fundamentally what drawback is there if force has a different measure for the cases of equilibrium and of retarded motion if one lays down clear ideas, and by the word 'force' means only the effect produced by the overcoming of the obstacle or that of resisting to it. Still, it must be admitted that the view of those who regard the force as a product of mass and velocity could stand a test not only for the case of equilibrium but also for the case of retarded motion, if in this second case one does not use the absolute quantity of the obstacles as a measure of the force but rather the sum of their resistances. For, there should be no doubt that this sum of resistances is proportional to the quantity of motion; as is generally acknowledged, the quantity of motion that the body loses in every instant is proportional to the product of resistance and the infinitesimal duration of the instant and that the sum of these products is evidently the total resistance. The whole difficulty is reduced thus to the problem of knowing whether one should measure the force by the absolute quantity of the obstacles or by the sum of their resistances. It would seem more natural to measure the force in the second manner because an obstacle is such only by virtue of its offering a resistance, and the correct expression for the obstacle which had been overcome is the sum of its resistances. Moreover, estimating the force in this way, one has the advantage of having a common measure for the case of equilibrium and of retarded motion. Nevertheless, as we cannot attach any precise meaning to the word 'force' except when limiting the term to express only the effect, I believe that one should leave it to every person to decide for himself in this matter, and the whole question will then consist only in a futile, metaphysical discussion, or in verbal dispute, not worthy to occupy the philosophers.

What has been said so far should suffice to clarify the matter to our readers. But a very simple consideration will convince them finally. Let us take a body with a simple tendency to move with a certain velocity, and let this tendency be opposed by an obstacle; let it move really and uniformly with that same velocity; let it finally begin moving with the selfsame velocity and then let this be consumed slowly and at the end be completely destroyed by any cause whatsoever: in all these cases the effect produced by the body is different, but the body itself which we have considered has nothing more to it in one case than in the other; only the action of the cause which produces the effect is differently applied. In the first case the effect is reduced to a simple tendency which cannot be properly measured with any precision because it does not result in any motion; in the second case the effect is the distance covered uniformly in a given time, and this effect is proportional to the velocity; in the third, the effect is the distance covered, until the total extinction of the motion, and this effect is proportional to the square of the velocity. These different effects are thus clearly products of the same cause; accordingly, those who say that the force is in one case proportional to the velocity and in another to the square of the same, could possibly be talking of the effect only, when expressing themselves in this way. This diversity of effects proceeding from one and the same cause can serve, by the way, to show us how little precision inheres in that quasi-axiom, so often used, about the proportionality of cause and effect. [. . .]

We believe to have demonstrated in this work that a body left to itself will persist eternally in its state of rest or uniform motion; we also believe to have shown that if that body tends to move at the same time along two sides of a parallelogram, the direction along which it will move will be that of the diagonal, and it will, so to say, choose it from all the others. Finally, we have demonstrated that all the laws of communication of motion between bodies are reducible to the laws of equilibrium, and that the laws of equilibrium themselves can be reduced to those of the equilibrium of two equal bodies tending to move in opposite directions with equal virtual velocities. In this last case the two bodies will destroy each other's motion; also, as a result of geometrical considerations a case of equilibrium results if the ratio of the two masses is equal to the inverse ratio of their velocities; the only thing still to be established is whether the case of equilibrium is unique, i.e. whether, if the masses are not in the inverse proportion of the velocities, one of the bodies will necessarily compel the other to move. But, it is easy to see that since there is a possible and necessary case of equilibrium, there is no need for the others: if this was not so, the laws of collision, which necessarily are reduced to the laws of equilibrium, would become indeterminate; but this should not occur, since a body, having collided with another, must necessarily bring forth a unique effect, an indispensable result of the existence and impenetrability of these bodies. Moreover, one can demonstrate the uniqueness of the law of equilibrium by another reasoning, but it is too mathematical to be given in this preliminary discourse, yet I shall endeavour to make it comprehensible in my treatise, to which I refer the reader.

From all these reflections it follows that the laws of Statics and of Mechanics, as expressed in this book, are those that follow from the existence of matter and motion. Experience proves us that these laws are in effect observed by the bodies that form our environment. Thus the laws of equilibrium and motion, those that observation has made us acquainted with, are truly necessary. A metaphysician would perhaps content himself with saying that it was the Creator's wisdom and the simplicity of His views not to have established other laws of equilibrium and motion than those which result from the very existence of the bodies and their mutual impenetrability; but we have come to believe in the necessity of abstaining from such reasoning as it seemed to us to lean on a very vague principle; the nature of the Supreme Being is too much hidden from us to recognize directly what does or does not conform with the designs of His wisdom; we can at best catch some glimpses of this wisdom in the observation of the laws of nature, while mathematical reasoning shows us the simplicity of these laws and when experience teaches us their application and extent.

It seems to me that these thoughts can serve to make us evaluate the demonstrations given by various philosophers of the laws of motion as being in accord with the principle of final causes, that is to say with the designs of the Author of Nature in establishing these laws. Such proofs can be convincing only insofar as they are preceded and supported by direct demonstrations and have been derived from principles which are within our reach; otherwise they could often lead us into error. It is for having followed that path, for having believed that it was the Creator's wisdom to conserve always the same quantity of motion in the universe, that Descartes was mistaken about the laws of collision. Those who imitate him run the risk of either being deceived like him, or taking for a general principle something that takes place only in special cases, or finally of regarding a purely mathematical consequence of some formula as a fundamental law of nature.

IMMANUEL KANT

The origins of the solar system

249

I have chosen a subject which is capable of exciting an unfavourable prejudice in a great number of my readers at the very outset, both on account of its own intrinsic difficulty, and also from the way they may regard it from the point of view of religion. To discover the system which binds together the great members of the creation in the whole extent of infinitude, and to derive the formation of the heavenly bodies themselves, and the origin of their movements, from the primitive state of nature by mechanical laws, seems to go far beyond the power of human reason. On the other hand, religion threatens to bring a solemn accusation against the audacity which would presume to ascribe to nature by itself results in which the immediate hand of the Supreme Being is rightly recognized; and it is troubled with concern, by finding in the ingenuity of such views an apology for atheism. I see all these difficulties well, and yet am not discouraged. I feel all the strength of the obstacles which rise before me, and yet I do not despair. I have ventured, on the basis of a slight conjecture, to undertake a dangerous expedition; and already I discern the promontories of new lands. Those who will have the boldness to continue the investigation will occupy them, and may have the satisfaction of designating them by their own names.

I did not enter on the prosecution of this undertaking until I saw myself in security regarding the duties of religion. My zeal was redoubled when at every step I saw the clouds disperse that appeared to conceal monsters behind their darkness; and when they were scattered I saw the glory of the Supreme Being break forth with the brightest splendour. As I now know that these efforts are free from everything that is reprehensible, I shall faithfully adduce all that well-disposed or even weak minds may find repellent in my scheme; and I am ready to submit to the judicial severity of the orthodox Areopagus with a frankness which is the mark of an honest conviction. The advocate of the faith may therefore be first allowed to make his reasons heard, in something like the following terms:

'If the structure of the world with all its order and beauty,' he says, 'is only an effect of matter left to its own universal laws of motion, and if the blind mechanics of the natural forces can evolve so glorious a product out of chaos, and can attain to such perfection of themselves, then the proof of the Divine Author which is drawn from the spectacle of the beauty of the universe wholly loses its force. Nature is thus sufficient for itself; the Divine government is unnecessary; Epicurus lives again in the midst of Christendom, and a profane philosophy tramples under foot the faith which furnishes the clear light needed to illuminate it.' [. . .]

I will therefore not deny that the theory of Lucretius, or his predecessors, Epicurus, Leucippus and Democritus, has much resemblance with mine. I assume, like these philosophers, that the first state of nature consisted in a universal diffusion of the primitive matter of all the bodies in space, or of the atoms of matter, as these philosophers have called them. Epicurus asserted a gravity or weight which forced these elementary particles to sink or fall; and this does not seem to differ much from Newton's Attraction, which I accept. He also gave them a certain deviation from the straight line in their falling movement, although he had absurd fancies regarding the causes and consequences of it. This deviation agrees in some degree with the alteration from the falling in a straight line, which we deduce from the repulsion of the particles. Finally, the vortices which arose from the disturbed motion, is also a theory of Leucippus and Democritus, and it will be also found in our scheme. So many points of affinity with a system which constituted the real theory of all denial of God in antiquity do not, however, draw my system into community with its errors. Something true will always be found even in the most nonsensical opinions that have ever obtained the consent of men. A false principle, or a couple of unconsidered conjunctive propositions, lead men away from the high footpath of truth by unnoticed by-paths into the abyss. Notwithstanding the similarity indicated, there yet remains an essential difference between the ancient cosmogony and that which I present, so that the very opposite consequences are to be drawn from mine.

The teachers of the mechanical production of the structure of the world referred to, derive all the order which may be perceived in it from mere chance which made the atoms to meet in such a happy concourse that they constituted a well-ordered whole. Epicurus had the hardihood to maintain that the atoms diverged from their straight motion without a cause, in order that they might encounter one another. All these theorizers pushed this absurdity so far that they even assigned the origin of all animated creatures to this blind concourse, and actually derived reason from the irrational. In my system, on the contrary, I find matter bound to certain necessary laws. Out of its universal dissolution and dissipation I see a beautiful and orderly whole quite naturally developing itself. This does not take place by accident, or of chance, but it is perceived that natural qualities necessarily bring it about. And are we not thereby moved to ask, why matter must just have had laws which aim at order and conformity? Was it possible that many things, each of which has its own nature independent of the others, should determine each other of themselves just in such a way that a well-ordered whole should arise therefore; and if they do this, is it not an undeniable proof of the community of their origin at first, which must have been a universal Supreme Intelligence, in which the natures of things were devised for common combined purposes?

Matter, which is the primitive constituent of all things, is therefore bound to certain laws, and when it is freely abandoned to these laws it must necessarily bring forth beautiful combinations. It has no freedom to deviate from this perfect plan. Since it is thus subject to a supremely wise purpose, it must necessarily have been put into such harmonious relationships by a First Cause ruling over it; and *there is a God, just because nature even in chaos cannot proceed otherwise than regularly and according to order.*

250

I assume that all the material of which the globes belonging to our solar system —all the planets and comets—consist, at the beginning of all things was decomposed into its primary elements, and filled the whole space of the universe in which the bodies formed out of it now revolve. This state of nature, when viewed in and by itself without any reference to a system, seems to be the very simplest that can follow upon nothing. At that time nothing had yet been formed. The construction of heavenly bodies at a distance from each other, their distances regulated by their attraction, their form arising out of the equilibrium of their collected matter, exhibit a later state. The state of nature which immediately bordered on the creation was as crude, as unformed, as possible. But even in the essential properties of the elements that constituted this chaos, there could be traced the mark of that perfection which they have derived from their origin, their essential character being a consequence of the eternal idea of the Divine Intelligence. The simplest and most general properties which seem to be struck out without design, the matter which appears to be merely passive and wanting form and arrangement, has in its simplest state a tendency to fashion itself by a natural evolution into a more perfect constitution. But the variety in the kinds of elements, is what chiefly contributes to the stirring of nature and to the formative modification of chaos, as it is by it that the repose which would prevail in a universal equality among the scattered elements is done away, so that the chaos begins to take form at the points where the more strongly attracting particles are. The kinds of this elementary matter are undoubtedly infinitely different, in accordance with the immensity which nature shows on all sides. Those elements, which are of greater specific density and force of attraction, and which of themselves occupy less room and are also rarer, would therefore be more scattered than the lighter kinds when the material of the world was equally diffused in space. [. . .]

From what has been said, it will appear that if a point is situated in a very large space where the attraction of the elements there situated acts more strongly than elsewhere, then the matter of the elementary particles scattered throughout the whole region will fall to that point. The first effect of this general fall is the formation of a body at this centre of attraction which, so to speak, grows from an infinitely small nucleus by rapid strides; and in the proportion in which this mass increases, it also draws with greater force the surrounding particles to unite with it. When the mass of this central body has grown so great that the velocity with which it draws the particles to itself from great distances, is bent sideways by the feeble degrees of repulsion with which they impede each other, and when it issues in lateral movements which are capable by means of the centrifugal force of encompassing the central body in an orbit, then there are produced whirls or vortices of particles, each of which by itself describes a curved line by the composition of the attracting force and the force of revolution that has been bent sideways. These kinds of orbits all intersect each other, for which their great dispersion in this space gives place. Yet these movements are in many ways in conflict with each other, and they naturally tend to bring one another to a uniformity, that is, into a state in which one movement is as little obstructive to the other as possible. This happens in two ways: first, by the particles limiting each other's movement till they all advance in one direction; and secondly, in this way, that the particles limit their vertical movements in virtue of which they are approaching the centre of attraction, till they are all moving horizontally, i.e. in parallel circles round the sun as their centre, no longer intersect each other, and by the centrifugal force becoming equal with the falling force they keep themselves constantly in free circular orbits at the distance at which they move. The result, finally, is that only those particles continue to move in this region of space which have acquired by their fall a velocity, and through the resistance of the other particles a direction, by which they can continue to maintain a *free circular movement*. In this state, when all the particles are moving in one direction and in parallel circles, i.e. in free circular movements carried on by the acquired propulsive forces around the central body, the conflict and the concourse of the elements is annulled, and everything is then in the state of the least reciprocal action. This state is the natural consequence which always ensues in the case of matter involved in conflicting movements. It is therefore clear that a great number of the scattered multitude of particles must attain to such exact determinate conditions through the resistance by which they seek to bring each other to this state; although a much greater multitude of them do not reach it and only serve to increase the mass of the central body into which they fall, as they cannot maintain themselves freely at the distance at which they are moving, but cross the circles of the nearer particles, and, finally, by their resistance lose all motion. This body in the centre of attraction which, in consequence of all this, has become the chief part of the planetary system by the mass of its collected matter, is the sun, although it has not yet that glow of flame which bursts out on its surface after its formation has become entirely complete.

ROGER BOSCOVICH

A continuum theory of atoms and forces

251

The primary elements of matter are in my opinion perfectly indivisible and non-extended points. They are so scattered in an immense vacuum that every two of them are separated from one another by a definite interval; this interval can be indefinitely increased or diminished, but can never vanish altogether without compenetration of the points themselves; for I do not admit as possible any immediate contact between them. On the contrary I consider that it is a certainty that, if the distance between two points of matter should become absolutely nothing, then the very same indivisible point of space, according to the usual idea of it, must be occupied by both together, and we have true compenetration in every way. Therefore indeed I do not admit the idea of vacuum interspersed amongst matter, but I consider that matter is interspersed in a vacuum and floats in it. [. . .]

I therefore consider that any two points of matter are subject to a determination to approach one another at some distances, and in an equal degree recede from one another at other distances. This determination I call 'force'; in the first case 'attractive', in the second case 'repulsive'; this term does not denote the mode of action, but the propensity itself, whatever its origin, of which the magnitude changes as the distances change; this is in accordance with a certain definite law, which can be represented by a geometrical curve or by an algebraical formula, and visualized in the manner customary with Mechanicians. [. . .]

Now the law of forces is of this kind; the forces are repulsive at very small

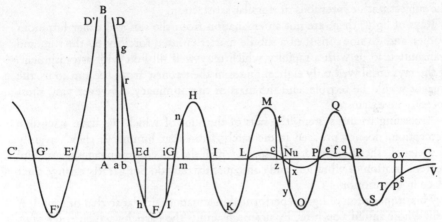

Graph of the forces about a point-like atom

distances, and become indefinitely greater and greater, as the distances are diminished indefinitely, in such a manner that they are capable of destroying any velocity, no matter how large it may be, with which one point may approach another, before ever the distance between them vanishes. When the distance between them is increased, they are diminished in such a way that at a certain distance, which is extremely small, the force becomes nothing. Then as the distance is still further increased, the forces are changed to attractive forces; these at first increase, then diminish, vanish, and become repulsive forces, which in the same way first increase, then diminish, vanish, and become once more attractive; and so on in turn, for a very great number of distances, which are all still very minute: until, finally, when we get to comparatively great distances, they begin to be continually attractive and approximately inversely proportional to the squares of the distances. This holds good as the distances are increased indefinitely to any extent, or at any rate until we get to distances that are far greater than all the distances of the planets and comets.

LEONHARD EULER

LETTERS TO A GERMAN PRINCESS

252

LETTER 134

Reflections on the analogy between colours and sounds

You will be pleased to recollect the objections I offered to the system of the emanation of light. They appear to me so powerful as completely to overturn that system. I have accordingly succeeded in my endeavours to convince certain natural philosophers of distinction, and they have embraced my sentiments of the subject with expressions of singular satisfaction.

Rays of light, then, are not an emanation from the sun and other luminous bodies, and do not consist of a subtile matter emitted forcibly by the sun, and transmitted to us with a rapidity which may well fill you with astonishment. If the rays employed only eight minutes in their course from the sun to us, the torrent would be terrible, and the mass of that luminary, however vast, must speedily be exhausted.

According to my system, the rays of the sun, of which we have a sensible perception, do not proceed immediately from that luminary; they are only particles of ether floating around us, to which the sun communicates nearer and nearer a motion of vibration, and consequently they do not greatly change their place in this motion.

This propagation of light is performed in a manner similar to that of sound. A bell, whose sound you hear, by no means emits the particles which enter your ears. You have only to touch it when struck to be assured that all its parts are in a very sensible agitation. This agitation immediately communicates itself to the more remote particles of air, so that all receive from it successively a similar motion of vibration, which, reaching the ear, excite in it the sensation of sound. The strings of a musical instrument put the matter beyond all doubt; you see them tremble, go and come. It is even possible to determine by calculation how often in a second each string vibrates; and this agitation, being communicated to the particles of air adjacent to the organ of hearing, the ear is struck by it precisely as often in a second. It is the perception of this tremulous agitation which constitutes the nature of sound. The greater the number of vibrations produced by the string in a second, the higher or sharper is the sound. Vibrations less frequent produce lower notes.

We find the circumstances which accompany the sensation of hearing, in a manner perfectly analogous, in that of sight.

The medium only and the rapidity of the vibrations differ. In sound, it is the air through which the vibrations of sonorous bodies are transmitted. But with respect to light, it is the ether, or that medium incomparably more subtile and more elastic than air, which is universally diffused wherever the air and grosser bodies leave interstices.

As often, then, as this ether is put into a state of vibration, and is transmitted to the eye, it excites in it the sentiment of vision, which is in that case nothing but a similar tremulous motion, whereby the small nervous fibres at the bottom of the eye are agitated.

253

LETTER 136

How opaque bodies are rendered visible

You will find no difficulty in the definition I have been giving of coloured bodies. The particles of their surface are always endowed with a certain degree of elasticity, which renders them susceptible of a motion of vibration, as a string is always susceptible of a certain sound; and it is the number of vibrations which these particles are capable of making in a second which determines the species of colour.

If the particles of the surface have not elasticity sufficient to admit of such agitation, the body must be black, this colour being nothing else but a deprivation of light, and all bodies from which no rays are transmitted to our eyes, appearing black.

I now come to a very important question, respecting which some doubts may be entertained. It may be asked: what is the cause of the motion of vibration which constitutes the colours of bodies?

Into the discovery of this, indeed, the whole is resolved; for as soon as the particles of bodies shall be put in motion, the ether diffused through the air will immediately receive a similar agitation, which, continued to our eyes, constitutes there that which we call *rays*, from which vision proceeds.

I remark, first, that the particles of bodies are not put in motion by an internal, but an external power, just as a string distended would remain for ever at rest, were it not put in motion by some external force. Such is the case of all bodies in the dark; for, as we see them not, it is a certain proof that they emit no rays, and that their particles are at rest. In other words, during the night bodies are in the same state with the strings of an instrument that is not touched, and which emit no sound; whereas bodies rendered visible may be compared to strings which emit sound.

And as bodies become visible as soon as they are illuminated, that is, as soon as the rays of the sun, or of some other luminous body, fall upon them, it must follow, that the same cause which illuminates them must excite their particles to generate rays, and to produce in our eyes the sensation of vision. The rays of light, then, falling upon a body, put its particles into a state of vibration.

This appears at first surprising, because on exposing our hands to the strongest light no sensible impression is made on them. It is to be considered, that the sense of touch is in us too gross to perceive these subtile and slight impressions; but that the sense of sight, incomparably more delicate, is powerfully affected by them. This furnishes an incontestable proof that the rays of light which fall upon a body possess sufficient force to act upon the minuter particles, and to communicate to them a tremulous agitation. And in this precisely consists the action necessary to explain how bodies, when illuminated, are put in a condition themselves to produce rays, by means of which they become visible to us. It is sufficient that bodies should be luminous or exposed to the light, in order to the agitation of their particles and thereby to their producing themselves rays which render them visible to us.

The perfect analogy between hearing and sight gives to this explanation the highest degree of probability. Let a harpsichord be exposed to a great noise, and you will see that not only the strings in general are put into a state of vibration, but you will hear the sound of each, almost as if it were actually touched. The mechanism of this phenomenon is easily comprehended, as soon as it is known that a string agitated is capable of communicating to the air the same motion of vibration which, transmitted to the ear, excites in it the sensation of the sound which that same string emits.

Now, as a string produces in the air such a motion, it follows that the air reciprocally acts on the string, and gives it a tremulous motion. And as a noise is capable of putting in motion the strings of a harpsichord, and of extracting sounds from them, the same thing must take place in the objects of vision.

Coloured bodies are similar to the strings of a harpsichord, and the different colours to the different notes, in respect of high and low. The light which falls on these bodies, being analogous to the noise to which the harpsichord is exposed, acts on the particles of their surface as that noise acts on the strings of the harpsichord; and these particles thus put in vibration will produce the rays which shall render the body visible.

This elucidation seems to me sufficient to dissipate every doubt relating to my theory of colours. I flatter myself, at least, that I have established the true principle of all colours, as well as explained how they become visible to us only by the light whereby bodies are illuminated, unless such doubts turn upon some other point which I have not touched upon.

254

LETTER 139

The true principle of nature on which are founded all the phenomena of electricity

The summary I have exhibited of the principle phenomena of electricity has no doubt excited a curiosity to know what occult powers of nature are capable of producing effects so surprising.

The greatest part of natural philosophers acknowledge their ignorance in this respect. They appear to be so dazzled by the endless variety of phenomena which every day present themselves, and by the singularly marvellous circumstances which accompany these phenomena, that they are discouraged from attempting an investigation of the true cause of them. They readily admit the existence of a subtile matter, which is the primary agent in the production of the phenomena, and which they denominate the electric fluid; but they are so embarrassed about determining its nature and properties, that this important branch of physics is rendered only more perplexed by their researches.

There is no room to doubt that we must look for the source of all the phenomena of electricity only in a certain fluid and subtile matter; but we have no need to go to the regions of imagination in quest of it. That subtile matter denominated *ether*, whose reality I have already endeavoured to demonstrate, is sufficient very naturally to explain all the surprising effects which electricity presents. I hope I shall be able to set this in so clear a light, that you shall be able to account for every electrical phenomenon, however strange an appearance it may assume.

The great requisite is to have a thorough knowledge of the nature of ether. The air which we breathe rises only to a certain height above the surface of the earth; the higher you ascend the more subtile it becomes, and at last it entirely ceases. We must not affirm that beyond the region of the air there is a perfect vacuum which occupies the immense space in which the heavenly bodies revolve. The rays of light, which are diffused in all directions from these heavenly bodies, sufficiently demonstrate that those vast spaces are filled with a subtile matter.

If the rays of light are emanations forcibly projected from luminous bodies, as some philosophers have maintained, it must follow that the whole space of the heavens is filled with these rays—nay, that they move through it with incredible rapidity. You have only to recollect the prodigious velocity with which the rays of the sun are transmitted to us. On this hypothesis, not only would there be no vacuum, but all space would be filled with a subtile matter, and that in a state of constant and most dreadful agitation.

But I think I have clearly proved that rays of light are no more emanations projected from luminous bodies than sound is from sonorous bodies. It is much more certain that rays of light are nothing else but a tremulous motion or agitation of a subtile matter, just as sound consists of a similar agitation excited in the air. And as sound is produced and transmitted by the air, light is produced and transmitted by that matter, incomparably more subtile, denominated ether, which consequently fills the immense space between the heavenly bodies.

Ether, then, is a medium proper for the transmission of rays of light: and this same quality puts us in a condition to extend our knowledge of its nature and properties. We have only to reflect on the properties of air, which render it adapted to the reception and transmission of sound. The principal cause is its elasticity or spring. You know that air has a power of expanding itself in all directions, and that it does expand the instant that obstacles are removed. The air is never at rest but when its elasticity is everywhere the same; whenever it is greater in one place than another the air immediately expands. We likewise discover by experiment that the more the air is compressed, the more its elasticity increases: hence the force of air-guns, in which the air, being very strongly compressed, is capable of discharging the ball with astonishing velocity. The contrary takes place when the air is rarefied: its elasticity becomes less in proportion as it is more rarefied, or diffused over a larger space.

On the elasticity of the air, then, relative to its density, depends the velocity of sound, which makes a progress of 1,142 feet in a second. If the elasticity of the air were increased, its density remaining the same, the velocity of sound would increase; and the same thing would take place if the air were more rare or less dense than it is, its elasticity being the same. In general, the more that any medium, similar to air, is elastic, and at the same time less dense, the more rapidly will the agitations excited in it be transmitted. And as light is transmitted so many thousand times more rapidly than sound, it must clearly follow that the ether, that medium whose agitations constitute light, is many thousand times more elastic than air, and, at the same time, many thousand times more rare or more subtile, both of these qualities contributing to accelerate the propagation of light.

Such are the reasons which lead us to conclude that ether is many thousand times more elastic and more subtile than air; its nature being in other respects similar to that of air, in as much as it is likewise a fluid matter, and susceptible of compression and of rarefaction. It is this quality which will conduct us to the explanation of all the phenomena of electricity.

255

LETTER 140

Continuation. Different nature of bodies relatively to electricity

Ether being a subtile matter and similar to air, but many thousand times more rare and more elastic, it cannot be at rest, unless its elasticity, or the force with which it tends to expand, be the same everywhere.

As soon as the ether in one place shall be more elastic than in another, which is the case when it is more compressed there, it will expand itself into the parts adjacent, compressing what it finds there till the whole is reduced to the same degree of elasticity. It is then in equilibrio, the equilibrium being nothing else but the state of rest, when the powers which have a tendency to disturb it counterbalance each other.

When, therefore, the ether is not in equilibrio the same thing must take place as in air, when its equilibrium is disturbed; it must expand itself from the place where its elasticity is greater towards that where it is less; but, considering its greater elasticity and subtility, this motion must be much more rapid than that of air. The want of equilibrium in the air produces wind, or the motion of that fluid from one place to another. There must therefore be produced a species of wind, but incomparably more subtile than that of air, when the equilibrium of the ether is disturbed, by which this last fluid will pass from places where it was more compressed and more elastic to those where it was less so.

This being laid down, I with confidence affirm that all the phenomena of electricity are a natural consequence of want of equilibrium in the ether, so that wherever the equilibrium of the ether is disturbed the phenomena of electricity must take place; consequently, electricity is nothing else but a derangement of the equilibrium of the ether.

In order to unfold all the effects of electricity, we must attend to the manner in which ether is blended and enveloped with all the bodies which surround us. Ether, in these lower regions, is to be found only in the small interstices which the particles of the air and of other bodies leave unoccupied. Nothing can be more natural than that the ether, from its extreme subtility and elasticity, should insinuate itself into the smallest pores of bodies which are impervious to air, and even into those of the air itself. You will recollect that all bodies, however solid they may appear, are full of pores; and many experiments incontestably demonstrate that these interstices occupy much more space than the solid parts; finally, the less ponderous a body is, the more it must be filled with these pores, which contain ether only. It is clear, therefore, that though the ether be thus diffused through the smallest pores of bodies, it must however be found in very great abundance in the vicinity of the earth.

You will easily comprehend that the difference of these pores must be very great, both as to magnitude and figure, according to the different nature of the bodies, as their diversity probably depends on the diversity of their pores. There must be, therefore, undoubtedly, pores more close, and which have less communication with others; so that the ether which they contain is likewise more confined, and cannot disengage itself but with great difficulty, though its elasticity may be much greater than that of the ether which is lodged in the adjoining pores. There must be, on the contrary, pores abundantly open, and of easy communication with the adjacent pores; in this case it is evident that the ether lodged in them can with less difficulty disengage itself than in the preceding; and if it is more or less elastic in these than in the others, it will soon recover its equilibrium.

In order to distinguish these two classes of pores, I shall denominate the first *close*, and the others *open*. Most bodies must contain pores of an intermediate species, which it will be sufficient to distinguish by the terms *more* or *less close*, and *more* or *less open*.

This being laid down, I remark, first, that if all bodies had pores perfectly close, it would be impossible to change the elasticity of the air contained in them; and even though the ether in some of these pores should have acquired, from whatever cause, a higher degree of elasticity than the others, it would always remain in that state, and never recover its equilibrium, from a total want of communication. In this case no change could take place in bodies; all would remain in the same state as if the ether were in equilibrio, and no phenomenon of electricity could be produced.

This would likewise be the case if the pores of all bodies were perfectly open; for then, though the ether might be more or less elastic in some pores than in others, the equilibrium would be instantly restored, from the entire freedom of communication—and that so rapidly that we should not be in a condition to remark the slightest change. For the same reason it would be impossible to disturb the equilibrium of the ether contained in such pores; as often as the equilibrium might be disturbed, it would be as instantaneously restored, and no sign of electricity would be discoverable.

The pores of all bodies being neither perfectly close nor perfectly open, it will always be possible to disturb the equilibrium of the ether which they contain; and when this happens, from whatever cause, the equilibrium cannot fail to re-establish itself; but this re-establishment will require some time, and this produces certain phenomena; and you will presently see, much to your satisfaction, that they are precisely the same which electrical experiments have discovered. It will then appear that the principles on which I am going to establish the theory of electricity are extremely simple, and at the same time absolutely incontrovertible.

256

LETTER 142

Of positive and negative electricity

You will easily comprehend, from what I have above advanced, that a body must become electrical whenever the ether contained in its pores becomes more or less elastic than that which is lodged in adjacent bodies. This takes place when a greater quantity of ether is introduced into the pores of such a body, or when part of the ether which it contained is forced out. In the former case, the ether becomes more compressed, and consequently more elastic; in the other, it becomes rarer, and loses it elasticity. In both cases it is no longer in equilibrio with that which is external; and the efforts which it makes to recover its equilibrium produce all the phenomena of electricity.

You see, then, that a body may become electric in two different ways, according as the ether contained in its pores becomes more or less elastic than that which is external; hence result two species of electricity: the one, by which the ether is rendered more elastic, or more compressed, is denominated *increased* or *positive electricity*; the other, in which the ether is less elastic, or more rarefied, is denominated *diminished* or *negative electricity*. [. . .]

PIERRE-SIMON de LAPLACE

The evolution of the solar system

257

The summary which we have given here of the history of astronomy indicates three clearly distinct periods, which, giving account of the phenomena, of the laws which govern them, and of the forces on which these laws depend, shows us the path which this science has followed and which should be followed by all the other natural sciences. The first period covers the observations of the astronomers before Copernicus, about the appearances of the celestial movements, and the theories which they had invented to explain the appearances and in order to make them follow their calculations. In the second period, Copernicus deduced from these appearances the motion of the earth around itself and around the sun and Kepler has discovered the laws of planetary motion. Finally, in the third period, Newton, leaning on these laws, worked his way up to the principle of universal gravitation, and the geometers, applying analysis to this principle, have derived all the astronomical phenomena and the numerous inequalities of the motions of the planets, satellites and comets. Thus, astronomy has become the solution of the great problems of mechanics, in which the elements of the celestial motions are the arbitrary constants. Astronomy has all the certainty that results from enormous numbers and from the variety of rigorously explained phenomena, and from the simplicity of the principle which by itself is enough to explain all those phenomena. Far from being afraid lest a new star belie the principle, we can state in advance that the motion will conform with the principle: this we have ourselves seen with regard to Uranus and the four planets which have been recently telescopically discovered, and every appearance of a comet has furnished a new proof.

Such is then, no doubt, the constitution of the solar system. The immense globe of the sun, the main focus of the diverse motions of the system, turns in twenty-five-and-a-half days around itself. Its surface is covered by an ocean of luminous matter; beyond this the planets, with their own satellites, move in almost circular orbits and in a plane only slightly inclined to the solar equator. The innumerable comets, after having approached the sun, go far away, to such distances that prove that the sun's empire extends much beyond the familiar limits of the planetary system. Not only does this star act on all these globes through its attraction by making them move around it, but it also pours out on them its light and heat. Its beneficial action blows to life the animals and plants that cover the earth, and analogy leads us to believe that it produces similar effects on the planets; for it is natural to think that matter, in which we see fertility developing in so many ways, will not be barren on such a big planet like Jupiter, which like the terrestrial globe has its days, nights and years, and one on which observations indicate changes which again suggest the action of very active forces. Man, created suitable for those temperatures which he enjoys on earth, would not be able, in all probability, to live on other planets; but should there not exist an infinity of living structures, related to different temperatures of the various globes of this universe? If the single difference in the elements and the climate is enough to cause such a variety of terrestrial products, how much more must differ those of the diverse planets and their satellites? Even the most active imagination cannot form an idea of that; but, to say the least, the existence of such living structures is rendered very plausible.

Although the elements of the planetary system are arbitrary, there are relations between them which can shed light on their origin. Considering them with attention one is astonished to see that all the planets move around the sun from west to east, and almost in the same plane: the satellites move around their planets in the same sense and again almost in the same plane; finally, the sun, the planets and all the satellites in which rotatory motion has been observed all turn around themselves in the same sense and almost in the plane of their translatory motion. The satellites exhibit in this regard a remarkable singularity: their rotatory motion is exactly equal to their orbital motion insofar that they always show the same hemisphere to their planet. This is what is observed at least in the cases of the moon, the four satellites of Jupiter, and the last satellite of Saturn—and these are the only satellites for which rotatory motion has been observed so far.

Such extraordinary phenomena cannot result from irregular causes. Submitting their probabilities to calculation one finds that there is two hundred thousand milliards to one against pure chance; this is a far greater probability than that of most historical events, which we do not doubt at all, and thus we should have at least as much confidence here that there is a fundamental cause that dictates the planetary motions.

Another equally remarkable phenomenon of the solar system is the small eccentricity of the orbits of the planets and satellites, while the orbits of the comets are very elongated; there is no intermediate eccentricity between the very large and very small ones in this system. We must again recognize here the effect of a regular cause: chance could not have given all the planets an almost circular orbit; it is the cause which has determined the motions of these bodies, which must have necessarily rendered them almost circular. Moreover it must be that the great eccentricity of the orbits of comets and the direction of their motions in all the senses are also necessary results; for, looking at the orbits of the retrograde comets, being inclined by more than 100° to the ecliptic, one finds that the mean inclination of the orbits of all the comets is very near 100°, as it must be, if these bodies have all been started on their way randomly. What is this fundamental cause? I will go into that in the note which ends this work, a hypothesis which seems to me to derive with great plausibility from the preceding phenomena, but one which I shall present with great hesitation that should inspire all those who do not have yet any results of their observations or calculations. [. . .]

I cannot refrain from observing here how much Newton has departed in this matter from the method of which he otherwise made such happy application. After the publication of his discoveries about the system of the world and of light, this great geometer, surrendering to considerations of a different sort, began investigating the motives of the Author of Nature to give the solar system that constitution which we have just described. In the general scholium at the end of the *Principia*, having described the singular phenomena of the motions of the planets and satellites being all in the same sense, almost in the same plane and in almost circular orbits, Newton adds:

> but it is not to be conceived that mere mechanical causes could give birth to so many regular motions, since the comets range over all parts of the heavens in very eccentric orbits. . . . This most beautiful system of the sun, planets, and comets, could only proceed from the counsel and dominion of an intelligent and powerful Being.

He repeats this same idea once more, at the end of the *Opticks*; he would have been convinced in it even more if he had known what we have just demonstrated, namely, that the conditions of the arrangement of planets and satellites are exactly those which assure their stability:

> blind Fate could never make all the Planets move one and the same way in Orbs concentrick, some Irregularities excepted, which may have risen from the mutual actions of Comets and Planets upon one another, and which will be apt to increase till this System wants a Reformation.

But could not this arrangement of the planets itself be a result of the laws of motion, and could not the Supreme Intelligence, which Newton makes to interfere, have made them all depend on a more general phenomenon? Such is, according to our conjecture, the phenomenon of nebulous matter, scattered in diverse heaps, in the immensity of the skies. Can one still assert that the conservation of the planetary system is part of the outlook of the Author of the Universe? The mutual attraction of the bodies of this system cannot alter its stability as Newton had supposed, but had there not been a fluid other than light in the heavenly spaces, its resistance and the decrease in the mass of the sun that its emission produces would, in the long run, destroy the arrangement of the planets, and in order to maintain it, a reform would no doubt be necessary. But so many extinct species of animals, in which M. Cuvier,[1] with rare sagacity, knew how to recognize organization among the numerous fossilized remains described by him, do not these indicate that there is a tendency in nature to change even things which seem most immutable? The grandeur and importance

1. George Cuvier (1769–1832), founder of comparative anatomy and of palaeontology. (Ed.)

of the solar system cannot make it escape the general law, for it is such only relative to our smallness and the system which seems to us so vast is nothing but a small point in the universe. If we review the history of progress of the human spirit and its errors, we shall see the final causes constantly receding to the limits of its knowledge. These causes, which Newton had extended to the limits of the solar system, were, even in his own times, taking place in the atmosphere in order to be able to explain the meteorites; these are thus, in the eyes of the philosophers, only expressions of our ignorance about the true causes.

In his controversy with Newton about the invention of the infinitesimal calculus, Leibniz had vigorously criticized the intervention of the Divinity to put the solar system in order. 'This,' he said, 'is to have very narrow and strange ideas about the wisdom and power of God.' Newton responded not less vigorously with a criticism of Leibniz's pre-established harmony, which he termed a 'perpetual miracle'. Posterity has not accepted these vain hypotheses, but it has given full justice to the mathematical works of these two great geniuses: the discovery of universal gravity and the efforts of its author to connect with it all the celestial phenomena will always remain an object of posterity's admiration and recognition.

258

In order to trace the causes of the original motion of the planetary system, we have at our disposal from the previous chapter the following five phenomena: the fact that all the planets move in the same sense and almost in the same plane, the motion of the satellites in the same sense as the planets, the fact that the rotatory motion of all these different bodies and of the sun is in the same sense and in the same as their forward motion, and again, almost in the same plane; the small eccentricity of the orbits of planets and satellites; finally the great eccentricity of the orbits of comets, although their inclination has been left to chance. [. . .]
Consideration of the planetary motions makes us think that, due to excessive heat, the sun's atmosphere originally extended beyond the orbits of all the planets and that it has then retreated successively to its present limits.

In the original state, where we shall now imagine the sun, it resembled those nebulae which the telescope shows us to be composed of a more or less brilliant core encircled by a nebulosity, which, on condensing at the surface of the core, transforms it into a star. If, by analogy, one imagines all the stars having been formed in this manner, one can imagine their state before that of nebulosity, and that one again preceded by another state in which the nebulous matter had been more and more diffuse, the core having been less and less luminous.

Thus going back as far as possible one arrives at a state of nebulosity so diffuse that one could scarcely subsume it under the existing things.

For a long time now, the particular arrangement of some stars which are visible even to the naked eye, has puzzled the philosophers. Michell[1] has already noted how impossible it would have been for the Pleiades, for example, to have been drawn together to such a small space which they indeed occupy, had it happened by pure chance: he concluded that this and similar groups of stars seen in the heaven are the effects of an original cause or of a general law of nature. These groups are a necessary result of the condensation of the nebulae into numerous cores; for it can be easily seen that the nebulous matter is incessantly being drawn together to these various cores, and they must form, in the long run, a group of stars very much like the Pleiades. Similarly, the condensation of a nebula into two cores will form two mutually very near stars, rotating around each other in a way which has already been observed for all double stars.

But how has the solar atmosphere determined the rotations and revolutions of the planets and the satellites? Had these bodies penetrated deeply into the solar atmosphere, its resistance would have made them fall into the sun. One can thus conjecture that the planets were formed on the successive limits of the atmosphere, through condensation of zones of vapour, and on cooling they had to be released by this atmosphere in the plane of its equator.

Let us recall the results we have given in the tenth chapter of this book. The solar atmosphere cannot extend indefinitely; its limit is at the point where the centrifugal force of the rotatory motion balances gravity; now, in proportion to the cooling which draws together the atmosphere and condenses the molecules which were carried within it on the surface of the star, the rotatory motion increases; for according to the area theorem the sum of areas covered by the radius vector of all the molecules of the sun and its atmosphere, and projected to the plane of its equator, is always the same. Thus the rotation must have become more rapid when the molecules got nearer the centre of the sun. The centrifugal force, resulting of this motion, becomes larger too, and thus the point where this force equals that of gravity moves nearer the centre. Supposing thus, that the atmosphere of the sun is extended in any one epoch to its possible limit, and this is a natural thing to accept, on cooling it will have to abandon those molecules which were located at that limit, and all those located on successive limits, each determined by the increasing rotation of the sun. These abandoned molecules will continue to circle around that star, since their centrifugal force has been balanced by gravity. But this balancing does not take place with regard to molecules of the atmosphere located on planes parallel to the plane of the solar equator; these have been drawn together in the atmosphere due to their weight, in proportion to its condensation, and they did not cease to belong to it; they did, however, come nearer the equator.

Let us now consider the vapour zones which have been abandoned successively. In all probability, through condensation and mutual attraction of their molecules, these zones had to form several concentric rings of vapour, circling

1. John Michell (1724–1793), British astronomer and physicist. (Ed.)

around the sun. The mutual friction of the molecules of each ring had to accelerate some and retard others until all acquired the same angular motion. Thus the actual velocity of the molecules further away from the centre of the star would be greater. There is another cause which had to contribute to this difference in velocities: the most distant molecules, those which through the cooling and condensation were drawn together to form the outer part of the ring, will always describe areas proportional to the times; this, because the central force which makes them move is always directed towards the star; now, the constancy of areas demands an increase of velocity in proportion to the extent of the drawing together. One can see that this same cause would make the velocity decrease for those molecules which were carried to the ring to form its inner part.

If all the molecules of a ring of vapours continued to condense without falling apart, they would form in the long run, a liquid or a solid ring. But the regularity which this formation requires in all the parts of the ring and in the procedure of cooling, must make such a formation exceedingly rare. Indeed, the solar system can show us only one single example, the rings of Saturn. Almost always each ring breaks up into several masses, which, cast away with almost equal speeds, continue to circle around the sun at the same distance. These masses had to assume a spheroidal shape with rotational motion directed in the same sense as their revolution, for their inner molecules had less real velocity than the outer ones; thus, they formed planets in a vaporous state. But if one of them were powerful enough to reunite through its attraction all the others, successfully, around its centre, the ring of vapours would then be transformed into a single, spheroidal mass of vapours, circling around the sun and rotating around itself in the same sense. This last case has been the most common; but the solar system demonstrates to us also the first case for the four small planets that move between Jupiter and Mars; this, at least, if we do not suppose with M. Olbers[1] that they had originally formed one single planet which exploded into several parts moving with different speeds. Now, if we follow the changes that a later cooling must produce in the vaporous planets, the formations of which we have suggested above, we shall see how a core develops in the centre of each and increases incessantly through the condensation of the atmosphere around it. In that state the planet resembles completely the sun and it is in the state of the nebulae which we have described; thus, the cooling must produce similar effects to those described at the various limits of its atmosphere, i.e. rings and satellites circling around its centre in the same sense as its revolution and rotating in the same sense around themselves. The regular distribution of the mass of Saturn's rings around its centre and in the plane of its equator is a natural consequence of this hypothesis, and without it, it would have been unexplainable: these rings seem to me to be ever valid proofs of the original extent of Saturn's atmosphere and of its successive shrinkage. [. . .]

1. Wilhelm Olbers (1758–1840), German astronomer. (Ed.)

A thorough examination of all the circumstances in this system, increases even more the probablility of our hypothesis. The original fluidity of the planets has already been indicated by the flattening of their shape and it conforms with the laws of mutual attraction between their molecules; it is proved even more strongly for the earth through the regular decrease[1] in the force of gravity when going from the equator towards the poles. This original fluid state, to which one has been conducted by astronomical phenomena, must also be manifest in the phenomena that natural history presents to us. But, in order to detect these, it is necessary to take into consideration the immense variety of combinations formed by the mixing of the terrestrial compounds in the vaporous state, when the decrease in temperature enables their elements to unite; moreover, one must consider the prodigious changes which must have been caused by this decrease in temperature in the interior and on the surface of the earth, in all its productions, in the constitution and pressure of the atmosphere, in the Ocean and in the bodies which are dissolved in it. Finally, we have to study those violent changes like volcanic eruptions which must have upset the regularity of those changes, in various epochs. If geological studies were conducted with such a point of view which would make them close and connected with astronomy, they would gain in many instances great precision and certitude.

Foundations of the theory of probability

259

All events, even those which on account of their insignificance do not seem to follow the great laws of nature, are a result of those laws just as necessarily as the revolutions of the sun. In ignorance of the ties which unite such events to the whole system of the universe, they have been made to depend upon final causes or upon hazard, according as to whether they occur and are repeated with regularity, or appear without regard to order; but these imaginary causes have gradually receded with the widening bounds of knowledge and disappear entirely before sound philosophy, which sees in them only the expression of our ignorance of the true causes.

Present events are connected with preceding ones by a tie based upon the evident principle that a thing cannot occur without a cause which produces it. This axiom, known by the name of *the principle of sufficient reason*, extends even to actions which are considered indifferent; the freest will is unable without a determinative motive to give them birth; if we assume two positions with exactly similar circumstances and find that the will is active in the one and inactive in the other, we say that its choice is an effect without a cause. It is then, says Leibniz, the blind chance of the Epicureans. The contrary opinion is an illusion of the mind, which, losing sight of the evasive reasons of the choice of the will in indifferent things, believes that choice is determined of itself and without motives.

1. This is obviously a slip: it must be read as 'increase'. (Ed.)

We ought then to regard the present state of the universe as the effect of its anterior state and as the cause of the one which is to follow. Given for one instant an intelligence which could comprehend all the forces by which nature is animated and the respective situation of the beings who compose it—an intelligence sufficiently vast to submit these data to analysis—it would embrace in the same formula the movements of the greatest bodies of the universe and those of the lightest atom; for it, nothing would be uncertain and the future, as the past, would be present to its eyes. The human mind offers, in the perfection which it has been able to give to astronomy, a feeble idea of this intelligence. Its discoveries in mechanics and geometry, added to that of universal gravity, have enabled it to comprehend in the same analytical expressions the past and future states of the system of the world. Applying the same method to some other objects of its knowledge, it has succeeded in referring to general laws observed phenomena and in foreseeing those which given circumstances ought to produce. All these efforts in the search for truth tend to lead it back continually to the vast intelligence which we have just mentioned, but from which it will always remain infinitely removed. This tendency, peculiar to the human race, is that which renders it superior to animals; and their progress in this respect distinguishes nations and ages and constitutes their true glory.

Let us recall that formerly, and at no remote epoch, an unusual rain or an extreme drought, a comet having in train a very long tail, the eclipses, the aurora borealis, and in general all the unusual phenomena were regarded as so many signs of celestial wrath. Heaven was invoked in order to avert their baneful influence. No one prayed to have the planets and the sun arrested in their courses; observation had soon made apparent the futility of such prayers. But as these phenomena, occurring and disappearing at long intervals, seemed to oppose the order of nature, it was supposed that Heaven, irritated by the crimes of the earth, had created them to announce its vengeance. Thus the long tail of the comet of 1456 spread terror through Europe, already thrown into consternation by the rapid successes of the Turks, who had just overthrown the Lower Empire. This star after four revolutions has excited among us a very different interest. The knowledge of the laws of the system of the world acquired in the interval had dissipated the fears begotten by the ignorance of the true relationship of man to the universe; and Halley, having recognized the identity of this comet with those of the years 1531, 1607, and 1682, announced its next return for the end of the year 1758 or the beginning of the year 1759. The learned world awaited with impatience this return which was to confirm one of the greatest discoveries that have been made in the sciences, and fulfil the prediction of Seneca when he said, in speaking of the revolutions of those stars which fall from an enormous height: 'The day will come when, by study pursued through several ages, the things now concealed will appear with evidence; and posterity will be astonished that truths so clear had escaped us.' Clairaut[1] then undertook to sub-

1. Alexis Clairaut (1713–1765), French mathematician and astronomer. (Ed.)

mit to analysis the perturbations which the comet had experienced by the action of the two great planets, Jupiter and Saturn; after immense calculations he fixed its next passage at the perihelion towards the beginning of April, 1759, which was actually verified by observation. The regularity which astronomy shows us in the movements of the comets doubtless exists also in all phenomena.

The curve described by a simple molecule of air or vapour is regulated in a manner just as certain as the planetary orbits; the only difference between them is that which comes from our ignorance.

Probability is relative, in part to this ignorance, in part to our knowledge. We know that of three or a greater number of events a single one ought to occur; but nothing induces us to believe that one of them will occur rather than the others. In this state of indecision it is impossible for us to announce their occurrence with certainty. It is, however, probable that one of these events, chosen at will, will not occur because we see several cases equally possible which exclude its occurrence, while only a single one favours it.

The theory of chance consists in reducing all the events of the same kind to a certain number of cases equally possible, that is to say, to such as we may be equally undecided about in regard to their existence, and in determining the number of cases favourable to the event whose probability is sought. The ratio of this number to that of all the cases possible is the measure of this probability, which is thus simply a fraction whose numerator is the number of favourable cases and whose denominator is the number of all the cases possible.

The preceding notion of probability supposes that, in increasing in the same ratio the number of favourable cases and that of all the cases possible, the probability remains the same. In order to convince ourselves let us take two urns, *A* and *B*, the first containing four white and two black balls, and the second containing only two white balls and one black one. We may imagine the two black balls of the first urn attached by a thread which breaks at the moment when one of them is seized in order to be drawn out, and the four white balls thus forming two similar systems. All the chances which will favour the seizure of one of the balls of the black system will lead to a black ball. If we conceive now that the threads which unite the balls do not break at all, it is clear that the number of possible chances will not change any more than that of the chances favourable to the extraction of the black balls; but two balls will be drawn from the urn at the same time; the probability of drawing a black ball from the urn *A* will then be the same as at first. But then we have obviously the case of urn *B* with the single difference that the three balls of this last urn would be replaced by three systems of two balls invariably connected.

When all the cases are favourable to an event the probability changes to certainty and its expression becomes equal to unity. Upon this condition certainty and probability are comparable, although there may be an essential difference between the two states of the mind when a truth is rigorously demonstrated to it, or when it still perceives a small source of error.

In things which are only probable the difference of the data, which each man has in regard to them, is one of the principal causes of the diversity of opinions which prevail in regard to the same objects. Let us suppose, for example, that we have three urns, *A, B, C,* one of which contains only black balls while the two others contain only white balls; a ball is to be drawn from the urn *C* and the probability is demanded that this ball will be black. If we do not know which of the three urns contains black balls only, so that there is no reason to believe that it is *C* rather than *B* or *A,* these three hypotheses will appear equally possible, and since a black ball can be drawn only in the first hypothesis, the probability of drawing it is equal to one third. If it is known that the urn *A* contains white balls only, the indecision then extends only to the urns *B* and *C,* and the probability that the ball drawn from the urn *C* will be black is one half. Finally this probability changes to certainty if we are assured that the urns *A* and *B* contain white balls only.

It is thus that an incident related to a numerous assembly finds various degrees of credence, according to the extent of knowledge of the auditors. If the man who reports it is fully convinced of it and if, by his position and character, he inspires great confidence, his statement, however extraordinary it may be, will have for the auditors who lack information the same degree of probability as an ordinary statement made by the same man, and they will have entire faith in it. But if some one of them knows that the same incident is rejected by other equally trustworthy men, he will be in doubt and the incident will be discredited by the enlightened auditors, who will reject it whether it be in regard to facts well averred or the immutable laws of nature.

It is to the influence of the opinion of those whom the multitude judges best informed and to whom it has been accustomed to give its confidence in regard to the most important matters of life that the propagation of those errors is due which in times of ignorance have covered the face of the earth. Magic and astrology offer us two great examples. These errors inculcated in infancy, adopted without examination, and having for a basis only universal credence, have maintained themselves during a very long time; but at last the progress of science has destroyed them in the minds of enlightened men, whose opinion consequently has caused them to disappear even among the common people, through the power of imitation and habit which had so generally spread them abroad. This power, the richest resource of the moral world, establishes and conserves in a whole nation ideas entirely contrary to those which it upholds elsewhere with the same authority. What indulgence ought we not then to have for opinions different from ours, when this difference often depends only upon the various points of view where circumstances have placed us! Let us enlighten those whom we judge insufficiently instructed; but first let us examine critically our own opinions and weigh with impartiality their respective probabilities.

The difference of opinions depends, however, upon the manner in which the influence of known data is determined. The theory of probabilities holds to considerations so delicate that is is not surprising that with the same data two persons arrive at different results, especially in very complicated questions. Let us examine now the general principles of this theory.

FIRST PRINCIPLE

The first of these principles is the definition itself of probability, which, as has been seen, is the ratio of the number of favourable cases to that of all the cases possible.

SECOND PRINCIPLE

But that supposes the various cases equally possible. If they are not so, we will determine first their respective possibilities, whose exact appreciation is one of the most delicate points of the theory of chance. Then the probability will be the sum of the possibilities of each favourable case. Let us illustrate this principle by an example.

Let us suppose that we throw into the air a large and very thin coin whose two large opposite faces, which we will call heads and tails, are perfectly similar. Let us find the probability of throwing heads at least one time in two throws. It is clear that four equally possible cases may arise, namely, heads at the first and at the second throw; heads at the first throw and tails at the second; tails at the first throw and heads at the second; finally, tails at both throws. The first three cases are favourable to the event whose probability is sought; consequently this probability is equal to $\frac{3}{4}$; so that it is a bet of three to one that heads will be thrown at least once in two throws.

We can count at this game only three different cases, namely, heads at the first throw, which dispenses with throwing a second time; tails at the first throw and heads at the second; finally, tails at the first and at the second throw. This would reduce the probability to $\frac{2}{3}$ if we should consider with d'Alembert these three cases as equally possible. But it is apparent that the probability of throwing heads at the first throw is $\frac{1}{2}$, while that of the two other cases is $\frac{1}{4}$, the first case being a simple event which corresponds to two events combined: heads at the first and at the second throw, and heads at the first throw, tails at the second. If we then, conforming to the second principle, add the possibility $\frac{1}{2}$ of heads at the first throw to the possibility $\frac{1}{4}$ of tails at the first throw and heads at the second, we shall have $\frac{3}{4}$ for the probability sought, which agrees with what is found in the supposition when we play the two throws. This supposition does not change at all the chance of that one who bets on this event; it simply serves to reduce the various cases to the cases equally possible.

THIRD PRINCIPLE

One of the most important points of the theory of probabilities and that which lends the most to illusions is the manner in which these probabilities increase or diminish by their mutual combination. If the events are independent of one another, the probability of their combined existence is the product of their respective probabilities. Thus the probability of throwing one ace with a single die is $\frac{1}{6}$; that of throwing two aces in throwing two dice at the same time is $\frac{1}{36}$. Each face of the one being able to combine with the six faces of the other, there are in fact thirty-six equally possible cases, among which one single case gives two aces. Generally the probability that a simple event in the same circumstances will occur consecutively a given number of times is equal to the probability of this simple event raised to the power indicated by this number. Having thus the successive powers of a fraction less than unity diminishing without ceasing, an event which depends upon a series of very great probabilities may become extremely improbable. Suppose then an incident be transmitted to us by twenty witnesses in such manner that the first has transmitted it to the second, the second to the third, and so on. Suppose again that the probability of each testimony be equal to the fraction $\frac{9}{10}$; that of the incident resulting from the testimonies will be less than $\frac{1}{8}$. We cannot better compare this diminution of the probability than with the extinction of the light of objects by the interposition of several pieces of glass. A relatively small number of pieces suffices to take away the view of an object that a single piece allows us to perceive in a distinct manner. The historians do not appear to have paid sufficient attention to this degradation of the probability of events when seen across a great number of successive generations; many historical events reputed as certain would be at least doubtful if they were submitted to this test.

In the purely mathematical sciences the most distant consequences participate in the certainty of the principle from which they are derived. In the applications of analysis to physics the results have all the certainty of facts or experiences. But in the moral sciences, where each inference is deduced from that which precedes it only in a probable manner, however probable these deductions may be, the chance of error increases with their number and ultimately surpasses the chance of truth in the consequences very remote from the principle.

SECTION V

Dalton to Mach

Section V: Dalton to Mach

INTRODUCTION

PERHAPS the most conspicuous advance made in the physical sciences in the nineteenth century was the progress of the non-mechanical disciplines of physics. As they reached the level of mechanics as regards quantification and mathematization, the first great syntheses of the physical branches of mechanics, thermodynamics, light and electromagnetism were possible; this led to two comprehensive statements, the law of conservation of energy and the second law of thermodynamics. Through these momentous achievements, by the second half of the nineteenth century the zenith of classical physics was attained. This largely determined the world-picture that prevailed for the following decades.

The main precursors of this advance were two important developments of the second half of the eighteenth century—the experiments which established the basic facts of electrostatics, and the beginnings of modern chemistry. Neither are represented by texts in this Anthology, the first because of lack of space, the second because no excerpts from works on chemistry are included except those relevant to the understanding of the development of atomic and molecular physics (as for instance Dalton's great work, Prout's hypothesis and Mendeleev's papers on the periodic system of the elements).

The beginnings of the modern theory of electricity are connected chiefly with the researches of Benjamin Franklin, Priestley, Coulomb, Galvani and Volta. They mark too the beginning of the decline of the age of 'subtle fluids'; here we include the aether, electric fluid, magnetic fluid, and 'caloric'. Of these, only the aether lived on till the end of the nineteenth century. The essential empirical data of electrical phenomena accumulated rather quickly, so that Poisson's mathematical theory of electrostatics had already been established in the first third of the nineteenth century, modelled after the analogous equations of classical mechanics and drawing upon concepts such as that of potential.

The history of the origins of modern chemistry is connected with the names of Gassendi and Boyle in the seventeenth century, with the revival of Epicurus' atomic doctrine, the corpuscular hypothesis of matter. The main originator of modern chemical practice and thought was Lavoisier who in 1789 introduced reliable methods of weighing and measuring and through whom the analytical balance became the most important tool of quantitative chemistry. Lavoisier demolished the hypothesis of the phlogiston, that hypothetical particle of negative mass introduced by Stahl early in the eighteenth century (this was a kind of revival of the Aristotelian concept of 'absolute lightness'). The modern concept of a chemical element began to evolve; Lavoisier already knew 23 of these, partly through specific reactions, but mainly by their differences in weight. In 1830 the number of known elements had grown to 54, particularly as a result of the researches of Dalton, Gay-Lussac, Avogadro, and Berzelius. Dalton's book, *A New System of Chemical Philosophy* (1808), is a landmark in the history of atomic theory. He stressed the importance of the relative weights of the ultimate particles in simple and complex bodies.

In 1815 and 1816, a dozen years after the first table of atomic masses had been drawn up by Dalton, two papers (originally anonymous) were published by the British physician William Prout. In these he claimed that the atomic weights of all the elements were multiples of that of hydrogen, the lightest element. Prout was certainly not in a position to verify his hypothesis with a satisfactory measure of exactitude. All the same, on the basis of the data available to him, and despite the uncertainty attached to the measuring methods of those years, he decided to publish his ideas. They anticipated the conclusions reached a century later in which the proton, the nucleus of the hydrogen atom, is an essential building-block of all atomic nuclei. His remarkable statement 'if the views we have ventured to advance be correct, we may almost consider the $\pi\rho\omega\tau\eta$ $\tilde{\upsilon}\lambda\eta$ of the ancients to be realized in hydrogen' is contained in the passages quoted in the Anthology. It was, however, challenged by many chemists. As late as 1860, at an international congress of chemistry in Karlsruhe, there was still much difference of opinion as to the most reliable method of determining relative atomic weights. One of the participants of this congress was Mendeleev, and the discussions influenced the direction of his subsequent researches which, after ten years of systematic work, led to the development of the Periodic Table of chemical elements. The idea that there existed a periodicity in the physical and chemical properties of the elements had been considered since the middle of the nineteenth century. Thus almost simultaneously with that of Mendeleev Lothar Meyer published his independent discovery. However, Mendeleev's priority is justified historically by his anticipation of the significance of his discovery as well as that of the whole problem of relative atomic weights for the subsequent development of the theory of matter. This is evident from the remarkable passage from his paper quoted in this Anthology. The discovery of spectroscopy by Kirchhoff and Bunsen (1859), which made the spectroscope a powerful tool for the investigation of the properties of matter, finally transformed the Periodic Law from a chapter in chemistry into a chapter in atomic physics.

The fascinating story of the development of electromagnetism from Ørsted's discovery (1820) to Hertz's experiments in the 1880s was preceded by the re-establishment of the wave theory of light by Young in Britain and by Fresnel in France. One of the central concepts of electromagnetism was that of the physical field, explaining electric and magnetic phenomena by means of contiguous action. The revival of the wave concept of light prepared the ground for this approach, based as it was on the continuum. Huygens' basic idea that the propagation of light is that of a state, not of a substance, and his explanation of its reflection and refraction by means of the wave model, were taken over in the first decades of the nineteenth century and developed on a higher level of experimental technique and mathematical algorithm. Naturally enough, it was accompanied by a criticism of Newton's corpuscular theory of light (although his corpuscles exhibited periodical 'fits') which for more than a hundred years had dominated the scientific scene and eclipsed Huygens' ideas. The fundamental facts of the interference, diffraction, and polarization of light rays found their natural explanation in the wave hypothesis. It is interesting to note that Young was helped in his theory of interference by Newton's remark that tidal waves, split into two trains on entering two different channels, would reinforce each other as well as be weakened after having united again. Of Fresnel's lucid expositions the most important one is perhaps his explanation of the generation of light waves from light-emitting particles, and the idea of the rapid changes of the plane of vibration of a wave due to the constantly changing directions of oscillation of the particles. The fundamental essence of a 'natural' ray of light, one consisting of a totality of polarized rays of different and rapidly changing planes of polarization, was thus clarified satisfactorily. The final stage of the wave theory of light was reached only after the completion of the theory of electromagnetism, the history of which will now briefly be outlined.

Ørsted's classical paper of 1820 on the deflection of a magnetic needle by an electric current convincingly established the connection between electricity and magnetism. It was obvious that what he called 'the conflict of electricity' takes place in the space surrounding the conductor through which the electric current flows, and that this 'conflict' can be represented by circles in any of the planes perpendicular to the conductor. Ørsted's discovery is referred to again and again by Ampère ('the Newton of electricity' as Maxwell called him). His systematic experiments on the mutual action of electric currents and on the effect of a current on a magnet laid the foundation for his great work on electrodynamics. A passage from Ampère's *Mathematical Theory of Electrodynamic Phenomena* exhibits the brilliant style characteristic of all his writings; it is most revealing for his philosophy of science, in particular for his attitude towards the relation of experiment, theory and the mathematical formulation of facts. It also contains an historical analysis of developments prior to his theory. Such historical surveys by great scientists (like that given by Maxwell fifty years after Ampère, and again that of Rutherford fifty years after him) impart to us a deeper insight into the history of scientific ideas than that which can be presented by a professional historian of science.

Ampère's electrodynamics was the precursor of the work of Faraday. With his remarkable imagination, ingenuity and skill Faraday can be regarded as the greatest experimental physicist of all time. From all the astounding wealth of his experimental researches only two passages are quoted in the Anthology, one related to his discovery of electromagnetic induction (1831) and the other concerning the action of magnetic fields on light. Nothing is quoted here of his discovery of the laws of electrolysis, a signal contribution to electrochemistry, and to the knowledge of the electrical structure of matter in general. On the other hand, extensive quotations are given from Faraday's papers on the essence of magnetic action, the significance of lines of magnetic force and on the concept of a physical field of force in general. Faraday's supreme power of expressing himself in terms of concrete pictures, that of describing his physical ideas in language that lends itself to immediate intuitive understanding, is all the more impressive as it is not obscured or diverted by abstract mathematical formulations. In this respect, his lack of mathematical background is a decidedly positive asset for every student of the history of science to whom the unfolding of a scientific idea is of primary interest. What emerge from the passages printed here are Faraday's conviction of the reality of the magnetic and electric fields and his conclusion that their associated lines of force are not mere mathematical abstractions but part and parcel of concrete physical reality. In the circumstances Faraday's criticism of the interpretation of the force of gravity as an action at a distance and his thoughts on the notion of a gravitational field surrounding massive bodies are most illuminating. In his view a satisfactory explanation of gravitation is not possible without the concept of a physical field of force analogous to that which he introduced for magnetism and electricity. His attempts—albeit with negative results—to discover experimentally some connection between gravity and electricity show how deeply he was concerned with this idea.

In an anthology which tries as much as possible to avoid any recourse to mathematical derivation and any mention of mathematical equations, it is impossible to do full justice to Maxwell's greatest achievement—his mathematical theory of electromagnetism. Despite the rapidly changing world-picture of science, his equations will always retain a major place in theoretical physics. The extensive passages from his most important papers quoted here should convey some idea of the greatness of Maxwell's conceptions even to those readers who have not the benefit of a mathematical background.

The excerpts throw into relief the development of Maxwell's essential train of thought, which in the course of a few years led to the final crystallization of his theory. First of all there is his elaboration of Faraday's concept of lines of force pictured by Maxwell as filaments or tubes. From the tension along such lines, and the pressure across them, there emerged an explanation in mechanical terms of the main features of magnetic and electric fields. Maxwell assumed the pressure across the lines of force to result from the centrifugal forces of vortices; their sense of revolution also determines the dipolar character of the lines forming the axes of the vortices. Moreover, the angular velocity of the revolution was assumed to be proportional to the intensity of the magnetic field. Some additional factor in the picture of adjacent tubes revolving in the same sense is required in order to turn it into a satisfactory mechanical model.

Maxwell reminds us of the 'idlers' which have to be placed between two wheels revolving in the same direction in order to prevent them from obstructing each other's motion. He therefore envisages layers of particles between the vortices to fulfil the same function. These particles play the part of electricity; as long as the magnetic field remains constant, they stay stationary between the vortices they separate, but as soon as the field strength changes, in other words as soon as the rotation of the vortices ceases to be uniform, the electrical particles are forced out of their positions to move across the direction of the magnetic field. Thus the essential phenomena of electromagnetic induction—namely the creation of electric currents and the concatenation of changing magnetic and electric fields—are described mechanically.

A few years later Maxwell published his final version of the dynamical theory of the electromagnetic field. The fact that in this paper he refrains from any further reference to his mechanical model and restricts himself to the exposition and interpretation of his electromagnetic equations is of an historical significance which can hardly be overestimated. Here, for the first time in the history of physics, a great scientist effectively concedes that mechanical analogies, however useful they may be as props for the scientific imagination, are no more than makeshifts which must not be regarded as true representations of physical reality. In view of the abstract nature of this reality only such mathematical descriptions as Maxwell's own equations can furnish an adequate representation.

However, Maxwell's great paper also contains convincing evidence of the fact that—in contradistinction to certain mechanical analogies—scientific thought may essentially be furthered by suitable *conceptual* analogies. The changing electric field between the plates of a capacitor as it is discharged by an electric current is described by him as a 'displacement current'. Hence he observes that only closed electric currents exist in nature, and consequently the changing electric field must be surrounded by a magnetic field in the same way as is an electrical conductor through which a current is flowing. Furthermore, Maxwell's identification of light as an electromagnetic phenomenon (his conclusion that 'light is an electromagnetic disturbance propagated according to electromagnetic laws') brought about a synthesis of three main fields of physics—electricity, magnetism, and light. Some twenty years later the experiments of Heinrich Hertz on electromagnetic waves showed that these waves indeed behave in every respect like light; thus was opened the first chapter of radio physics.

During the two centuries from the time of Huygens to Maxwell's electromagnetic theory of light, the aether was supposed to be the medium in which the propagation of light waves took place. Physicists clung to the obvious analogy between the propagation of sound waves in an elastic medium such as air and that of light in an aetherial one. To this substance they attributed properties according to the changing level of advancing physical knowledge. However, this did not lead as could have been hoped to the emergence of an unambiguous picture of this hypothetical medium. Even one of its simplest properties, that of its state of motion, was subject to conflicting theories—is the aether at rest with regard to a moving body, is it partially carried along with it, or, finally, is it completely swept along by the body moving in it? The first suggestion would seem to be confirmed by Bradley's discovery in 1727 of the aberration of light; the second agrees with Fizeau's experiments on the propagation of light in flowing water; the third could serve as an explanation for the surprising result of Michelson's celebrated interferometric experiments. Instead of detecting the expected influence of the 'aether wind' of 30 km/s arising from the earth's orbital motion, his investigation had a 'null effect'. Another explanation of the experimental result requires of course the negation of the existence of the aether as a physical substratum. This was one of the starting-points for Einstein's theory of relativity.

A synthesis of an even more universal nature than that of Maxwell's theory was achieved during the nineteenth century with the first and second laws of thermodynamics. The first law is nothing else than the law of conservation of energy, initially formulated in 1847 by Helmholtz and later expounded by him and others in a series of papers. The history of this law can be traced back to two roots.

One was the conclusion arrived at by physicists in the early nineteenth century, that the conception of heat as a substance (in other words, the hypothesis of the 'caloric') was untenable; experiments such as that of Rumford furnished conclusive evidence of this. Scientists then reverted to a conception which had been part of the atomic theories of earlier times, namely that heat derives from the movements of the ultimate particles of matter. If heat could thus be expressed in the mechanical terms of *vis viva* or of kinetic energy, it should be possible to find the numerical equivalence of a certain quantity of heat with a certain quantity of mechanical work. This train of thought led to the brilliant experiments of Joule by which the 'mechanical equivalent of heat' was established.

The second root of the law was the centuries-old search for an entity which, among the multitude of quantities undergoing changes and transformations in the variegated processes of nature, would exhibit the important property of permanence or constancy. This entity was found to be the energy of a physical system (Helmholtz first used for it the term 'force'). The universal significance of energy was thrown into relief by the mechanical theory of heat and by the fact that ultimately other forms of energy, such as electrical and magnetic energies, or even the energy of living organisms, can be expressed in terms of heat or mechanical energy. Helmholtz's assertion that in a given system the energy is always conserved was confirmed experimentally. This was so too in all subsequent investigations, including those of atomic physics. It is in fact the positive counterpart of the centuries-old lack of success in the search for a *perpetuum mobile*. The vast importance of the law of conservation of energy for the unification of our world-picture needs hardly to be emphasized. This

Anthology also contains excerpts from another of Helmholtz's papers (written in English in the original). Here, in the context of an exposition of the law, he rejects the notion of a 'vital energy' for the explanation of organic processes. Indeed, up to now, biology has no need for any recourse to 'vitalism' and can rely on its theories of purely physico-chemical concepts.

Of even greater significance for the physical world-picture and its philosophical implications is the second law of thermodynamics, discovered in 1824 by Carnot. In substance it means that, although the total energy of any system through all its transformations is conserved, there exists a definite sense of direction in these transformations so that there is a tendency for energy of a higher degree of order (such as mechanical energy of heat at high temperature) to be degraded into energy of a lower degree of order (such as heat at low temperature). The first step in the conceptual development leading to the final clarification and formulation of this law was Carnot's ingenious thought-experiment in which his ideal engine transformed heat into mechanical energy. (In the discussion, by the way, he still used the concept of a 'heat substance'.) It turned out that only a fraction of the heat put into the engine could be transformed into mechanical work, the size of the fraction depending only on the temperature difference between the two reservoirs of heat indispensable for its working. The remainder of the heat is dissipated in the form of low-grade heat. The second step was Clausius' introduction of the concept of the entropy of a system. This can be represented as a mathematical quantity, thus allowing for a mathematico-physical treatment of the problem. The fact that in general the entropy of a system always increases is an indication of that unilateral sense of direction of physical events which, in contrast to the strict reversibility of purely mechanical processes, makes it possible to distinguish physically between past and future. Clausius' famous formulation that heat passes by itself only from higher to lower temperatures is equivalent to the statement that each creation of a source of high-temperature energy (namely each decrease of entropy) is possible only with such an investment of work that, on balance, the whole process will lead to an overall increase of entropy.

Statistical mechanics and the kinetic theory of heat as developed by Maxwell, Gibbs and Boltzmann marked the third step in the crystallization of the entropy law. In this stage it became clear that the second law is of a probabilistic character, in contrast to the strictly causal nature of the first law. This distinction, however, is not valid any more in our contemporary physics, which has come to the conclusion that all processes of nature are based on primary probabilities. In this connection a passage from an account of Maxwell's theory of heat is of interest. Here he introduces his famous 'demon' as the symbol of a mechanism by which a decrease of entropy might be made possible. Maxwell's demon was 'exorcised' in 1929 by Szillard after Boltzmann had previously generalized the concept of entropy as a measure of the disorder of a system. Furthermore, he had succeeded in quantitatively connecting entropy and probability, reasoning that the higher the entropy of a system, in other words the greater its disorder, the higher the probability of its state. The distinction between the earlier and the later states of a system can be reduced to the difference between the levels of order in that system. The situation corresponding to higher disorder (namely the one of higher probability) must be attributed to a later time. Boltzmann's speculations on the entropy of the universe as a whole (assuming that one could regard that as a closed system) are highly suggestive, in view of his probabilistic approach to the problem. One can avoid the notion of the 'heat death' of the universe by supposing statistical fluctuations on a cosmic scale which are able to decrease the entropy of vast regions, such as of distant clusters of galaxies. However, as Boltzmann points out, conscious beings living in that region would not be able to realize this, since they would always necessarily attribute a state of greater entropy to a later time, and thus their sense of past and future would be reversed in comparison with ours.

Finally, two passages from Mach are quoted in this chapter, both coming from his extremely illuminating book on the historical development of mechanics. One of them, on the 'Economy of Science', is a characteristic example of the positivistic spirit which reached its climax during the nineteenth century and of which Mach himself was an eminent exponent. The other passage criticizes Newton's concept of inertia relative to absolute space, and points out that the idea of inertial motion must be bound up with the assumption of the existence of other masses in space. This latter in particular considerably influenced the thinking of Einstein as it led to his general theory of relativity.

V: FROM DALTON TO MACH

JOHN DALTON

The ultimate particles and their relative weights

260

There are three distinctions in the kinds of bodies, or three states, which have more especially claimed the attention of philosophical chemists; namely, those which are marked by the terms *elastic fluids, liquids, and solids*. A very familiar instance is exhibited to us in water, of a body, which, in certain circumstances, is capable of assuming all the three states. In steam we recognize a perfectly elastic fluid, in water, a perfect liquid, and in ice a complete solid. These observations have tacitly led to the conclusion which seems universally adopted, that all bodies of sensible magnitude, whether liquid or solid, are constituted of a vast number of extremely small particles, or atoms of matter bound together by a force of attraction, which is more or less powerful according to circumstances, and which as it endeavours to prevent their separation, is very properly called in that view, *attraction of cohesion*; but as it collects them from a dispersed state (as from steam into water) it is called, *attraction of aggregation*, or more simply, *affinity*. Whatever names it may go by, they still signify one and the same power. It is not my design to call in question this conclusion, which appears completely satisfactory; but to shew that we have hitherto made no use of it, and that the consequence of the neglect, has been a very obscure view of chemical agency, which is daily growing more so in proportion to the new lights attempted to be thrown upon it.

The opinions I more particularly allude to, are those of Berthollet[1] on the Laws of chemical affinity; such as that chemical agency is proportional to the mass, and that in all chemical unions, there exist insensible gradations in the proportions of the constituent principles. The inconsistence of these opinions, both with reason and observation, cannot, I think, fail to strike every one who takes a proper view of the phenomena.

Whether the ultimate particles of a body, such as water, are all alike, that is, of the same figure, weight, etc., is a question of some importance. From what is known, we have no reason to apprehend a diversity in these particulars: if it does exist in water, it must equally exist in the elements constituting water, namely, hydrogen and oxygen. Now it is scarcely possible to conceive how the aggregates of dissimilar particles should be so uniformly the same. If some of the particles of water were heavier than others, if a parcel of the liquid on any occasion were constituted principally of these heavier particles, it must be supposed to affect the specific gravity of the mass, a circumstance not known.

1. Claude Berthollet (1748–1822), French chemist, one of the founders of the concept of chemical affinity. He recognized the significance of mass proportions in chemical reactions. (Ed.)

Similar observations may be made on other substances. Therefore we may conclude that *the ultimate particles of all homogeneous bodies are perfectly alike in weight, figure, etc.* In other words, every particle of water is like every other particle of water; every particle of hydrogen is like every other particle of hydrogen, etc. [. . .]

When any body exists in the elastic state, its ultimate particles are separated from each other to a much greater distance than in any other state; each particle occupies the centre of a comparatively large sphere, and supports its dignity by keeping all the rest, which by their gravity, or otherwise are disposed to encroach up it, at a respectful distance. When we attempt to conceive the *number* of particles in an atmosphere, it is somewhat like attempting to conceive the number of stars in the universe; we are confounded with the thought. But if we limit the subject, by taking a given volume of any gas, we seem persuaded that, let the divisions be ever so minute, the number of particles must be finite; just as in a given space of the universe, the number of stars and planets cannot be infinite.

Chemical analysis and synthesis go no farther than to the separation of particles one from another, and to their reunion. No new creation or destruction of matter is within the reach of chemical agency. We might as well attempt to introduce a new planet into the solar system, or to annihilate one already in existence, as to create or destroy a particle of hydrogen. All the changes we can produce, consist in separating particles that are in a state of cohesion or combination, and joining those that were previously at a distance.

In all chemical investigations, it has justly been considered an important object to ascertain the relative *weights* of the simples which constitute a compound. But unfortunately the inquiry has terminated here; whereas from the relative weights in the mass, the relative weights of the ultimate particles or atoms of the bodies might have been inferred, from which their number and weight in various other compounds would appear, in order to assist and to guide future investigations, and to correct their results. Now it is one great object of this work, to shew the importance and advantage of ascertaining *the relative weights of the ultimate particles, both of simple and compound bodies, the number of simple elementary particles which constitute one compound particle, and the number of less compound particles which enter into the formation of one more compound particle.*

WILLIAM PROUT

261

On the relation between the specific gravities of bodies in their gaseous state and the weights of their atoms

The author of the following essay submits it to the public with the greatest diffidence; for though he has taken the utmost pains to arrive at the truth, yet he has not that confidence in his abilities as an experimentalist as to induce him to dictate to others far superior to himself in chemical acquirements and fame. He trusts, however, that its importance will be seen, and that some one will undertake to examine it, and thus verify or refute its conclusions. If these should be proved erroneous, still new facts may be brought to light, or old ones better established, by the investigation; but if they should be verified, a new and interesting light will be thrown upon the whole science of chemistry.

It will perhaps be necessary to premiss that the observations about to be offered are chiefly founded on the doctrine of volumes as first generalized by M. Gay-Lussac;[1] and which, as far as the author is aware at least, is now universally admitted by chemists. [. . .]

The following tables exhibit a general view of the above results, and at the same time the proportions, both in volume and weight, in which they unite with oxygen and hydrogen: also the weights of other substances, which have not been rigidly examined, are here stated from analogy. [. . .]

On a general review of the tables, we may notice,

(1) That all the elementary numbers, hydrogen being considered as 1, are divisible by 4, except carbon, azote, and barytium, and these are divisible by 2, appearing therefore to indicate that they are modified by a higher number than that of unity or hydrogen. Is the other number 16, or oxygen? And are all substances compounded of these two elements? [. . .]

There is an advantage in considering the volume of hydrogen equal to the atom, as in this case the specific gravities of most, or perhaps all, elementary substances (hydrogen being 1) will either exactly coincide with, or be some multiple of, the weights of their atoms; whereas if we make the volume of oxygen unity, the weights of the atoms of most elementary substances, except oxygen, will be double that of their specific gravities with respect to hydrogen. [. . .]

If the views we have ventured to advance be correct, we may almost consider the $\pi\rho\acute{\omega}\tau\eta$ $\ddot{v}\lambda\eta$ of the ancients to be realized in hydrogen; an opinion, by the by, not altogether new. If we actually consider this to be the case, and further consider the specific gravities of bodies in their gaseous state to represent the number of volumes condensed into one; or, in other words, the number of the absolute weight of a single volume of the first matter ($\pi\rho\acute{\omega}\tau\eta$ $\ddot{v}\lambda\eta$) which they contain, which is extremely probable, multiples in weight must always indicate multiples in volume, and vice versa; and the specific gravities, or absolute weights of all bodies in a gaseous state, must be multiples of the specific gravity or absolute weight of the first matter ($\pi\rho\acute{\omega}\tau\eta$ $\ddot{v}\lambda\eta$), because all bodies in a gaseous state which unite with one another unite with reference to their volume.

1. Louis Gay-Lussac (1778–1850), French chemist and physicist. One of his discoveries was the law concerning the volumes of reacting gaseous compounds. (Ed.)

TABLE I: ELEMENTARY SUBSTANCES

Name	Sp. gr. hydr. being 1.	Wt. of atom, 2 vols. hydr. being 1.	Wt. of atom, oxygen being 10.	Wt. of atom, oxygen being 10, from experiment.	Sp. gr. atmospheric air being 1.	Sp. gr. atmospheric air being 1, from experiment.	Wt. in grs. of 100 cub. inches. Barom. 30, Therm. 60.	Wt. in grs. of 100 cub. inches from exper.
Hydrogen	1	1	1·25	1·32	·06944	·073	2·118	2·23
Carbon	6	6	7·5	7·54	·4166	—	12·708	—
Azote	14	14	17·5	17·54	·9722	·969	29·652	29·56
Phosphorus	14	14	17·5	17·4	·9722	—	29·652	—
Oxygen	16	8	10	10	1·1111	1·104	33·888	33·672
Sulphur	16	16	20	20	1·1111	—	33·888	—
Calcium	20	20	25	25·46	1·3888	—	42·36	—
Sodium	24	24	30	29·1	1·6666	—	50·832	—
Iron	28	28	35	34·5	1·9444	—	59·302	—
Zinc	32	32	40	41	2·222	—	67·777	—
Chlorine	36	36	45	44·1	2·5	2·483	76·248	—
Potassium	40	40	50	49·1	2·7777	—	84·72	—
Barytium	70	70	87·5	87	4·8611	—	148·26	—
Iodine	124	124	155	156·21	8·6111	—	262·632	—

Table 1. This, as well as the other tables, will be easily understood. In the first column we have the specific gravities of the different substances in a gaseous state, hydrogen being 1: and if we suppose the volume to be 47·21435 cubic inches, the numbers will at the same time represent the number of grains which this quantity of each gas will weigh. In the third column are the corrected numbers, the atom of oxygen being supposed, according to Dr Thomson, Dr Wollaston, &c., to be 10: and in the fourth, the same, as obtained by experiment, are stated, to show how nearly they coincide.

AUGUSTIN FRESNEL

The wave theory of light: interference, diffraction, polarization

262

7. It seems to me that the theory of vibrations suits the phenomena better than Newton's theory, and if this theory has not given a satisfactory explanation of

refraction, this might result from the fact that from this point of view light has not yet been studied profoundly enough. The hypothesis is simple and one feels that it should be fertile in consequences, but these can be obtained only with difficulty.

8. The strongest objection to this theory is that it has been based on a comparison between light and sound; but nothing indicates that one could draw an exact comparison between the vibrations of air, a heavy fluid, and the vibrations of caloric, a tenuous fluid of which it has borrowed its elasticity. The motion of light is infinitely more rapid than that of air; its motion should then deviate much less from its original direction[1] if no obstacle hinders it; for light on meeting a body can be reflected, refracted or diffracted, like sound.

9. An objection, the only one to which I find any difficulty in giving an answer, led me to occupy myself with projected shadows. To treat the phenomenon in its greatest simplicity I have diminished as much as possible the dimensions of the point of light and observed at the same time that the shadows cast were not sharply defined. Yet they should have been, had light propagated only in its original direction. One sees that it spreads into the shadow and it is difficult to assign the point where it stops, the limit of the angle of bending. I could perceive some light up to the middle of the shadow of a two-centimeter-wide ruler, when looking at it directly with a strong magnifying glass. [. . .]

263

Before giving my attention to the numerous phenomena, all covered by the name: diffraction, I believe it is necessary to give some general consideration to the two theoretical systems which have been so far advocated by scientists. Newton had held that light-molecules, emitted by the bodies which give us light, arrived directly at our eyes and then through their impact produced the sense of vision. Descartes, Hooke, Huygens, Euler, had thought that light resulted from the vibrations of a universal, extremely tenuous fluid, agitated by the rapid movements of the particles of the luminous body in much the same manner as air is perturbed by the vibrations of sonorous bodies; it is of a sort that on this theory it is not the molecules of the fluid in contact with the luminous bodies which reach the organ of vision, but solely the motion which has been impressed upon them.

1. The first hypothesis has the advantage of leading to more obvious results, for mechanical analysis is more easily applicable to it. The second theory, on the contrary, presents in this respect great difficulties. But in choosing a theory, one must pay attention only to the simplicity of the hypothesis; those of the calculation should have no weight in the balance of probabilities. Nature is not embarrassed by the difficulties of analysis—it has avoided only the complications of the means. It may seem that it has tried to achieve a lot with very little: this

1. It is possible that even in vacuum the motion of a beam of light may cause others in oblique direction to its own, and that these motions more feeble and of different nature are insensible to the eye, where the extent of sensation is much less than that of the ears.

is a principle which the perfection of physical science reinforces in every new proof. Astronomy, the honour of human spirit, provides an especially striking confirmation; all the laws of Kepler have been brought, by the genius of Newton, under the single law of gravitation, which has seemed lately to explain and even to discover the most complicated and least apparent perturbations of the planetary motions.

2. If one has sometimes lost the way when trying to simplify the elements of a science, it is only because one has established a theory before having assembled a great enough number of facts; such a hypothesis, which is very simple when considering only one class of phenomena, necessitates many more hypotheses when trying to escape the narrow circle in which one has locked oneself. If nature has intended to produce maximum effect with a minimum of causes, it is in the collection of its laws that it should resolve this great problem. No doubt, it is difficult to discover the basis of this admirable economy, i.e. the simplest causes of the phenomena, considered under a point of view, even so extended. But if this general principle of the philosophy of the physical sciences does not immediately lead to the recognition of truth, it can nonetheless guide the efforts of the human spirit in eliminating those theories which put the phenomena under too many different causes, and in making it adopt preferentially those which, relying on the smallest number of hypotheses, prove to be most fertile in consequences.

3. In agreement with this, that theory which makes light consist in the vibrations of a universal fluid has great advantage above the emission theory. This theory allows us to form a conception of the way in which light is susceptible to so many modifications. I do not mean here those momentary changes that light experiences while traversing the bodies, and which can always be attributed to the nature of these media, but I wish to speak of those permanent modifications which light carries along and which impress on it new characteristics. One assumes that a fluid, a collection of an infinity of moving molecules mutually interdependent, is susceptible of many different modifications due to the relative motions which are impressed on them. The vibrations of the air and the sensations which these produce on the organ of hearing will serve a remarkable example.

In the theory of emission, on the contrary, the course of each light-molecule being independent of the others, the number of different modifications to which they are susceptible seems extremely limited. One can add a rotatory motion to that of translation, but that is all. As to oscillatory motions, this could be conceivable only in media which would keep them up by an unequal action of their parts on the different sides of the light-molecules, supposedly having different properties. When this action ceases, the oscillations must cease also, or to be transformed into rotatory motion. Thus, rotatory motion and the different faces of one and the same light-molecule are the only mechanical resources of the emission theory to represent all the permanent modifications of light.

This would seem very inadequate if we consider the multitude of phenomena which is offered to us in optics. One will become even more convinced when reading the *Mathematical and Experimental Treatise on Physics* of M. Biot, in which the major consequences of Newton's theory are explained with many details and with great clarity. One will see that in order to account for the phenomena, one must accumulate for each light-particle a great number of different modifications, often too difficult to bring them all into accord.

4. According to the theory of waves the infinite variety of rays of different colour which constitute white light derives simply from a difference in the lengths of the light-waves, in the same manner as the various musical tones derive from the difference in lengths of sound waves. In the Newtonian theory one cannot attribute the diversity of colours or of the sensations produced on the organ of vision to the differences in mass or initial velocity of the light-molecules, for this would bring the result that dispersion would have to be always proportional to refraction, and experience proves the contrary. Thus it must *necessarily* be admitted that the molecules of differently coloured rays are not of the same nature. Thus there are as many different light-molecules as there are colours of different nuances in the solar spectrum.

5. Having explained reflection and refraction through attractive and repulsive forces, emanating from the surface of bodies, Newton, in order to explain the phenomena of coloured rings, imagined that for the light-molecules fits of easy reflection and easy transmission change periodically at equal intervals. It was natural to suppose that these intervals, like the speed of light, are always the same in the same medium, and that, consequently, when the incidence is more oblique, the traversed path having increased, the diameter of the rings will diminish. Experience teaches, on the contrary, that the diameter of the rings increases with the obliquity of incidence and Newton was obliged to conclude that the fits augment in length, and that in a much larger proportion than did the traversed path. He had to expect also to find the fits longer in media in which light propagates with greater speed, which according to him are the denser bodies; for it was natural to suppose that their durations would remain isochronous in different media. Experience has proved him wrong: he recognized that the thickness of the layer of air and water, for example, which reflect the same colour at vertical incidence, is exactly in the ratio of the sinus of incidence to the sinus of refraction, for the passage of light from air to water; and precisely this is one of the most striking confirmations of the wave theory. He had thus to assume that the length of the fits was in inverse ratio to the speed of light or, which comes to the same, that the duration of the fits diminishes in the same ratio as that with which the square of the speed increases.

Thus, the theory of emission is so little sufficient to explain the phenomena that each new phenomenon necessitates a new hypothesis. [. . .]

7. Not only is the hypothesis of the fits implausible because of its complexity and difficulty to reconcile it with the facts in its own consequences, but it does

not suffice even for the explanation of the phenomena of coloured rings, for the sake of which it had been invented. [. . .] It is very natural to assume that this phenomenon results of the influence that light, reflected from the second surface of the layer of air, exercises on that light which is reflected from the first, and this influence varies with the difference of traversed paths. Thus, the coloured rings lead to the principle of mutual influence of light-rays, like the phenomena of diffraction, even though it does not demonstrate that with the same evidence.

8. In the wave theory this principle is a consequence of the fundamental hypothesis. One imagines, in fact, that when two systems of light waves tend to produce absolutely opposing motions at the same points of space, they have to weaken mutually and even be destroyed completely, if the two impulsions are equal, and the oscillations must, on the contrary, reinforce each other if they are in the same direction. The intensity of the light will then depend on the respective positions of the two systems of waves, or, and this comes to the same, on the difference of the traversed paths if they emanate from a common source. In the contrary case the disturbances which the vibrations from two illuminated points necessarily experience, and which must succeed each other with great rapidity, will not take place simultaneously and in the same manner because these are independent; consequently the effects of the influence of two wave systems which are produced, changing each instant, cannot be perceived by the eye. [. . .]

10. The multiplicity and complexity of its hypotheses is not the only defect of the emission theory. Even if we admitted all that what I have just disproved, I will show in the continuation of this memoir that one cannot give a complete explanation of the phenomena, and that only the wave theory can account for all those phenomena which occur in the diffraction of light.

264

If the polarization of a ray of light consists in all its vibrations taking place in the same direction, it follows from my hypothesis on the generation of light-waves that a ray of light, emanating from a single centre of oscillation, will, at every given instant, be polarized in a given plane. Then, an instant later, the direction of the motion changes and with it changes the plane of polarization; and these changes succeed each other with the same rapidity as the perturbations of the vibrations of the light-emitting particle; so that even if one could separate the light emitted by this particle from the light which emanates from the other luminous points, one would not see any polarization whatsoever.

If one considers now the effect produced by the reunion of all the waves which emanate from the various points of a light-emitting body, one will see that at each instant, and at a given point of the ether, the resultant of all the motions crossing at that point will be in a given direction, but that this direction will vary from one instant to another. Thus, a ray of light can be considered as the reunion, or rather the rapid succession of groups of waves polarized in all the directions. When viewing the matter this way, the act of polarization does not consist in creating transverse motions but in decomposing them into two invariable directions, perpendicular to each other, and in separating the two components one from the other; thus, in each of these directions the oscillatory motions will operate always in the same plane.

265 HANS CHRISTIAN ØRSTED

Experiments on the effect of the electric conflict on a
magnetic needle

The first experiments respecting the subject which I mean at present to explain, were made by me last winter, while lecturing on electricity, galvanism, and magnetism, in the University. It seemed demonstrated by these experiments that the magnetic needle was moved from its position by the galvanic apparatus, but that the galvanic circle must be complete, and not open, which last method was tried in vain some years ago by very celebrated philosophers. But as these experiments were made with a feeble apparatus, and were not, therefore, sufficiently conclusive, considering the importance of the subject, I associated myself with my friend Esmarck to repeat and extend them by means of a very powerful galvanic battery, provided by us in common. [. . .]

The galvanic apparatus which we employed consists of 20 copper troughs, the length and height of each of which was 12 inches; but the breadth scarcely exceeded $2\frac{1}{2}$ inches. Every trough is supplied with two plates of copper, so bent that they could carry a copper rod, which supports the zinc plate in the water of the next trough. The water of the troughs contained $\frac{1}{60}$th of its weight of sulphuric acid, and an equal quantity of nitric acid. The portion of each zinc plate sunk in the water is a square whose side is about 10 inches in length. A smaller apparatus will answer provided it be strong enough to heat a metallic wire red hot.

The opposite ends of the galvanic battery were joined by a metallic wire, which, for shortness sake, we shall call the *uniting conductor*, or the *uniting wire*. To the effect which takes place in this conductor and in the surrounding space, we shall give the name of the *conflict of electricity*.

Let the straight part of this wire be placed horizontally above the magnetic needle, properly suspended, and parallel to it. If necessary, the uniting wire is bent so as to assume a proper position for the experiment. Things being in this state, the needle will be moved, and the end of it next the negative side of the battery will go westward.

If the distance of the uniting wire does not exceed three-quarters of an inch from the needle, the declination of the needle makes an angle of about 45°. If the distance is increased, the angle diminishes proportionally. The declination likewise varies with the power of the battery. [. . .]

The effect of the uniting wire passes to the needle through glass, metals, wood, water, resin, stoneware, stones; for it is not taken away by interposing plates of glass, metal or wood. Even glass, metal, and wood, interposed at once, do not destroy, and indeed scarcely diminish the effect. The disc of the electrophorus, plates of porphyry, a stone-ware vessel, even filled with water, were interposed with the same result. We found the effects unchanged when the needle was included in a brass box filled with water. It is needless to observe that the transmission of effects through all these matters has never before been observed in electricity and galvanism. The effects, therefore, which take place in the conflict of electricity are very different from the effects of either of the electricities.

If the uniting wire be placed in a horizontal plane under the magnetic needle, all the effects are the same as when it is above the needle, only they are in an opposite direction; for the pole of the magnetic needle next the negative end of the battery declines to the east.

That these facts may be more easily retained, we may use this formula—the pole *above* which the *negative* electricity enters is turned to the *west*; *under* which, to the *east*. [. . .]

A brass needle, suspended like a magnetic needle, is not moved by the effect of the uniting wire. Likewise needles of glass and of gum lac remain unacted on.

We may now make a few observations towards explaining these phenomena.

The electric conflict acts only on the magnetic particles of matter. All non-magnetic bodies appear penetrable by the electric conflict, while magnetic bodies, or rather their magnetic particles, resist the passage of this conflict. Hence they can be moved by the impetus of the contending powers.

It is sufficiently evident from the preceding facts that the electric conflict is not confined to the conductor, but dispersed pretty widely in the circumjacent space.

From the preceding facts we may likewise collect that this conflict performs circles; for without this condition, it seems impossible that the one part of the uniting wire, when placed below the magnetic pole, should drive it towards the east, and when placed above it towards the west; for it is the nature of a circle that the motions in opposite parts should have an opposite direction. Besides, a motion in circles, joined with a progressive motion, according to the length of the conductor, ought to form a conchoidal or spiral line; but this, unless I am mistaken, contributes nothing to explain the phenomena hitherto observed.

All the effects on the north pole above-mentioned are easily understood by supposing that negative electricity moves in a spiral line bent towards the right, and propels the north pole, but does not act on the south pole. The effects on the south pole are explained in a similar manner, if we ascribe to positive electricity a contrary motion and power of acting on the south pole, but not upon the north. The agreement of this law with nature will be better seen by a repetition of the experiments than by a long explanation. The mode of judging of the experiments will be much facilitated if the course of the electricities in the uniting wire be pointed out by marks or figures.

I shall merely add to the above that I have demonstrated in a book published five years ago that heat and light consist of the conflict of the electricities. From the observations now stated, we may conclude that a circular motion likewise occurs in these effects. This I think will contribute very much to illustrate the phenomena to which the appellation of polarization of light has been given.

ANDRÉ MARIE AMPÈRE

266

On the interaction between two electric currents

Electromotive action manifests itself in two sorts of effects which I believe that I must distinguish by a precise definition. I shall call the first *electric tension* and the other *electric current*. The first is observed when two bodies, between which the electromotive action takes place, are separated from each other by non-conducting bodies at each point of their surface except at those where it is established; the second is where these, on the contrary, form part of a circuit of conducting bodies which communicate at points on their surface different from those at which the action is produced. In the first case the effect of this action is to put the two bodies or the two systems of bodies, between which it takes place, in two states of tension of which the difference is constant while the action is constant, when for example, it is produced by contact of two different substances of different nature; this difference varies, on the contrary, with the cause which has produced it if it has occurred as a result of friction or pressure.

The first case is the only one which can take place when the electromotive action develops between different parts of the same non-conducting body. Tourmaline offers a good example when it changes its temperature.

In the second case there is no more electric tension, light bodies are not sensibly attracted, and the ordinary electrometer cannot serve any more to indicate what is going on in the bodies; meanwhile the electromotive action continues to act; for, if, for example water, an acid, an alkali or solution of salt form part of the circuit, the bodies are decomposed, especially if, as it has been known for a long time, the electromotive action is constant; and, besides, as has been discovered by M. Ørsted, when electromotive action is produced by the contact of two metals, the magnetic needle is deflected from its direction when it is placed near any portion of the circuit; but these effects cease, water is not any more decomposed, the needle returns to its ordinary position, the moment when the circuit is broken, when the tension is re-established and when the light bodies are again drawn together, which proves that the tensions are not the cause of the decomposition of water nor of the changes in the direction of the magnetic needle as discovered by Ørsted. Evidently, this second case is the only one that can take place when the electromotive action develops between the different parts of one and the same conducting body. The consequences of Ørsted's experiments as deduced in this *Memorandum* will lead us to recognize the existence of this condition in the only case where it still has to be admitted.

Let us now see where lies the difference between these two orders of entirely different phenomena, one of which consists in the tension and the attractions or repulsions, phenomena well known for a long time, and the other in the decomposition of water and a great number of other substances, in the change of direction of the needle and in a sort of attractions and repulsions which are completely different from the ordinary electric attractions and repulsions which I believe to have been the first to recognize and which I have designated *attractions and repulsions of electric currents* in order to distinguish them from the others. When there is no conducting continuity between one of the bodies or one of the systems of bodies in which the electromotive action develops, to the other, and when these bodies themselves are conductors like in a voltaic pile, one could conceive of this action only as a constant carrying of positive electricity into the one body and negative electricity into the other; in the first moment when nothing opposes the effect which the actions tend to produce the two electricities accumulate, each in that part of the total system towards which it is being carried; but this action stops when the difference of the electric tensions gives to their mutual attraction, which tends to reunite them, sufficient force to make equilibrium with electromotive action. Then everything remains in that state, except the loss of electricity which takes place slowly towards the non-conducting body, air for example, which breaks the circuit; for it seems that no bodies exist which are perfect insulators. As far as the leakage is taking place the tension diminishes, but as it diminishes, the mutual attraction of the two electricities will no longer be in equilibrium with the electromotive action, this last force, in case it is constant, will again carry positive electricity to the one side and negative to the other, and tensions will be re-established. It is this state of a system of electromotive and conducting bodies which I call electric tension. It is known that it exists in the two halves of the system, be it after we have separated them, be it even in the case when they remain in contact after the electromotive action has ceased, provided that it has arisen by friction or pressure between bodies which are not both conductors. In these two cases the tensions diminish gradually through the leakage of electricity of which we have spoken.

But when the two bodies or the two systems of bodies between which the electromotive action takes place are also connected by conducting bodies between which there is no other electromotive action equal and opposite to the first, which would maintain the state of electrical equilibrium, and consequently the tension resulting from it, these tensions would disappear or at least become very small, and phenomena would be produced which we indicated as characteristic of the second case. But as nothing else has been changed in the arrangement of the bodies between which the electromotive action has developed, one cannot doubt that it continues to act, and as the mutual attraction of the two electricities, measured by the difference in electric tension which has become zero or has considerably diminished, can no longer come to equilibrium with this action, there is general agreement that it continues to carry the two electricities in the two directions where it has been carrying it previously; in such a way that a double current is the result, one of positive electricity, the other of negative electricity, starting out in opposite directions from the points where the electromotive action arises, and going out to reunite in that part of the circuit which is opposite to those where it started. The currents of which I am talking are accelerated until the inertia of the electric fluids and the resistance which they encounter because of the imperfections of even the best conductors, make equilibrium with the electromotive force, after which they continue indefinitely with a constant velocity, so long as this force keeps the same intensity, but they cease on the instant the circuit has been broken. It is this state of electricities in a series of conducting bodies which I will call for the sake of brevity the *electric current*; and, as I shall have to talk all the time of two opposite directions in which the electricities move, I shall understand every time when this problem is being dealt with, to avoid tedious repetition after the words, 'the sense of the electric current' the words 'of the positive electricity'; thus, like in the case of voltaic pile the expression: 'direction of the electric current in the pile', will designate the direction from where the hydrogen is liberated in the decomposition of water to the end where the oxygen is obtained; and this expression: 'the direction of the electric current in a conductor, which established the contact between the two ends of the pile' will designate the direction which, on the contrary, goes from the end where the oxygen is produced towards the end where the hydrogen is developed. In order to cover the two cases by a single definition one can say that the direction of the current is that one which is followed by hydrogen and saline bases when water or a saline substance form part of a circuit and as decomposed by the current, whether, in a voltaic pile, these substances are part of the conductor or are interposed between those pairs of which the voltaic pile is built up. [...]

But the differences which I have mentioned are not the only ones which distinguish these two states of electricity. I have discovered some much more remarkable ones, when arranging in parallel directions two straight pieces of two conducting wires joining the ends of two voltaic piles; the one was fixed, the other suspended from two points and made very mobile by a counterweight, could easily approach or move away from the first while keeping in parallel with it. I have then observed that letting a current pass at the same time through both wires, they are mutually attracted when the currents are in the same direction, and repel each other when the currents are in opposite senses.

Now the attractions and repulsions of electric currents differ essentially from those which are produced by electricity at rest; first, they cease, as chemical decomposition does, the instant we break the circuit of the conducting bodies; second, in the ordinary electric attractions and repulsions, opposite species of electricity attract, and those of the same name repel each other; in the attractions and repulsions of electric currents it is exactly the contrary, it is when the two wires are placed parallel in such a manner that the ends of the same name are on the same side and very near each other that attraction occurs, and there is repulsion when in the two conductors, being always parallel, the currents are in opposite directions, in such a way that the ends of the same name are at the greatest possible distance from each other. Third, in the case when attraction occurs, and when it is strong enough to bring the movable conductor in contact with the fixed one, they remain attached to each other like two magnets, and do not separate after that, as it happens when two conducting bodies which are drawn together because they are electrified, the one positively, the other negatively, come in contact. Finally, and it seems that this last circumstance derives from the same cause as does the preceding: two electric currents are attracted or repelled in vacuum exactly as they are in air; this is again contrary to what is observed in the mutual action of ordinary electrified conducting bodies.

267

On the interaction between an electric current and a magnet

The first idea that occurred to me when I wanted to search for the causes of the new phenomena discovered by M. Ørsted is, that the order in which one discovers two facts, has nothing to do with the consequences of the analogies which they present. Thus we were able to suppose that before knowing that the magnetic needle has a constant direction south to north, one has already known its property that it is deflected by an electric current into a position perpendicular to the current, in a manner that always the same pole of the needle is carried to the left of the current, and that one has only later discovered its property to turn towards the north that of its poles which is to the left of the current; would it not be the simplest idea and the one which presents itself immediately to him who wants to explain the constant direction of the needle, to admit that there is in the earth an electric current in such a direction that the north would be to the left of a man who, lying on its surface with his face towards the needle, would receive this current in the direction from his feet towards his head, and to conclude from this that it takes place from east to west in a direction perpendicular to the magnetic meridian? This hypothesis becomes even more plausible if one pays more attention to the totality of all known facts; this current, if it exists, should be compared to that which I showed to be acting on the magnetic needle in the pile, as it is directed from the copper end to the zinc end when one establishes a conductor between them, and which would occur even if the pile, forming a closed circuit, would be reunited by an element similar to the others; for there is probably nothing on our globe which resembles a continuous and homogeneous conductor; but the various materials of which it is composed are precisely those in the case of a voltaic pile, formed by elements arranged at random, and which, coming back to itself, would be formed as a continuous chain around the earth. Elements thus arranged would, no doubt, give less electric energy than they would have given had they been in a periodically regulated order; but it would be necessary that they be expressly so arranged that, in a series of different substances forming a closed circuit around the Earth, there should be no current in one or the other sense. [. . .]

Now, if the electric currents are the cause of the deflecting action of the Earth, electric currents will also be the cause of the action of one magnet on the other, from which it follows that a magnet should be considered an assemblage of electric currents, which all flow in planes perpendicular to its axis, directed in such a way that the southern pole of the magnet, which is turned towards the north, is to the right of these currents while always to the left of a current placed outside the magnet, and which faces it in a parallel direction; or rather these currents are established first in the magnet along the shortest closed circuits, be it from left to right or from right to left, and then the line perpendicular to the planes of the currents becomes the axis of the magnet and its ends become the poles. Thus, at each of the poles of the magnet the electric currents of which it is composed are directed along concentric closed circuits; I have imitated that arrangement as much as it was possible with an electric current by bending the conducting wire into a spiral; this spiral was formed with a brass wire and ended in two rectilinear portions of the same wire which were enclosed in two tubes of glass in order that they should not touch each other and could be attached to the two ends of the pile.

According to the direction in which one lets the current flow in that spiral, it is in effect strongly attracted or repelled by the pole of a magnet which is placed in such a way that the direction of its axis is perpendicular to the plane of the spiral, depending on whether the electric currents of the spiral and of the magnetic pole are in the same direction or in opposite directions. On replacing the magnet by another spiral, in which the current is in the same direction as that of the magnet, the same attractions and repulsions occur; it is in this way that I discovered that two electric currents attract each other when they are in the same direction and repel each other in the other case.

On replacing then, in the experiment of the mutual action between the pole of a magnet and a current in a metallic wire bent into a spiral, that spiral by another magnet, one has again the same effects, be it attraction or repulsion, in conformity with the law of known phenomena of a magnet; thus it is evident that all the circumstances of these last phenomena are a necessary result of the arrangement of the electric currents in a manner, as I have just made clear, in which these currents attract or repel each other. I have constructed another apparatus in which the conducting wire is wound in a helix around a glass tube; according to my theory of these phenomena, this conductor should, if an electric current is led through it, act similarly to a magnetic needle or bar in all the circumstances where these act on other bodies or are subjected to terrestrial magnetism. I have already observed a part of the effects which I expected of the behaviour of a conductor wound in a helix, and I do not doubt that the more one will vary the experiments founded on the analogy which the theory establishes between this instrument and a magnetic bar, the more proofs one will obtain that the existence of the electric currents in the magnet is the unique cause of all magnetic phenomena.

268

On the mathematical theory of electrodynamical phenomena

The period in the history of science which is marked by the works of Newton is not only one of the most important discoveries man has made of the causes of the grand phenomena of nature; it is also the period in which the human spirit has opened up a new road in the sciences, with the object to study these phenomena.

So far, one has been almost exclusively searching for the causes in the impulsion of an unknown fluid which drags the material particles along the direction of its own particles; and wherever one has seen a revolving motion, one considered it a vortex rotating in the same sense.

Newton has taught us that such sorts of motion must, like all motions that we encounter in nature, be reduced by calculation to forces which are always acting between any two material particles along the direction of the line joining them, in such a manner that the action which one of them exercises on the other is equal and opposite to the action which this other exercises in the same time on the first, and that consequently, supposing the two particles invariably connected, no motion can result from their mutual action. It is this law which, having been confirmed by now by all observations and calculations, Newton has expressed in the last of his three axioms which he put at the beginning of the *Philosophiae Naturalis Principia Mathematica*. But it was not sufficient to have risen to such a high conception; it was necessary to find according to what law these forces vary with the respective **position** of the particles between which they act, or, what comes to the same, to express their value by a formula.

Newton was far from thinking that such a law could be invented by starting with more or less plausible abstract considerations. He established that it had to be derived from observed facts, or rather, from those empirical laws which, like those of Kepler, are only generalized results of a great many facts.

To observe the facts, while varying the circumstances as much as possible to support this first work by precise measurements, in order to deduce from them general laws uniquely founded on experience, and to deduce from these laws, independently of any hypothesis on the nature of the forces which produce these phenomena, the mathematical value of these forces, that is to say the formulas which represent them—this was the road followed by Newton. This course has been adopted in general in France by those scientists to whom physics owes the immense progress it has made lately, and this is the road I have followed in all my investigations on electrodynamic phenomena. I have relied only on experience to establish the laws of these phenomena, and from it I have deduced the formula which alone can represent the forces to which they are due. I made no attempt to investigate the cause itself of these forces, being convinced that all research of this sort must be preceded by purely experimental knowledge of the laws and by the determination of the value of the elementary forces, uniquely deduced from these laws, the direction of which forces is that of the line connecting the material points between which the forces are exercised. It is for this reason that I have avoided speaking of the ideas I might have had about the nature of the causes of the forces which emanate from voltaic conductors, as far as they cannot be found among the notes which accompany my 'Summary exposition of some new Electromagnetic Experiments as having been conducted by several Physicists since March 1821' which I read at a public meeting of the Académie des Sciences on 18 April, 1822; one can find what I said in these notes on page 215 of my *Collection of Electrodynamic Observations*. It does not seem that this course, the only one which can lead to results independent of all hypotheses, has been preferred by the physicists of the rest of Europe as it has been in France; and an illustrious scientist, the first to have seen that the poles of a magnet are displaced by the action of a conducting wire in a direction perpendicular to that of the wire, has concluded that the electrical material turns around it and pushes the poles in the sense of its motion, exactly as Descartes had made the material of his vortices turn in the sense of the planetary revolutions. Guided by the principles of Newtonian philosophy, I have reduced the phenomena observed by M. Ørsted, as one has to do with regard to all the phenomena that nature offers us, to forces which are acting always along the lines joining the particles between which they are exercised; and if I have established that the same disposition or the same motion of electricity which exists in the conducting wire is the one that takes place also around the magnetic particles, it was certainly not in order to make them act by impulsion in the manner of a vortex, but in order to calculate according to my formula the forces which act between these particles and those of a conductor or another magnet, along the lines joining each two of the particles where the action is considered mutual, and in order to show that the results of the calculation are completely verified: first by experiments I have conducted, and by those which one owes to M. Pouillet for the precise determination of the position where the moving conductor must be found in order to stay in equilibrium while it is under the influence of another conductor or a magnet; second, by the agreement between the results of the laws which Coulomb[1] and M. Biot[2] have deduced from their experiments, the first relating to the mutual action of two magnets and the other to the action between the magnet and a conducting wire.

The chief advantage in formulas which are derived in such a manner directly from some general facts, given by a sufficient number of observations so that they certainly could not be contested, is to stay independent of hypotheses which could have helped their authors in their investigations, as well as of those which could be substituted later on. The expression of universal attraction deduced from the laws of Kepler does not depend at all on the hypotheses which some authors have tried to make about the mechanical cause which they wanted to assign to it. The theory of heat rests truly on general facts, given directly by observation; and the equation deduced from these facts, being confirmed by the agreement between the results derived from the equation and those given by the experiment, must be accepted as expressing the true laws of the propagation of heat both by those who attribute this to radiation of calorific molecules and by those who try to explain the same phenomenon by the vibrations of a fluid spread in space; it is only necessary that the first show how the equation with which we deal results from their way of looking at things, and that the second group deduce it from general formulas of vibratory motions; this, not in order to add certainty to the equation but in order that their hypotheses may stand. That physicist who has not taken sides in this matter admits this equation as the exact representation of the facts without troubling himself with the manner in which it could be derived from one or the other of the explanations about which we were talking; and if new phenomena and new calculations came to demonstrate that the effects of heat cannot be really explained except according to the theory of vibrations, that great physicist[3] who has given this equation for the first time, and who has created it in order to apply to the object of his investigations a new means of integration, will not be less the author of the mathematical theory of heat than Newton is of the theory of planetary motions, even though this has not been demonstrated as completely by him as it has been since by his successors.

It is the same with the formula by which I have represented the electromagnetic action. Whatever the physical causes by which one wishes to explain the phenomena produced by this action, the formula which I obtained will always remain an expression of the facts. If one succeeds in deducing it from one of those considerations by which so many other phenomena have been explained, like attractions in the inverse ratio to the square of distance, which become unnoticable at an appreciable distance from the particles between which they act, the vibrations of a fluid spread in space, etc., one has come a step forward in that

1. Charles Augustin de Coulomb (1736–1806), French physicist, discoverer of the well-known law of electrostatics now named after him. (Ed.)
2. Jean Biot (1774–1862), French physicist, stated in 1820, together with Savart, the mathematical law of the magnetic action of an electrical current. (Ed.)
3. Jean Baptist Joseph Fourier (1768–1830), French physicist and mathematician. Among other famous discoveries he developed the mathematical theory of heat conduction. (Ed.)

part of physics; but this investigation, with which I have not yet occupied myself, although I recognize its importance, shall never change the results of my work, for to be in agreement with the facts, the adopted hypothesis must always agree with the formula which represents those facts completely.

SADI CARNOT

269

Reflections on the motive power of fire

Every one knows that heat can produce motion. That it possesses vast motive power no one can doubt, in these days when the steam-engine is everywhere so well known.

To heat also are due the vast movements which take place on the earth. It causes the agitations of the atmosphere, the ascension of clouds, the fall of rain and of other precipitations, the currents of water on the surface of the globe, and of which man has thus far employed but a small portion. Even earthquakes and volcanic eruptions are the result of heat.

From this immense reservoir we may draw the moving force necessary for our purposes. Nature, in providing us with combustibles on all sides, has given us the power to produce, at all times and in all places, heat and the impelling power which is the result of it. To develop this power, to appropriate it to our uses, is the object of heat-engines. [. . .]

The question has often been raised whether the motive power of heat is unbounded, whether the possible improvements in steam-engines have an assignable limit—a limit which the nature of things will not allow to be passed by any means whatever; or whether, on the contrary, these improvements may be carried on indefinitely. We have long sought, and are seeking today, to ascertain whether there are in existence agents preferable to the vapour of water for developing the motive power of heat; whether atmospheric air, for example, would not present in this respect great advantages. We propose now to submit these questions to a deliberate examination.

The phenomenon of the production of motion by heat has not been considered from a sufficiently general point of view. We have considered it only in machines the nature and mode of action of which have not allowed us to take in the whole extent of application of which it is susceptible. In such machines the phenomenon is, in a way, incomplete. It becomes difficult to recognize its principles and study its laws.

In order to consider in the most general way the principle of the production of motion by heat, it must be considered independently of any mechanism or any particular agent. It is necessary to establish principles applicable not only to steam-engines but to all imaginable heat-engines, whatever the working substance and whatever the method by which it is operated.

Machines which do not receive their motion from heat, those which have for a motor the force of men or of animals, a waterfall, an air current, etc., can be studied even to their smallest details by the mechanical theory. All cases are foreseen, all imaginable movements are referred to these general principles, firmly established, and applicable under all circumstances. This is the character of a complete theory. A similar theory is evidently needed for heat-engines. We shall have it only when the laws of physics shall be extended enough, generalized enough, to make known beforehand all the effects of heat acting in a determined manner on any body.

We will suppose in what follows at least a superficial knowledge of the different parts which compose an ordinary steam-engine; and we consider it unnecessary to explain what are the furnace, boiler, steam-cylinder, piston, condenser, etc.

The production of motion in steam-engines is always accompanied by a circumstance on which we should fix our attention. This circumstance is the re-establishing of equilibrium in the caloric; that is, its passage from a body in which the temperature is more or less elevated, to another in which it is lower. What happens in fact in a steam-engine actually in motion? The caloric developed in the furnace by the effect of the combustion traverses the walls of the boiler, produces steam, and in some way incorporates itself with it. The latter carrying it away, takes it first into the cylinder, where it performs some function, and from thence into the condenser, where it is liquefied by contact with the cold water which it encounters there. Then, as a final result, the cold water of the condenser takes possession of the caloric developed by the combustion. It is heated by the intervention of the steam as if it had been placed directly over the furnace. The steam is here only a means of transporting the caloric. It fills the same office as in the heating of baths by steam, except that in this case its motion is rendered useful.

We easily recognize in the operations that we have just described the re-establishment of equilibrium in the caloric, its passage from a more or less heated body to a cooler one. The first of these bodies, in this case, is the heated air of the furnace; the second is the condensing water. The re-establishment of equilibrium of the caloric takes place between them, if not completely, at least partially, for on the one hand the heated air, after having performed its function, having passed round the boiler, goes out through the chimney with a temperature much below that which it had acquired as the effect of combustion; and on the other hand, the water of the condenser, after having liquefied the steam, leaves the machine with a temperature higher than that with which it entered.

The production of motive power is then due in steam-engines not to an actual consumption of caloric, but *to its transportation from a warm body to a cold body*, that is, to its re-establishment of equilibrium—an equilibrium considered as destroyed by any cause whatever, by chemical action such as combustion, or by any other. We shall see shortly that this principle is applicable to any machine set in motion by heat.

According to this principle, the production of heat alone is not sufficient to give birth to the impelling power: it is necessary that there should also be cold; without it, the heat would be useless. And in fact, if we should find about us only bodies as hot as our furnaces, how can we condense steam? What should we do with it if once produced? We should not presume that we might discharge it into the atmosphere, as is done in some engines; the atmosphere would not receive it. It does receive it under the actual condition of things, only because it fulfils the office of a vast condenser, because it is at a lower temperature; otherwise it would soon become fully charged, or rather would be already saturated.

Wherever there exists a difference of temperature, wherever it has been possible for the equilibrium of the caloric to be re-established, it is possible to have also the production of impelling power. Steam is a means of realizing this power, but it is not the only one. All substances in nature can be employed for this purpose, all are susceptible of changes of volume, of successive contractions and dilatations, through the alternation of heat and cold. All are capable of overcoming in their changes of volume certain resistances, and of thus developing the impelling power. A solid body—a metallic bar for example—alternately heated and cooled increases and diminishes in length, and can move bodies fastened to its ends. A liquid alternately heated and cooled increases and diminishes in volume, and can overcome obstacles of greater or less size, opposed to its dilatation. An aeriform fluid is susceptible of considerable change of volume by variations of temperature. If it is enclosed in an expansible space, such as a cylinder provided with a piston, it will produce movements of great extent. Vapours of all substances capable of passing into a gaseous condition, as of alcohol, of mercury, of sulphur, etc., may fulfil the same office as vapour of water. The latter, alternately heated and cooled, would produce motive power in the shape of permanent gases, that is, without ever returning to a liquid state. Most of these substances have been proposed, many even have been tried, although up to this time perhaps without remarkable success.

We have shown that in steam-engines the motive power is due to a re-establishment of equilibrium in the caloric; this takes place not only for steam-engines, but also for every heat-engine—that is, for every machine of which caloric is the motor. Heat can evidently be a cause of motion only by virtue of the changes of volume or of form which it produces in bodies.

These changes are not caused by uniform temperature, but rather by alternations of heat and cold. Now to heat any substance whatever requires a body warmer than the one to be heated; to cool it requires a cooler body. We supply caloric to the first of these bodies that we may transmit it to the second by means of the intermediary substance. This is to re-establish, or at least to endeavour to re-establish, the equilibrium of the caloric.

It is natural to ask here this curious and important question: Is the motive power of heat invariable in quantity, or does it vary with the agent employed to realize it as the intermediary substance, selected as the subject of action of the heat?

It is clear that this question can be asked only in regard to a given quantity of caloric, the difference of the temperatures also being given. We take, for example, one body A kept at a temperature of 100° and another body B kept at a temperature of 0°, and ask what quantity of motive power can be produced by the passage of a given portion of caloric (for example, as much as is necessary to melt a kilogram of ice) from the first of these bodies to the second. We inquire whether this quantity of motive power is necessarily limited, whether it varies with the substance employed to realize it, whether the vapour of water offers in this respect more or less advantage than the vapour of alcohol, of mercury, a permanent gas, or any other substance. We will try to answer these questions, availing ourselves of ideas already established.

We have already remarked upon this self-evident fact, a fact which at least appears evident as soon as we reflect on the changes of volume occasioned by heat: *wherever there exists a difference of temperature, motive power can be produced.* Reciprocally, wherever we can consume this power, it is possible to produce a difference of temperature, it is possible to occasion destruction of equilibrium in the caloric. Are not percussion and the friction of bodies actually means of raising their temperature, of making it reach spontaneously a higher degree than that of the surrounding bodies, and consequently of producing a destruction of equilibrium in the caloric, where equilibrium previously existed? It is a fact proved by experience, that the temperature of gaseous fluids is raised by compression and lowered by rarefaction. This is a sure method of changing the temperature of bodies, and destroying the equilibrium of the caloric as many times as may be desired with the same substance. The vapour of water employed in an inverse manner to that in which it is used in steam-engines can also be regarded as a means of destroying the equilibrium of the caloric. To be convinced of this we need to observe closely the manner in which motive power is developed by the action of heat on vapour of water. Imagine two bodies A and B, kept each at a constant temperature, that of A being higher than that of B. These two bodies, to which we can give or from which we can remove the heat without causing their temperatures to vary, exercise the functions of two unlimited reservoirs of caloric. We will call the first the furnace and the second the refrigerator.

If we wish to produce motive power by carrying a certain quantity of heat from the body A to the body B we shall proceed as follows:

(1) To borrow caloric from the body A to make steam with it—that is, to make this body fulfil the function of a furnace, or rather of the metal composing the boiler in ordinary engines—we here assume that the steam is produced at the same temperature as the body A.

(2) The steam having been received in a space capable of expansion, such as a cylinder furnished with a piston, to increase the volume of this space, and consequently also that of the steam. Thus rarefied, the temperature will fall spontaneously, as in all elastic fluids; now assume that the rarefaction may be continued to the point where the temperature becomes precisely that of the body B.

(3) To condense the steam by putting it in contact with the body *B*, and at the same time exerting on it a constant pressure until it is entirely liquefied. The body *B* fills here the place of the injection-water in ordinary engines, with this difference, that it condenses the vapour without mingling with it, and without changing its own temperature.

The operations which we have just described might have been performed in an inverse direction and order. There is nothing to prevent forming vapour with the caloric of the body *B*, and at the temperature of that body, compressing it in such a way as to make it acquire the temperature of the body *A*, finally condensing it by contact with this latter body, and continuing the compression to complete liquefaction.

By our first operations there would have been at the same time production of motive power and transfer of caloric from the body *A* to the body *B*. By the inverse operations there is at the same time expenditure of motive power and return of caloric from the body *B* to the body *A*. But if we have acted in each case on the same quantity of vapour, if there is produced no loss either of motive power or caloric, the quantity of motive power produced in the first place will be equal to that which would have been expended in the second, and the quantity of caloric passed in the first case from the body *A* to the body *B* would be equal to the quantity which passes back again in the second from the body *B* to the body *A*; so that an indefinite number of alternative operations of this sort could be carried on without in the end having either produced motive power or transferred caloric from one body to the other.

Now if there existed any means of using heat preferable to those which we have employed, that is, if it were possible by any method whatever to make the caloric produce a quantity of motive power greater than we have made it produce by our first series of operations, it would suffice to divert a portion of this power in order by the method just indicated to make the caloric of the body *B* return to the body *A* from the refrigerator to the furnace, to restore the initial conditions, and thus to be ready to commence again an operation precisely similar to the former; and so on: this would be not only perpetual motion, but an unlimited creation of motive power without consumption either of caloric or of any other agent whatever. Such a creation is entirely contrary to ideas now accepted, to the laws of mechanics and of sound physics. It is inadmissible. We should then conclude that *the maximum of motive power resulting from the employment of steam is also the maximum of motive power realizable by any means whatever.* [. . .]

According to established principles at the present time, we can compare with sufficient accuracy the motive power of heat to that of a waterfall. Each has a maximum that we cannot exceed, whatever may be, on the one hand, the machine which is acted upon by the water, and whatever, on the other hand, the substance acted upon by the heat. The motive power of a waterfall depends on its height and on the quantity of the liquid; the motive power of heat depends also

393

on the quantity of caloric used, and on what may be termed, on what in fact we will call, the *height of its fall*, that is to say, the difference of temperature of the bodies between which the exchange of caloric is made. In the waterfall the motive power is exactly proportional to the difference of level between the higher and lower reservoirs. In the fall of caloric the motive power undoubtedly increases with the difference of temperature between the warm and the cold bodies; but we do not know whether it is proportional to this difference.

JAMES PRESCOTT JOULE

270

On the mechanical equivalent of heat

In accordance with the pledge I gave the Royal Society some years ago, I have now the honour to present it with the results of the experiments I have made in order to determine the mechanical equivalent of heat with exactness. I will commence with a slight sketch of the progress of the mechanical doctrine, endeavouring to confine myself, for the sake of conciseness, to the notice of such researches as are immediately connected with the subject. I shall not therefore be able to review the valuable labours of Mr Forbes and other illustrious men, whose researches on radiant heat and other subjects do not come exactly within the scope of the present memoir.

For a long time it had been a favourite hypothesis that heat consists of 'a force or power belonging to bodies', but it was reserved for Count Rumford[1] to make the first experiments decidedly in favour of that view. That justly celebrated natural philosopher demonstrated by his ingenious experiments that the very great quantity of heat excited by the boring of cannon could not be ascribed to a change taking place in the calorific capacity of the metal; and he therefore concluded that the motion of the borer was communicated to the particles of metal, thus producing the phenomena of heat. 'It appears to me,' he remarks, 'extremely difficult, if not quite impossible, to form any distinct idea of anything capable of being excited and communicated in the manner the heat was excited and communicated in these experiments, except it be motion.'

One of the most important parts of Count Rumford's paper, though one to which little attention has hitherto been paid, is that in which he makes an estimate of the quantity of mechanical force required to produce a certain amount of heat. Referring to his third experiment, he remarks that the 'total quantity of ice-cold water which, with the heat actually generated by friction, and accumulated in $2^h 30^m$, might have been heated 180°, or made to boil,= 26·58 lb.' In the next page he states that 'the machinery used in the experiment could easily be carried round by the force of one horse (though, to render the

1. Benjamin Thompson, Count Rumford (1753–1814), American physicist, well known for his experiments on the creation of heat by friction. (Ed.)

work lighter, two horses were actually employed in doing it)'. Now the power of a horse is estimated by Watt at 33,000 foot-pounds per minute, and therefore if continued for two hours and a half will amount to 4,950,000 foot-pounds, which, according to Count Rumford's experiment, will be equivalent to 26·58 lb of water raised 180°. Hence the heat required to raise a lb of water 1° will be equivalent to the force represented by 1,034 foot-pounds. This result is not very widely different from that which I have deduced from my own experiment related in this paper, viz. 772 foot-pounds; and it must be observed that the excess of Count Rumford's equivalent is just such as might have been anticipated from the circumstance, which he himself mentions, that 'no estimate was made of the heat accumulated in the wooden box, nor of that dispersed during the experiment'.

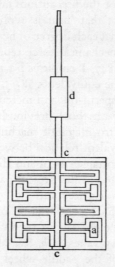

Joule's apparatus for the determination of the mechanical equivalent of heat[1]

About the end of the last century Sir Humphrey Davy communicated a paper to Dr Beddoes's *West Country Contributions*, entitled 'Researches on Heat, Light, and Respiration', in which he gave ample confirmation to the views of Count Rumford. By rubbing two pieces of ice against one another in the vacuum of an air-pump, part of them was melted, although the temperature of the receiver was kept below the freezing-point. This experiment was the more decisively in favour of the doctrine of the immateriality of heat, inasmuch as the capacity of ice for heat is much less than that of water. It was therefore with good reason that Davy drew the inference that 'the immediate cause of the phenomena of heat is motion, and the laws of its communication are precisely the same as the laws of the communication of motion'. [. . .]

1. The axis is made to revolve by the transmission of the motion of a sinking weight. (Ed.).

From the explanation given by Count Rumford of the heat arising from the friction of solids, one might have anticipated, as a matter of course, that the evolution of heat would also be detected in the friction of liquid and gaseous bodies. Moreover there were many facts, such as, for instance, the warmth of the sea after a few days of stormy weather, which had long been commonly attributed to fluid friction. Nevertheless the scientific world, preoccupied with the hypothesis that heat is a substance, and following the deductions drawn by Pictet from experiments not sufficiently delicate, have almost unanimously denied the possibility of generating heat in that way. The first mention, so far as I am aware, of experiments in which the evolution of heat from fluid friction is asserted was in 1842 by R. Mayer,[1] who states that he has raised the temperature of water from 12°C to 13°C by agitating it, without, however, indicating the quantity of force employed, or the precautions taken to secure a correct result. In 1843 I announced the fact that 'heat is evolved by the passage of water through narrow tubes', and that each degree of heat per lb of water required for its evolution in this way a mechanical force represented by 770 foot-pounds. Subsequently, in 1845 and 1847, I employed a paddle-wheel to produce the fluid friction, and obtained the equivalents 781·5, 782·1 and 787·6 respectively from the agitation of water, sperm-oil, and mercury. Results so closely coinciding with one another, and with those previously derived from experiments with elastic fluids and the electro-magnetic machine, left no doubt on my mind as to the existence of an equivalent relation between force and heat; but still it appeared of the highest importance to obtain that relation with still greater accuracy. This I have attempted in the present paper. [. . .]

Pl.II. fig. 69 represents a vertical and fig. 70 a horizontal plan of the apparatus employed for producing the friction of water, consisting of a brass paddle-wheel furnished with eight sets of revolving arms, *a, a,* &c., working between four sets of stationary vanes, *b, b,* &c., affixed to a framework also in sheet brass. The brass axis of the paddle-wheel worked freely, but without shaking, on its bearings at *c c,* and at *d* was divided into two parts by a piece of boxwood intervening, so as to prevent the conduction of heat in that direction.

Pl.II. fig. 71 represents the copper vessel into which the revolving apparatus was firmly fitted: it had a copper lid, the flange of which, furnished with a very thin washer of leather saturated with white-lead, could be screwed perfectly water-tight to the flange of the copper vessel. In the lid there were two necks, *a, b,* the former for the axis to revolve in without touching, the latter for the insertion of the thermometer. [. . .]

The method of experimenting was simply as follows: The temperature of the frictional apparatus having been ascertained and the weights wound up with the assistance of the stand *h,* the roller was refixed to the axis. The precise height of the weights above the ground having then been determined by means of the

---

1. Robert v. Mayer (1814–1878), German physican and physicist. One of the first discoverers of the equivalence of heat and work. (Ed.)

graduated slips of wood k, k, the roller was set at liberty and allowed to revolve until the weights reached the flagged floor of the laboratory, after accomplishing a fall of about 63 inches. The roller was then removed to the stand, the weights wound up again, and the friction renewed. After this had been repeated twenty times, the experiment was concluded with another observation of the temperature of the apparatus. The mean temperature of the laboratory was determined by observations made at the commencement, middle, and termination of each experiment.

Previously to, or immediately after, each of the experiments I made trial of the effect of radiation and conduction of heat to or from the atmosphere in depressing or raising the temperature of the frictional apparatus. In these trials the position of the apparatus, the quantity of water contained by it, the time occupied, the method of observing the thermometers, the position of the experimenter, in short everything, with the exception of the apparatus being at rest, was the same as in the experiments in which the effect of friction was observed. [. . .]

I will therefore conclude by considering it as demonstrated by the experiments contained in this paper,—

1st. *That the quantity of heat produced by the friction of bodies, whether solid or liquid, is always proportional to the quantity of force expended.* And,

2nd. *That the quantity of heat capable of increasing the temperature of a pound of water (weighed in vacuo, and taken at between 55° and 60°) by 1° Fahr. requires for its evolution the expenditure of a mechanical force represented by the fall of 772 lb. through the space of one foot.*

HERMANN HELMHOLTZ

271

On the conservation of force

This discourse is meant primarily for physicists; with this in view I prefer to present the gist of the matter in the form of a physical presupposition, free of any philosophical considerations. Moreover, I will explore the physical implications of this hypothesis in the various branches of physics and compare those with the empirical laws of the phenomena of nature.

The task of deriving our propositions can be attempted in two different ways: using as a premiss either the theorem that no interaction between any combination of bodies can yield unlimited force, or the presupposition that all operations of nature can be traced back to attractive or repulsive forces, the intensity of which is a function solely of the distance between the interacting particles. That the two premisses are in fact identical will be shown at the beginning of the present lecture. In addition, these presuppositions have an even greater significance for the final and real goal of physical natural sciences, a significance which I shall endeavour to demonstrate in this introduction. [. . .]

Science treats the objects of the external world in the light of one of two possible abstractions: first it notes their existence disregarding their interactions and their influence on our senses; as such it designates them simply as matter. The existence of matter *per se* has no effect on us; we note its distribution in space and its quantity (mass), which is being considered immutable forever. We are not entitled to ascribe qualitative differences to matter as such, because when we talk about different materials we perceive those differences only in their different effects, i.e. in the forces they activate.

Thus, matter as such cannot undergo any other change but spatial, i.e. motion. But objects of nature are not effectless, on the contrary, we realize their existence only through their influence on our senses, deducing that those effects must have been caused by some objects. If we wish to apply the concept of matter to reality we have to attribute to matter those qualities which we have abstracted from it in a second abstraction, namely the ability to cause effects, or in other words to operate forces. It is obvious that the concepts of matter and forces are inseparable when applied to nature. An object of pure matter would be indifferent to the world around it, as it would never cause a change in the world, nor would it affect our senses. A pure force would be something which is existent and non-existent, because that which is existent we call matter. It is equally erroneous to consider matter as real but force as a mere concept, not corresponding to anything real; both are abstractions of reality, formed in exactly the same way: we realize matter through the very forces it exercises.

We have now seen that natural phenomena ought to be traced back to immutable ultimate causes turning out to be time-independent forces. In science, units of matter that activate such time-independent forces and have indestructible qualities we call chemical elements. If we picture to ourselves the world reduced to elements with unchangeable qualities, the only imaginable changes in such a system will be spatial ones: i.e. motion. The external circumstances that could modify the effects of the forces must be of a spatial character and thus the forces can only be forces of motion, determined by the spatial relationships.

To be precise: phenomena of nature must be reduced to motions of material particles with unchangeable motive forces which depend solely on the spatial conditions.

Motion is change of spatial arrangements; such arrangements can be perceived only in limited space and not when seen in undifferentiated empty space. Thus, in reality, motion can occur only as a change of spatial relationships between at least two material particles; motive force can be referred to as the cause of this change only if at least two bodies are present at the time when spatial relationships undergo change and can thus be defined as the tendency of masses to change their mutual positions.

Furthermore, the force that two masses exercise on each other must be reduced to all the forces that all particles of the two masses exert on all the others; this is the reason why mechanics treats of forces between material points, i.e. points of space filled with matter. But points have no other spatial relationship with each other than their mutual distance, as the direction of a line connecting two points can be determined only against at least two other points. If these points exercise a motive force on each other this can cause a change in their mutual distance, i.e. it must be a repulsive or attractive force; this follows from the law of sufficient reason.

The forces of interaction between two masses must be determined according to their strength and direction if the configuration is known. But two points define only one direction, that of the line connecting them; thus the interaction forces must act along this line and their intensity must be a function of the distance.

Finally, it becomes evident that the task of physical natural sciences is to reduce phenomena of Nature to immutable forces of attraction and repulsion, the intensity of which will be determined solely by distance. The solubility of this problem is the condition for a complete understanding of Nature. That this limitation of the concept of motive force has been ignored by analytical mathematical mechanics has two reasons: this discipline did not have in the past enough insight into its own fundamental presuppositions; on the other hand it had the ambition to calculate the resultant effect of numerous forces of motions, even in cases where the reduction of this resultant to its individual components was not feasible. And yet, a substantial part of the general principles of mechanics treating the motions of systems of mass-particles is applicable only for the case where the interacting forces of the system are immutable forces of repulsion or attraction; namely the principle of virtual velocities, that of the conservation of motion of the centre of gravity, conservation of the main plane of rotation and the moment of rotation of free systems, that of the conservation of living force. In terrestrial conditions, of all these, only the first and the last are applicable, while the others refer to ideally free systems. We will show that the first is actually a special case of the last and thus the last-mentioned principle is the most important result of our whole discussion.

Theoretical natural science, if it does not want to stop halfway on its course towards understanding Nature, must be in accord with these postulates about the nature of simple forces and the implications thereof. It will have accomplished its task when the reduction of the phenomena to simple forces is achieved and when it becomes demonstrable that this is the only possible reduction permitted by the phenomena. When we reach that point we will have found the necessary conceptual framework in which to understand Nature and then we can ascribe to our conception the merit of showing objective truth.

On the application of the law of the conservation of force to organic nature

The most important progress in natural philosophy by which the present century is distinguished, has been the discovery of a general law which embraces and rules all the various branches of physics and chemistry. This law is of as much importance for the highest speculations on the nature of forces, as for immediate and practical questions in the construction of machines. This law at present is commonly known by the name of 'the principle of conservation of force'. It might be better perhaps to call it, with Mr Rankine,[1] 'the conservation of energy', because it does not relate to that which we call commonly *intensity* of force; it does not mean that the intensity of the natural forces is constant: but it relates more to the whole amount of power which can be gained by any natural process, and by which a certain amount of work can be done. For example: if we apply this law to gravity, it does not mean, what is strictly and undoubtedly true, that the intensity of the gravity of any given body is the same as often as the body is brought back to the same distance from the centre of the earth. Or with regard to the other elementary forces of nature—for example, chemical force: when two chemical elements come together, so that they influence each other, either from a distance or by immediate contact, they will always exert the same force upon each other—the same force both in intensity and in its direction and in its quantity. This other law indeed is true; but it is not the same as the principle of conservation of force. We may express the meaning of the law of conservation of force by saying, that every force of nature when it effects any alteration, loses and exhausts its faculty to effect the same alteration a second time. But while, by every alteration in nature, that force which has been the the cause of this alteration is exhausted, there is always another force which gains as much power of producing new alterations in nature as the first has lost. Although, therefore, it is the nature of all inorganic forces to become exhausted by their own working, the power of the whole system in which these alterations take place is neither exhausted nor increased in quantity, but only changed in form. Some special examples will enable you better to understand this law than any general theories. We will begin with gravity; that most general force, which not only exerts its influence over the whole universe, but which at the same time gives the means of moving to a great number of our machines. Clocks and smaller machines, you know, are often set in motion by a weight. The same is really the case with water-mills. Water-mills are driven by falling water; and it is the gravity, the weight of the falling water, which moves the mill. Now you know that by water-mills, or by a falling weight, every machine can be put in

---

1. William Rankine (1820–1872), Scottish engineer, developed the first theory of the steam engine, based on the mechanical theory of heat. (Ed.)

motion; and that by these motive powers every sort of work can be done which can be done at all by any machine. You see, therefore, that the weight of a heavy body, either solid or fluid, which descends from a higher place to a lower place is a motive power, and can do every sort of mechanical work. Now if the weight has fallen down to the earth, then it has the same amount of gravity, the same intensity of gravity; but its power to move, its power to work, is exhausted; it must become again raised before it can work anew. In this sense, therefore, I say that the faculty of producing new work is exhausted—is lost; and this is true of every power of nature when this power has produced alteration Hence, therefore, the faculty of producing work, of doing work, does not depend upon the intensity of gravity. The intensity of gravity may be the same, the weight may be in a higher position or in a lower position, but the power to work may be quite different. The power of a weight to work, or the amount of work which can be produced by a weight, is measured by the product of the height to which it is raised and the weight itself. Therefore our common measure is foot-pound; that is, the product of the number of feet and the number of pounds. Now we can by the force of a falling weight raise another weight; as, for example, the falling water in a water-mill may raise the weight of a hammer. Therefore it can be shown that the work of the raised hammer expressed in foot-pounds, that is, the weight of the hammer multiplied by the height expressed in feet to which it is raised, that this amount of work cannot be greater than the product of the weight of water which is falling down, and the height from which it fell down. Now we have another form of motive power, of mechanical motive power; that is, velocity. The velocity of any body in this sense, if it is producing work, is called *vis viva*, or living force, of that body. You will find many examples of it. Take the ball of a gun. If it is shot off, and has a great velocity, it has an immense power of destroying; and if it has lost its velocity, it is quite a harmless thing. The great power it has depends only on its velocity. In the same sense, the velocity of the air, the velocity of the wind, is motive power; for it can drive windmills, and by the machinery of the windmills it can do every kind of mechanical work. Therefore you see that also velocity in itself is a motive force.

Take a pendulum which swings to and fro. If the pendulum is raised to the side, the weight is raised up; it is a little higher than when it hangs straightly down, perpendicular. Now if you let if fall, and it comes to its position of equilibrium, it has gained a certain velocity. Therefore, at first, you had motive power in the form of a raised weight. If the pendulum comes again to the position of equilibrium, you have motive power in the form of *vis viva*, in the form of velocity, and then the pendulum goes again to the other side, and it ascends again till it loses its velocity; then again, *vis viva* or velocity is changed into elevation of the weight: so you see in every pendulum that the power of a raised weight can be changed into velocity, and the velocity into the power of a raised weight. These two are equivalent.

Then take the elasticity of a bent spring. It can do work, it can move machines or watches. The cross-bow contains such springs. These springs of the watch and cross-bow are bent by the force of the human arm, and they become in that way reservoirs of mechanical power. The mechanical power which is communicated to them by the force of the human arm, afterwards is given out by a watch during the next day. It is spent by degrees to overpower the friction of the wheels. By the cross-bow, the power is spent suddenly. If the instrument is shot off, the whole amount of force which is communicated to the spring is then again communicated to the shaft, and gives it a great *vis viva*.

Now the elasticity of air can be a motive power in the same way as the elasticity of solid bodies; if air is compressed, it can move other bodies; let us take the air-gun; there the case is quite the same as with the cross-bow. The air is compressed by the force of the human arm; it becomes a reservoir of mechanical power; and if it is shot off, the power is communicated to the ball in the form of *vis viva*, and the ball has afterwards the same mechanical power as is communicated to the ball of a gun loaded with powder.

The elasticity of compressed gases is also the motive power of the mightiest of our engines, the steam-engine; but there the case is different. The machinery is moved by the force of the compressed vapours, but the vapours are not compressed by the force of the human arm, as in the case of the compressed air-gun. The compressed vapours are produced immediately in the interior of the boiler by the heat which is communicated to the boiler from the fuel.

You see, therefore, that in this case the heat comes in the place of the force of the human arm, so that we learn by this example, that heat is also a motive power. This part of the subject, the equivalence of heat as a motive power, with mechanical power, has been that branch of this subject which has excited the greatest interest, and has been the subject of deep research.

It may be considered as proved at present, that if heat produces mechanical power, that is, mechanical work, a certain amount of heat is always lost.On the other hand, heat can be also produced by mechanical power, namely, by friction and the concussion of unelastic bodies. You can bring a piece of iron into a high temperature, so that it becomes glowing and luminous, by only beating it continuously with a hammer. Now, if mechanical power is produced by heat, we always find that a certain amount of heat is lost; and this is proportional to the quantity of mechanical work produced by that heat. We measure mechanical work by foot-pounds, and the amount of heat we measure by the quantity of heat which is necessary to raise the temperature of one pound of water by one degree, taking the centigrade scale. The equivalent of heat has been determined by Mr Joule, of Manchester. He found that one unit of heat, or that quantity of heat which is necessary for raising the temperature of a pound of water one degree centigrade, is equivalent to the mechanical work by which the same mass of water is raised to $423\frac{1}{2}$ metres, or, 1,389 English feet. This is the mechanical equivalent of heat.

Hence, if we produce so much heat as is necessary for raising the temperature of one pound of water by one degree, then we must apply an amount of mechanical work equal to raising one pound of water 1,389 English feet, and lose it for gaining again that heat.

By these considerations, it is proved, that heat cannot be a ponderable matter, but that it must be a motive power, because it is converted into motion or into mechanical power, and can be either produced by motion or mechanical power. Now, in the steam-engine we find that heat is the origin of the motive power, but the heat is produced by burning fuel, and therefore the origin of the motive power is to be found in the fuel, that is, in the chemical forces of the fuel, and in the oxygen with which the fuel combines.

You see from this, that the chemical forces can produce mechanical work, and can be measured by the same units and by the same measures as any other mechanical force. We may consider the chemical forces as attractions, in this instance, as attraction of the carbon of the fuel for the oxygen of the air; and if this attraction unites the two bodies, it produces mechanical work just in the same way as the earth produces work, if it attracts a heavy body. Now the conservation of force, of chemical force, is of great importance for our subject today, and it may be expressed in this way. If you have any quantity of chemical materials, and if you cause them to pass from one state into a second state, in any way, so that the amount of the materials at the beginning, and the amount of the materials at the end of this process be the same, then you will have always the same amount of work, of mechanical work or its equivalent, done during this process. Neither more nor less work can be done by the process. Commonly, no mechanical work in the common sense is done by chemical force, but usually it produces only heat; hence the amount of heat produced by any chemical process must be independent of the way in which that chemical process goes on. The way may be determined by the will of the experimenter as he likes. [. . .]

We must consider the living bodies under the same point of view, and see how it stands with them. Now if you compare the living body with a steam-engine, then you have the completest analogy. The living animals take in food that consists of inflammable substances, fat and the so-called hydrocarbons, as starch and sugar, and nitrogenous substances, as albumen, flesh, cheese, and so on. Living animals take in these inflammable substances and oxygen; the oxygen of the air, by respiration. Therefore, if you take, in the place of fat, starch, and sugar, coals or wood, and the oxygen of the air, you have the substances in the steam-engine. The living bodies give out carbonic acid and water; and then if we neglect very small quantities of more complicated matters which are too small to be reckoned here, they give up their nitrogen in the form of urea. Now let us suppose that we take an animal on one day, and on any day afterwards; and let us suppose that this animal is of the same weight the first day and the second day, and that its body is composed quite in the same way on both days. During the time—the interval of time—between these two days the animal has taken in food and oxygen, and has given out carbonic acid, water, and urea. Therefore, a certain quantity of inflammable substance, of nutriment, has combined with oxygen, and has produced nearly the same substances, the same combinations, which would be produced by burning the food in an open fire, at least, fat, sugar, starch, and so on; and those substances which contained no nitrogen would give us quite in the same way carbonic acid and water, if they are burnt in the open fire, as if they are burnt in the living body; only the oxidation in the living body goes on more slowly. The albuminous substances would give us the same substances, and also nitrogen, as if they were burnt in the fire. You may suppose, for making both cases equal, that the amount of urea which is produced in the body of the animal, may be changed without any very great development of heat, into carbonate of ammonia, and carbonate of ammonia may be burnt, and gives nitrogen, water, and carbonic acid. The amount of heat which would be produced by burning urea into carbonic acid and nitrogen, would be of no great value when compared with the great quantity of heat which is produced by burning the fat, the sugar, and the starch. Therefore we can change a certain amount of food into carbonic acid, water, and nitrogen, either by burning the whole in the open fire, or by giving it to living animals as food, and burning afterwards only the urea. In both cases we come to the same result. [. . .]

Let us now consider what consequences must be drawn when we find that the laws of animal life agree with the law of the conservation of force, at least as far as we can judge at present regarding this subject. As yet we cannot prove that the work produced by living bodies is an exact equivalent of the chemical forces which have been set into action. It is not yet possible to determine the exact value of either of these quantitites so accurately as will be done ultimately; but we may hope that at no distant time it may be possible to determine this with greater accuracy. There is no difficulty opposed to this task. Even at present I think we may consider it as extremely probable that the law of the conservation of force holds good for living bodies.

Now we may ask, what follows from this fact as regards the nature of the forces which act in the living body?

The majority of the physiologists in the last century, and in the beginning of this century, were of the opinion that the processes in living bodies were determined by one principal agent which they chose to call the 'vital principle'. The physical forces in the living body they supposed could be suspended or again set free at any moment, by the influence of the vital principle; and that by this means this agent could produce changes in the interior of the body, so that the health of the body would be thereby preserved or restored.

Now the conservation of force can exist only in those systems in which the forces in action (like all forces of inorganic nature) have always the same intensity and direction if the circumstances under which they act are the same. If it were possible to deprive any body of its gravity, and afterwards to restore its gravity, then indeed we should have the perpetual motion. Let the weight come down as long as it is heavy; let it rise if its gravity is lost; then you have produced mechanical work from nothing. Therefore this opinion that the chemical or mechanical power of the elements can be suspended, or changed, or removed in the interior of the living body, must be given up if there is complete conservation of force.

There may be other agents acting in the living body, than those agents which act in the inorganic world; but those forces, as far as they cause chemical and mechanical influences in the body, must be quite of the same character as inorganic forces, in this at least, that their effects must be ruled by necessity, and must be always the same, when acting in the same conditions, and that there cannot exist any arbitrary choice in the direction of their actions.

This is that fundamental principle of physiology which I mentioned in the beginning of this discourse.

Still at the beginning of this century physiologists believed that it was the vital principle which caused the processes of life, and that it detracted from the dignity and nature of life, if anybody expressed his belief that the blood was driven through the vessels by the mechanical action of the heart, or that respiration took place according to the common laws of the diffusion of gases.

The present generation, on the contrary, is hard at work to find out the real causes of the processes which go on in the living body. They do not suppose that there is any other difference between the chemical and the mechanical actions in the living body, and out of it, than can be explained by the more complicated circumstances and conditions under which these actions take place; and we have seen that the law of the conservation of force legitimizes this supposition. This law, moreover, shows the way in which this fundamental question, which has excited so many theoretical speculations, can be really and completely solved by experiment.

RUDOLF CLAUSIUS

Heat cannot by itself pass from a colder to a warmer body

273

1. When I wrote my *First Memoir on the Mechanical Theory of Heat*, two different views were entertained relative to the deportment of heat in the production of mechanical work. One was based on the old and widely spread notion, that heat is a peculiar substance, which may be present in greater or less quantity in a body, and thus determine the variations of temperature. Conformably with this notion was the opinion that, although heat could change its mode of distribution by passing from one body into another, and could further exist in different conditions, to which the terms *latent* and *free* were applied, yet the quantity of

heat in the whole mass could neither increase nor diminish, since matter can neither be created nor destroyed.

Upon this view is based the paper published by S. Carnot, in the year 1824, wherein machines driven by heat are subjected to a general theoretical treatment. Carnot, in investigating more closely the circumstances under which moving force can be produced by heat, found that in all cases there is a passage of heat from a body of higher into one of a lower temperature; as in the case of a steam engine where, by means of steam, heat passes from the fire or from a body of very high temperature, to the condenser, a space containing bodies of lower temperature. He compared this manner of producing work with that which occurs when a mass of water falls from a higher to a lower level, and consequently, in correspondence with the expression '*une chute d'eau*', he described the fall of heat from a higher to a lower temperature as '*une chute du calorique*'.

Regarding the subject from this point of view, he lays down the theorem that the magnitude of the work produced always bears a certain general relation to the simultaneous transfer of heat, i.e. to the quantity of heat which passes over, and to the temperatures of the bodies between which the transfer takes place, and that this relation is independent of the nature of the substances through which the production of work and the transfer of heat are effected. His proof of the necessity of such a relation is based on the axiom *that it is impossible to create a moving force out of nothing,* or in other words, *that perpetual motion is impossible.*

The other view above referred to is that heat is not invariable in quantity; but that when mechanical work is produced by heat, heat must be consumed, and that, on the contrary, by the expenditure of work a corresponding quantity of heat can be produced. This view stands in immediate connection with the new theory respecting the nature of heat, according to which heat is not a substance but a motion. Since the end of the last century various writers, amongst whom Rumford, Davy, and Seguin may be mentioned, have accepted this theory; but it is only since 1842 that Mayer of Heilbronn, Colding of Copenhagen, and Joule of Manchester examined the theory more closely, founded it, and established with certainty the law of the equivalence of heat and work.

According to this theory, the causal relation involved in the process of the production of work by heat is quite different from that which Carnot assumed. Mechanical work ensues from the conversion of existing heat into work, just in the same manner as, by the ordinary laws of mechanics, force is overcome, and work thereby produced, by motion which already exists; in the latter case the motion suffers a loss, in *vis viva*, equivalent to the work done, so that we may say that the *vis viva* of motion has been converted into work. Carnot's comparison therefore, in accordance with which the production of work by heat corresponds to the production of work by the falling of a mass of water—and, in fact, the fall of a certain quantity of heat from a higher to a lower temperature may be regarded as a cause of the work produced, was no longer admissible according to modern views. On this account it was thought that one of two alternatives must necessarily be accepted; either Carnot's theory must be retained and the modern view rejected, according to which heat is consumed in the production of work, or, on the contrary, Carnot's theory must be rejected and the modern view adopted.

2. When at the same period I entered on the investigation of this subject, I did not hesitate to accept the view that heat must be consumed in order to produce work. Nevertheless I did not think that Carnot's theory, which had found in Clapeyron[1] a very expert analytical expositor, required total rejection; on the contrary, it appeared to me that the theorem established by Carnot, after separating one part and properly formulizing the rest, might be brought into accordance with the more modern law of the equivalence of heat and work, and thus be employed together with it for the deduction of important conclusions. The theorem of Carnot thus modified was treated by me in the second part of the above-cited memoir, in the first part of which I had considered the law of the equivalence of heat and work.

In my later memoirs I succeeded in establishing simpler and at the same time more comprehensive theorems by pursuing further the same considerations which had led me to the first modification of Carnot's theorem. I will not now enter, however, upon these extensions of the theory, but will limit myself for the present to the question how, in accordance with the law of the equivalence of heat and work, the necessity can be demonstrated of the other theorem in its modified form.

The axiom employed by Carnot in the proof of his theorem, and which consists in the impossibility of creating moving force, or, more properly expressed, mechanical work of nothing, could no longer be employed in establishing the modified theorem. In fact, since in the latter it is already assumed that to produce mechanical work an equivalent amount of heat must be consumed, it follows that the supposition of the creation of work is altogether out of the question, no matter whether a transfer of heat from a warm to a colder body does or does not accompany the consumption of heat.

On the other hand, I found that another and, in my opinion, a more certain basis can be secured for the proof by reversing the sequence of reasoning pursued by Carnot, and by accepting as an axiom a theorem, in a somewhat modified form, which may be regarded as a consequence of his assumptions.

In fact, after establishing from the axiom that work cannot be produced from nothing, the theorem that in order to produce work a corresponding quantity of heat must be transferred from a warmer to a colder body, Carnot to be consistent could not but conclude that, in order to transfer heat from a colder to a warmer body, work must be expended. Although we must now abandon the argument which led to this result, and notwithstanding the fact that the result itself in its original form is not quite admissible, it is nevertheless manifest that

1. Émile Clapeyron (1799–1864), French engineer and physicist, developed analytically Carnot's basic ideas on the theory of heat. (Ed.)

an essential difference exists between the transfer of heat from a warmer to a colder body and the transfer from a colder to a warmer, since the first may take place spontaneously under circumstances which render the latter impossible.

On investigating the subject more closely, and taking into consideration the known properties and actions of heat, I came to the conviction that the difference in question had its origin in the nature of heat itself, inasmuch as by its very nature it must tend to equalize existing differences of temperature. Heat accordingly incessantly strives to pass from warmer to colder bodies, and a passage in a contrary direction can only take place under circumstances where simultaneously another quantity of heat passes from a warmer to a colder body, or when some change occurs which has the peculiarity of not being reversible without causing on its part such a transfer from a warmer to a colder body. This change which simultaneously takes place is consequently to be regarded as the equivalent of that transfer of heat from a colder to a warmer body, so that it cannot be said that the transfer has taken place *of itself* (*von selbst*).

I thought it permissible, therefore, to lay down the axiom, that *Heat cannot of itself pass from a colder to a warmer body*, and to employ it in demonstrating the second fundamental theorem of the mechanical theory of heat.

MICHAEL FARADAY

Electromagnetic induction

274

10. Two hundred and three feet of copper wire in one length were coiled round a large block of wood; another two hundred and three feet of similar wire were interposed as a spiral between the turns of the first coil, and metallic contact everywhere prevented by twine. One of these helices was connected with a galvanometer, and the other with a battery of one hundred pairs of plates four inches square, with double coppers, and well charged. When the contact was made, there was a sudden and very slight effect at the galvanometer, and there was also a similar slight effect when the contact with the battery was broken. But whilst the voltaic current was continuing to pass through the one helix, no galvanometrical appearances nor any effect like induction upon the other helix could be perceived, although the active power of the battery was proved to be great, by its heating the whole of its own helix, and by the brilliancy of the discharge when made through charcoal.

34. Another arrangement was then employed connecting the former experiments on volta-electric induction (6–26) with the present. A combination of helices like that already described (6) was constructed upon a hollow cylinder of pasteboard: there were eight lengths of copper wire, containing altogether 220 feet; four of these helices were connected end to end, and then with the galvanometer (7); the other intervening four were also connected end to end, and the battery of one hundred pairs discharged through them. In this form

the effect on the galvanometer was hardly sensible (11), though magnets could be made by the induced current (13). But when a soft iron cylinder seven eighths of an inch thick, and twelve inches long, was introduced into the pasteboard tube, surrounded by the helices, then the induced current affected the galvanometer powerfully, and with all the phenomena just described (30). It possessed also the power of making magnets with more energy, apparently, than when no iron cylinder was present.

36. Similar effects were then produced by *ordinary magnets*: thus the hollow helix just described (34) had all its elementary helices connected with the galvanometer by two copper wires, each five feet in length; the soft iron cylinder was introduced into its axis; a couple of bar magnets, each twenty-four inches long, were arranged with their opposite poles at one end in contact, so as to resemble a horse-shoe magnet, and then contact made between the

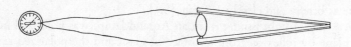

Electromagnetic induction: Faraday's apparatus

other poles and the ends of the iron cylinder, so as to convert it for the time into a magnet (fig. 2.): by breaking the magnetic contacts, or reversing them, the magnetism of the iron cylinder could be destroyed or reversed at pleasure.

60. Whilst the wire is subject to either volta-electric or magneto-electric induction, it appears to be in a peculiar state; for it resists the formation of an electrical current in it, whereas, if in its common condition, such a current would be produced; and when left uninfluenced it has the power of originating a current, a power which the wire does not possess under common circumstances. This electrical condition of matter has not hitherto been recognized, but it probably exerts a very important influence in many if not most of the phenomena produced by currents of electricity. For reasons which will immediately appear (71), I have, after advising with several learned friends, ventured to designate it as the *electro-tonic* state.

275

The action of magnets on light

2146. I have long held an opinion, almost amounting to conviction, in common I believe with many other lovers of natural knowledge, that the various forms under which the forces of matter are made manifest have one common origin; or, in other words, are so directly related and mutually dependent, that they are convertible, as it were, one into another, and possess equivalents of power in their action. In modern times the proofs of their convertibility have been accumulated to a very considerable extent, and a commencement made of the determination of their equivalent forces.

2147. This strong persuasion extended to the powers of light, and led, on a former occasion, to many exertions, having for their object the discovery of the direct relation of light and electricity, and their mutual action in bodies subject jointly to their power; but the results were negative and were afterwards confirmed, in that respect, by Wartmann.

2148. These ineffectual exertions, and many others which were never published, could not remove my strong persuasion derived from philosophical considerations; and, therefore, I recently resumed the inquiry by experiment in a most strict and searching manner, and have at last succeeded in *magnetizing and electrifying a ray of light, and in illuminating a magnetic line of force*. These results, without entering into the detail of many unproductive experiments, I will describe as briefly and clearly as I can. [. . .]

2150. A ray of light issuing from an Argand lamp, was polarized in a horizontal plane by reflexion from a surface of glass, and the polarized ray passed through a Nichol's eye-piece revolving on a horizontal axis, so as to be easily examined by the latter. Between the polarizing mirror and the eye-piece two powerful electro-magnetic poles were arranged, being either the poles of a horse-shoe magnet, or the contrary poles of two cylinder magnets; they were separated from each other about 2 inches in the direction of the line of the ray, and so placed, that, if on the same side of the polarized ray, it might pass near them; or if on contrary sides, it might go between them, its direction being always parallel, or nearly so, to the magnetic lines of force (2149). After that, any transparent substance placed between the two poles, would have passing through it, both the polarized ray and the magnetic lines of force at the same time and in the same direction. [. . .]

2152. A piece of this glass, about 2 inches square and 0·5 of an inch thick, having flat and polished edges, was placed as a *diamagnetic* (2149) between the poles (not as yet magnetized by the electric current), so that the polarized ray should pass through its length; the glass acted as air, water, or any other indifferent substance would do; and if the eye-piece were previously turned into such a position that the polarized ray was extinguished, or rather the image produced by it rendered invisible, then the introduction of this glass made no alteration in that respect. In this state of circumstances the force of the electro-magnet was developed, by sending an electric current through its coils, and immediately the image of the lamp-flame became visible, and continued so as long as the arrangement continued magnetic. On stopping the electric current, and so causing the magnetic force to cease, the light instantly disappeared; these phenomena could be renewed at pleasure, at any instant of time, and upon any occasion, showing a perfect dependence of cause and effect. [. . .]

2154. The character of the force thus impressed upon the diamagnetic is that of *rotation*; for when the image of the lamp-flame has thus been rendered visible, revolution of the eye-piece to the right or left, more or less, will cause its extinction; and the further motion of the eye-piece to the one side or other of this position will produce the reappearance of the light, and that with complementary tints, according as this further motion is to the right- or left-hand.

276

On the physical lines of magnetic force

Many powers act manifestly at a distance; their physical nature is incomprehensible to us: still we may learn much that is real and positive about them, and amongst other things something of the condition of the space between the body acting and that acted upon, or between the two mutually acting bodies. Such powers are presented to us by the phenomena of gravity, light, electricity, magnetism, &c. These when examined will be found to present remarkable differences in relation to their respective lines of forces; and at the same time that they establish the existence of real physical lines in some cases, will facilitate the consideration of the question as applied especially to magnetism.

When two bodies, *a*, *b*, gravitate towards each other, the line in which they act is a straight line, for such is the line which either would follow if free to move. The attractive force is not altered, either in *direction* or *amount*, if a third body is made to act by gravitation or otherwise upon either or both of the two first. A balanced cylinder of brass gravitates to the earth with a weight exactly the same, whether it is left like a pendulum freely to hang towards it, or whether it is drawn aside by other attractions or by tension, whatever the amount of the latter may be. A new gravitating force may be exerted upon *a*, but that does not in the least affect the amount of power which it exerts towards *b*. We have no evidence that *time* enters in any way into the exercise of this power, whatever the distance between the acting bodies, as that from the sun to the earth, or from star to star. We can hardly conceive of this force in one particle by itself; it is when two or more are present that we comprehend it: yet in gaining this idea we perceive no difference in the character of the power in the different particles; all of the same kind are *equal*, *mutual* and *alike*. In the case of gravitation, no effect which sustains the idea of an independent or physical line of force is presented to us; and as far as we at present know, the line of gravitation is merely an ideal line representing the direction in which the power is exerted.

Take the Sun in relation to another force which it exerts upon the earth, namely its illuminating or warming power. In this case rays (which are lines of force) pass across the intermediate space; but then we may affect these lines by different media applied to them in their course. We may alter their direction either by reflection or refraction; we may make them pursue curved or angular courses. We may cut them off at their origin and then search for and find them before they have attained their object. They have a relation to *time*, and occupy

8 minutes in coming from the sun to the earth: so that they may exist independently either of their source or their final home, and have in fact a clear distinct physical existence. They are in extreme contrast with the lines of gravitating power in this respect; as they are also in respect of their condition at their terminations. The two bodies terminating a line of gravitating force are alike in their actions in every respect, and so the line joining them has like relations in both directions. The two bodies at the terminals of a ray are utterly unlike in action; one is a source, the other a destroyer of the line; and the line itself has the relation of a stream flowing in one direction. In these two cases of gravity and radiation, the difference between an abstract and a physical line of force is immediately manifest.

Turning to the case of Static Electricity we find here attractions (and other actions) at a distance as in the former cases; but when we come to compare the attraction with that of gravity, very striking distinctions are presented which immediately affect the question of a physical line of force. In the first place, when we examine the bodies bounding or terminating the lines of attraction, we find them as before, mutually and equally concerned in the action; but they are not alike: on the contrary, though each is endued with a force which speaking generally is of the like nature, still they are in such contrast that their actions on a third body in a state like either of them are precisely the reverse of each other—what the one attracts the other repels; and the force makes itself evident at one of those manifestations of power endued with a dual and antithetical condition. Now with all such dual powers, attraction cannot occur unless the two conditions of force are present and in face of each other through the lines of force. Another essential limitation is that these two conditions must be exactly equal in amount; not merely to produce the effects of attraction, but in every other case; for it is impossible so to arrange things that there shall be present or be evolved more electric power of the one kind than of the other. Another limitation is that they must be in physical relation to each other; and that when a positive and a negative electrified surface are thus associated, we cannot cut off this relation except by transferring the forces of these surfaces to equal amounts of the contrary forces provided elsewhere. Another limitation is that the power is definite in amount. If a ball *a* be charged with 10 of positive electricity, it may be made to act with that amount of power of another ball *b* charged with 10 of negative electricity; but if 5 of its power be taken up by a third ball *c* charged with negative electricity, then it can only act with 5 of power on ball *a*, and that ball must find or evolve 5 of positive power elsewhere: this is quite unlike what occurs with gravity, a power that presents us with nothing dual in its character. Finally, the electric force acts in curved lines. If a ball be electrified positively and insulated in the air, and a round metallic plate be placed about 12 or 15 inches off, facing it and uninsulated, the latter will be found, by the necessity mentioned above, in a negative condition; but it is not negative only on the side facing the ball, but on the other or outer face also, as may be shown by a carrier applied there, or by a strip of gold or silver leaf hung against that outer face. Now the power affecting this face does not pass through the uninsulated plate, for the thinnest gold leaf is able to stop the inductive action, but round the edges of the face, and therefore acts in curved lines. All these points indicate the existence of physical lines of electric force:—the absolutely essential relation of positive and negative surfaces to each other, and their dependence on each other contrasted with the known mobility of the forces, admit of no other conclusion. The action also in curved lines must depend upon a physical line of force. And there is a third important character of the force leading to the same result, namely its affection by media having different specific inductive capacities.

When we pass to Dynamic Electricity the evidence of physical lines of force is far more patent. A voltaic battery having its extremities connected by a conducting medium, has what has been expressively called a current of force running round the circuit, but this current is an axis of power having equal and contrary forces in opposite directions. It consists of lines of force which are compressed or expanded according to the transverse action of the conductor, which changes in direction with the form of the conductor, which are found in every part of the conductor, and can be taken out from any place by channels properly appointed for the purpose; and nobody doubts that they are physical lines of force.

Finally as regards a Magnet, which is the object of the present discourse. A magnet presents a system of forces perfect in itself, and able, therefore, to exist by its own mutual relations. It has the dual and antithetic character belonging to both static and dynamic electricity; and this is made manifest by what are called its polarities, i.e. by the opposite powers of like kind found at and towards its extremities. These powers are found to be absolutely equal to each other; one cannot be changed in any degree as to amount without an equal change of the other; and this is true when the opposite polarities of a magnet are not related to each other, but to the polarities of other magnets. The polarities, or the *northness* and *southness* of a magnet are not only related to each other, through or within the magnet itself, but they are also related externally to opposite polarities (in the manner of static electric induction), or they cannot exist; and this external relation involves and necessitates an exactly equal amount of the new opposite polarities to which those of the magnet are related. So that if the force of a magnet *a* is related to that of another magnet *b*, it cannot act on a third magnet *c* without being taken off from *b*, to an amount proportional to its action on *c*. The lines of magnetic force are shown by the moving wire to exist both within and outside of the magnet; also they are shown to be closed curves passing in one part of their course through the magnet; and the amount of those within the magnet at its equator is exactly equal in force to the amount in any section including the whole of those on the outside. The line of force outside a magnet can be affected in their direction by the use of various media placed in their course. A magnet can in no way be procured having only one magnetism, or even the smallest excess of northness or southness one over the other. When the polarities of a magnet are not related externally to the forces of other magnets, then they are related to each other, i.e. the northness and southness of an isolated magnet are externally dependent on and sustained by each other.

Now all these facts, and many more, point to the existence of physical lines of force external to the magnets as well as within. They exist in curved as well as in straight lines; for if we conceive of an isolated straight bar-magnet, or more especially of a round disc of steel magnetized regularly, so that its magnetic axis shall be in one diameter, it is evident that the polarities must be related to each other externally by curved lines of force; for no straight line can at the same time touch two points having northness and southness. Curved lines of force can, as I think, only consist with physical lines of force.

The phenomena exhibited by the moving wire confirm the same conclusion. As the wire moves across the lines of force, a current of electricity passes or tends to pass through it, there being no such current before the wire is moved. The wire when quiescent has no such current, and when it moves it need not pass into places where the magnetic force is greater or less. It may travel in such a course that if a magnetic needle were carried through the same course it would be entirely unaffected magnetically, i.e. it would be a matter of absolute indifference to the needle whether it were moving or still. Matters may be so arranged that the wire when still shall have the same diamagnetic force as the medium surrounding the magnet, and so in no way cause disturbance of the lines of force passing through both; and yet when the wire moves, a current of electricity shall be generated in it. The mere fact of motion cannot have produced this current: there must have been a state or condition around the magnet and sustained by it, within the range of which the wire was placed; and this state shows the physical constitution of the lines of magnetic force.

What this state is, or upon what it depends, cannot as yet be declared. It may depend upon the aether, as a ray of light does, and an association has already been shown between light and magnetism. It may depend upon a state of tension, or a state of vibration, or perhaps some other state analogous to the electric current, to which the magnetic forces are so intimately related. Whether it of necessity requires matter for its sustentation will depend upon what is understood by the term matter. If that is to be confined to ponderable or gravitating substances, then matter is not essential to the physical lines of magnetic force any more than to a ray of light or heat; but if in the assumption of an aether we admit it to be a species of matter, then the lines of force may depend upon some function of it. Experimentally mere space is magnetic; but then the idea of such mere space must include that of the aether, when one is talking on that belief; or if hereafter any other conception of the state or condition of space rises up, it must be admitted into the view of that, which just now in relation to experiment is called mere space. On the other hand it is, I think, an ascertained fact, that ponderable matter is not essential to the existence of physical lines of magnetic force.

Thoughts on ray vibrations

You are aware of the speculation which I some time since uttered respecting that view of the nature of matter which considers its ultimate atoms as centres of force, and not as so many little bodies surrounded by forces, the bodies being considered in the abstract as independent of the forces and capable of existing without them. In the latter view, these little particles have a definite form and a certain limited size; in the former view such is not the case, for that which represents size may be considered as extending to any distance to which the lines of force of the particle extend: the particle indeed is supposed to exist only by these forces, and where they are it is. The consideration of matter under this view gradually led me to look at the lines of force as being perhaps the seat of the vibrations of radiant phenomena.

Another consideration bearing conjointly on the hypothetical view both of matter and radiation, arises from the comparison of the velocities with which the radiant action and certain powers of matter are transmitted. The velocity of light through space is about 190,000 miles in a second; the velocity of electricity is, by the experiments of Wheatstone, shown to be as great as this, if not greater: the light is supposed to be transmitted by vibrations through an aether which is, so to speak, destitute of gravitation, but infinite in elasticity; the electricity is transmitted through a small metallic wire, and is often viewed as transmitted by vibrations also. That the electric transference depends on the forces or powers of the matter of the wire can hardly be doubted, when we consider the different conductibility of the various metallic and other bodies; the means of affecting it by heat or cold; the way in which conducting bodies by combination enter into the constitution of non-conducting substances, and the contrary; and the actual existence of one elementary body, carbon, both in the conducting and non-conducting state. The power of electric conduction (being a transmission of force equal in velocity to that of light) appears to be tied up in and dependent upon the properties of the matter, and is, as it were, existent in them.

I suppose we may compare together the matter of the aether and ordinary matter (as, for instance, the copper of the wire through which the electricity is conducted), and consider them as alike in their essential constitution; i.e. either as both composed of little nuclei, considered in the abstract as matter, and of force or power associated with these nuclei, or else both consisting of mere centres of force, according to Boscovich's theory and the view put forth in my speculation; for there is no reason to assume that the nuclei are more requisite in the one case than in the other. It is true that the copper gravitates and the aether does not, and that therefore the copper is ponderable and the aether is not; but that cannot indicate the presence of nuclei in the copper more than in the aether, for of all powers of matter gravitation is the one in which the force extends to the greatest possible distance from the supposed nucleus, being infinite in relation to the size of the latter, and reducing that nucleus to a mere centre of force. The smallest atom of matter on the earth acts directly on the smallest atom of matter in the sun, though they are 95,000,000 of miles apart; further, atoms which, to our knowledge, are at least nineteen times that distance, and indeed, in cometary masses, far more, are in a similar way tied together by the lines of force extending from and belonging to each. What is there in the condition of the particles of the supposed aether, if there be even only *one* such particle between us and the sun, that can in subtility and extent compare to this?

Let us not be confused by the *ponderability* and *gravitation* of heavy matter, as if they proved the presence of the abstract nuclei; these are due not to the nuclei, but to the force superadded to them, if the nuclei exist at all; and, if the *aether* particles be without this force, which according to the assumption is the case, then they are more material, in the abstract sense, than the matter of this our globe; for matter, according to the assumption, being made up of nuclei and force, the aether particles have in this respect proportionately more of the nucleus and less of the force.

On the other hand, the infinite elasticity assumed as belonging to the particles of the aether, is as striking and positive a force of it as gravity is of ponderable particles, and produces in its way effects as great; in witness whereof we have all the varieties of radiant agency as exhibited in luminous, calorific, and actinic phenomena.

Perhaps I am in error in thinking the idea generally formed of the aether is that its nuclei are almost infinitely small, and that such force as it has, namely its elasticity, is almost infinitely intense. But if such be the received notion, what then is left in the aether but force or centres of force? As gravitation and solidity do not belong to it, perhaps many may admit this conclusion; but what are gravitation and solidity? certainly not the weight and contact of the abstract nuclei. The one is the consequence of an *attractive* force, which can act at distances as great as the mind of man can estimate or conceive; and the other is the consequence of a *repulsive* force, which forbids for ever the contact or touch of any two nuclei; so that these powers or properties should not in any degree lead those persons who conceive of the aether as a thing consisting of force only, to think any otherwise of ponderable matter, except that is has more and other *forces* associated with it than the aether has.

In experimental philosophy we can, by the phenomena presented, recognize various kinds of lines of force; thus there are the lines of gravitating force, those of electro-static induction, those of magnetic action, and others partaking of a dynamic character might be perhaps included. The lines of electric and magnetic action are by many considered as exerted through space like the lines of gravitating force. For my own part, I incline to believe that when there are intervening particles of matter (being themselves only centres of force), they take part in carrying on the force through the line, but that when there are none, the line proceeds through space. Whatever the view adopted respecting them may be, we can, at all events, affect these lines of force in a manner which may be conceived as partaking of the nature of a shake or lateral vibration. For suppose two bodies, *A B*, distant from each other and under mutual action, and therefore connected by lines of force, and let us fix our attention upon one resultant of force having an invariable direction as regards space; if one of the bodies should move in the least degree right or left, or if its power be shifted for a moment within the mass (neither of these cases being difficult to realize if *A* and *B* be either electric or magnetic bodies), then an effect equivalent to a lateral disturbance will take place in the resultant upon which we are fixing our attention; for, either it will increase in force whilst the neighbouring resultants are diminishing, or it will fall in force as they are increasing.

It may be asked, what lines of force are there in nature which are fitted to convey such an action and supply for the vibrating theory the place of the aether? I do not pretend to answer this question with any confidence; all I can say is, that I do not perceive in any part of space, whether (to use the common phrase) vacant or filled with matter, anything but forces and the lines in which they are exerted. The lines of weight or gravitating force are, certainly, extensive enough to answer in this respect any demand made upon them by radiant phenomena; and so, probably, are the lines of magnetic force: and then who can forget that Mossotti has shown that gravitation, aggregation, electric force, and the electro-chemical action may all have one common connection or origin; and so, in their actions at a distance, may have in common that infinite scope which some of these actions are known to possess?

The view which I am so bold to put forth considers, therefore, radiation as a high species of vibration in the lines of force which are known to connect particles and also masses of matter together. It endeavours to dismiss the aether, but not the vibrations. The kind of vibration which, I believe, can alone account for the wonderful, varied, and beautiful phenomena of polarization, is not the same as that which occurs on the surface of disturbed water, or the waves of sound in gases or liquids, for the vibrations in these cases are direct, or to and from the centre of action, whereas the former are lateral. It seems to me, that the resultant of two or more lines of force is an apt condition for that action which may be considered as equivalent to a *lateral* vibration; whereas a uniform medium, like the aether, does not appear apt, or more apt than air or water.

The occurrence of a change at one end of a line of force easily suggests a consequent change at the other. The propagation of light, and therefore probably of all radiant action, occupies *time*; and, that a vibration of the line of force should account for the phenomena of radiation, it is necessary that such vibration should occupy time also. I am not aware whether there are any data by which it has been, or could be ascertained whether such a power as gravitation acts without occupying time, or whether lines of force being already in existence, such a lateral disturbance of them at one end as I have suggested above, would require time, or must of necessity be felt instantly at the other end.

As to that condition of the lines of force which represents the assumed high elasticity of the aether, it cannot in this respect be deficient: the question here seems rather to be, whether the lines are sluggish enough in their action to render them equivalent to the aether in respect of the time known experimentally to be occupied in the transmission of radiant force.

The aether is assumed as pervading all bodies as well as space: in the view now set forth, it is the forces of the atomic centres which pervade (and make) all bodies, and also penetrate all space. As regards space, the difference is, that the aether presents successive parts or centres of action, and the present supposition only lines of action; as regards matter, the difference is, that the aether lies between the particles and so carries on the vibrations, whilst as respects the supposition, it is by the lines of force between the centres of the particles that the vibration is continued. As to the difference in intensity of action within matter under the two views, I suppose it will be very difficult to draw any conclusion, for when we take the simplest state of common matter and that which most nearly causes it to approximate to the condition of the aether, namely the state of rare gas, how soon do we find in its elasticity and the mutual repulsion of its particles, a departure from the law, that the action is inversely as the square of the distance!

*Magnetic action: critique of former theories and Faraday's
own explanation*

278

3300. Within the last three years I have been bold enough, though only as an experimentalist, to put forth new views of magnetic action in papers having for titles, 'On Lines of Magnetic Force', and 'On Physical Lines of Magnetic Force'. The first paper was simply an attempt to give, for the use of experimentalists and others, a correct expression of the dual nature, amount and direction of the magnetic power both within and outside of magnets, apart from any assumption regarding the character of the source of the power; that the mind, in reasoning forward towards new developments and discoveries, might be free from the bondage and deleterious influence of assumptions of such a nature (3075–3243). The second paper was a speculation respecting the possible physical nature of the force, as existing outside of the magnet as well as within it, and within what are called magnetic bodies, and was expressly described as being entirely hypothetical in its character (3243).

3301. There are at present two, or rather three general hypotheses of the physical nature of magnetic action. First, that of aethers, carrying with it the idea of fluxes or currents, and this Euler has set forth in a simple manner to the unmathematical philosopher in his Letters;—in that hypothesis the magnetic fluid or aether is supposed to move in streams through magnets, and also the space and substances around them. Then there is the hypothesis of two magnetic fluids, which being present in all magnetic bodies, and accumulated at the poles of a magnet, exert attractions and repulsions upon portions of both fluids at a distance, and so cause the attractions and repulsions of the distant bodies containing them. Lastly, there is the hypothesis of Ampère, which assumes the existence of electrical currents round the particles of magnets, which currents, acting at a distance upon other particles having like currents, arranges them in the masses to which they belong, and so renders such masses subject to the magnetic action. Each of these ideas is varied more or less by different philosophers, but the three distinct expressions of them which I have just given will suffice for my present purpose. My physico-hypothetical notion does not go so far in assumption as the second and third of these ideas, for it does not profess to say how the magnetic force is originated or sustained in a magnet; it falls in rather with the first view, yet does not assume so much. Accepting the magnet as a centre of power surrounded by lines of force, which, as representants of the power, are now justified by mathematical analysis (3302), it views these lines as *physical* lines of power, essential both to the existence of the force within the magnet, and to its conveyance to, and exertion upon, magnetic bodies at a distance. Those who entertain in any degree the aether notion might consider these lines as currents, or progressive vibrations, or as stationary undulations, or as a state of tension. For many reasons they should be contemplated round a wire carrying an electric current, as well as when issuing from a magnetic pole.

3302. The attention of two very able men and eminent mathematicians has fallen upon my proposition to represent the magnetic power by lines of magnetic force; and it is to me a source of great gratification and much encouragement to find that they affirm the truthfulness and generality of the method of representation. [. . .]

3303. The encouragement I derive from this appreciation by mathematicians of the mode of figuring to one's self the magnetic forces by lines, emboldens me to dwell a little more upon the further point of the true but unknown natural magnetic action. Indeed, what we really want, is not a variety of different methods of representing the forces, but the one true physical signification of that which is rendered apparent to us by the phenomena, and the laws governing them. Of the two assumptions most usually entertained at present, magnetic fluids and electric currents, *one* must be wrong, perhaps *both* are; and I do not perceive that the mathematician, even though he may think that each contains a higher principle than any I have advanced, can tell the true from the false, or say that either is true. Neither of these views could have led the mind to the phenomena of diamagnetism, and I think not to the magnetic rotation of light; and I suppose that if the question of the possibility of diamagnetic phenomena could have been asked beforehand, a mathematician, guided by either hypothesis, must have denied that possibility. The notion that I have introduced complicates the matter still further, for it is inconsistent with either of the former views, so long as they depend exclusively upon action at a distance without inter-mediation; and yet in the form of lines of force it represents magnetic actions truly in all that is not hypothetical. So that there are now three fundamental notions, and *two* of them at least must be impossible, i.e. untrue.

3304. It is evident, therefore, that our physical views are very doubtful; and I think good would result from an endeavour to shake ourselves loose from such preconceptions as are contained in them, that we may contemplate for a time the force as much as possible in its purity. [. . .]

3305. If we could tell the *disposition* of the force of a magnet, first at the place of its origin, and next in the space around, we should then have attained to a very important position in the pursuit of our subject; and if we could do that, assuming little or nothing, then we should be in the very best condition for carrying the pursuit further. Supposing that we imagine the magnet a sort of sun (as there is every reason to believe that the sun is a magnet) polarized, with antithetical powers, ever filling all space around it with its curved beams, as either the sun or a candle fills space with luminous rays; and supposing that such a view takes equal position with either of the two former views in representing truly the disposition of the forces, and that mathematical consideration cannot at present decide which of the three views is either above or inferior to its co-rivals; it surely becomes necessary that physical reasoning should be brought to bear upon the subject as largely as possible. For if there be such physical lines of magnetic force as correspond (in having a real existence) to the rays of light, it does not seem so very impossible for experiment to touch *them*; and it must be very important to obtain an answer to the inquiry respecting their existence, especially as the answer is likely enough to be in the affirmative. I therefore purpose, without asserting anything regarding the physical hypothesis of the magnet more strongly than before (3299), to call the attention of experimenters, in a somewhat desultory manner, to the subject again, both as respects the deficiency of the present physical views and the possible existence of lines of physical force, concentrating the observations I may have to make about a few points—as *polarity*, *duality*, &c., as occasion may best serve; and I am encouraged to make this endeavour by the following considerations. 1. The confirmation by mathematicians of the truthfulness of the abstract lines of force in representing the direction and amount of the magnetic power;—2. My own personal advantageous use of the lines on numerous occasions (3174);—3. The close analogy of the magnetic force and the other dual powers, either in the static or dynamic state, and especially of the magnet with the voltaic battery or any other sustaining source of an electric current;—4. Euler's idea of magnetic aethers or circulating fluids;—5. The strong conviction expressed by Sir Isaac Newton, that even gravity cannot be carried on to produce a distant effect except by some interposed agent fulfilling the conditions of a physical line of force;—6. The example of the conflict and final experimental settlement of the two theories of light.

The nature of gravitation

279

It is, probably, of great importance that our thoughts should be stirred up at this time to a reconsideration of the general nature of physical force, and especially to those forms of it which are concerned in actions at a distance. These are, by the dual powers, connected very intimately with those which occur at insensible distances; and it is to be expected that the progress which physical science has made in latter times will enable us to approach this deep and difficult subject with far more advantage than any possessed by philosophers at former periods. At present we are accustomed to admit action at sensible distances, as of one magnet upon another, or of the sun upon the earth, as if such admission were itself a perfect answer to any inquiry into the nature of the physical means which cause distant bodies to affect each other; and the man who hesitates to admit the sufficiency of the answer, or of the assumption on which it rests, and asks for a more satisfactory account, runs some risk of appearing ridiculous or ignorant before the world of science. Yet Newton, who did more than any other man in demonstrating the law of action of distant bodies, including amongst such the sun and Saturn, which are 900 millions of miles apart, did not leave the subject without recording his well-considered judgement, that the mere attraction of distant portions of matter was not a sufficient or satisfactory thought for a philosopher. That gravity should be innate, inherent, and essential to matter, so that one body may act upon another at a distance through a vacuum, without the mediation of anything else, by and through which their action and force may be conveyed from one to another, is, he says, to him a great absurdity. Gravity must be caused by an agent, acting constantly according to certain laws; but whether this agent be material or immaterial he leaves to the consideration of his readers. This is the onward-looking thought of one, who by his knowledge and like quality of mind, saw in the diamond an unctuous substance coagulated, when as yet it was known but as a transparent stone, and foretold the presence of a combustible substance in water a century before water was decomposed or hydrogen discovered; and I cannot help believing that the time is near at hand, when his thought regarding gravity will produce fruit:—and, with that impression, I shall venture a few considerations upon what appears to me the insufficiency of the usually accepted notions of gravity, and of those forces generally, which are supposed to act at a distance, having respect to the modern and philosophic view of the conservation and indestructibility of force.

The notion of the gravitating force is, with those who admit Newton's law, but go with him no further, that matter attracts matter with a strength which is inversely as the square of the distance. Consider, then, a mass of matter (or a particle), for which present purpose the sun will serve, and consider a globe like one of the planets, as our earth, either created or taken from distant space and placed near the sun as our earth is;—the attraction of gravity is then exerted, and we say that the sun attracts the earth, and also that the earth attracts the sun. But if the sun attracts the earth, that force of attraction must either arise *because* of the presence of the earth near the sun; or it must have *pre-existed* in the sun when the earth was not there. If we consider the first case, I think it will be exceedingly difficult to conceive that the sudden presence of the earth, 95 millions of miles from the sun, and having no previous physical connection with it, nor any physical connection caused by the mere circumstance of juxtaposition, should be able to raise up in the sun a power having no previous existence. As respects gravity, the earth must be considered as inert, previously, as the sun; and can have no more inducing or affecting power over the sun than the sun over it: both are assumed to be *without* power in the beginning of the case; how then can that power arise by their mere approximation or coexistence? That a body without force should raise up force in a body at a distance from it, is too hard to imagine; but it is harder still, if that can be possible, to accept the idea when we consider that it includes the *creation of force*. Force may be opposed by force, may be diverted, directed partially or exclusively, may even be converted, as far as we understand the matter, disappearing in one form to reappear in another; but it cannot be created or annihilated, or truly suspended, i.e. rendered existent without action or without its equivalent action. The conservation of power is now a thought deeply impressed upon the minds of philosophic men; and I think that, as a body, they admit that the creation or annihilation of force is equally impossible with the creation or annihilation of matter. But if we conceive the sun existing alone in space, exerting no force of gravitation exterior to it; and then conceive another sphere in space having like conditions, and that the two are brought towards each other; if we assume, that by their mutual presence each causes the other to act,—this is to assume not merely a creation of power, but a *double creation*, for both are supposed to rise from a previously inert to a powerful state. On their dissociation they, by the assumption, pass into the powerless state again, and this would be equivalent to the *annihilation* of force. It will be easily understood, that the case of the sun or the earth, or of any one of two or more acting bodies, is reciprocal;—and also that the variation of attraction, with any degree of approach or separation of the bodies, involves the same result of creation or annihilation of power as the creation or annihilation (which latter is only the total removal) of either of the acting bodies would do.

Such, I think, must be the character of the conclusion, if it be supposed that the attraction of the sun upon the earth arises *because* of the presence of the earth, and the attraction of the earth upon the sun, because of the presence of the sun; there remains the case of the power, or the efficient source of the power, having pre-existed in the sun (or the earth) *before* the earth (or the sun) was in presence. In the latter view it appears to me that, consistently with the conservation of force, one of three sub-cases must occur: either the gravitating force of the sun, when directed upon the earth, must be removed in an equivalent degree from some other bodies, and when taken off from the earth (by the disappearance of the latter) be disposed of on some other bodies;—or else it must take up some *new* form of power when it ceases to be gravitation, and consume some other form of power when it is developed as gravitation;—or else it must be *always* existing around the sun through infinite space. The first sub-case is not imagined by the usual hypothesis of gravitation, and will hardly be supposed probable; for, if it were true, it is scarcely possible that the effects should not have been observed by astronomers, when considering the motions of the planets in different positions with respect to each other and the sun. Moreover, gravitation is not assumed to be a dual power, and in them only as yet have such removals been observed by experiment or conceived by the mind. The second sub-case, or that of a new or another form of power, is also one which has never been imagined by others, in association with the theory of gravity. I made some endeavours, experimentally, to connect gravity with electricity, having this very object in view (Phil. Trans. 1851, p. 1); but the results were entirely negative. The view, if held for a moment, would imply that not merely the sun, but all matter, whatever its state, would have extra powers set up in it, if removed in any degree from gravitation; that the particles of a comet at its perihelion would have changed in character, by the conversion of some portion of their molecular force into the increased amount of gravitating force which they would then exert; and that at its aphelion, this extra gravitating force would have been converted back into some other kind of molecular force, having either the former or a new character: the conversion either way being to a perfectly equivalent degree. One could not even conceive of the diffusion of a cloud of dust, or its concentration into a stone, without supposing something of the same kind to occur; and I suppose that nobody will accept the idea as possible. The third sub-case remains, namely that the power is always existing around the sun and through infinite space, whether secondary bodies be there to be acted upon by gravitation or not; and not only around the sun, but around every particle of matter which has existence. This case of a constant necessary condition to action in space, when as respects the sun the earth is *not* in place, and of a certain gravitating action as the result of that previous condition when the earth *is* in place, I can conceive, consistently, as I think, with the conservation of force: and I think the case is that which Newton looked at in gravity; is, in philosophical respects, the same as that admitted by all in regard to light, heat, and radiant phenomena; and (in a sense even more general and extensive) is that now driven upon our attention in an especially forcible and instructive manner, by the phenomena of electricity and magnetism, because of their dependence on dual forms of power.

JAMES CLERK MAXWELL

280

The theory of molecular vortices applied to magnetic
phenomena

In all phenomena involving attractions or repulsions, or any forces depending on the relative position of bodies, we have to determine the *magnitude* and *direction* of the force which would act on a given body, if placed in a given position.

In the case of a body acted on by the gravitation of a sphere, this force is inversely as the square of the distance, and in a straight line to the centre of the sphere. In the case of two attracting spheres, or of a body not spherical, the magnitude and direction of the force vary according to more complicated laws. In electric and magnetic phenomena, the magnitude and direction of the result-ant force at any point is the main subject of investigation. Suppose that the direc-tion of the force at any point is known, then, if we draw a line so that in every part of its course it coincides in direction with the force at that point, this line may be called a *line of force*, since it indicates the direction of the force in every part of its course.

By drawing a sufficient number of lines of force, we may indicate the direc-tion of the force in every part of the space in which it acts.

Thus if we strew iron filings on paper near a magnet, each filing will be magnetized by induction, and the consecutive filings will unite by their opposite poles, so as to form fibres, and these fibres will *indicate* the direction of the lines of force. The beautiful illustration of the presence of magnetic force afforded by this experiment, naturally tends to make us think of the lines of force as something real, and as indicating something more than the mere resultant of two forces, whose seat of action is at a distance, and which do not exist there at all until a magnet is placed in that part of the field. We are dissatisfied with the explanation founded on the hypothesis of attractive and repellent forces directed towards the magnetic poles, even though we may have satisfied ourselves that the phenomenon is in strict accordance with that hypothesis, and we cannot help thinking that in every place where we find these lines of force, some physical state or action must exist in sufficient energy to produce the actual phenomena.

My object in this paper is to clear the way for speculation in this direction, by investigating the mechanical results of certain states of tension and motion in a medium, and comparing these with the observed phenomena of magnetism and electricity. By pointing out the mechanical consequences of such hypotheses, I hope to be of some use to those who consider the phenomena as due to the action of a medium, but are in doubt as to the relation of this hypothesis to the experimental laws already established, which have generally been expressed in the language of other hypotheses. [. . .]

I propose now to examine magnetic phenomena from a mechanical point of view, and to determine what tensions in, or motions of, a medium are capable of producing the mechanical phenomena observed. If, by the same hypothesis, we can connect the phenomena of magnetic attraction with electro-magnetic phenomena and with those of induced currents, we shall have found a theory which, if not true, can only be proved to be erroneous by experiments which will greatly enlarge our knowledge of this part of physics. [. . .]

If we observe the lines of force between two magnets, as indicated by iron filings, we shall see that whenever the lines of force pass from one pole to another, there is *attraction* between these poles; and where the lines of force from the poles avoid each other and are dispersed into space, the poles *repel* each other, so that in both cases they are drawn in the direction of the resultant of the lines of force.

It appears therefore that the stress in the axis of a line of magnetic force is a *tension*, like that of a rope. [. . .]

Let us now suppose that the phenomena of magnetism depend on the existence of a tension in the direction of the lines of force, combined with a hydrostatic pressure; or in other words, a pressure greater in the equatorial than in the axial direction: the next question is, what mechanical explanation can we give of this inequality of pressures in a fluid or mobile medium? The explanation which most readily occurs to the mind is that the excess of pressure in the equatorial direction arises from the centrifugal force of vortices or eddies in the medium having their axes in directions parallel to the lines of force.

This explanation of the cause of the inequality of pressures at once suggests the means of representing the dipolar character of the line of force. Every vortex is essentially dipolar, the two extremities of its axis being distinguished by the direction of its revolution as observed from those points.

We also know that when electricity circulates in a conductor, it produces lines of magnetic force passing through the circuit, the direction of the lines depending on the direction of the circulation. Let us suppose that the direction of revolution of our vortices is that in which vitreous electricity must revolve in order to produce lines of force whose direction within the circuit is the same as that of the given lines of force.

We shall suppose at present that all the vortices in any one part of the field are revolving in the same direction about axes nearly parallel, but that in passing from one part of the field to another, the direction of the axes, the velocity of rotation, and the density of the substance of the vortices are subject to change. We shall investigate the resultant mechanical effect upon an element of the medium, and from the mathematical expression of this resultant we shall deduce the physical character of its different component parts.

281

The theory of molecular vortices applied to electric currents

We have already shewn that all the forces acting between magnets, substances capable of magnetic induction, and electric currents, may be mechanically accounted for on the supposition that the surrounding medium is put into such a state that at every point the pressures are different in different directions, the direction of least pressure being that of the observed lines of force, and the difference of greatest and least pressures being proportional to the square of the intensity of the force at that point.

Such a state of stress, if assumed to exist in the medium, and to be arranged according to the known laws regulating lines of force, will act upon the magnets, currents, &c., in the field with precisely the same resultant forces as those calculated on the ordinary hypothesis of direct action at a distance. This is true independently of any particular theory as to the *cause* of this state of stress, or the mode in which it can be sustained in the medium. We have therefore a satisfactory answer to the question, 'Is there any mechanical hypothesis as to the condition of the medium indicated by lines of force, by which the observed resultant forces may be accounted for?' The answer is, the lines of force indicate the direction of *minimum pressure* at every point of the medium.

The second question must be, 'What is the mechanical cause of this difference of pressure in different directions?' We have supposed, in the first part of this paper, that this difference of pressure is caused by molecular vortices, having their axes parallel to the lines of force.

We also assumed, perfectly arbitrarily, that the direction of these vortices is such that, on looking along a line of force from south to north, we should see the vortices revolving in the direction of the hands of a watch.

We found that the velocity of the circumference of each vortex must be proportional to the intensity of the magnetic force, and that the density of the substance of the vortex must be proportional to the capacity of the medium for magnetic induction.

We have as yet given no answers to the questions, 'How are these vortices set in rotation?' and 'Why are they arranged according to the known laws of lines of force about magnets and currents?' These questions are certainly of a higher order of difficulty than either of the former; and I wish to separate the suggestions I may offer by way of provisional answer to them, from the mechanical deductions which resolved the first question, and the hypothesis of vortices which gave a probable answer to the second.

We have, in fact, now come to inquire into the physical connection of these vortices with electric currents, while we are still in doubt as to the nature of electricity, whether it is one substance, two substances, or not a substance at all, or in what way it differs from matter, and how it is connected with it.

We know that the lines of force are affected by electric currents, and we know the distribution of those lines about a current; so that from the force we can determine the amount of the current. Assuming that our explanation of the lines of force by molecular vortices is correct, why does a particular distribution of vortices indicate an electric current? A satisfactory answer to this question would lead us a long way towards that of a very important one, 'What is an electric current?'

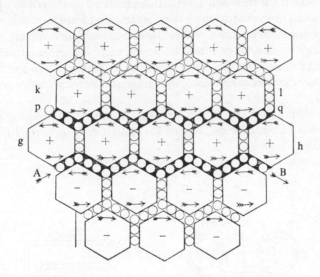

The molecular vortices according to Maxwell

I have found great difficulty in conceiving of the existence of vortices in a medium, side by side, revolving in the same direction about parallel axes. The contiguous portions of consecutive vortices must be moving in opposite directions; and it is difficult to understand how the motion of one part of the medium can coexist with, and even produce, an opposite motion of a part in contact with it.

The only conception which has at all aided me in conceiving of this kind of motion is that of the vortices being separated by a layer of particles, revolving each on its own axis in the opposite direction to that of the vortices, so that the contiguous surfaces of the particles and of the vortices have the same motion.

In mechanism, when two wheels are intended to revolve in the same direction, a wheel is placed between them so as to be in gear with both, and this wheel is called an 'idle wheel'. The hypothesis about the vortices which I have to suggest is that a layer of particles, acting as idle wheels, is interposed between each vortex and the next, so that each vortex has a tendency to make the neighbouring vortices revolve in the same direction with itself. [. . .]

It appears therefore that, according to our hypothesis, an electric current is represented by the transference of the movable particles interposed between the neighbouring vortices. We may conceive that these particles are very small compared with the size of a vortex, and that the mass of all the particles together is inappreciable compared with that of the vortices, and that a great many vortices, with their surrounding particles, are contained in a single complete molecule of the medium. The particles must be conceived to roll without sliding between the vortices which they separate, and not to touch each other, so that, as long as they remain within the same complete molecule, there is no loss of energy by resistance. When, however, there is a general transference of particles in one direction, they must pass from one molecule to another, and in doing so, may experience resistance, so as to waste electrical energy and generate heat. [. . .]

In Plate VIII, p. 488, fig. 1, let the vertical circle *EE'* represent an electric current flowing from copper *C* to zinc *Z* through the conductor *EE'*, as shewn by the arrows.

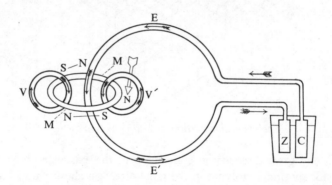

The theory of molecular vortices applied to an electric current

Let the horizontal circle *MM'* represent a line of magnetic force embracing the electric circuit, the north and south directions being indicated by the lines *SN* and *NS*.

Let the vertical circles *V* and *V'* represent the molecular vortices of which the line of magnetic force is the axis. *V* revolves as the hands of a watch, and *V'* the opposite way.

It will appear from this diagram, that if *V* and *V'* were contiguous vortices, particles placed between them would move downwards; and that if the particles were forced downwards by any cause, they would make the vortices revolve as in the figure. We have thus obtained a point of view from which we may regard the relation of an electric current to its lines of force as analogous to the relation of a toothed wheel or rack to wheels which it drives.[. . .]

We may now recapitulate the assumptions we have made, and the results we have obtained.

1. Magneto-electric phenomena are due to the existence of matter under certain conditions of motion or of pressure in every part of the magnetic field, and not to direct action at a distance between the magnets or currents. The substance producing these effects may be a certain part of ordinary matter, or it may be an aether associated with matter. Its density is greatest in iron, and least in diamagnetic substances; but it must be in all cases, except that of iron, very rare, since no other substance has a large ratio of magnetic capacity to what we call a vacuum.

2. The condition of any part of the field, through which lines of magnetic force pass, is one of unequal pressure in different directions, the direction of the lines of force being that of least pressure, so that the lines of force may be considered lines of tension.

3. This inequality of pressure is produced by the existence in the medium of vortices or eddies, having their axes in the direction of the lines of force, and having their direction of rotation determined by that of the lines of force.

We have supposed that the direction was that of a watch to a spectator looking from south to north. We might with equal propriety have chosen the reverse direction, as far as known facts are concerned, by supposing resinous electricity instead of vitreous to be positive. The effect of these vortices depends on their density, and on their velocity at the circumference, and is independent of their diameter. The density must be proportional to the capacity of the substance for magnetic induction, that of the vortices in air being one. The velocity must be very great, in order to produce so powerful effects in so rare a medium.

The size of the vortices is indeterminate, but is probably very small as compared with that of a complete molecule of ordinary matter.

4. The vortices are separated from each other by a single layer of round particles, so that a system of cells is formed, the partitions being these layers of particles, and the substance of each cell being capable of rotating as a vortex.

5. The particles forming the layer are in *rolling contact* with both the vortices which they separate, but do not rub against each other. They are perfectly free to roll between the vortices and so to change their place, provided they keep within one *complete molecule* of the substance; but in passing from one molecule to another they experience resistance, and generate irregular motions, which constitute heat. These particles, in our theory, play the part of electricity. Their motion of translation constitutes an electric current, their rotation serves to transmit the motion of the vortices from one part of the field to another, and the tangential pressures thus called into play constitute electromotive force. The conception of a particle having its motion connected with that of a vortex by perfect rolling contact may appear somewhat awkward. I do not bring it

forward as a mode of connection existing in nature, or even as that which I would willingly assent to as an electrical hypothesis. It is, however, a mode of connection which is mechanically conceivable, and easily investigated, and it serves to bring out the actual mechanical connections between the known electromagnetic phenomena; so that I venture to say that any one who understands the provisional and temporary character of this hypothesis, will find himself rather helped than hindered by it in his search after the true interpretation of the phenomena.

The action between the vortices and the layers of particles is in part tangential; so that if there were any slipping or differential motion between the parts in contact, there would be a loss of the energy belonging to the lines of force, and a gradual transformation of that energy into heat. Now we know that the lines of force about a magnet are maintained for an indefinite time without any expenditure of energy; so that we must conclude that wherever there is tangential action between different parts of the medium, there is no motion of slipping between those parts. We must therefore conceive that the vortices and particles roll together without slipping; and that the interior strata of each vortex receive their proper velocities from the exterior stratum without slipping, that is, the angular velocity must be the same throughout each vortex.

The only process in which electromagnetic energy is lost and transformed into heat, is in the passage of electricity from one molecule to another. In all other cases the energy of the vortices can only be diminished when an equivalent quantity of mechanical work is done by magnetic action.

6. The effect of an electric current upon the surrounding medium is to make the vortices in contact with the current revolve so that the parts next to the current move in the same direction as the current. The parts furthest from the current will move in the opposite direction; and if the medium is a conductor of electricity, so that the particles are free to move in any direction, the particles touching the outside of these vortices will be moved in a direction contrary to that of the current, so that there will be an induced current in the opposite direction to the primary one.

If there were no resistance to the motion of the particles, the induced current would be equal and opposite to the primary one, and would continue as long as the primary current lasted, so that it would prevent all action of the primary current at a distance. If there is a resistance to the induced current, its particles act upon the vortices beyond them, and transmit the motion of rotation to them, till at last all the vortices in the medium are set in motion with such velocites of rotation that the particles between them have no motion except that of rotation, and do not produce currents.

In the transmission of the motion from one vortex to another, there arises a force between the particles and the vortices, by which the particles are pressed in one direction and the vortices in the opposite direction. We call the force acting on the particles the electromotive force. The reaction on the vortices is equal and opposite, so that the electromotive force cannot move any part of the medium as a whole, it can only produce currents. When the primary current is stopped, the electromotive forces all act in the opposite direction.

7. When an electric current or a magnet is moved in presence of a conductor, the velocity of rotation of the vortices in any part of the field is altered by that motion. The force by which the proper amount of rotation is transmitted to each vortex, constitutes in this case also an electromotive force, and, if permitted, will produce currents.

8. When a conductor is moved in a field of magnetic force, the vortices in it and in its neighbourhood are moved out of their places, and are changed in form. The force arising from these changes constitutes the electromotive force on a moving conductor, and is found by calculation to correspond with that determined by experiment.

We have now shewn in what way electromagnetic phenomena may be imitated by an imaginary system of molecular vortices. Those who have been already inclined to adopt an hypothesis of this kind, will find here the conditions which must be fulfilled in order to give it mathematical coherence and a comparison, so far satisfactory, between its necessary results and known facts. Those who look in a different direction for the explanation of the facts, may be able to compare this theory with that of the existence of currents flowing freely through bodies, and with that which supposes electricity to act at a distance with a force depending on its velocity, and therefore not subject to the law of conservation of energy.

The facts of electromagnetism are so complicated and various, that the explanation of any number of them by several different hypotheses must be interesting, not only to physicists, but to all who desire to understand how much evidence the explanation of phenomena lends to the credibility of a theory, or how far we ought to regard a coincidence in the mathematical expression of two sets of phenomena as an indication that these phenomena are of the same kind. We know that partial coincidences of this kind have been discovered; and the fact that they are only partial is proved by the divergence of the laws of the two sets of phenomena in other respects. We may chance to find, in the higher parts of physics, instances of more complete coincidence, which may require much investigation to detect their ultimate divergence.

282

A dynamical theory of the electromagnetic field

1. The most obvious mechanical phenomenon in electrical and magnetical experiments is the mutual action by which bodies in certain states set each other in motion while still at a sensible distance from each other. The first step, therefore, in reducing these phenomena into scientific form, is to ascertain the magnitude and direction of the force acting between the bodies, and when it is

found that this force depends in a certain way upon the relative position of the bodies and on their electric or magnetic condition, it seems at first sight natural to explain the facts by assuming the existence of something either at rest or in motion in each body, constituting its electric or magnetic state, and capable of acting at a distance according to mathematical laws.

In this way mathematical theories of statical electricity, of magnetism, of the mechanical action between conductors carrying currents, and of the induction of currents have been formed. In these theories the force acting between the two bodies is treated with reference only to the condition of the bodies and their relative position, and without any express consideration of the surrounding medium.

These theories assume, more or less explicitly, the existence of substances the particles of which have the property of acting on one another at a distance by attraction or repulsion. [. . .]

2. The mechanical difficulties, however, which are involved in the assumption of particles acting at a distance with forces which depend on their velocities are such as to prevent me from considering this theory as an ultimate one, though it may have been, and may yet be useful in leading to the coordination of phenomena.

I have therefore preferred to seek an explanation of the fact in another direction by supposing them to be produced by actions which go on in the surrounding medium as well as in the excited bodies, and endeavouring to explain the action between distant bodies without assuming the existence of forces capable of acting directly at sensible distances.

3. The theory I propose may therefore be called a theory of the *Electromagnetic Field*, because it has to do with the space in the neighbourhood of the electric or magnetic bodies, and it may be called a *Dynamical* Theory, because it assumes that in that space there is matter in motion, by which the observed electromagnetic phenomena are produced.

4. The electromagnetic field is that part of space which contains and surrounds bodies in electric or magnetic conditions.

It may be filled with any kind of matter, or we may endeavour to render it empty of all gross matter, as in the case of Geissler's tubes and other so-called vacua.

There is always, however, enough of matter left to receive and transmit the undulations of light and heat, and it is because the transmission of these radiations is not greatly altered when transparent bodies of measurable density are substituted for the so-called vacuum, that we are obliged to admit that the undulations are those of an aethereal substance, and not of the gross matter, the presence of which merely modifies in some way the motion of the aether.

We have therefore some reason to believe, from the phenomena of light and heat, that there is an aethereal medium filling space and permeating bodies, capable of being set in motion and of transmitting that motion from one part to another, and of communicating that motion to gross matter so as to heat it and affect it in various ways.

5. Now the energy communicated to the body in heating it must have formerly existed in the moving medium, for the undulations had left the source of heat son_e time before they reached the body, and during that time the energy must have been half in the form of motion of the medium and half in the form of elastic resilience. From these considerations Professor W. Thomson[1] has argued that the medium must have a density capable of comparison with that of gross matter, and has even assigned an inferior limit to that density.

6. We may therefore receive, as a datum derived from a branch of science independent of that with which we have to deal, the existence of a pervading medium, of small but real density, capable of being set in motion, and of transmitting motion from one part to another with great, but not infinite, velocity.

Hence the parts of this medium must be so connected that the motion of one part depends in some way on the motion of the rest; and at the same time these connections must be capable of a certain kind of elastic yielding, since the communication of motion is not instantaneous, but occupies time.

The medium is therefore capable of receiving and storing up two kinds of energy, namely, the 'actual' energy depending on the motions of its parts, and 'potential' energy, consisting of the work which the medium will do in recovering from displacement in virtue of its elasticity.

The propagation of undulations consists in the continual transformation of one of these forms of energy into the other alternately, and at any instant the amount of energy in the whole medium is equally divided, so that half is energy of motion, and half is elastic resilience.

7. A medium having such a constitution may be capable of other kinds of motion and displacement than those which produce the phenomena of light and heat, and some of these may be of such a kind that they may be evidenced to our senses by the phenomena they produce.

8. Now we know that the luminiferous medium is in certain cases acted on by magnetism; for Faraday discovered that when a plane polarized ray traverses a transparent diamagnetic medium in the direction of the lines of magnetic force produced by magnets or currents in the neighbourhood, the plane of polarization is caused to rotate.

This rotation is always in the direction in which positive electricity must be carried round the diamagnetic body in order to produce the actual magnetization of the field. [. . .]

9. We may now consider another phenomenon observed in the electromagnetic field. When a body is moved across the lines of magnetic force it experiences what is called an electromotive force; the two extremities of the body tend to become oppositely electrified, and an electric current tends to flow

1. Sir William Thomson, later Lord Kelvin (1824–1907), English physicist, famous for his researches in thermodynamics and electromagnetism. (Ed.)

through the body. When the electromotive force is sufficiently powerful, and is made to act on certain compound bodies, it decomposes them, and causes one of their components to pass towards one extremity of the body, and the other in the opposite direction.

Here we have evidence of a force causing an electric current in spite of resistance, electrifying the extremities of a body in opposite ways; a condition which is sustained only by the action of the electromotive force, and which, as soon as that force is removed, tends, with an equal and opposite force, to produce a counter current through the body and to restore the original electrical state of the body; and finally, if strong enough, tearing to pieces chemical compounds and carrying their compoents in opposite directions, while their natural tendency is to combine, and to combine with a force which can generate an electromotive force in the reverse direction.

This, then, is a force acting on a body caused by its motion through the electromagnetic field, or by changes occurring in that field itself; and the effect of the force is either to produce a current and heat the body, or to decompose the body, or, when it can do neither, to put the body in a state of electric polarization,—a state of constraint in which opposite extremities are oppositely electrified, and from which the body tends to relieve itself as soon as the disturbing force is removed.

10. According to the theory which I propose to explain, this 'electromotive force' is the force called into play during the communication of motion from one part of the medium to another, and it is by means of this force that the motion of one part causes motion in another part. When electromotive force acts on a conducting circuit, it produces a current, which, as it meets with resistance, occasions a continual transformation of electrical energy into heat, which is incapable of being restored again to the form of electrical energy by any reversal of the process.

11. But when electromotive force acts on a dielectric it produces a state of polarization of its parts similar in distribution to the polarity of the parts of a mass of iron under the influence of a magnet, and like the magnetic polarization, capable of being described as a state in which every particle has its opposite poles in opposite conditions.

In a dielectric under the action of electromotive force, we may conceive that the electricity in each molecule is so displaced that one side is rendered positively and the other negatively electrical, but that the electricity remains entirely connected with the molecule, and does not pass from one molecule to another. The effect of this action on the whole dielectric mass is to produce a general displacement of electricity in a certain direction. This displacement does not amount to a current, because when it has attained to a certain value it remains constant, but it is the commencement of a current, and its variations constitute currents in the positive or the negative direction according as the displacement is increasing or decreasing. In the interior of the dielectric there is no indication of electrification, because the electrification of the surface of any molecule is neutralized by the opposite electrification of the surface of the molecules in contact with it; but at the bounding surface of the dielectric, where the electrification is not neutralized, we find the phenomena which indicate positive or negative electrification.

The relation between the electromotive force and the amount of electric displacement it produces depends on the nature of the dielectric, the same electromotive force producing generally a greater electric displacement in solid dielectrics, such as glass or sulphur, than in air.

12. Here, then, we perceive another effect of electromotive force, namely, electric displacement, which according to our theory is a kind of elastic yielding to the action of the force, similar to that which takes place in structures and machines owing to the want of perfect rigidity of the connections. [. . .]

15. It appears therefore that certain phenomena in electricity and magnetism lead to the same conclusion as those of optics, namely, that there is an aethereal medium pervading all bodies, and modified only in degree by their presence; that the parts of this medium are capable of being set in motion by electric currents and magnets; that this motion is communicated from one part of the medium to another by forces arising from the connections of those parts; that under the action of these forces there is a certain yielding depending on the elasticity of these connections; and that therefore energy in two different forms may exist in the medium, the one form being the actual energy of motion of its parts, and the other being the potential energy stored up in the connections, in virtue of their elasticity.

16. Thus, then, we are led to the conception of a complicated mechanism capable of a vast variety of motion, but at the same time so connected that the motion of one part depends, according to definite relations, on the motion of other parts, these motions being communicated by forces arising from the relative displacement of the connected parts, in virtue of their elasticity. Such a mechanism must be subject to the general laws of Dynamics, and we ought to be able to work out all the consequences of its motion, provided we know the form of the relation between the motions of the parts.

17. We know that when an electric current is established in a conducting circuit, the neighbouring part of the field is characterized by certain magnetic properties, and that if two circuits are in the field, the magnetic properties of the field due to the two currents are combined. Thus each part of the field is in connection with both currents, and the two currents are put in connection with each other in virtue of their connection with the magnetization of the field. The first result of this connection that I propose to examine, is the induction of one current by another, and by the motion of conductors in the field.

The second result, which is deduced from this, is the mechanical action between conductors carrying currents. The phenomenon of the induction of currents has been deduced from their mechanical action by Helmholtz and Thomson. I have followed the reverse order, and deduced the mechanical action from the laws of induction.[. . .]

18. I then apply the phenomena of induction and attraction of currents to the exploration of the electromagnetic field, and the laying down of systems of lines of magnetic force which indicate its magnetic properties. By exploring the same field with a magnet, I shew the distribution of its equipotential magnetic surfaces, cutting the lines of force at right angles.

In order to bring these results within the power of symbolical calculation, I then express them in the form of the General Equations of the Electromagnetic Field. These equations express—

(A) The relation between electric displacement, true conduction, and the total current, compounded of both.

(B) The relation between the lines of magnetic force and the inductive coefficients of a circuit, as already deduced from the laws of induction.

(C) The relation between the strength of a current and its magnetic effects, according to the electromagnetic system of measurement.

(D) The value of the electromotive force in a body, as arising from the motion of the body in the field, the alteration of the field itself, and the variation of electric potential from one part of the field to another.

(E) The relation between electric displacement, and the electromotive force which produces it.

(F) The relation between an electric current, and the electromotive force which produces it.

(G) The relation between the amount of free electricity at any point, and the electric displacements in the neighbourhood.

(H) The relation between the increase or diminution of free electricity and the electric currents in the neighbourhood.

There are twenty of these equations in all, involving twenty variable quantities.

19. I then express in terms of these quantities the intrinsic energy of the Electromagnetic Field as depending partly on its magnetic and partly on its electric polarization at every point.

From this I determine the mechanical force acting, 1st, on a movable conductor, carrying an electric current; 2ndly, on a magnetic pole; 3rdly, on an electrified body. [. . .]

20. The general equations are next applied to the case of a magnetic disturbance propagated through a non-conducting field, and it is shewn that the only disturbances which can be so propagated are those which are transverse to the direction of propagation, and that the velocity of propagation is the velocity V found from experiments such as those of Weber,[1] which expresses the number of electrostatic units of electricity which are contained in one electromagnetic unit.

1. Wilhem Weber (1804–1891), German physicist, developed the absolute system of electrical units and based it on his precision measurements. (Ed.)

This velocity is so nearly that of light, that it seems we have strong reason to conclude that light itself (including radiant heat, and other radiations if any) is an electromagnetic disturbance in the form of waves propagated through the electromagnetic field according to electromagnetic laws. [. . .]

96. [. . .] The velocity of light in air, by M. Fizeau's[1] experiments, is

$$V = 314,858,000;$$

according to the more accurate experiments of M. Foucault,[2]

$$V = 298,000,000.$$

The velocity of light in the space surrounding the earth, deduced from the coefficient of aberration and the received value of the radius of the earth's orbit, is

$$V = 308,000,000.$$

97. Hence the velocity of light deduced from experiment agrees sufficiently well with the value of V deduced from the only set of experiments we as yet possess. The value of V was determined by measuring the electromotive force with which a condenser of known capacity was charged, and then discharging the condenser through a galvanometer, so as to measure the quantity of electricity in it in electromagnetic measure. The only use made of light in the experiment was to see the instruments. The value of V found by M. Foucault was obtained by determining the angle through which a revolving mirror turned, while the light reflected from it went and returned along a measured course. No use whatever was made of electricity or magnetism.

The agreement of the results seems to shew that light and magnetism are affections of the same substance, and that light is an electromagnetic disturbance propagated through the field according to electromagnetic laws.

A historical survey of theories on action at a distance

283

If we are ever to discover the laws of nature, we must do so by obtaining the most accurate acquaintance with the facts of nature, and not by dressing up in philosophical language the loose opinions of men who had no knowledge of the facts which throw most light on these laws. And as for those who introduce aetherial, or other media, to account for these actions, without any direct evidence of the existence of such media, or any clear understanding of how the media do their work, and who fill all space three and four times over with aethers

1. Armand Hippolyte Louis Fizeau (1819-1896), French physicist, known in particular for the first terrestrial determination of the velocity of light. (Ed.)
2. Léon Foucault (1819-1868), French physicist, famous for his pendulum experiment to demonstrate the axial rotation of the earth. Carried out precision experiments on the velocity of light. (Ed.)

of different sorts, why the less these men talk about their philosophical scruples about admitting action at a distance the better.

If the progress of science were regulated by Newton's first law of motion, it would be easy to cultivate opinions in advance of the age. We should only have to compare the science of today with that of fifty years ago; and by producing, in the geometrical sense, the line of progress, we should obtain the science of fifty years hence.

The progress of science in Newton's time consisted in getting rid of the celestial machinery with which generations of astronomers had encumbered the heavens, and thus 'sweeping cobwebs off the sky'.

Though the planets had already got rid of their crystal spheres, they were still swimming in the vortices of Descartes. Magnets were surrounded by effluvia, and electrified bodies by atmospheres, the properties of which resemble in no respect those of ordinary effluvia and atmospheres.

When Newton demonstrated that the force which acts on each of the heavenly bodies depends on its relative position with respect to the other bodies, the new theory met with violent opposition from the advanced philosophers of the day, who described the doctrine of gravitation as a return to the exploded method of explaining everything by occult causes, attractive virtues, and the like.

Newton himself, with that wise moderation which is characteristic of all his speculations, answered that he made no pretence of explaining the mechanism by which the heavenly bodies act on each other. To determine the mode in which their mutual action depends on their relative position was a great step in science, and this step Newton asserted that he had made. To explain the process by which this action is effected was a quite distinct step, and this step Newton, in his *Principia*, does not attempt to make.

But so far was Newton from asserting that bodies really do act on one another at a distance, independently of anything between them, that in a letter to Bentley, which has been quoted by Faraday in this place, he says:

'It is inconceivable that inanimate brute matter should, without the mediation of something else, which is not material, operate upon and affect other matter without mutual contact, as it must do if gravitation, in the sense of Epicurus, be essential and inherent in it. . . . That gravity should be innate, inherent, and essential to matter, so that one body can act upon another at a distance, through a vacuum, without the mediation of anything else, by and through which their action and force may be conveyed from one to another, is to me so great an absurdity, that I believe no man who has in philosophical matters a competent faculty of thinking can ever fall into it.'

Accordingly, we find in his *Optical Queries*, and in his letters to Boyle, that Newton had very early made the attempt to account for gravitation by means of the pressure of a medium, and that the reason he did not publish these investigations 'proceeded from hence only, that he found he was not able, from experiment and observation, to give a satisfactory account of this medium, and the manner of its operation in producing the chief phenomena of nature'.

The doctrine of direct action at a distance cannot claim for its author the discoverer of universal gravitation. It was first asserted by Roger Cotes, in his preface to the *Principia*, which he edited during Newton's life. According to Cotes, it is by experience that we learn that all bodies gravitate. We do not learn in any other way that they are extended, movable, or solid. Gravitation, therefore, has as much right to be considered an essential property of matter as extension, mobility, or impenetrability.

And when the Newtonian philosophy gained ground in Europe, it was the opinion of Cotes rather than that of Newton that became most prevalent, till at last Boscovich propounded his theory, that matter is a congeries of mathematical points, each endowed with the power of attracting or repelling the others according to fixed laws. In his world, matter is unextended, and contact is impossible. He did not forget, however, to endow his mathematical points with inertia. In this some of the modern representatives of his school have thought that he 'had not quite got so far as the strict modern view of "matter" as being but an expression for modes or manifestations of "force"'.

But if we leave out of account for the present the development of the ideas of science, and confine our attention to the extension of its boundaries, we shall see that it was most essential that Newton's method should be extended to every branch of science to which it was applicable—that we should investigate the forces with which bodies act on each other in the first place, before attempting to explain *how* that force is transmitted. No men could be better fitted to apply themselves exclusively to the first part of the problem, than those who considered the second part quite unnecessary.

Accordingly Cavendish, Coulomb, and Poisson, the founders of the exact sciences of electricity and magnetism, paid no regard to those old notions of 'magnetic effluvia' and 'electric atmospheres', which had been put forth in the previous century, but turned their undivided attention to the determination of the law of force, according to which electrified and magnetized bodies attract or repel each other. In this way the true laws of these actions were discovered and this was done by men who never doubted that the action took place at a distance, without the intervention of any medium, and who would have regarded the discovery of such a medium as complicating rather than as explaining the undoubted phenomena of attraction.

We have now arrived at the great discovery by Ørsted of the connection between electricity and magnetism. Ørsted found that an electric current acts on a magnetic pole, but that it neither attracts it nor repels it, but causes it to move round the current. He expressed this by saying that 'the electric conflict acts in a revolving manner'.

The most obvious deduction from this new fact was that the action of the current on the magnet is not a push-and-pull force, but a rotatory force, and accordingly many minds were set a-speculating on vortices and streams of aether whirling round the current.

But Ampère, by a combination of mathematical skill and experimental ingenuity, first proved that two electric currents act on one another, and then analysed this action into the resultant of a system of push-and-pull forces between the elementary parts of these currents.

The formula of Ampère, however, is of extreme complexity, as compared with Newton's law of gravitation, and many attempts have been made to resolve it into something of greater apparent simplicity.

I have no wish to lead you into a discussion of any of these attempts to improve a mathematical formula. Let us turn to the independent method of investigation employed by Faraday in those researches in electricity and magnetism which have made this Institution one of the most venerable shrines of science.

No man ever more conscientiously and systematically laboured to improve all his powers of mind than did Faraday from the very beginning of his scientific career. But whereas the general course of scientific method then consisted in the application of the ideas of mathematics and astronomy to each new investigation in turn, Faraday seems to have had no opportunity of acquiring a technical knowledge of mathematics, and his knowledge of astronomy was mainly derived from books.

Hence, though he had a profound respect for the great discovery of Newton, he regarded the attraction of gravitation as a sort of sacred mystery, which, as he was not an astronomer, he had no right to gainsay or to doubt, his duty being to believe it in the exact form in which it was delivered to him. Such a dead faith was not likely to lead him to explain new phenomena by means of direct attractions.

Besides this, the treatises of Poisson and Ampère are of so technical a form, that to derive any assistance from them the student must have been thoroughly trained in mathematics, and it is very doubtful if such a training can be begun with advantage in mature years.

Thus Faraday, with his penetrating intellect, his devotion to science, and his opportunities for experiments, was debarred from following the course of thought which had led to the achievements of the French philosophers, and was obliged to explain the phenomena to himself by means of a symbolism which he could understand, instead of adopting what had hitherto been the only tongue of the learned.

This new symbolism consisted of those lines of force extending themselves in every direction from electrified and magnetic bodies, which Faraday in his mind's eye saw as distinctly as the solid bodies from which they emanated.

The idea of lines of forces and their exhibition by means of iron filings was nothing new. They had been observed repeatedly, and investigated mathematically as an interesting curiosity of science. But let us hear Faraday himself, as he introduces to his reader the method which in his hands became so powerful.

'It would be a voluntary and unnecessary abandonment of most valuable aid if an experimentalist, who chooses to consider magnetic power as represented by lines of magnetic force, were to deny himself the use of iron filings. By their employment he may make many conditions of the power, even in complicated cases, visible to the eye at once, may trace the varying direction of the lines of force and determine the relative polarity, may observe in which direction the power is increasing or diminishing, and in complex systems may determine the neutral points, or places where there is neither polarity nor power, even when they occur in the midst of powerful magnets. By their use probable results may be seen at once, and many a valuable suggestion gained for future leading experiments.'

284

Experiment on lines of force

In this experiment each filing becomes a little magnet. The poles of opposite names belonging to different filings attract each other and stick together, and more filings attach themselves to the exposed poles, that is, to the end of the row of filings. In this way the filings, instead of forming a confused system of dots over the paper, draw together, filing to filing, till long fibres of filings are formed, which indicate by their direction the lines of force in every part of the field.

The mathematicians saw in this experiment nothing but a method of exhibiting at one view the direction in different places of the resultant of two forces, one directed to each pole of the magnet; a somewhat complicated result of the simple law of force.

But Faraday, by a series of steps as remarkable for their geometrical definiteness as for their speculative ingenuity, imparted to his conception of these lines of force a clearness and precision far in advance of that with which the mathematicians could then invest their own formulae.

In the first place, Faraday's lines of force are not to be considered merely as individuals, but as forming a system, drawn in space in a definite manner so that the number of the lines which pass through an area, say of one square inch, indicates the intensity of the force acting through the area. Thus the lines of force become definite in number. The strength of a magnetic pole is measured by the number of lines which proceed from it; the electrotonic state of a circuit is measured by the number of lines which pass through it.

In the second place, each individual line has a continuous existence in space and time. When a piece of steel becomes a magnet, or when an electric current begins to flow, the lines of force do not start into existence each in its own place, but as the strength increases new lines are developed within the magnet or current, and gradually grow outwards, so that the whole system expands from within, like Newton's rings in our former experiment. Thus every line of force preserves its identity during the whole course of its existence, though its shape and size may be altered to any extent.

I have no time to describe the methods by which every question relating to the forces acting on magnets or on currents, or to the induction of currents in conducting circuits, may be solved by the consideration of Faraday's lines of force. In this place they can never be forgotten. By means of this new symbolism, Faraday defined with mathematical precision the whole theory of electromagnetism, in language free from mathematical technicalities, and applicable to the most complicated as well as the simplest cases. But Faraday did not stop here. He went on from the conception of geometrical lines of force to that of physical lines of force. He observed that the motion which the magnetic or electric force tends to produce is invariably such as to shorten the lines of force and to allow them to spread out laterally from each other. He thus perceived in the medium a state of stress, consisting of a tension, like that of a rope, in the direction of the lines of force, combined with a pressure in all directions at right angles to them.

This is quite a new conception of action at a distance, reducing it to a phenomenon of the same kind as that action at a distance which is exerted by means of the tension of ropes and the pressure of rods. When the muscles of our bodies are excited by that stimulus which we are able in some unknown way to apply to them, the fibres tend to shorten themselves and at the same time to expand laterally. A state of stress is produced in the muscle, and the limb moves. This explanation of muscular action is by no means complete. It gives no account of the cause of the excitement of the state of stress, nor does it even investigate those forces of cohesion which enable the muscles to support this stress. Nevertheless, the simple fact, that it substitutes a kind of action which extends continuously along a material substance for one of which we know only a cause and an effect at a distance from each other, induces us to accept it as a real addition to our knowledge of animal mechanics.

For similar reasons we may regard Faraday's conception of a state of stress in the electromagnetic field as a method of explaining action at a distance by means of the continuous transmission of force, even though we do not know how the state of stress is produced.

But one of Faraday's most pregnant discoveries, that of the magnetic rotation of polarized light, enables us to proceed a step farther. The phenomenon, when analysed into its simplest elements, may be described thus: —Of two circularly polarized rays of light, precisely similar in configuration, but rotating in opposite directions, that ray is propagated with the greater velocity which rotates in the same direction as the electricity of the magnetizing current.

It follows from this, as Sir W. Thomson has shewn by strict dynamical reasoning, that the medium when under the action of magnetic force must be in a state of rotation—that is to say, that small portions of the medium, which we may call molecular vortices, are rotating, each on its own axis, the direction of this axis being that of the magnetic force.

Here, then, we have an explanation of the tendency of the lines of magnetic force to spread out laterally and to shorten themselves. It arises from the centrifugal force of the molecular vortices.

The mode in which electromotive force acts in starting and stopping the vortices is more abstruse, though it is of course consistent with dynamical principles.

We have thus found that there are several different kinds of work to be done by the electromagnetic medium if it exists. We have also seen that magnetism has an intimate relation to light, and we know that there is a theory of light which supposes it to consist of the vibrations of a medium. How is this luminiferous medium related to our electromagnetic medium?

It fortunately happens that electromagnetic measurements have been made from which we can calculate by dynamical principles the velocity of propagation of small magnetic disturbances in the supposed electromagnetic medium.

This velocity is very great, from 288 to 314 millions of metres per second, according to different experiments. Now the velocity of light, according to Foucault's experiments, is 298 millions of metres per second. In fact, the different determinations of either velocity differ from each other more than the estimated velocity of light does from the estimated velocity of propagation of small electromagnetic disturbance. But if the luminiferous and the electromagnetic media occupy the same place, and transmit disturbances with the same velocity, what reason have we to distinguish the one from the other? By considering them as the same, we avoid at least the reproach of filling space twice over with different kinds of aether.

Besides this, the only kind of electromagnetic disturbances which can be propagated through a non-conducting medium is a disturbance transverse to the direction of propagation, agreeing in this respect with what we know of that disturbance which we call light. Hence, for all we know, light also may be an electromagnetic disturbance in a non-conducting medium. If we admit this, the electromagnetic theory of light will agree in every respect with the undulatory theory, and the work of Thomas Young[1] and Fresnel will be established on a firmer basis than ever, when joined with that of Cavendish[2] and Coulomb by the key-stone of the combined sciences of light and electricity—Faraday's great discovery of the electromagnetic rotation of light.

The vast interplanetary and interstellar regions will no longer be regarded as waste places in the universe, which the Creator has not seen fit to fill with the symbols of the manifold order of His kingdom. We shall find them to be already full of this wonderful medium; so full, that no human power can remove

1. Thomas Young (1773–1829), English physicist, developed the wave theory of light and thus explained the phenomena of interference. (Ed.)
2. Henry Cavendish (1731–1810), English physicist and chemist. His many important discoveries include the verification of the inverse square law in electrostatics. (Ed.)

it from the smallest portion of space, or produce the slightest flaw in its infinite continuity. It extends unbroken, from star to star; and when a molecule of hydrogen vibrates in the dog-star, the medium receives the impulses of these vibrations; and after carrying them in its immense bosom for three years, delivers them in due course, regular order, and full tale into the spectroscope of Mr Huggins, at Tulse Hill.

But the medium has other functions and operations besides bearing light from man to man, and from world to world, and giving evidence of the absolute unity of the metric system of the universe. Its minute parts may have rotatory as well as vibratory motions, and the axes of rotation form those lines of magnetic force which extend in unbroken continuity into regions which no eye has seen, and which, by their action on our magnets, are telling us in language not yet interpreted, what is going on in the hidden underworld from minute to minute and from century to century.

And these lines must not be regarded as mere mathematical abstractions. They are the directions in which the medium is exerting a tension like that of a rope, or rather, like that of our own muscles. The tension of the medium in the direction of the earth's magnetic force is in this country one grain weight on eight square feet. In some of Dr Joule's experiments, the medium has exerted a tension of 200 lb weight per square inch.

But the medium, in virtue of the very same elasticity by which it is able to transmit the undulations of light, is also able to act as a spring. When properly wound up, it exerts a tension, different from the magnetic tension, by which it draws oppositely electrified bodies together, produces effects through the length of telegraph wires, and when of sufficient intensity, leads to the rupture and explosion called lightning.

These are some of the already discovered properties of that which has often been called vacuum, or nothing at all. They enable us to resolve several kinds of action at a distance into actions between contiguous parts of a continuous substance. Whether this resolution is of the nature of explication or complication, I must leave to the metaphysicians.

The application of statistics to molecules

285

The study of molecules has developed a method of its own, and it has also opened up new views of nature. [. . .]

As long as we have to deal with only two molecules, and have all the data given us, we can calculate the result of their encounter; but when we have to deal with millions of molecules, each of which has millions of encounters in a second, the complexity of the problem seems to shut out all hope of a legitimate solution.

The modern atomists have therefore adopted a method which is, I believe,

444

new in the department of mathematical physics, though it has long been in use in the section of Statistics. When the working members of Section F get hold of a report of the Census, or any other document containing the numerical data of Economic and Social Science, they begin by distributing the whole population into groups, according to age, income-tax, education, religious belief, or criminal convictions. The number of individuals is far too great to allow of their tracing the history of each separately, so that, in order to reduce their labour within human limits, they concentrate their attention on a small number of artificial groups. The varying number of individuals in each group, and not the varying state of each individual, is the primary datum from which they work.

This, of course, is not the only method of studying human nature. We may observe the conduct of individual men and compare it with that conduct which their previous character and their present circumstances, according to the best existing theory, would lead us to expect. Those who practise this method endeavour to improve their knowledge of the elements of human nature in much the same way as an astronomer corrects the elements of a planet by comparing its actual position with that deduced from the received elements. The study of human nature by parents and schoolmasters, by historians and statesmen, is therefore to be distinguished from that carried on by registrars and tabulators, and by those statesmen who put their faith in figures. The one may be called the historical, and the other the statistical method.

The equations of dynamics completely express the laws of the historical method as applied to matter, but the application of these equations implies a perfect knowledge of all the data. But the smallest portion of matter which we can subject to experiment consists of millions of molecules, not one of which ever becomes individually sensible to us. We cannot, therefore, ascertain the actual motion of any one of these molecules; so that we are obliged to abandon the strict historical method, and to adopt the statistical method of dealing with large groups of molecules.

The data of the statistical method as applied to molecular science are the sums of large numbers of molecular quantities. In studying the relations between quantities of this kind, we meet with a new kind of regularity, the regularity of averages, which we can depend upon quite sufficiently for all practical purposes, but which can make no claim to that character of absolute precision which belongs to the laws of abstract dynamics.

Thus molecular science teaches us that our experiments can never give us anything more than statistical information, and that no law deduced from them can pretend to absolute precision.

Maxwell's 'demon'

286

One of the best established facts in thermodynamics is that it is impossible in a system enclosed in an envelope which permits neither change of volume nor passage of heat, and in which both the temperature and the pressure are everywhere the same, to produce any inequality of temperature or of pressure without the expenditure of work. This is the second law of thermodynamics, and it is undoubtedly true so long as we can deal with bodies only in the mass and have no power of perceiving or handling the separate molecules of which they are made up. But if we conceive a being whose faculties are so sharpened that he can follow every molecule in its course, such a being, whose attributes are still as essentially finite as our own, would be able to do what is at present impossible to us. For we have seen that the molecules in a vessel full of air at uniform temperature are moving with velocities by no means uniform though the mean velocity of any great number of them, arbitrarily selected, is almost exactly uniform. Now let us suppose that such a vessel is divided into two portions A and B, by a division in which there is a small hole, and that a being, who can see the individual molecules, opens and closes this hole, so as to allow only the swifter molecules to pass from A to B, and only the slower ones to pass from B to A. He will thus, without expenditure of work, raise the temperature of B and lower that of A, in contradiction to the second law of thermodynamics.

GUSTAV KIRCHHOFF and ROBERT BUNSEN

287

Chemical analysis by observations of the spectrum

It is a well-known characteristic of some substances that when placed in a flame certain bright lines appear in their spectrum. These lines open a method of qualitative analysis, extending the field of chemical reactions, and also leading to the solution of problems which have previously been considered inaccessible. We shall confine ourselves to the task of developing these methods for the alkali metals and alkaline earths and showing its usefulness by several examples.

The above-mentioned lines become more conspicuous the higher the temperature of the flame and the smaller its luminous power. The gas lamp described by one of us elsewhere has a high-temperature flame and very low luminous power; it is thus preferentially suitable for experiments with substances exhibiting those strange bright lines. [. . .]

Those investigators who, due to repeated observations, are well acquainted with individual spectra need not even measure exactly the individual lines; the colour, relative position, shape and shading and profiles of the lines are characteristic features, amply sufficient for orientation in the spectrum even for the untrained eye. These features of the lines are analogous to those characteristically different looking precipitates which are used as means of identification.

Exactly as we should characterize the precipitate as gelatinous, powdery, cheesy, granular or crystalline, so the spectral lines exhibit individual features like being sharply defined, fading away on one or both sides, gradually or unequally or in their relative width. Furthermore, exactly as only those precipitates serve for identification purposes which still come out after the precipitating substances have been considerably diluted, so only those lines will be considered in spectral analysis that appear even for the smallest amounts of test substance at not too high temperatures. As far as these features are considered, the two methods are of equal value. On the other hand, spectral analysis has a peculiarity which makes it preferable to any other analytical method; this has to do with the colouring which serves as one of the means of identification: most of the precipitates, which serve as identification means, are white and only a few show some colouring. The colouring of these last is not stable and varies with the thickness of the precipitate layer. The smallest quantity of impurity suffices to change the colouring and thus make the substance unrecognizable. As a result, slight colour differences can on no account serve as a basis for identification. In spectral analysis, on the other hand, the coloured lines stay unchanged and are not influenced by the presence of impurities; the position of the lines in the spectrum is a chemical characteristic of such fundamental nature as the atomic weight of the substance and can be measured with astronomical exactitude. Of special significance is the fact that spectral analysis removes any former limits on the characteristics which serve as chemical identification marks. Spectral analysis promises valuable information about the distribution and arrangements of geological formations. Even the few experiments reported here reveal the unexpected fact that not only potassium and sodium, but also lithium and strontium, albeit in small quantities, belong to the most widely spread elements on the earth.

Of no less importance should spectral analysis prove to be in the discovery of new elements. If there are substances that are so rare in nature that our old means of analysis are of no value in their discovery, it is reasonable to hope that these will be discovered by their flame-spectra even if they are available in such small quantities that would elude any other chemical means. That, indeed, such previously unknown elements exist, we have had ample opportunity to convince ourselves. Relying on the unquestionable results of spectral analysis we feel fully confident that in addition to potassium, sodium and lithium, there exists a fourth metal of the alkali-group; this element gives a simple and characteristic spectrum, very much like that of lithium. With our spectral apparatus it exhibits only two lines: a weak blue that almost coincides with the strontium-δ line, and another blue, lying not much further to the violet end of the spectrum, and it rivals the lithium line in intensity and sharpness.

Not only is spectral analysis, as we believe to have shown, a means of wonderful simplicity to discover the smallest traces of an element on the earth, but it opens up a new field for chemical research which was up till now hidden from us: this field of research exceeds the confines of the earth and indeed of the whole solar system. With this method, if we want to analyse an incandescent gas, it is enough to see it; this being so it should be applicable to the analysis of the solar atmosphere and to that of the brighter fixed stars. In these cases, however, a modification is necessary: one of us has theoretically proved in the essay 'The relationship between the emissivity and absorptivity of heat and light of bodies' that the spectrum of an incandescent gas will be reversed, i.e. if behind the line-emitting substance there is a strong-intensity light-source with a continuous spectrum, the bright lines of the element will turn into dark ones. This implies that the spectrum of the sun with its dark lines is nothing but a reversal of the spectrum of the sun's atmosphere. In this case a chemical analysis of the sun's atmosphere means merely seeking out those substances which, when put in a flame, emit bright lines that coincide with the dark lines in the solar spectrum.

DMITRI IVANOVICH MENDELEEV

288

The periodic law of chemical elements

When studying the properties of the elements with the view of drawing conclusions and forming chemical predictions, we should give the same consideration to the properties of the group to which the given element belongs as to their individual characteristics. Only on the ground of such comparative studies and on basis of measurable properties can we draw general conclusions on the properties of the elements.

Such a property is atomic weight, and it will serve as such for a long time to come. Our conception of atomic weight (as the smallest part of an element which occurs in a molecule of one of its compounds) has gained an unshakable firmness mainly through applying Avogadro's and Ampère's law and through the labours of Laurent, Gerhardt, Regnault, Rose and Cannizzaro; we can boldly assert that the concept of atomic weight will stay unchanged through all the changes in the theoretical framework of the chemists.

The very name of this concept (atomic weight) presupposes the hypothesis of atomic structure of matter; but we are not dealing here with the problem of designations but with concepts which have to be expressed. Mechanical-physical considerations can enlarge our chemical knowledge only through a comparative ordering of the elements according to their atomic weight. Thus it seems to me that it is a most profitable procedure to investigate the dependence of the properties of the elements on their atomic weights. In this I see one of the major tasks of future chemistry. As for theoretical understanding, this seems to me of as much importance as studying the relationships between isomers. I shall endeavour in the following to demonstrate this dependence of the properties of the elements on their atomic weights, and especially of their ability to form various compounds. [. . .]

Since 1868, when my work, *The Foundations of Chemistry*, appeared for the first time, I have been engaged in trying to solve this problem. Here I wish to present the results of this investigation. [. . .]

I shall call *periodic law* the mutual relations between the properties of the elements and their atomic weights, which are applicable to all elements; this law needs to be developed further. The relations mentioned here have the form of a periodic function.

(1) THE ESSENCE OF THE PERIODIC LAW

It has been observed a long time ago that the elements with high atomic number are analogues to elements with much smaller atomic number. Claus has shown that *Os, Ir, Pt* with an approximate atomic weight of 195 possess properties similar to those of *Ru, Rh, Pd* with a much smaller atomic weight (of approximately 105). Marignac has pointed out the analogy between $Ta=182$ and $W=184$ on the one hand, and $Nb=94$ and $Mo=96$ on the other. *Au* and *Hg* correspond to the lighter analogues *Ag* and *Cd* as well as to the even lighter *Cu* and *Zn*. Caesium and Baryum are analogues of Potassium and Calcium, and many similar comparisons lead to the expectation that all the elements can be arranged according to their atomic weights. In this process one suddenly realizes the surprising simplicity of the mutual relations between the elements. As a proof we shall for example arrange all the light elements with atomic weights 7 to 36 in an arithmetic series according to their atomic weight:

$$Li=7; \ Be=9,4; \ B=11; \ C=12; \ N=14; \ O=16; \ F=19$$
$$Na=23; \ Mg=24; \ Al=27,3; \ Si=28; \ P=31; \ S=32; \ Cl=35,5$$

As it is seen the character of the elements changes regularly and gradually with growing atomic weight, and it does so periodically, i.e. in both rows it changes in the same way, so that the corresponding members are analogues: *Na* and *Li, Mg* and *Be, C* and *Si, O* and *S* and so on. Thus the corresponding members of both rows form the same kind of compounds, and they possess the same valency (as it is usually called).

OBSERVATIONS TO TABLE I [*see page 450*]

For the sake of brevity we have rounded off the atomic weight numbers; for in most of the cases there is no certainty whatever as to the correctness of the first decimals and not even of the units. A question mark in front of the chemical symbol of an element means that no fixed place has been allocated to it as yet, due to insufficient research. A question mark after the atomic weight means that the data about the number are insufficient, i.e. the equivalent of the element could not have been ascertained so far. Several atomic weights have been changed in the table, according to the law of periodicity (See Chapter 5): thus Tellurium stands as 125 as required by the law of periodicity and not as 128 as given by Berzelius *et al.*

TABLE I

Typical elements							
			K =39	*Rb* = 85	*Cs*=133	—	—
			Ca =40	*Sr* = 87	*Ba*=137	—	—
			—	?*Yt* = 88?	?*Di*=138?	*Er* =178?	—
			Ti =48?	*Zr* = 90	*Ce*=140?	?*La* =180?	*Th*=231
			V =51	*Nb* = 94	—	*Ta*=182	—
			Cr = 52	*Mo*= 96	—	*W* =184	*U* =240
			Mn=55	—	—	—	—
			Fe =56	*Ru* =104	—	*Os* =195?	—
			Co =59	*Rh* =104	—	*Ir* =197	—
			Ni =59	*Pd* =106	—	*Pt* =198?	—
H=1	*Li*= 7	*Na* =23	*Cu* =63	*Ag* =108	—	*Au*=199?	—
	Be= 9.4	*Mg*=24	*Zn* =65	*Cd* =112	—	*Hg*=200	—
	B =11	*Al* =27.3	—	*In* =113	—	*Tl* =204	—
	C =12	*Si* =28	—	*Sn* =118	—	*Pb* =207	—
	N=14	*P* =31	*As* =75	*Sb* =122	—	*Bi* =208	—
	O=16	*S* =32	*Se* =78	*Te* =125?	—	—	—
	F =19	*Cl* =35.5	*Br* =80	*J* =127	—	—	—

[...]

On the application of the periodic law for the determination of the properties of not-yet-discovered elements

These developments make it clear that the law of periodicity enables us to draw conclusions about unknown properties of elements, the analogous atoms of which are known. Furthermore, we find that many elements are missing in tables I and II. These tables show clearly the periodic relationship between the elements and we would expect to find the missing ones in specified places in them. I will now describe the properties of some of these expected elements in order to show the correctness of the law of periodicity, although the actual proof lies in the future. Moreover, prediction of properties of unknown elements makes their detection possible, predicting the possible reactions of their compounds. [...]

On the application of the periodic law to the correction of the values of atomic weights

Everybody knows the fate of Prout's hypothesis that the atomic weights of the elements are integer multiples of the weight of hydrogen. There could have been no doubt whatsoever that this hypothesis overstated the facts, after exact investigations had shown that there are atomic weights which contain fractions, and after Stas[1] has shown that there are no rational fractions among them, despite Marignac's[2] splendid critical remarks in support of the hypothesis. It seems to me

1. Jean Servais Stas (1813–1891), Belgian chemist, known for his determinations of atomic weights which contradicted Prout's hypothesis. (Ed.)
2. Gallissard de Marignac (1817–1894), Swiss chemist, supported Prout's hypothesis in his polemics with Stas. (Ed.)

that there are no sufficient reasons in favour of accepting this hypothesis. Even granted that the matter constituting the elements be completely homogeneous, there is no reason to suppose that n parts of the weight of an element, or n atoms, will yield the same n parts after transmutation into an atom of another element, i.e. that the atom of the second element will be n times heavier than the first. One can regard the law of conservation of weight as a special case of the law of conservation of force or of movement. Surely weight depends on a special kind of movement of matter, and there is no reason to deny the possibility of a transmutation of these movements into chemical energy or some other form of movement during the formation of elementary atoms. Up to now we have seen that there exists an intimate and even historical connection between two phenomena which are observed in elements, namely the constancy of their atomic weight, and the impossibility of their decomposition. Thus in case a known element would be decomposed or a new one would be formed, these phenomena could well be accompanied by a decrease or increase in weight. In this way one also could explain to a certain extent the difference in chemical energy of various elements. In expressing this idea the only thing I have in mind is that there might be some possibility to reconcile both the view of the compound nature of the elements, tacitly held by many chemists, and the rejection of Prout's hypothesis. [. . .]

All that has been said above did not aim at erecting a closed system, for I know there are still several corrections and additions necessary. Yet, it seems to me that I can assume that if the chemists follow the reliable directions described above, they will reach their aim much sooner. Daring hypotheses have something attractive about them, and sometimes they even constitute a temporary progress; but more often these hypotheses lead to false conclusions and they sink into oblivion, especially if they are not supported by scientific laws. This is so because the establishment of such laws must also imply the subsequent task of the scientific endeavour.

LUDWIG BOLTZMANN

289

Application of the second law to the universe as a whole

Is the apparent irreversibility of all known natural processes consistent with the idea that all natural events are possible without restriction? Is the apparent unidirectionality of time consistent with the infinite extent or cyclic nature of time? He who tries to answer these questions in the affirmative sense must use as a model of the world a system whose changes with time are determined by equations in which the positive and negative directions of time are equivalent, and by means of which the appearance of irreversibility over long periods of time is explicable by some special assumption. But this is precisely what happens in the atomic view of the world.

One can think of the world as a mechanical system of an enormously large number of constituents, and of an immensely long period of time, so that the dimensions of that part containing our own 'fixed stars' are minute compared to the extension of the universe; and times that we call eons are likewise minute compared to such a period. Then in the universe, which is in thermal equilibrium throughout and therefore dead, there will occur here and there relatively small regions of the same size as our galaxy (we call them single worlds) which, during the relatively short time of eons, fluctuate noticeably from thermal equilibrium, and indeed the state probability in such cases will be equally likely to increase or decrease. For the universe, the two directions of time are indistinguishable, just as in space there is no up or down. However, just as at a particular place on the earth's surface we call 'down' the direction towards the centre of the earth, so will a living being in a particular time interval of such a single world distinguish the direction of time towards the less probable state from the opposite direction (the former towards the past, the latter towards the future). By virtue of this terminology, such small isolated regions of the universe will always find themselves 'initially' in an improbable state. This method seems to me to be the only way in which one can understand the second law—the heat death of each single world—without a unidirectional change of the entire universe from a definite initial state to a final state.

Obviously no one would consider such speculations as important discoveries or even—as did the ancient philosophers—as the highest purpose of science. However, it is doubtful that one should despise them as completely idle. Who knows whether they may not broaden the horizon of our circle of ideas, and by stimulating thought, advance the understanding of the facts of experience?

That in nature the transition from a probable to an improbable state does not take place as often as the converse, can be explained by assuming a very improbable initial state of the entire universe surrounding us, in consequence of which an arbitrary system of interacting bodies will in general find itself initially in an improbable state. However, one may object that here and there a transition from a probable to an improbable state must occur and occasionally be observed. To this the cosmological considerations just presented give an answer. From the numerical data on the inconceivably great rareness of transition from a probable to a less probable state in observable dimensions during an observable time, we see that such a process within what we have called an individual world—in particular, our individual world—is so unlikely that its observability is excluded.

In the entire universe, the aggregate of all individual worlds, there will, however, in fact occur processes going in the opposite direction. But the beings who observe such processes will simply reckon time as proceeding from the less probable to the more probable states, and it will never be discovered whether they reckon time differently from us, since they are separated from us by eons of time and spatial distances $10^{10^{10}}$ times the distance of Sirius—and moreover their language has no relation to ours.

Very well, you may smile at this; but you must admit that the model of the world developed here is at least a possible one, free of inner contradiction, and also a useful one, since it provides us with many new viewpoints. It also gives an incentive, not only to speculation, but also to experiments (for example on the limit of divisibility, the size of the sphere of action, and the resulting deviations from the equations of hydrodynamics, diffusion, and heat conduction) which are not stimulated by any other theory.

290

Statistical mechanics

At present it appears that we cannot define infinity in any other way than as the limit of ever-increasing finite quantities; at least, until now, no one has been able to fashion any other concept of infinity that can be grasped. Thus if we wish to arrive at a picture of the continuum which can be expressed in words, then first we must necessarily imagine a large finite number of particles which are endowed with certain qualities, and then investigate the behaviour of the sum total of such particles. Now certain properties of this sum total can approach a definite limit when the number of particles is ever increased and their size is ever decreased. It can then be maintained about these properties that they are properties of a continuum, and this is, in my opinion, the only definition, free of contradictions, of a continuum endowed with definite properties.

Therefore, the question of whether matter is composed of atoms, or whether it is continuous, is reduced to the question of whether the observed properties of matter are most accurately represented either by the properties of the particles, when the number of such particles is very large but finite, or by the limits of these properties, when the number of particles is ever increasing. To be sure, this does not answer the ancient philosophical question, but we are at least saved from the tendency to resolve it in an ambiguous and unprofitable way. It is the same thought process by which we first investigate the properties of a finite sum and then let the number of terms increase enormously; and when, as it often happens, the differential equations themselves are made the starting-point of a mathematico-physical theory, these equations are nothing else than the abbreviated expression of exactly the same thought process in algebraic symbols.

In any case we cannot think of the terms of this sum total, which we chose as the image of material bodies, as being always absolutely at rest, because otherwise there could be no motion at all, nor can we think of them as being relatively at rest in one and the same body, because otherwise we could not give an account of the nature of fluids. Furthermore, no attempt has yet been made to think of them as other than subject to the general laws of mechanics. Therefore we wish to choose for the explanation of Nature the sum total of an extraordinarily large number of very small basic individual units which are in continuous motion and are subject to the laws of mechanics. Against this an objection has been raised, which we can suitably make the starting-point for the considerations which constitute the final goal of this lecture. The basic equations of mechanics do not alter their form in the least if only the sign of time is reversed in them. All purely mechanical events can therefore take place exactly the same in one sense as in the other, that is, in the sense of increasing time exactly in the same way as in the sense of decreasing time. But we notice even in daily life that the future and the past do in no way coincide as completely as do the directions right and left; that rather they are clearly different from each other.

This is expressed more precisely by the so-called second law of the mechanical theory of heat. This law states that if any system of bodies is isolated and not influenced by other bodies, then we can always predict in which direction any change of state of this system takes place. That is to say, that we can determine a certain function of the state of all bodies, i.e. the entropy, which is of such a nature that every change of this state can only proceed in the sense involving an increase of this function, so that with increasing time it can only increase. This law, to be sure, is only arrived at by means of abstraction, as the Galilean Principle; for it is impossible to remove a system of bodies completely from the influence of all other bodies. But since, combined with the other laws, it has always yielded correct results until now, we maintain that it is correct, exactly as is the case with the Galilean Principle.

It follows from this law that every isolated system of bodies must always tend towards a certain final state for which the entropy is a maximum. It may seem surprising to find as the consequence of this law that the whole universe necessarily moves to a final state in which all events cease, but this consequence is self-evident if the world is considered finite and subject to the second law. If the world is considered infinite, then the conceptual difficulties discussed arise again if one does not think of infinity as a mere limit.

Since absolutely nothing analogous to the second law exists in the differential equations of mechanics themselves, this law can only be represented mechanically by assumptions about the initial conditions. In order to find the suitable assumptions, we have to bear in mind that we must presume, for the sake of explanation of apparently continuous bodies, that of each species of atoms, or, more generally, mechanical individuals, an exceedingly great number must be present in the most diverse initial conditions. In order to treat this assumption mathematically, a special science has been invented, whose aim is not to find the motion of a single mechanical system, but rather to find the properties of a complex of very many mechanical systems, which start from the great diversity of initial conditions. The credit for having systematized this science, having presented it in a comprehensive treatise, and having given it a characteristic name belongs to one of the great American scientists, perhaps the greatest, as far as purely abstract thinking and purely theoretical research is concerned—

Willard Gibbs,[1] the recently deceased professor of Yale College. He called this science statistical mechanics. It is divided into two parts. The first part investigates the conditions under which the externally apparent properties of a complex of very many mechanical individuals do not change at all, in spite of vigorous motion of the individuals; this first part I should like to call statistical statics. The second part calculates the gradual changes of these externally visible properties, when the first condition is not met; this might be called statistical dynamics. The wide vistas opened to us when we think of the application of this science not only to mechanical bodies but also to the statistics of living creatures, of human society, sociology, etc., can only be indicated here briefly.

An exposition of the details of this science would only be possible with the help of mathematical formulae, in a series of lectures. Besides mathematical difficulties the theory is not free from difficulties in principle, because it is based on the theory of probability. Now this theory is, indeed, precisely as exact as all other branches of mathematics, as soon as the concept of equal probability is given. But this concept, being a fundamental concept, cannot in turn be derived, but must rather be considered as something given. This is like the formulae of the method of least squares which too follow rigorously only under certain assumptions about the equal probability of elementary errors. This difficulty in principle explains why even the most simple result of statistical statics, the proof of Maxwell's distribution law of the velocities of gas molecules, is still being contested.

The propositions of statistical mechanics are strict consequences of the assumptions made and they will always remain true, as do all well-founded mathematical propositions. However, their application to events in nature is the prototype of a physical hypothesis. For if we start from the simplest basic assumptions about equal probability, we find laws for the relationships of aggregates of very many individuals, completely analogous to those which experience shows for the behaviour of the material world. Translatory or rotatory visible motion must be gradually transformed into invisible motion of the smallest particles: that is, into heat motion. As Helmholtz says characteristically: ordered motion is progressively transformed into disordered motion; the mixture of several substances as well as of several temperatures, that is, of regions of more or less vigorous molecular motion, must become ever more uniform. That this mixture was not yet complete at the beginning, that the world started rather from a very improbable initial condition, this can be counted among the fundamental hypotheses of the whole theory, and it can be said that the reason for this is just as little known as the reason why the world in general is precisely so and not otherwise. But one can adopt still another point of view. States where the components of a mixture are largely separated,

1. Josiah Willard Gibbs (1839–1903), American mathematician and physicist, one of the founders of statistical mechanics. (Ed.)

respectively those of large temperature differences, are, according to the theory, not absolutely impossible, but rather only extremely improbable, although to an incomprehensibly high degree.

Thus if we only assume a world sufficiently large, then, according to the very laws of probability, there will occur regions of the dimensions of the heavens of the fixed stars with a completely improbable distribution of states. The temporal course of both their formation and dissolution will be one-sided. If therefore there are intelligent beings in such a region, they must gain exactly the same impression of time as we have, although the temporal course for the universe as a whole will not be one-sided. The theory developed here, however, boldly transcends experience, but it has precisely that property which all such theories should have because it shows us the facts of experience in a completely new light and stimulates further thinking and research. For in contrast to the first law, the second law appears as a mere probability law, as Gibbs already stated in the seventies of the last century.

I have not avoided philosophical questions here, in the firm hope that concerted cooperation between philosophy and science will add new nourishment to each of these branches of learning; more than this—that only thus a truly consistent system of thought can be reached. If Schiller said to the scientists and the philosophers of his age, 'Enmity be between you; it is still too early for an alliance', I do not disagree with him; I just believe that now the time for the alliance has come.

ALBERT ABRAHAM MICHELSON

291

The ether

The velocity of light is so enormously greater than anything with which we are accustomed to deal that the mind has some little difficulty in grasping it. A bullet travels at the rate of approximately half a mile a second. Sound, in a steel wire, travels at the rate of three miles a second. From this—if we agree to except the velocities of the heavenly bodies—there is no intermediate step to the velocity of light, which is about 186,000 miles a second. We can, perhaps, give a better idea of this velocity by saying that light will travel around the world seven times between two ticks of a clock.

Now, the velocity of wave propagation can be seen, without the aid of any mathematical analysis, to depend on the elasticity of the medium and its density; for we can see that if a medium is highly elastic the disturbance would be propagated at a great speed. Also, if the medium is dense the propagation would be slower than if it were rare. It can easily be shown that if the elasticity were represented by E, and the density by D, the velocity would be represented by the square root of E divided by D. So that, if the density of the medium which propagates light waves were as great as the density of steel, the elasticity, since the velocity of light is some 60,000 times as great as that of the propagation of sound in a steel wire, must be 60,000 squared times as great as the elasticity of steel. Thus, this medium which propagates light vibrations would have to have an elasticity of the order of 3,600,000,000 times the elasticity of steel. Or, if the elasticity of the medium were the same as that of steel, the density would have to be 3,600,000,000 times as small as that of steel, that is to say, roughly speaking, about 50,000 times as small as the density of hydrogen, the lightest known gas. Evidently, then, a medium which propagates vibrations with such an enormous velocity must have an enormous high elasticity or abnormally low density. In any case, its properties would be of an entirely different order from the properties of the substances with which we are accustomed to deal, so that it belongs in a category by itself.

Another course of reasoning which leads to this same conclusion—namely, that this medium is not an ordinary form of matter, such as air or gas or steel— is the following: Sound is produced by a bell under a receiver of an air pump. When the air has the same density inside the receiver as outside, the sound reaches the ear of an observer without difficulty. But when the air is gradually pumped out of the receiver, the sound becomes fainter and fainter until it ceases entirely. If the same thing were true of light, and we exhausted a vessel in which a source of light—an incandescent lamp, for example—had been placed, then, after a certain degree of exhaustion was reached, we ought to see the light less clearly than before. We know, however, that the contrary is the case, i.e. that the light is actually brighter and clearer when the exhaustion of the receiver has been carried to the highest possible degree. The probabilities are enormously against the conclusion that light is transmitted by the very small quantity of residual gas. There are other theoretical reasons, into which we will not enter. [...]

But these are not the only paradoxes connected with the medium which transmits light. There was an observation made by Bradley a great many years ago, for quite another purpose. He found that when we observe the position of a star by means of the telescope, the star seems shifted from its actual position, by a certain small angle called the angle of aberration. He attributed this effect to the motion of the earth in its orbit, and gave an explanation of the phenomenon which is based on the corpuscular theory and is apparently very simple. We will give this explanation, notwithstanding the fact that we know the corpuscular theory to be erroneous.

Let us suppose a raindrop to be falling vertically and an observer to carrying, say, a gun, the barrel being as nearly vertical as he can hold it. If the observer is not moving and the raindrop falls in the centre of the upper end of the barrel, it will fall centrally through the lower end. Suppose, however, that the observer is in motion in the direction *bd*; the raindrop will still fall exactly vertically, but if the gun advances laterally while the raindrop is within the barrel, it strikes against the side. In order to make the raindrop move centrally along the axis of the barrel, it is evidently necessary to incline the gun at an angle such as *bad*. The gun barrel is now pointing, apparently, in the wrong direction, by an angle whose tangent is the ratio of the velocity of the observer to the velocity of the raindrop.

According to the undulatory theory, the explanation is a trifle more complex; but it can easily be seen that, if the medium we are considering is motionless and the gun barrel represents a telescope, and the waves from the star are moving in the direction *ad*, they will be concentrated at a point which is in the axis of

The aberration of light

the telescope, unless the latter is in motion. But if the earth carrying the telescope is moving with a velocity something like twenty miles a second, and we are observing the stars in a direction approximately at right angles to the direction of that motion, the light from the star will not come to a focus on the axis of the telescope, but will form an image in a new position, so that the telescope appears to be pointing in the wrong direction. In order to bring the image on the axis of the instrument, we must turn the telescope from its position through an angle whose tangent is the ratio of the velocity of the earth in its orbit to the velocity of light. The velocity of light is, as before stated, 186,000 miles a second— 200,000 in round numbers—and the velocity of the earth in its orbit is roughly twenty miles a second. Hence the tangent of the angle of aberration would be measured by the ratio of 1 to 10,000. More accurately, this angle is 20″.445. The limit of accuracy of the telescope, as was pointed out in several of the preceding lectures, is about one-tenth of a second; but, by repeating these measurements under a great many variations in the conditions of the problem, this limit may be passed, and it is practically certain that this number is correct to the second decimal place.

When this variation in the apparent position of the stars was discovered, it was accounted for correctly by the assumption that light travels with a finite velocity, and that, by measuring the angle of aberration, and knowing the speed of the earth in its orbit, the velocity of light could be found. This velocity has since been determined much more accurately by experimental means, so that now we use the velocity of light to deduce the velocity of the earth and the radius of its orbit. [. . .]

Maxwell [. . .] further indicated that if we made two measurements of the velocity of light, one in the direction in which the earth is travelling in its orbit, and one in a direction at right angles to this, then the time it takes light to pass over the same length of path is greater in the first case than in the second.

We can easily appreciate the fact that the time is greater in this case, by considering a man rowing in a boat, first in a smooth pond and then a river. If he rows at the rate of four miles an hour, for example, and the distance between the stations is twelve miles, then it would take him three hours to pull there and three to pull back—six hours in all. This is his time when there is no current. If there is a current, suppose at the rate of one mile an hour, then the time it would take to go from one point to the other would be, not 12 divided by 4, but 12 divided by $4+1$, i.e. 2.4 hours. In coming back the time would be 12 divided by $4-1$, which would be 4 hours, and this added to the other time equals 6.4 instead of 6 hours. It takes him longer, then, to pass back and forth when the medium is in motion than when the medium is at rest. We can understand, then, that it would take light longer to travel back and forth in the direction of the motion of the earth. The difference in the time is, however, so exceedingly small, being of the order of 1 in 100,000,000, that Maxwell considered it practically hopeless to attempt to detect it.

In spite of this apparently hopeless smallness of the quantities to be observed, it was thought that the minuteness of the light waves might again come to our rescue. As a matter of fact, an experiment was devised for detecting this small quantity. The conditions which the apparatus must fulfil are rather complex. The total distance travelled must be as great as possible, something of the order of one hundred million waves, for example. Another condition requires that we be able to interchange the direction without altering the adjustment by even the one hundred-millionth part. Further, the apparatus must be absolutely free from vibration.

The problem was practically solved by reflecting part of the light back and forth a number of times and then returning it to its starting-point. The other path was at right angles to the first, and over it the light made a similar series of excursions, and was also reflected back to the starting-point. This starting-point was a separating plane in an interferometer, and the two paths at right angles were the two arms of an interferometer. Notwithstanding the very considerable difference in path, which must involve an exceedingly high order

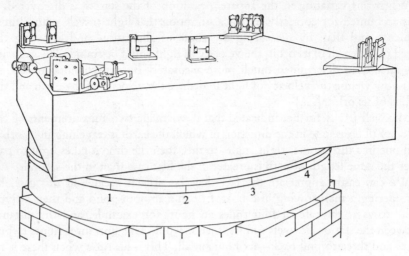

Michelson's interferometer

of accuracy in the reflecting surfaces and a constancy of temperature in the air between, it was possible to see fringes and to keep them in position for several hours at a time.

These conditions having been fulfilled, the apparatus was mounted on a stone support, about four feet square and one foot thick, and this stone was mounted on a circular disc of wood which floated in a tank of mercury. The resistance to motion is thus exceedingly small, so that by a very slight pressure on the circumference the whole could be kept in slow and continuous rotation. It would take, perhaps, five minutes to make one single turn. With this slight motion there is practically no oscillation; the observer has to follow around and at intervals to observe whether there is any displacement of the fringes.

It was found that there was no displacement of the interference fringes, so that the result of the experiment was negative and would, therefore, show that there is still a difficulty in the theory itself; and this difficulty, I may say, has not yet been satisfactorily explained. [. . .]

The experiment is to me historically interesting, because it was for the solution of this problem that the interferometer was devised. I think it will be admitted that the problem, by leading to the invention of the interferometer, more than compensated for the fact that this particular experiment gave a negative result.

HEINRICH HERTZ

292

On the relationship between light and electricity

Talking of the relationship between light and electricity, the layman will immediately think of electric light. [. . .] The physicist, on the other hand, will think of subtle interactions between the two forces, like the rotation of the plane of polarization by a current, or the change in resistance of a wire under the influence of light. However, these are not examples of direct interaction between light and electricity; here a medium acts between the two forces: ponderable matter. Accordingly, these phenomena will not occupy us here, either. There are other, more intimate connections between the two forces.

My proposition is that light is an electrical phenomenon; this is true for light as such or any special sort of light: of the sun, of a candle, of a glowworm. Remove electricity from the world and all light will be quenched: remove the luminiferous ether from the universe and electric and magnetic forces will not propagate in space.

This is what we assert. It is not an invention of today or yesterday, it has a longer history; even more: its history serves as the justification for holding to it. My own experiments in this field constitute only one link in a long chain. It is of the chain as a whole that I wish to speak to you, not about the individual links. It is not easy to treat these matters simply and correctly at the same time: the events we are going to describe take place in empty space—in free ether; they are intangible for the hand, inaudible for the ear, invisible for the eye; only to the inner perception and conceptual associations are they accessible, and not at all to sense description. We shall do our best to associate the new concepts with the existing views and conceptions. Let us recall what we know for certain about light and electricity separately and only then try to seek out the relationship between them. [. . .]

The suggestion that electrical forces can exist independently of their sources came into early conflict with the ruling theories of electricity. In the same way the accepted theory of optics refused to consider the idea that light could be any other than elastical in nature. Any attempt to investigate these suggestions more thoroughly seemed an idle speculation. How deeply we must admire the happy genius of the man who knew how to combine two such far-lying conjectures, recognizing that they mutually supported each other and that in reality they were two conclusions of one and the same theory, a theory, the inner probability of which could not have been foreseen or predicted! This man was the Englishman Maxwell. His work, published in 1865, is known as *The Electromagnetic Theory of Light*. One cannot study this amazing theory without realizing that here the mathematical expressions live a life of their own, exhibiting an insight more penetrating and more clever than we are, nay more clever than the man who invented them. [. . .]

If one disputes Maxwell's theory of electricity, there is no logical ground to accept his theory of light either. Contrarily, if one clings to the notion that light is a phenomenon of elastic nature, Maxwell's theory of electricity cannot be held by him. But if one entered this edifice free of any preconceived ideas, he would then see how the parts support each other like the stones of an arch: the whole would seem to constitute a bridge spanning the gulf of the unknown and connecting the two banks of the known.

The difficulties of the theory were immediately overcome only by a few of Maxwell's disciples. Once one penetrated its world of ideas one became perforce an ardent supporter wishing to study the new theory's presuppositions and investigate its implications. Experimental proof had to rely for a long time to come on individual assertions, that is to say, on information external to the theory.

I have just compared Maxwell's theory to an arch spanning the abyss of the unknown; let me carry this analogy further and say that for a long time anything that turned out to serve the strengthening of the arch came in the form of additional support to the two buttressing pillars on the banks. Thus the arch was strong enough to support itself but it had too large a span for further building up in height. For that purpose special pillars were necessary; these had to stand on firm ground and to come as support to the middle of the arch. Such a pillar would have been analogous to the proof that light indeed had direct magnetic or electric effects: it could then support directly the optical wing of the edifice while indirectly supporting the electrical wing. Another pillar could have been the proof that indeed there exist waves of electric or magnetic forces, propagating in the same way as do waves of light. This pillar, on the contrary, would support directly the electric part and indirectly the optical part. A harmonic completion of the construction task will necessitate the erection of both pillars, but as a first condition one of them must suffice. The first of these problems has not yet been found amenable for intensive study, but the other has been solved and put on a strong foundation. [. . .]

Let a physicist be given some tuning-forks and several resonators, and let him be asked to demonstrate that the propagation of sound-waves takes time; he will have no difficulties even if at his disposal is only an enclosure of the size of a room. He will put one of the tuning-forks at any given point of the room and then listen with a resonator to the intensity of the sound at various points; he will show that at some points the wave-intensity is very weak and that at these points one vibration was annihilated by another one, which, though it had left the source later, had gone a shorter way and happened to reach the same point. Whenever the case is that the shorter path needs less time to be covered, the propagation is in time and not immediate; thus, the demonstration is complete. Now, our acoustician will show that these silent points follow each other periodically, at equal distances; out of this he calculates the wavelength and if he knows the period of the tuning-fork, he can now find the velocity of the sound-waves. Exactly the same procedure should be followed with our electrical vibrations. In place of the tuning-fork we put the vibrating conductor; in place of the resonator we put a wire in the form of a broken loop which we shall call the electrical resonator. We will notice now that in the resonator sparks occur if it is placed at some special points in the room, while at others nothing occurs. Again we see how the 'dead points' follow each other in periodic succession— propagation in time has been demonstrated, wavelength has become a measurable quantity. The question occurs whether the waves are transverse or longitudinal. Let us hold our wire in two different positions at the same place in the wave; it will respond at one position and fail to do so at the other; indeed, we need nothing more than that: the question has been answered, the waves are transverse. We are asked what their velocity is: we multiply the measured wavelength by the calculated period of vibrations and get a velocity which is very much like the speed of light. [. . .]

Our experiments belonged to those higher levels where the pass between electricity and optics lies. We can go even a few steps further and try to enter the province of conventional optics. It might prove to be advisable to leave the theory out altogether. There are many friends of Nature who are all deeply interested in the essence of light, who could probably understand simple experiments, but to whom Maxwell's theory is a book closed with seven seals. Besides, economy of science demands that wherever a straight path is available, the crooked one should be abandoned. If it is feasible to produce phenomena of light directly with electrical means, we do not need theory as a mediator: the relationship emerges out of the very experiments. Such experiments can be devised. Let us place the conductor, which generates the vibrations, in the focal plane of a big concave mirror. This keeps the waves together which radiate from the mirror in the form of a powerful beam. We cannot directly feel or observe this beam, but its effects become noticeable when it excites sparks in a conductor which it encounters on its way. Such sparks become visible if we arm the conductor with an electrical resonator: in all other aspects this beam is just like a light-beam: we can send it in various directions through turning the mirror and we can show its rectilinear propagation if we follow its path. If we put conductors in the path of the beam these will obstruct its way and cast shadows, but the beam is not annihilated by them: it is reflected; if we now follow the reflected beam we find that this follows the laws of reflection of light. It can be also refracted exactly as light: in order to refract a beam of light we let the beam pass through a prism which diverts it from its original direction. We do the same here with the same effect. True, for electric beams we have to take into consideration the dimensions of the waves and choose a very large prism; thus a cheap material is required like pitch or asphalt. Finally, we can detect such characteristics of electric beams which have been observed hitherto only with light-beams, namely phenomena of polarizability. Inserting a grid of suitable structure in the way of our beam, sparks will light up or be quenched in our resonator obeying the same geometrical laws which govern illumination or darkening of the plane of vision of a polarizer after a crystal plate has been inserted into the polarizing apparatus.

So far the experiments. Having put up these, we have penetrated into the province of the science of light. Planning and describing these experiments we were thinking optically, not electrically. We do not see any more currents flowing in the conductors or particles of electricity accumulating; we only see waves in the air, crossing each other, disintegrating, uniting, strengthening or weakening each other.

We have left the province of electrical phenomena and arrived step by step to that of purely optical phenomena. We have gone now over the pass and the road will slope down before it levels out again. The senses have perceived, natural understanding has comprehended the relationship between light and electricity—a relationship surmised and foreseen by theory. From the high point of the pass, an insight into both provinces has been gained, and both provinces seem now bigger than we have previously imagined. Optical waves are no longer limited to ether waves of the size of fractions of a millimeter; these cover now waves which are measured in decimetres, metres and kilometres. And yet, all this huge territory seems a tiny appendage of the province of electricity which embraces now more than ever before. We recognize electricity at a thousand places where nobody had a certain knowledge of its presence previously. In every flame, in every radiating atom we see an electric process. Even if a body does not radiate light but only heat, we know it to be a seat of electricity.

Electricity has come to embrace the whole of Nature. It even touches us personally: we have an electrical organ in our body: the eye. This we see if we look down into the world of the particulars. Not less rewarding will it be to look up at the mountain peaks of the general goals of science. Here the problem of action-at-a-distance comes to mind. Is there such a thing at all? Once we had believed that there existed several. Now only one has been left to us: gravitation. Might we be deceived in this too? The laws that govern gravitation make it suspicious. In another direction looms the great problem of the nature of electricity; it seems hiding behind the more specific problems of the nature of magnetic and electric forces in space. Linked to these emerges the major riddle of the nature and qualities of the ether: that medium which fills all space; we inquire about its structure, about its motion or state of rest, its endlessness or limitation. It appears now that this last question overrides all the others in importance: that knowledge of the ether may not only clarify the nature of the once so-called ponderable bodies, but the nature of matter itself, and its most innate properties: gravity and inertia.

ERNST MACH

Critique of Newton's absolute space

293

Let us now examine the point on which Newton, apparently with sound reasons, rests his distinction of absolute and relative motion. If the earth is affected with an *absolute* rotation about its axis, centifugal forces are set up in the earth: it assumes an oblate form, the acceleration of gravity is diminished at the equator, the plane of Foucault's pendulum rotates, and so on. All these phenomena disappear if the earth is at rest and the other heavenly bodies are affected with absolute motion round it, such that the same *relative* rotation is produced. This is, indeed, the case, if we start *ab initio* from the idea of absolute space. But if we take our stand on the basis of facts, we shall find we have knowledge only of *relative* spaces and motions. *Relatively*, not considering the unknown and neglected medium of space, the motions of the universe are the same whether we adopt the Ptolemaic or the Copernican mode of view. Both views are, indeed, equally *correct*; only the latter is more simple and more *practical*. The universe is not *twice* given, with an earth at rest and an earth in motion; but only *once*, with its *relative* motions, alone determinable. It is, accordingly, not permitted us to say how things would be if the earth did not rotate. We may interpret the one case that is given us, in different ways. If, however, we so interpret it that we come into conflict with experience, our interpretation is simply wrong. The principles of mechanics can, indeed, be so conceived, that even for relative rotations centrifugal forces arise.

Newton's experiment with the rotating vessel of water simply informs us, that the relative rotation of the water with respect to the sides of the vessel produces *no* noticeable centrifugal force, but that such forces *are* produced by its relative rotation with respect to the mass of the earth and the other celestial bodies. No one is competent to say how the experiment would turn out if the sides of the vessel increased in thickness and mass till they were ultimately several leagues thick. The one experiment only lies before us, and our business is, to bring it into accord with the other facts known to us, and not with the arbitrary fictions of our imagination.

294

The economy of science

1. It is the object of science to replace, or *save*, experiences, by the reproduction and anticipation of acts in thought. Memory is handier than experience, and often answers the same purpose. This economical office of science, which fills its whole life, is apparent at first glance; and with its full recognition all mysticism in science disappears. Science is communicated by instruction, in order that

one man may profit by the experience of another and be spared the trouble of accumulating it for himself; and thus, to spare posterity, the experiences of whole generations are stored up in libraries.

Language, the instrument of this communication, is itself an economical contrivance. Experiences are analysed, or broken up, into simpler and more familiar experiences, and then symbolized at some sacrifice of precision. The symbols of speech are as yet restricted in their use within national boundaries, and doubtless will long remain so. But written language is gradually being metamorphosed into an ideal universal character. It is certainly no longer a mere transcript of speech. Numerals, algebraic signs, chemical symbols, musical notes, phonetic alphabets, may be regarded as parts already formed of this universal character of the future; they are, to some extent, decidedly conceptual, and of almost general international use. [. . .]

2. In the reproduction of facts in thought, we never reproduce the facts in full, but only that side of them which is important to us, moved to this directly or indirectly by a practical interest. Our reproductions are invariably abstractions. Here again is an economical tendency.

Nature is composed of sensations as its elements. Primitive man, however, first picks out certain compounds of these elements—those namely that are relatively permanent and of greater importance to him. The first and oldest words are names of 'things'. Even here, there is an abstractive process, an abstraction from the surroundings of the things, and from the continual small changes which these compound sensations undergo, which being practically unimportant are not noticed. No inalterable thing exists. The thing is an abstraction, the name a symbol, for a compound of elements from whose changes we abstract. The reason we assign a single word to a whole compound is that we need to suggest all the constituent sensations at once. When, later, we come to remark the changeableness, we cannot at the same time hold fast to the idea of the thing's permanence, unless we have recourse to the conception of a thing-in-itself, or other such like absurdity. Sensations are not signs of things; but, on the contrary, a thing is a thought-symbol for a compound sensation of relative fixedness. Properly speaking the world is not composed of 'things' as its elements, but of colours, tones, pressures, spaces, times, in short what we ordinarily call individual sensations.

The whole operation is a mere affair of economy. In the reproduction of facts, we begin with the more durable and familiar compounds, and supplement these later with the unusual by way of corrections. Thus, we speak of a perforated cylinder, of a cube with bevelled edges, expressions involving contradictions, unless we accept the view here taken. All judgements are such amplifications and corrections of ideas already admitted. [. . .]

We must admit, therefore, that there is no result of science which in point of principle could not have been arrived at wholly without methods. But, as a matter of fact, within the short span of a human life and with man's limited powers of memory, any stock of knowledge worthy of the name is unattainable except by the *greatest* mental economy. Science itself, therefore, may be regarded as a minimal problem, consisting of the completest possible presentment of facts with the *least possible expenditure of thought*.

7. The function of science, as we take it, is to replace experience. Thus, on the one hand, science must remain in the province of experience, but, on the other, must hasten beyond it, constantly expecting confirmation, constantly expecting the reverse. Where neither confirmation nor refutation is possible, science is not concerned. Science acts and acts only in the domain of *uncompleted* experience. Exemplars of such branches of science are the theories of elasticity and of the conduction of heat, both of which ascribe to the smallest particles of matter only such properties as observation supplies in the study of the larger portions. The comparison of theory and experience may be farther and farther extended, as our means of observation increase in refinement.

Experience alone, without the ideas that are associated with it, would forever remain strange to us. Those ideas that hold good throughout the widest domains of research and that supplement the greatest amount of experience, are the *most scientific*. The principle of continuity, the use of which everywhere pervades modern inquiry, simply prescribes a mode of conception which conduces in the highest degree to the economy of thought. [. . .]

9. Yet not all the prevalent scientific theories originated so naturally and artlessly. Thus, chemical, electrical and optical phenomena are explained by atoms. But the mental artifice atom was not formed by the principle of continuity; on the contrary, it is a product especially devised for the purpose in view. Atoms cannot be perceived by the senses; like all substances, they are things of thought. Furthermore, the atoms are invested with properties that absolutely contradict the attributes hitherto observed in bodies. However well fitted atomic theories may be to reproduce certain groups of facts, the physical inquirer who has laid to heart Newton's rules will only admit those theories as *provisional* helps, and will strive to attain, in some more natural way, a satisfactory substitute.

SECTION VI

Planck to Pauli

Section VI: Planck to Pauli

INTRODUCTION

THE nineteenth century saw the completion of the world-picture dominated by classical physics; the highlights of this culmination of the classical world-view were the definitive formulation of Newton's mechanics, the discovery of the first and second laws of thermodynamics, the growth of the theory of electromagnetism—including that of light—and the development of statistical mechanics based on the classical conception of causality. But certain unresolvable problems associated with the classical world-picture led to its profound modification, during the first three decades of the present century, by the establishment of two new systems of physical thought—those of relativity and quantum physics.

The Theory of Relativity—the achievement of Albert Einstein—was a triumph of theoretical reasoning and epistemological analysis which put an end to a situation that had become increasingly intriguing and problematic during the course of the nineteenth century. Einstein began by giving a new evaluation and interpretation of known facts, and of earlier attempts at their explanation—among them that of Lorentz, whose important researches had extended Maxwell's theory to include the motion of electrons in an electromagnetic field. Einstein's main starting-point for the Special Theory of Relativity was an interpretation of the failure of the many attempts which had been made to detect the motion of the earth relative to the aether—for example, the null result of the Michelson–Morley experiments. Why did the velocity of light—and, indeed, of every electromagnetic signal—remain constant and be neither added to, nor subtracted from, that of the source of the light? By an ingenious mental turn-about, Einstein transformed this puzzling problem into a postulate: the velocity of light, he suggested, was a universal constant independent of the state of motion of the observer or that of the light source. This postulate, apparently incompatible with the assumptions about the kinematics of relative motion on which Newtonian mechanics was based—that is, with the classical postulate of relativity—Einstein combined with the latter to form a new principle of relativity. The resulting mathematical equations (the 'Lorentz transformations') contain not only the velocity of a moving body relative to the observer, but also c, the velocity of light; c thus became a quantity of universal significance in that it entered into not only optical or electromagnetic observations, but also those of mechanical phenomena. In fact, c had now to be regarded as the upper limit of all physically meaningful velocities—that is, of the velocities of physical signals or of material objects. But whereas the velocity of, for example, a sound signal depends on the state of motion of the source of sound with respect to the elastic medium, the velocity of electromagnetic radiation had now to be taken as independent of that of the medium in which it was supposedly propagated. The reason for this, according to the theory of relativity, was that there is no such medium, and the universal constancy of c has to be regarded as the property of a physical magnitude which has the role of an absolute, invariant entity.

The absoluteness of c in relativistic physics replaced that of space and time in Newtonian physics. Since Einstein, we know that lengths and times have no absolute values, but that their values are relative to the state of motion of the observer, and that this relativity arises from the fact that the velocity of light is constant and finite. An infinite velocity of light would reduce Einstein's universe to the classical Newtonian world in which, in particular, the absoluteness of time is shown in the assumption of an absolute simultaneity of physical events, independent of the observer of such events. In the general theory of relativity, Einstein went further, and generalized the absoluteness of c into the absoluteness of the 'interval', the spatio-temporal distance between two events, which now assumed a universal significance independent of the observer. In the words of Hermann Minkowski (written in 1908): 'From now on space and time by themselves shall be completely reduced to shadows, and only a sort of union of them both shall retain an existence of its own.'

From a philosophical point of view, the relativization of the concept of simultaneity was certainly the most revolutionary of all the profound changes brought about by the theory of relativity. Another radical change, however, was in that of the concept of mass, which had been supposed to be a constant for a given body, giving a measure of its resistance to acceleration. As, now, according to the theory of relativity, no body with a finite mass could reach the velocity of light, it followed that its resistance to acceleration—that is, its mass— must approach infinity as its velocity approached c. Only the mass of a body at rest with respect to some observer, the so-called rest mass, could now be regarded as a constant. In the equations derived by Einstein, the product of this rest mass with the square of the velocity of light $(m_0 c^2)$ appeared as a magnitude defining the energy of a body at rest in empty space; such a body, according to classical concepts, would have neither kinetic nor potential energy. The full significance of the equivalence of mass and energy was realized and confirmed only after the advent of modern atomic physics, as were the other major consequences of the special theory of relativity, such as the relativity of time and the dependence of mass on velocity already mentioned above.

In addition to the fundamental enrichment of our knowledge, the theory of relativity contributed to the unification of our world-picture in that it succeeded in removing several inconsistencies which still troubled the physicists of the nineteenth century, and in explaining the well-known phenomena of optics and electromagnetism in terms of a single, self-consistent set of presuppositions. In its totality, the theory of relativity represents the classical example of an all-embracing system of thought which, *a posteriori*, appears to us as elementary and convincing in its structural simplicity.

A few passages from the beginning of Einstein's famous paper of 1905 are quoted in the Anthology in order to give some idea of his fundamental approach, based on a critique of the meaning of measurement and on an analysis of the role of the measuring observer. Einstein, in this paper, put forward all the important results described above despite restricting his considerations to observers in a state of inertial motion relative to each other. This amounted to a considerable idealization of the physical world as we know it, for observers on the earth are accelerated with respect to extra-terrestrial phenomena because the earth, as do the other planets, moves with a velocity changing in both direction and size, as well as rotating on its axis. In other words, the special theory of relativity took no account of gravitation and the acceleration due to it. Some passages from the beginning of Einstein's paper of 1916, also quoted in this Anthology, indicate the train of thought which led him to this further generalization: 'The laws of physics must be of such a nature that they apply to systems of reference of *any kind* of motion.' To illustrate this, Einstein pointed out that Mach had already emphasized 'an inherent epistemological defect' in classical mechanics, which was still present in the theory of special relativity. Both assumed that the visible effects of a rotatory motion (for example, centrifugal forces or Coriolis forces) were due to a rotation relative to empty space; but this was merely a fictitious cause, and not an observable entity. Einstein then went on to discuss the nature of an accelerating force of a much more universal nature—the force of gravitation. Rightly, we attribute gravitation to the presence of masses—for example, the acceleration of a freely falling body to the presence of the earth. In this context Einstein gave, for the first time, an interpretation of the basic fact that (within a small region, at any rate) the acceleration due to gravity was the same for all bodies, irrespective of their mass. Could we not 'create' the effects of gravitation by going over to another system of reference? The answer was given by Einstein's famous thought experiment: a man removed from the influence of any gravitational field, and resting in a lift which was accelerated uniformly in the direction from the floor to the ceiling, would observe that all objects which he released from his hands were 'falling' downwards at the same rate, irrespective of their mass; whereas any other man, in a state of rest or inertial motion outside the lift, would explain his observations as due to the law of inertia—for him, the released objects would remain at rest in their initial positions. It seemed, then, as though, in certain cases, the concepts of inertia and gravitation could be interchanged in the interpretation of observed facts—in other words, in certain cases inertia and gravitation could be regarded as equivalent physical entities.

These considerations of Einstein's were his point of departure for a bold and ingenious attempt to abolish, for the important case of gravitation, the Newtonian dichotomy between inertia and force. In fact, the general theory of relativity did away with the concept of gravitation as a force, and explained its effects, such as the planetary motions, as inertial motions in a spatio-temporal universe of non-Euclidean metric. It is possible to translate the logic of this new and revolutionary conception into the language of mathematics. In the special theory of relativity, the spatio-temporal interval between two events is made up of elements of space and time, and can be represented by a well-known geometrical theorem: 'the square on the hypotenuse of a right-angled triangle is equal to the sum of the squares on the other two sides.' The Pythagorean theorem in this form applies, of course, to Euclidean geometry. However, it can be generalized by multiplying the squares of the elements of space and time, whose sum makes up the interval, by suitable coefficients which on the one hand enable the theorem to function as the geometrical expression of a non-Euclidean geometry, and on the other can be interpreted as gravitational potentials.

This geometrization of the gravitational field is of very great epistemological significance, in that it creates a close and satisfying interdependence between space and matter. In terms of it, the geometry of a certain region of space has to be regarded as determined mainly by the distribution of matter in that region, in such a way that a region sufficiently far away from any matter assumes the Euclidean metric. On the other hand, the orbits of bodies moving freely in any region are determined by its geometry, and are geodetic lines—that is, the non-Euclidean analogues of straight lines in Euclidean space—and all free motions are thus generalized inertial motions. The enormous conceptual advantages of the general theory of relativity are achieved at the price of a considerable degree of abstraction. But one should not forget that the apparently concrete character of established theories is to a large extent determined by psychological factors—that is, by the fact that we have got used to them.

In his paper on geometry and experience, the first passages of which are quoted in this Anthology, Einstein clearly and beautifully points out that, in contrast to our former naïve view, the geometry of the physical world can not be taken as given *a priori*, but that it can be determined only by empirical means. The excerpts from another paper of his, on the aether and relativity, show that the theory of general relativity returned, in a kind of way, to the concept of an aether, though not as a 'subtle fluid' but as an expression of the new conception of space as an active physical agent determining the mechanical behaviour of the bodies moving in it.

Finally, it should be noted that the general theory of relativity, besides explaining more satisfactorily all the facts known to classical mechanics, optics, and electromagnetism, yielded, as unexpected by-products, the prediction of three small but important effects—the perihelion movement of Mercury, the red-shift of light emitted in a gravitational field, and the bending of light rays in their passage through such a field. The prediction and explanation of such phenomena, which otherwise could not have been explained but by hypotheses *ad hoc,* naturally increase our confidence in a new theory. The revolutionary cosmological consequences of the theory of relativity, first discussed by Einstein himself, were further elucidated by the progress of astrophysics from the thirties of this century onwards. This extremely interesting branch of physical science, which is still in a state of flux and might well have repercussions on some fundamental concepts of the theory, is not represented in the Anthology.

The conceptual development of quantum theory, the other great system of physical thought which originated in the twentieth century, was not as straightforward as was that of relativity. The main reason for this was that the clarification, evolution, and integration of quantum theory into a coherent body of knowledge was bound up with an accumulation of basic facts whose acquisition had to await the development of experimental techniques. It was only in the second half of the nineteenth century that the technique of producing higher vacua improved sufficiently to make possible the generation and observation of electrically charged particles moving in electric and magnetic fields. Spectroscopic techniques made considerable advances during the same period, not only in the construction of spectrographs but also in the production of instruments for measuring the intensities of radiations extending into the infra-red and ultraviolet regions. In the last decade of the century radioactive radiation and X-rays were discovered, as was the electron, whose characteristic properties—its charge and mass—were later determined more and more precisely.

A remarkably illuminating retrospective survey of these developments is contained in an excellent article by Ernest Rutherford, entitled *The Electric Structure of Matter,* several extracts from which are given in this Anthology. Rutherford, one of the fathers of atomic physics, whose experimental genius reached the level of Faraday's, also gave an account in this article of his own fundamental researches into the spontaneous transformation of radioactive atoms, which led to the theory of radioactive decay established by him and Soddy. He also described briefly his famous observations of the deflection of α-particles in their passage through thin metallic foils, which finally gave the clue to the structure of the atom—a very small nucleus, in which practically the whole of the mass of the atom was concentrated, and a surrounding envelope of electrons, orders of magnitude larger than the dimensions of the nucleus.

These discoveries, however, were made in the decade following the birth of the quantum hypothesis in 1900. When Planck first published it, it had the character of an *ad hoc* hypothesis, enabling him to derive a mathematical formula conforming to the observed spectral distribution of the electromagnetic (heat and light) radiation emitted by a body which absorbs practically all the radiation falling on it (that is, neither transmitting nor reflecting it). The assumption of classical physics, derived originally from the theory of sound, that the particles of which a macroscopic body was made up, supposed to be small 'vibrators' or 'oscillators', were absorbing or emitting radiation in continuously varying quantities, had to be abandoned. Planck assumed that such oscillators could absorb or emit energy only in discrete quanta proportional to the frequency of oscillation, the factor of proportionality being Planck's quantum of action. This specialized and particular hypothesis turned out to be but the starting-point for a comprehensive theory which, far more than relativity, overthrew a centuries-old world-picture. Planck's essay of 1913, selections from which are quoted in the Anthology, not only gives a popular but strictly scientific account of his hypothesis; it also explains it in the context of a lucid account of the historical and philosophical background, accompanied by some penetrating observations on the significance of change in physical theories and on the necessity for belief in a physical reality.

Following Planck's paper of 1900, a further step of great importance for the establishment of quantum physics as a comprehensive theory was a paper published by Einstein in 1905. In this he introduced the hypothesis of light quanta (or, as they were later called, photons) as a complement, indispensable for the explanation of certain facts concerning the interaction of light with matter, to the hypothesis of light waves. Einstein's argument in the first part of this paper gives a clear example of the vital role played by conceptual analogies in physical thought. He pointed out that, while statistical mechanics had replaced the phenomenological concept of a gas, or of matter in general, as a continuous entity by the particle concept (a suggestion already evident in Daniel Bernoulli's explanation of gas pressure as the result of the impacts of particles on the walls of the containing vessel), Maxwell's electromagnetic theory of radiation made use exclusively of continuous spatial functions—applying, for instance, the continuous wave picture to the explanation of optical phenomena. He argued that in certain cases this picture could well be insufficient to explain satisfactorily the observed phenomena, and should be replaced by that of a sum of discrete light quanta interacting with electrons, atoms, and so on. Einstein illustrated the value of this picture in understanding the photo-electric effect. This, he maintained, could be explained as the sum of individual processes, in which each light quantum of a given energy—that is, of a given frequency—hitting a body did a certain amount of work, equal to that energy, on an electron in the body. Part of the energy would be used up in releasing the electron from the body and transferring it to its surface, and the rest would be converted into kinetic energy of the free electron. In this paper of Einstein's, the necessity of using two contradictory models—that of a particle and that of a wave—for explaining different phenomena in microphysics was emphasized for the first time.

Eight years later Bohr, who was then working in Rutherford's laboratory in Cambridge, made a further significant contribution to the elucidation of quantum phenomena in his important paper of 1913. Only the introduction and concluding summary of this paper are quoted, omitting all the elaborate calculations based on that model of the hydrogen atom proposed by Bohr on the basis of Rutherford's theory of atomic structure. This classical paper, together with

Bohr's retrospective review of a dozen years later, are an excellent illustration of scientific thought at a turning-point in the history of ideas. In order to explain the line spectra of the elements, which, because of their great stability and sharpness, had defied the attempts of physicists to draw exact parallels between acoustic and electromagnetic oscillators, Bohr developed a model which combined elements of a classical picture with a revolutionary deviation from classical concepts. On the one hand, he thought of the atom as a microphysical analogue of the solar system—that is, as an analogue of one of the paradigms of Newtonian mechanics—while on the other, he abandoned the principles of Maxwellian electromagnetism by postulating that electrons in their 'stationary' orbits in the atom did not emit any electromagnetic radiation. He thus succeeded in extending the quantum hypothesis by applying it to the structure of the atom: atoms, and micro-systems in general, were now held to exist only in discrete states of energy, any transition from one of these states to another being associated with the emission or absorption of a quantum of light.

In his historical review, Bohr gave an account of some of the fundamental experiments which confirmed his hypothesis, as well as of a wealth of theoretical and experimental detail which lent further support to the 'classical' quantum theory. Among others, he mentioned Einstein's paper of 1917, in which a satisfactory explanation of the exchange of energy between an atom and the electromagnetic radiation surrounding it, involving quantization of the atomic energy states as well as of the radiation energy, was given using statistical considerations. He also touched upon the attempts, which were to become desperate during the late twenties of this century, to effect a reconciliation between the wave and quantum theories of light. The state of perplexity which prevailed among the physicists during those years is reflected in a paper (written by Bohr, Kramers and Slater) which suggested that the law of conservation of energy was of a statistical character—that is, that the energy balance of atomic processes was preserved only if taken as an average over a sum of many such processes. This hypothesis, however, which was supposed to save the classical wave theory of light—that is, the idea of a continuous propagation of electromagnetic energy, despite the obviously discrete and sudden exchanges of energy observed in quantum phenomena—turned out to be shortlived; it was refuted by striking experiments which demonstrated the strict validity of the law of conservation of energy for each individual process.

The vacillations of those years coincided with the transition from the interim period of 'classical' quantum theory to the final phase of quantum mechanics. From our vantage-point of a perspective of several decades it is easy to understand the shortcomings of that intermediate stage of quantum physics, with its contradictions and its failure to account theoretically for certain empirical data in atomic physics. The root of the failure is to be found in the compromise between classical ideas and the new quantum approach. An important overture to this new phase was Louis de Broglie's theory of the wave-nature of matter. His ingenious suggestion is a further instance of the fruitfulness, at certain crucial moments in the history of scientific thought, of the drawing of parallels. Radiation, so de Broglie argued, exhibited the dual character of continuous waves and discrete quanta, so could the same feature not be found in matter? In other words: could not elementary particles of matter, such as protons or electrons, have wave as well as particle aspects? In fact this idea, developed in the first of his two papers represented in this Anthology, was confirmed by experiments demonstrating the wave characteristics of a beam of electrons passed through a crystal.

The excerpts from the second paper, de Broglie's Nobel lecture of 1929, indicate the great success which the generalization of his idea achieved in the form of Schrödinger's and Dirac's wave mechanics. Indeed, by far the most significant aspect of Schrödinger's wave equation of the hydrogen atom was that it established a mathematical method which not only overcame the shortcomings of the earlier approach to the problem, but also provided the most adequate and convenient procedure for dealing with other quantum phenomena. It can be said without exaggeration that the Schrödinger and Dirac equations, the latter of which took relativistic postulates into account, are still, after more than forty years, the cornerstone of the success of theoretical physics in giving a comprehensive description of a whole complex of empirical data in the realm of atomic and nuclear phenomena. The more recent achievements of the theoreticians are still of a rather fragmentary and provisional character.

The final establishment of quantum mechanics resulted from the penetrating and passionate discussions of the leading physicists concerning the interpretation of the new wave mechanics. An echo of those discussions is to be found in the final passage of the excerpt from de Broglie's Nobel lecture. They began with the question, whether Schrödinger's wave function had to be taken literally or statistically. In other words: did this mathematical expression indicate a return to a continuous description of atomic events with, for example, an electron in the field of a nucleus being regarded as really 'smeared out' as a matter wave; or had these waves to be interpreted as 'probability waves' which allowed only for a statistical evaluation of the probability of finding an electron in a given place? While nobody doubted the logical consistency of the new theory some physicists, among them even Einstein, could not reconcile themselves to the thought that the statistical approach—that is, the idea of the rule of primary probabilities in nature—had to be accepted as final.

However, the overwhelming majority of physicists accepted the new and radical view, which amounted to a complete break with the concepts of classical physics, including those of relativity. A decisive step in this direction resulted from Heisenberg's thought experiments, which elucidated the nature of experiments involving atomic events and led to the establishment of his uncertainty relations. It turned out that, in contrast to the situation in macro-physics, so-called 'conjugate variables', such as the position and velocity of a particle, could not in principle be determined simultaneously with absolute certainty. The theoretical limits of precision were expressed mathematically by the relation which stated that the product of the uncertainties of position and velocity would always be at least of the order of h, Planck's quantum of action. Heisenberg's relations thus highlight the essential difference between quantum and classical mechanics, a difference which would disappear if we assumed the value of h to be zero. The strict determinism of classical mechanics, presupposing the possibility of predicting every individual event, rested on two assumptions:

1. Each event could be described by a differential equation representing the event, for example, the motion of a body, as a function of certain causes.

2. The state of the body at a given moment—that is, the initial conditions, involving its position and velocity at that moment, could be exactly determined. This latter assumption, however, would be valid only if the interference of the observer carrying out the determination were to be neglected. Quantum mechanics, dealing with situations in which the means of observation and the observed object were of the same order of magnitude, revealed that such interference could not be ignored or eliminated and that it was of such a nature as to preclude exact determination, the consequence being that the statistical approach had to be accepted as primary in physical science.

The methodological analysis provoked by quantum physics led, therefore, to conclusions that were far more radical than those of the theory of relativity. Relativistic arguments had already taken into account the role of the observer, as he had to use signals propagated with the finite velocity c, whereas classical mechanics had assumed their velocity to be infinite. But the epistemological position of the theory of relativity with regard to the nature of an observation remained the same as that of classical physics. In quantum mechanics, on the other hand, observation means interaction: physical events, on this view, consist of a series of interactions of an indeterminate nature between an observer, including his measuring apparatus, and the observed objects.

The far reaching philosophical consequences of this situation were clearly and beautifully expounded in several articles by Bohr and Heisenberg, passages from which are included in the Anthology. One important result to emerge from this epistemological revolution was that the fact of interaction necessitated the inclusion of the observer in the description of a physical phenomenon. Thus the dual (wave/particle) nature of light or matter was explained as a consequence of the mutually exclusive experimental arrangements used for the observation and description of phenomena in which one or the other concept was involved. This, the so-called 'Copenhagen interpretation' of quantum mechanics, prevails today, despite many attempts to refute it. Niels Bohr has coined the concept of 'complementarity' as a central notion of indeterministic physics and has shown, moreover, that complementarity, and the associated idea of an intrinsic interrelation between subject and object, has a universal significance transcending the field of the physical sciences and playing an important role in many areas of human thought and action.

This introduction would be incomplete without some mention of Pauli, whose exclusion principle was an essential contribution to the development of quantum physics in the third decade of this century. No examples of his scientific writings are included here, but I have quoted a short passage from one of his philosophical articles. Pauli's views on the meaning of physical phenomena and reality are characteristic of the anti-positivistic trend, already discernible in Planck's essays, which began to emerge at the beginning of the century. They embody the *Credo* of one of the greatest scientists of his generation, who saw the contemporary revolution in physical thought as an integral part of a new world-view.

VI: PLANCK TO PAULI

MAX PLANCK
The quantum hypothesis
295

The initial impulse towards revision and modification of a physical theory comes almost always from the discovery of one or more facts which do not fit into the present framework of the theory. Facts always furnish the Archimedean point from which even the most ponderous theory may be lifted off its hinges. In that sense nothing is more interesting to the real Theoretician than a fact which directly contradicts a hitherto universally accepted theory, for it is just here that his real work begins.

Now what are we to do in a case like this? Only one thing is certain; something must be changed in the accepted theory, and in such a manner that it agrees with the new fact, but it is often a difficult and complicated question at what point of the theory the correction is to be applied, for one fact is insufficient to furnish a theory. A theory, indeed, consists as a rule of a whole series of theorems connected with each other. It may be compared to a complicated organism, whose separate parts fit together so intimately that any interference at one place is felt in various other places, sometimes far removed. Wherefore since every conclusion of theory results from the cooperation of several theorems, it follows that, as a rule, several theorems may be made responsible for each failure of the theory, and there are generally several possibilities of finding the way out. Usually, the question is eventually reduced to a conflict between two or three propositions which hitherto have found a place in the theory, but of which one must be abandoned in face of the new fact. The conflict lasts often for years or decades; and its final decision not only means the destruction of the defeated theorem, but also quite naturally—and this is specially important—a corresponding confirmation and elevation of the victorious constituent theorems which survive.

And now we must note the extremely important and remarkable result that in all this war and conflict it is just the great Physical Principles which have held the field—such as the Principle of the Conservation of Energy, the Principle of the Conservation of Momentum, the Principle of Least Action, and the chief laws of Thermodynamics. Their importance has thus been considerably increased; while, on the other hand, the theorems which have succumbed in the fight are those on which theoretical developments were based tacitly, either because they seemed so self-evident that it was not, as a rule, considered necessary to mention them, or because they were forgotten. In general, then, one may assert that the most recent development of theoretical physics is marked by the victory of the great Physical Principles over certain deeply-rooted and yet merely habitual assumptions and conceptions.

To illustrate these statements, I may adduce some of those theorems which have hitherto been used without any hesitation as the self-evident foundations of any theory, but which, in the light of new facts, have proved untenable, or extremely doubtful, in face of the general principles of Physics. I mention three: The Invariability of Chemical Atoms; The Mutual Independence of Space and Time; and The Continuity of all Dynamical Effects. [. . .]

The third of the above theories concerns the Continuity of all Dynamical Effects. This was formerly taken for granted as the basis of all physical theories, and, in close correspondence with Aristotle, was condensed into the well-known dogma—*Natura non facit saltus.* But even in this venerable stronghold of Physical Science present-day investigation has made a considerable breach. This time it is the principles of Thermodynamics with which that theorem has been brought into collision by new facts, and unless all signs are misleading, the days of its validity are numbered. Nature does indeed seem to make jumps—and very extraordinary ones. As an illustration, let me make an instructive comparison.

Let us imagine a sheet of water in which strong winds have produced high waves. Even after the total cessation of the wind, the waves will be maintained for some time and will pass from one shore to the other. But there will be a certain characteristic change in them. During their impact on the shore, or on other solid obstacles, the energy of motion of the longer and coarser waves is converted to an ever greater extent into the energy of motion of shorter and slighter waves; and this process will continue until at last the waves have become so small and their motion so slight that they are quite lost to view. That is the familiar transmutation of visible motion into heat, of molar into molecular, of ordered into disordered motion; for in ordered motion many neighbouring molecules have a common velocity, whilst in disordered motion every molecule has its separate and separately directed velocity.

This process of disintegration or subdivision does not proceed indefinitely, but finds its natural limit in the size of the atoms. For the motion of a single atom by itself is always an ordered one, since the separate parts of an atom all move with the same common velocity. The larger the atoms, the less can the total energy of motion be subdivided. So far, everything is perfectly clear, and the Classical Theory is in excellent agreement with experience.

But now let us take another and quite analogous process, not dealing with water waves but with waves of light and heat. Let us assume that rays emitted by a brightly glowing body are collected by suitable mirrors into a completely enclosed hollow space, and that they are continually thrown to and fro between the reflecting walls of that space. Here also there will be a gradual transmutation of the energy of radiation from longer waves to shorter waves, from ordered radiation to disordered radiation. The longer and coarser waves correspond to the infra-red rays, and the shorter and slighter waves correspond to the ultraviolet rays of the spectrum. Hence, according to the Classical Theory, we must expect the total energy of radiation to concentrate itself upon the ultra-violet portion of the spectrum; or, in other words, we must expect the infra-red and the visible rays to disappear gradually and convert themselves ultimately into invisible ultra-violet or chemical rays.

But of such a phenomenon no trace can be discovered in Nature. The conversion sooner or later attains a perfectly definite and assignable limit, and after that, the radiation-conditions remain stable in every respect.

In order to reconcile this fact with the Classical Theory the most varied experiments have already been made, but the result has always been that the contradiction went too deep into the roots of the Theory to leave them unhurt. So again nothing remains but to re-examine the foundations of the Theory. And again we must admit that the principles of Thermodynamics have shown themselves to be unshakable. For the only method so far found to promise a complete solution of the riddle depends directly upon the two laws of Thermodynamics; though it combines with them a new and peculiar hypothesis, which, if we utilize the two illustrations above mentioned, can be expressed somewhat as follows:

In the case of the Water waves, the disintegration of the energy of motion is limited by the fact that the atoms hold the energy together, in a way, each atom representing a certain finite material Quantum which can only move as a whole. In the same sort of way certain processes must be at work in the case of light and heat rays, although they are quite of an immaterial nature, which shall hold together the energy of radiation in definite finite Quanta, and shall unite it the more strongly the shorter the waves and the quicker therefore the frequency of the oscillations.

In what way we are to conceive the nature of quanta of a purely dynamical nature, we cannot yet say for certain. Possibly such quanta might be accounted for if each source of radiation can only emit energy when that energy attains at least a certain minimum value; just as a rubber pipe, into which air is gradually compressed, bursts and scatters its contents only when the elastic energy in it attains a certain quantity.

In any case, the hypothesis of Quanta has led to the idea that there are changes in Nature which do not occur continuously but in an explosive manner. I need hardly remind you that this view has become much more conceivable since the discovery and investigation of Radio-Active Phenomena. Besides, all difficulties connected with detailed explanation are at present overshadowed by the circumstance that the Quantum Hypothesis has yielded results which are in closer agreement with radiation-measurements than are all previous theories.

Moreover, if it is a good sign for a new hypothesis that it is found applicable to regions for which it was not originally devised, then the Quantum Hypothesis can surely claim a favourable testimony. I shall only refer to a quite particularly striking point. Since we have succeeded in liquefying Air, Hydrogen, and

Helium, a new field of activity has been opened for experimental investigation in the region of low temperatures, and in this region a number of new and in some ways highly surprising results have been obtained. To heat a piece of copper from $-250°$ to $-249°$, i.e. by one degree, we do not require the same quantity of heat as for heating it from $0°$ to $1°$, but a quantity about thirty times less. If we took the original temperature of the copper still lower, the corresponding quantity of heat would turn out many times smaller, without any assignable limit. This fact not only runs counter to our habitual ideas, but also is out of harmony with the demands of the Classical Theory. For although for more than 100 years we have learnt to disintinguish between temperature and quantity of heat, yet we were led by the Kinetic Theory of Matter to suppose that these two quantities, even if not strictly proportional to each other, preserved at all events a sensibly parallel course.

The Quantum Hypothesis has entirely cleared up this difficulty, and in addition has yielded another result of high importance, viz. that the forces controlling the thermal oscillations in a solid body are of the same kind as those which control its elastic oscillations. With the help of the Quantum Hypothesis, therefore, we can now calculate quantitatively the thermal energy of a monatomic substance at various temperatures, from its elastic properties—a performance which was far beyond the reach of the Classical Theory. Hence arise a number of further questions which appear very strange at first sight—for instance, whether perhaps the vibrations of a tuning-fork are not absolutely continuous but are broken up into quanta. It is true that in acoustic vibrations the energy quanta will be extremely small, on account of their relatively low frequency. Thus, in the middle *a*, they would amount to only 3 quadrillionths of a unit of work in absolute mechanical measure. It would be just as little necessary to alter the Theory of Elasticity on that account as on account of the quite analogous circumstance that it treats *matter* as perfectly continuous, whereas it is really constituted atomistically, i.e. according to quanta. But fundamentally the revolutionary aspect of the new conception must be clear to everybody; and although the nature of dynamical quanta still remains somewhat puzzling, yet, in view of the facts now known, it is difficult to doubt their existence, in some form or other. For whatever we can measure must exist.

Thus in the light of recent investigation, the Physical representation of the Universe exhibits an ever more intimate correspondence between its various features, and also manifests a certain peculiar structure whose refinement was hidden to the less trained eye and therefore remained concealed. But ever the question arises: What is the significance of this progress in fundamental conceptions for the satisfaction of our thirst for knowledge? Do we approach one step nearer to a real knowledge of Nature by the refining of our world-image? To this fundamental question let us devote a brief consideration. It is not as if anything essentially new could be said in this region, already traversed by manifold and endless speculation, but that while on this point modern views are often diametrically opposed, yet everyone who takes a deep interest in the real aims of Science must necessarily take up some position.

Thirty-five years ago, Hermann von Helmholtz in this very place expounded the view that our perceptions never give us an image of, but at most a message from, the external world. For every attempt fails to demonstrate any kind of similarity between the nature of the external impression and the nature of the corresponding sensation; all conceptions which we make for ourselves of the external world only reflect our own sensations in the last resort. Is there any sense, therefore, in opposing to our consciousness an independent 'intrinsic Nature'? Are not indeed all so-called 'Laws of Nature' essentially but more or less effective rules by means of which we summarize the temporal course of our sensations as accurately and conveniently as possible? If that were so, then not only common-sense but exact Science would have been fundamentally at fault from the beginning. For it is impossible to deny that the whole evolution of Physical knowledge up to now has aimed towards the completest fundamental division between the happenings of external Nature and the processes of human perception. The way out of this embarrassing difficulty is seen as soon as we go one step further along this line of thought. Let us suppose that a Physical representation of the Universe had been found which fulfils all our demands, and therefore one that can completely and accurately represent all laws of Nature empirically known; still that that image even remotely resembles 'real' Nature, can in no way be proven. But this assertion has another side to it, which is generally too little emphasized: for, in exactly the same sense, the much bolder assertion that the proposed image represents real Nature in all points with absolute fidelity cannot be in any way refuted. For the first step in such a dis-proof would be the ability to assert anything with certainty concerning real Nature, and that, as everybody agrees, is absolutely excluded.

We see that an immense gulf yawns here, into which no Science can ever penetrate. The filling of this gulf is a function not of pure reason, but of practical reason—it is a matter of common-sense.

Just as a given cosmic scheme cannot be scientifically established, so we may also be assured that it will survive every attack so long as it agrees with itself and with the facts of experience. But we must not fall into the error of supposing that it is possible to advance, even in the most exact of all Sciences, without the help of any world-image, i.e. without any unprovable hypotheses. Even in Physics, the phrase holds good that 'There is no Salvation without Faith'—at least a faith in a certain reality outside ourselves. It is this confident faith which guides the advancing creative impulse, this it is which gives the necessary support to the groping imagination, this which alone can raise the spirit depressed by failure and inspire it to new efforts. An observer who does not allow himself to be led in his work by any hypothesis, however cautious and provisional, renounces beforehand all deeper understanding of his own results. Whoever rejects faith in the reality of atoms and electrons, or the electromagnetic nature of Light-waves, or the identity of Heat and Motion, can never be found guilty of a logical or empirical contradiction, but he will find it difficult from his standpoint to advance Physical knowledge.

It is true that faith alone does nothing. As the history of all Science shows, it is liable also to lead astray and to issue in narrowness and fanaticism. If it is to be a reliable guide, it must constantly be tested by the laws of thought and by experience which in the last resort can only be furnished by conscientious and often laborious self-denying solitary work. There is no Prince of Science who is not willing, in case of necessity, to do menial work, whether in the laboratory, the library, in the open air, or at the writing-desk. It is just this hard struggle which ripens and purifies the cosmic view. Only he who has in his own body gone through the process can fully realize its meaning and importance.

ALBERT EINSTEIN

296

On the electrodynamics of moving bodies

I. DEFINITION OF SIMULTANEITY

Let us take a system of coordinates in which the equations of Newtonian mechanics hold good. In order to render our presentation more precise and to distinguish this system of coordinates verbally from others which will be introduced hereafter, we call it the 'stationary system'.

If a material point is at rest relatively to this system of coordinates, its position can be defined relatively thereto by the employment of rigid standards of measurement and the methods of Euclidean geometry, and can be expressed in Cartesian coordinates.

If we wish to describe the *motion* of a material point, we give the values of its coordinates as functions of the time. Now we must bear carefully in mind that a mathematical description of this kind has no physical meaning unless we are quite clear as to what we understand by 'time'. We have to take into account that all our judgements in which time plays a part are always judgements of *simultaneous events*. If, for instance, I say, 'That train arrives here at 7 o'clock', I mean something like this: 'The pointing of the small hand of my watch to 7 and the arrival of the train are simultaneous events.'

It might appear possible to overcome all the difficulties attending the definition of 'time' by substituting 'the position of the small hand of my watch' for 'time'. And in fact such a definition is satisfactory when we are concerned with defining a time exclusively for the place where the watch is located; but it is no longer satisfactory when we have to connect in time series of events occurring at different places, or—what comes to the same thing—to evaluate the times of events occurring at places remote from the watch.

We might, of course, content ourselves with time values determined by an observer stationed together with the watch at the origin of the coordinates, and coordinating the corresponding positions of the hands with light signals, given out by every event to be timed, and reaching him through empty space. But this coordination has the disadvantage that it is not independent of the standpoint of the observer with the watch or clock, as we know from experience. We arrive at a much more practical determination along the following line of thought.

If at the point A of space there is a clock, an observer at A can determine the time values of events in the immediate proximity of A by finding the positions of the hands which are simultaneous with these events. If there is at the point B of space another clock in all respects resembling the one at A, it is possible for an observer at B to determine the time values of events in the immediate neighbourhood of B. But it is not possible without further assumption to compare, in respect of time, an event at A with an event at B. We have so far defined only an 'A time' and a 'B time'. We have not defined a common 'time' for A and B, for the latter cannot be defined at all unless we establish *by definition* that the 'time' required by light to travel from A to B equals the 'time' it requires to travel from B to A. Let a ray of light start at the 'A time' t_A from A towards B, let it at the 'B time' t_B be reflected at B in the direction of A, and arrive again at A at the 'A time' t'_A.

In accordance with this definition the two clocks synchronize if

$$t_B - t_A = t'_A - t_B.$$

We assume that this definition of synchronism is free from contradictions, and possible for any number of points; and that the following relations are universally valid:

(i) If the clock at B synchronizes with the clock at A, the clock at A synchronizes with the clock at B.

(ii) If the clock at A synchronizes with the clock at B and also with the clock at C, the clocks at B and C also synchronize with each other.

Thus with the help of certain imaginary physical experiments we have settled what is to be understood by synchronous stationary clocks located at different places, and have evidently obtained a definition of 'simultaneous', or 'synchronous', and of 'time'. The 'time' of an event is that which is given simultaneously with the event by a stationary clock located at the place of the event. This clock being synchronous, and indeed synchronous for all time determinations, with a specified stationary clock.

In agreement with experience we further assume the quantity

$$\frac{2AB}{t'_A - t_A} = c$$

to be a universal constant—the velocity of light in empty space.

It is essential to have time defined by means of stationary clocks in the stationary system, and the time now defined being appropriate to the stationary system we call it 'the time of the stationary system'.

2. ON THE RELATIVITY OF LENGTHS AND TIMES

The following reflections are based on the principle of relativity and on the principle of the constancy of the velocity of light. These two principles we define as follows:

1. The laws by which the states of physical systems undergo change are not affected, whether these changes of state be referred to the one or the other of two systems of coordinates in uniform translatory motion.

2. Any ray of light moves in the 'stationary' system of coordinates with the determined velocity c, whether the ray be emitted by a stationary or by a moving body. Hence

$$\text{velocity} = \frac{\text{light path}}{\text{time interval}}$$

where time interval is to be taken in the sense of the definition in 1.

Let there be given a stationary rigid rod; and let its length be l as measured by a measuring-rod which is also stationary. We now imagine the axis of the rod lying along the axis of x of the stationary system of coordinates, and that a uniform motion of parallel translation with velocity v along the axis of x in the direction of increasing x is then imparted to the rod. We now inquire as to the length of the moving rod, and imagine its length to be ascertained by the following two operations:

(a) The observer moves together with the given measuring-rod and the rod to be measured, and measures the length of the rod directly by superposing the measuring-rod, in just the same way as if all three were at rest.

(b) By means of stationary clocks set up in the stationary system and synchronizing in accordance with 1, the observer ascertains at what points of the stationary system the two ends of the rod to be measured are located at a definite time. The distance between these two points, measured by the measuring-rod already employed, which in this case is at rest, is also a length which may be designated 'the length of the rod'.

In accordance with the principle of relativity the length to be discovered by the operation (a)—we will call it 'the length of the rod in the moving system'— must be equal to the length l of the stationary rod.

The length to be discovered by the operation (b) we will call 'the length of the (moving) rod in the stationary system'. This we shall determine on the basis of our two principles, and we shall find that it differs from l.

Current kinematics tacitly assumes that the lengths determined by these two operations are precisely equal, or in other words, that a moving rigid body at the epoch *t* may in geometrical respects be perfectly represented by *the same* body *at rest* in a definite position.

We imagine further that at the two ends *A* and *B* of the rod, clocks are placed which synchronize with the clocks of the stationary system, that is to say that their indications correspond at any instant to the 'time of the stationary system' at the places where they happen to be. These clocks are therefore 'synchronous in the stationary system'.

We imagine further that with each clock there is a moving observer, and that these observers apply to both clocks the criterion established in 1 for the synchronization of two clocks. Let a ray of light depart from *A* at the time t_A, let it be reflected at *B* at the time t_B, and reach *A* again at the time t'_A. Taking into consideration the principle of the constancy of the velocity of light we find that

$$t_B - t_A = \frac{r_{AB}}{c - v} \quad \text{and} \quad t'_A - t_B = \frac{r_{AB}}{c + v}$$

where r_{AB} denotes the length of the moving rod—measured in the stationary system. Observers moving with the moving rod would thus find that the two clocks were not synchronous, while observers in the stationary system would declare the clocks to be synchronous.

So we see that we cannot attach any *absolute* signification to the concept of simultaneity, but that two events which, viewed from a system of coordinates, are simultaneous, can no longer be looked upon as simultaneous events when envisaged from a system which is in motion relatively to that system.

297

The foundation of the general theory of relativity

I. OBSERVATIONS ON THE SPECIAL THEORY OF RELATIVITY

The special theory of relativity is based on the following postulate, which is also satisfied by the mechanics of Galileo and Newton.

If a system of coordinates *K* is chosen so that, in relation to it, physical laws hold good in their simplest form, the *same* laws also hold good in relation to any other system of coordinates *K'* moving in uniform translation relatively to *K*. This postulate we call the 'special principle of relativity'. The word 'special' is meant to intimate that the principle is restricted to the case when *K'* has a motion of uniform translation relatively to *K*, but that the equivalence of *K'* and *K* does not extend to the case of non-uniform motion of *K'* relatively to *K*.

Thus the special theory of relativity does not depart from classical mechanics through the postulate of relativity, but through the postulate of the constancy of the velocity of light *in vacuo*, from which, in combination with the special principle of relativity, there follow, in the well-known way, the relativity of simultaneity, the Lorentzian transformation, and the related laws for the behaviour of moving bodies and clocks.

The modification to which the special theory of relativity has subjected the theory of space and time is indeed far-reaching, but one important point has remained unaffected. For the laws of geometry, even according to the special theory of relativity, are to be interpreted directly as laws relating to the possible relative positions of solid bodies at rest; and, in a more general way, the laws of kinematics are to be interpreted as laws which describe the relations of measuring bodies and clocks. To two selected material points of a stationary rigid body there always corresponds a distance of quite definite length, which is independent of the locality and orientation of the body, and is also independent of the time. To two selected positions of the hands of a clock at rest relatively to the privileged system of reference there always corresponds an interval of time of a definite length, which is independent of place and time. We shall soon see that the general theory of relativity cannot adhere to this simple physical interpretation of space and time.

2. THE NEED FOR AN EXTENSION OF THE POSTULATE OF RELATIVITY
In classical mechanics, and no less in the special theory of relativity, there is an inherent epistemological defect which was, perhaps for the first time, clearly pointed out by Ernst Mach. We will elucidate it by the following example: Two fluid bodies of the same size and nature hover freely in space at so great a distance from each other and from all other masses that only those gravitational forces need to be taken into account which arise from the interaction of different parts of the same body. Let the distance between the two bodies be invariable, and in neither of the bodies let there be any relative movements of the parts with respect to one another. But let either mass, as judged by an observer at rest relatively to the other mass, rotate with constant angular velocity about the line joining the masses. This is a verifiable relative motion of the two bodies. Now let us imagine that each of the bodies has been surveyed by means of measuring instruments at rest relatively to itself, and let the surface of S_1 prove to be a sphere, and that of S_2 an ellipsoid of revolution. Thereupon we put the question—What is the reason for this difference in the two bodies? No answer can be admitted as epistemologically satisfactory, unless the reason given is an *observable fact of experience*. The law of causality has not the significance of a statement as to the world of experience, except when *observable facts* ultimately appear as causes and effects.

Newtonian mechanics does not give a satisfactory answer to this question. It pronounces as follows: The laws of mechanics apply to the space R_1, in respect to which the body S_1 is at rest, but not to the space R_2, in respect to which the body S_2 is at rest. But the privileged space R_1 of Galileo, thus introduced, is a merely *fictitious* cause, and not a thing that can be observed. It is therefore clear that Newton's mechanics does not really satisfy the requirement of causality in the case under consideration, but only apparently does so, since it makes the fictitious cause R_1 responsible for the observable difference in the bodies S_1 and S_2.

The only satisfactory answer must be that the physical system consisting of S_1 and S_2 reveals within itself no imaginable cause to which the differing behaviour of S_1 and S_2 can be referred. The cause must therefore lie *outside* this system. We have to take it that the general laws of motion, which in particular determine the shapes of S_1 and S_2, must be such that the mechanical behaviour of S_1 and S_2 is partly conditioned, in quite essential respects, by distant masses which we have not included in the system under consideration. These distant masses and their motions relative to S_1 and S_2 must then be regarded as the seat of the causes (which must be susceptible to observation) of the different behaviour of our two bodies S_1 and S_2. They take over the role of the fictitious cause R_1. Of all imaginable spaces R_1, R_2, etc., in any kind of motion relatively to one another, there is none which we may look upon as privileged *a priori* without reviving the above-mentioned epistemological objection. *The laws of physics must be of such a nature that they apply to systems of reference in any kind of motion.* Along this road we arrive at an extension of the postulate of relativity.

In addition to this weighty argument from the theory of knowledge, there is a well-known physical fact which favours an extension of the theory of relativity. Let K be a Galilean system of reference, i.e. a system relatively to which (at least in the four-dimensional region under consideration) a mass, sufficiently distant from other masses, is moving with uniform motion in a straight line. Let K' be a second system of reference which is moving relatively to K in *uniformly accelerated* translation. Then, relatively to K', a mass sufficiently distant from other masses would have an accelerated motion such that its acceleration and direction of acceleration are independent of the material composition and physical state of the mass.

Does this permit an observer at rest relatively to K' to infer that he is on a 'really' accelerated system of reference? The answer is in the negative; for the above-mentioned relation of freely movable masses to K' may be interpreted equally well in the following way. The system of reference K' is unaccelerated, but the space-time territory in question is under the sway of a gravitational field, which generated the acceleration motion of the bodies relatively to K'.

This view is made possible for us by the teaching of experience as to the existence of a field of force, namely, the gravitational field, which possesses the remarkable property of imparting the same acceleration to all bodies. The mechanical behaviour of bodies relatively to K' is the same as presents itself to experience in the case of systems which we are wont to regard as 'stationary' or as 'privileged'. Therefore, from the physical standpoint, the assumption readily suggests itself that the systems K and K' may both with equal right be looked upon as 'stationary', that is to say, they have an equal title as systems of reference for the physical description of phenomena.

It will be seen from these reflections that in pursuing the general theory of relativity we shall be led to a theory of gravitation, since we are able to 'produce' a gravitational field merely by changing the system of coordinates. It will also be obvious that the principle of the constancy of the velocity of light *in vacuo* must be modified, since we easily recognize that the path of a ray of light with respect to K' must in general be curvilinear, if with respect to K light is propagated in a straight line with a definite constant velocity.

298

Geometry and experience

One reason why mathematics enjoys special esteem, above all other sciences, is that its laws are absolutely certain and indisputable, while those of all other sciences are to some extent debatable and in constant danger of being overthrown by newly discovered facts. In spite of this, the investigator in another department of science would not need to envy the mathematician if the laws of mathematics referred to objects of our mere imagination, and not to objects of reality. For it cannot occasion surprise that different persons should arrive at the same logical conclusions when they have already agreed upon the fundamental laws (axioms), as well as the methods by which other laws are to be deduced therefrom. But there is another reason for the high repute of mathematics, in that it is mathematics which affords the exact natural sciences a certain measure of security, to which without mathematics they could not attain.

At this point an enigma presents itself which in all ages has agitated inquiring minds. How can it be that mathematics, being after all a product of human thought which is independent of experience, is so admirably appropriate to the objects of reality? Is human reason, then, without experience, merely by taking thought, able to fathom the properties of real things?

In my opinion the answer to this question is, briefly, this: As far as the laws of mathematics refer to reality, they are not certain; and as far as they are certain, they do not refer to reality. It seems to me that complete clearness as to this state of things first became common property through that new departure in mathematics which is known by the name of mathematical logic or 'Axiomatics'. The progress achieved by axiomatics consists in its having neatly separated the logical-formal from its objective or intuitive content; according to axiomatics the logical-formal alone forms the subject-matter of mathematics, which is not concerned with the intuitive or other content associated with the logical-formal.

Let us for a moment consider from this point of view any axiom of geometry, for instance, the following: Through two points in space there always passes one and only one straight line. How is this axiom to be interpreted in the older sense and in the more modern sense?

The older interpretation: Every one knows what a straight line is, and what a point is. Whether this knowledge springs from an ability of the human mind or from experience, from some collaboration of the two or from some other source, is not for the mathematician to decide. He leaves the question to the philosopher. Being based upon this knowledge, which precedes all mathematics, the axiom stated above is, like all other axioms, self-evident, that is, it is the expression of a part of this *a priori* knowledge.

The more modern interpretation: Geometry treats of entities which are denoted by the words 'straight line', 'point', etc. These entities do not take for granted any knowledge or intuition whatever, but they presuppose only the validity of the axioms, such as the one stated above, which are to be taken in a purely formal sense, i.e. as void of all content of intuition or experience. These axioms are free creations of the human mind. All other propositions of geometry are logical inferences from the axioms (which are to be taken in the nominalistic sense only). The matter of which geometry treats is first defined by the axioms. Schlick in his book on epistemology has therefore characterized axioms very aptly as 'implicit definitions'.

This view of axioms, advocated by modern axiomatics, purges mathematics of all extraneous elements, and thus dispels the mystic obscurity which formerly surrounded the principles of mathematics. But a presentation of its principles thus clarified makes it also evident that mathematics as such cannot predicate anything about perceptual objects or real objects. In axiomatic geometry the words 'point', 'straight line', etc., stand only for empty conceptual schemata. That which gives them substance is not relevant to mathematics.

Yet on the other hand it is certain that mathematics generally, and particularly geometry, owes its existence to the need which was felt of learning something about the relations of real things to one another. The very word 'geometry', which, of course, means earth-measuring, proves this. For earth-measuring has to do with the possibilities of the disposition of certain natural objects with respect to one another, namely, with parts of the earth, measuring-lines, measuring-wands, etc. It is clear that the system of concepts of axiomatic geometry alone cannot make any assertions as to the relations of real objects of this kind, which we will call practically-rigid bodies. To be able to make such assertions, geometry must be stripped of its merely logical-formal character by the coordination of real objects of experience with the empty conceptual framework of axiomatic geometry. To accomplish this, we need only add the proposition: Solid bodies are related, with respect to their possible dispositions, as are bodies in Euclidean geometry of three dimensions. Then the propositions of Euclid contain affirmations as to the relations of practically-rigid bodies.

Geometry thus completed is evidently a natural science; we may in fact regard it as the most ancient branch of physics. Its affirmations rest essentially on induction from experience, but not on logical inferences only. We will call this completed geometry 'practical geometry', and shall distinguish it in what follows from 'purely axiomatic geometry'. The question whether the practical geometry of the universe is Euclidean or not has a clear meaning, and its answer can only be furnished by experience. All linear measurement in physics is practical geometry in this sense, so too is geodetic and astronomical linear measurement, if we call to our help the law of experience that light is propagated in a straight line, and indeed in a straight line in the sense of practical geometry.

I attach special importance to the view of geometry which I have just set forth, because without it I should have been unable to formulate the theory of relativity. Without it the following reflection would have been impossible: In a system of reference rotating relatively to an inert system, the laws of disposition of rigid bodies do not correspond to the rules of Euclidean geometry on account of the Lorentz[1] contraction; thus if we admit non-inert systems we must abandon Euclidean geometry. The decisive step in the transition to general co-variant equations would certainly not have been taken if the above interpretation had not served as a stepping-stone. If we deny the relation between the body of axiomatic Euclidean geometry and the practically-rigid body of reality, we readily arrive at the following view, which was entertained by that acute and profound thinker, H. Poincaré:[2] Euclidean geometry is distinguished above all other imaginable axiomatic geometrics by its simplicity. Now since axiomatic geometry by itself contains no assertions as to the reality which can be experienced, but can do so only in combination with physical laws, it should be possible and reasonable—whatever may be the nature of reality—to retain Euclidean geometry. For if contradictions between theory and experience manifest themselves, we should rather decide to change physical laws than to change axiomatic Euclidean geometry. If we deny the relation between the practically-rigid body and geometry, we shall indeed not easily free ourselves from the convention that Euclidean geometry is to be retained as the simplest. Why is the equivalence of the practically-rigid body and the body of geometry —which suggests itself so readily—denied by Poincaré and other investigators? Simply because under closer inspection the real solid bodies in nature are not rigid, because their geometrical behaviour, that is, their possibilities of relative disposition, depend upon temperature, external forces, etc. Thus the original, immediate relation between geometry and physical reality appears destroyed, and we feel impelled towards the following more general view, which characterizes Poincaré's standpoint. Geometry (G) predicates nothing about the rela-

---

1. Hendrik Anton Lorentz (1853–1928), Dutch physicist. He extended Maxwell's equations by including the movement of the electron in the electromagnetic field. The 'Lorentz contraction' was a hypothesis introduced to explain the null-effect of the Michelson experiment. (Ed.)
2. Henri Poincaré (1854–1912), French mathematican and physicist. In addition to his important researches in mathematics and celestial mechanics he made essential contributions to the philosophy of science. He emphasized the role of conventions in scientific method. (Ed.)

tions of real things, but only geometry together with the purport (P) of physical laws can do so. Using symbols, we may say that only the sum of (G)+(P) is subject to the control of experience. Thus (G) may be chosen arbitrarily, and also parts of (P); all these laws are conventions. All that is necessary to avoid contradictions is to choose the remainder of (P) so that (G) and the whole of (P) are together in accord with experience. Envisaged in this way, axiomatic geometry and the part of natural law which has been given a conventional status appear as epistemologically equivalent.

Sub specie aeterni Poincaré, in my opinion, is right. The idea of the measuring-rod and the idea of the clock coordinated with it in the theory of relativity do not find their exact correspondence in the real world. It is also clear that the solid body and the clock do not in the conceptual edifice of physics play the part of irreducible elements, but that of composite structures, which may not play any independent part in theoretical physics. But it is my conviction that in the present stage of development of theoretical physics these ideas must still be employed as independent ideas; for we are still far from possessing such certain knowledge of theoretical principles as to be able to give exact theoretical constructions of solid bodies and clocks.

Further, as to the objection that there are no really rigid bodies in nature, and that therefore the properties predicated of rigid bodies do not apply to physical reality—this objection is by no means so radical as might appear from a hasty examination. For it is not a difficult task to determine the physical state of a measuring-rod so accurately that its behaviour relatively to other measuring-bodies shall be sufficiently free from ambiguity to allow it to be substituted for the 'rigid' body. It is to measuring-bodies of this kind that statements as to rigid bodies must be referred.

All practical geometry is based upon a principle which is accessible to experience, and which we will now try to realize. We will call that which is enclosed between two boundaries, marked upon a practically-rigid body, a tract. We imagine two practically-rigid bodies, each with a tract marked out on it. These two tracts are said to be 'equal to one another' if the boundaries of the one tract can be brought to coincide permanently with the boundaries of the other. We now assume that:

If two tracts are found to be equal once and anywhere, they are equal always and everywhere.

Not only the practical geometry of Euclid, but also its nearest generalization, the practical geometry of Riemann,[1] and therewith the general theory of relativity, rest upon this assumption. Of the experimental reasons which warrant this assumption I will mention only one. The phenomenon of the propagation of light in empty space assigns a tract, namely, the appropriate path of light, to

1. Bernhard Riemann (1826–1866), German mathematician. His paper 'On the hypotheses which form the foundation of geometry' introduced a new approach to geometry which included Euclidean as well as non-Euclidean geometries as special cases. (Ed.)

each interval of local time, and conversely. Thence it follows that the above assumption for tracts must also hold good for intervals of clock-time in the theory of relativity. Consequently it may be formulated as follows: If two ideal clocks are going at the same rate at any time and at any place (being then in immediate proximity to each other), they will always go at the same rate, no matter where and when they are again compared with each other at one place. If this law were not valid for real clocks, the proper frequencies for the separate atoms of the same chemical element would not be in such exact agreement as experience demonstrates. The existence of sharp spectral lines is a convincing experimental proof of the above-mentioned principle of practical geometry. This is the ultimate foundation in fact which enables us to speak with meaning of the mensuration, in Riemann's sense of the word, of the four-dimensional continuum of space-time.

The question whether the structure of this continuum is Euclidean, or in accordance with Riemann's general scheme, or otherwise, is, according to the view which is here being advocated, properly speaking a physical question which must be answered by experience, and not a question of a mere convention to be selected on practical grounds. Riemann's geometry will be the right thing if the laws of disposition of practically-rigid bodies are transformable into those of the bodies of Euclid's geometry with an exactitude which increases in proportion as the dimensions of the part of space-time under consideration are diminished.

299

Aether and relativity

The space-time theory and the kinematics of the special theory of relativity were modelled on the Maxwell–Lorentz theory of the electromagnetic field. This theory therefore satisfies the conditions of the special theory of relativity, but when viewed from the latter it acquires a novel aspect. For if K be a system of coordinates relatively to which the Lorentzian ether is at rest, the Maxwell–Lorentz equations are valid primarily with reference to K. But by the special theory of relativity the same equations without any change of meaning also hold in relation to any new system of coordinates K' which is moving in uniform translation relatively to K. Now comes the anxious question: Why must I in the theory distinguish the K system above all K' systems, which are physically equivalent to it in all respects, by assuming that the ether is at rest relatively to the K system? For the theoretician such an asymmetry in the theoretical structure, with no corresponding asymmetry in the system of experience, is intolerable. If we assume the ether to be at rest relatively to K, but in motion relatively to K', the physical equivalence of K and K' seems to me from the logical standpoint, not indeed downright incorrect, but nevertheless inacceptable.

The next position which it was possible to take up in face of this state of things appeared to be the following. The ether does not exist at all. The electromagnetic fields are not states of a medium, and are not bound down to any bearer, but they are independent realities which are not reducible to anything else, exactly like the atoms of ponderable matter. This conception suggests itself the more readily as, according to Lorentz's theory, electromagnetic radiation, like ponderable matter, brings impulse and energy with it, and as, according to the special theory of relativity, both matter and radiation are but special forms of distributed energy, ponderable mass losing its isolation and appearing as a special form of energy.

More careful reflection teaches us, however, that the special theory of relativity does not compel us to deny ether. We may assume the existence of an ether; only we must give up ascribing a definite state of motion to it, i.e. we must by abstraction take from it the last mechanical characteristic which Lorentz had still left it. [. . .]

Certainly, from the standpoint of the special theory of relativity, the ether hypothesis appears at first to be an empty hypothesis. In the equations of the electromagnetic field there occur, in addition to the densities of the electric charge, *only* the intensities of the field. The career of electromagnetic processes *in vacuo* appears to be completely determined by these equations, uninfluenced by other physical quantities. The electromagnetic fields appear as ultimate, irreducible realities, and at first it seems superfluous to postulate a homogeneous, isotropic ether-medium, and to envisage electromagnetic fields as states of this medium.

But on the other hand there is a weighty argument to be adduced in favour of the ether hypothesis. To deny the ether is ultimately to assume that empty space has no physical qualities whatever. The fundamental facts of mechanics do not harmonize with this view. For the mechanical behaviour of a corporeal system hovering freely in empty space depends not only on relative positions (distances) and relative velocities, but also on its state of rotation, which physically may be taken as a characteristic not appertaining to the system in itself. In order to be able to look upon the rotation of the system, at least formally, as something real, Newton objectivizes space. Since he classes his absolute space together with real things, for him rotation relative to an absolute space is also something real. Newton might no less well have called his absolute space 'Ether'; what is essential is merely that besides observable objects, another thing, which is not perceptible, must be looked upon as real, to enable acceleration or rotation to be looked upon as something real.

It is true that Mach tried to avoid having to accept as real something which is not observable by endeavouring to substitute in mechanics a mean acceleration with reference to the totality of the masses in the universe in place of an acceleration with reference to absolute space. But inertial resistance opposed to relative acceleration of distant masses presupposes action at a distance; and as the modern physicst does not believe that he may accept this action at a distance, he comes back once more, if he follows Mach, to the ether, which has to serve as medium for the effects of inertia. But this conception of the ether to which we are led by Mach's way of thinking differs essentially from the ether as conceived by Newton, by Fresnel, and by Lorentz. Mach's ether not only *conditions* the behaviour of inert masses, but *is also conditioned* in its state by them.

Mach's idea finds its full development in the ether of the general theory of relativity. According to this theory the metrical qualities of the continuum of space-time differ in the environment of different points of space-time, and are partly conditioned by the matter existing outside of the territory under consideration. This space-time variability of the reciprocal relations of the standards of space and time, or, perhaps, the recognition of the fact that 'empty space' in its physical relation is neither homogeneous nor isotropic, compelling us to describe its state by ten functions (the gravitation potentials $g_{\mu\nu}$), has, I think, finally disposed of the view that space is physically empty. But therewith the conception of the ether has again acquired an intelligible content, although this content differs widely from that of the ether of the mechanical undulatory theory of light. The ether of the general theory of relativity is a medium which is itself devoid of *all* mechanical and kinematical qualities, but helps to determine mechanical (and electromagnetic) events. [. . .]

Recapitulating, we may say that according to the general theory of relativity space is endowed with physical qualities; in this sense, therefore, there exists an ether. According to the general theory of relativity space without ether is unthinkable; for in such space there not only would be no propagation of light, but also no possibility of existence for standards of space and time (measuring-rods and clocks), nor therefore any space-time intervals in the physical sense. But this ether may not be thought of as endowed with the quality characteristic of ponderable media, as consisting of parts which may be tracked through time. The idea of motion may not be applied to it.

300

On a heuristic point of view about the creation and conversion of light

There exists an essential formal difference between the theoretical pictures physicists have drawn of gases and other ponderable bodies and Maxwell's theory of electromagnetic processes in so-called empty space. Whereas we assume the state of a body to be completely determined by the positions and velocities of an, albeit very large, still finite number of atoms and electrons, we use for the determination of the electromagnetic state in space continuous spatial functions, so that a finite number of variables cannot be considered to be sufficient to fix completely the electromagnetic state in space. According to

Maxwell's theory, the energy must be considered to be a continuous function in space for all purely electromagnetic phenomena, thus also for light, while according to the present-day ideas of physicists the energy of a ponderable body can be written as a sum over the atoms and electrons. The energy of a ponderable body cannot be split into arbitrarily many, arbitrarily small parts, while the energy of a light ray, emitted by a point source of light is according to Maxwell's theory (or in general according to any wave theory) of light distributed continuously over an ever increasing volume.

The wave theory of light which operates with continuous functions in space has been excellently justified for the representation of purely optical phenomena and it is unlikely ever to be replaced by another theory. One should, however, bear in mind that optical observations refer to time averages and not to instantaneous values and notwithstanding the complete experimental verification of the theory of diffraction, reflexion, refraction, dispersion, and so on, it is quite conceivable that a theory of light involving the use of continuous functions in space will lead to contradictions with experience, if it is applied to the phenomena of the creation and conversion of light.

In fact, it seems to me that the observations on 'black-body radiation', photoluminescence, the production of cathode rays by ultraviolet light and other phenomena involving the emission or conversion of light can be better understood on the assumption that the energy of light is distributed discontinously in space. According to the assumption considered here, when a light ray starting from a point is propagated, the energy is not continuously distributed over an ever increasing volume, but it consists of a finite number of energy quanta, localized in space, which move without being divided and which can be absorbed or emitted only as a whole.

In the following, I shall communicate the train of thought and the facts which led me to this conclusion, in the hope that the point of view to be given may turn out to be useful for some research workers in their investigations. [. . .]

ON THE PRODUCTION OF CATHODE RAYS BY ILLUMINATION OF SOLIDS

The usual idea that the energy of light is continuously distributed over the space through which it travels meets with especially great difficulties when one tries to explain photoelectric phenomena, as was shown in the pioneering paper by Mr. Lenard.

According to the idea that the incident light consists of energy quanta with an energy $h\nu$,[1] one can picture the production of cathode rays by light as follows. Energy quanta penetrate into a surface layer of the body, and their energy is at least partly transformed into electron kinetic energy. The simplest

---

1. In the original text the proportionality factor of ν is given as $R\beta/N$. R is the universal gas constant, equal to kN ($k=$ Boltzmann's constant, $N=$ Avogadro's number), and $\beta=h/k$. This factor can thus be replaced by h, Planck's constant. (Ed.)

picture is that a light quantum transfers all of its energy to a single electron, we shall assume that that happens. We must, however, not exclude the possibility that electrons only receive part of the energy from light quanta. An electron obtaining kinetic energy inside the body will have lost part of its kinetic energy when it has reached the surface. Moreover, we must assume that each electron on leaving the body must produce work P, which is characteristic for the body. Electrons which are excited at the surface and at right angles to it will leave the body with the greatest normal velocity. The kinetic energy of such electrons is

$$h\nu - P \; [\ldots]$$

As far as I can see, our ideas are not in contradiction to the properties of the photoelectric action observed by Mr Lenard. If every energy quantum of the incident light transfers its energy to electrons independently of all other quanta, the velocity distribution of the electrons, that is, the quality of the resulting cathode radiation, will be independent of the intensity of the incident light; on the other hand, *ceteris paribus*, the number of electrons leaving the body should be proportional to the intensity of the incident light.

ERNEST RUTHERFORD

301

The electric structure of matter

In order to view the extensive territory which has been conquered by science in this interval, it is desirable to give a brief summary of the state of knowledge of the constitution of matter at the beginning of this epoch. Ever since its announcement by Dalton the atomic theory has steadily gained ground, and formed the philosophic basis for the explanation of the facts of chemical combination. In the early stages of its application to physics and chemistry it was unnecessary to have any detailed knowledge of the dimensions or structure of the atom. It was only necessary to assume that the atoms acted as individual units, and to know the relative masses of the atoms of the different elements. In this next stage, for example, in the kinetic theory of gases, it was possible to explain the main properties of gases by supposing that the atoms of the gas acted as minute perfectly elastic spheres. During this period, by the application of a variety of methods, many of which were due to Lord Kelvin, rough estimates had been obtained of the absolute dimensions and mass of the atoms. These brought out the minute size and mass of the atom and the enormous number of atoms necessary to produce a detectable effect in any kind of measurement. From this arose the general idea that the atomic theory must of necessity for ever remain unverifiable by direct experiment, and for this reason it was suggested by one school of thought that the atomic theory should be banished from the teaching of chemistry, and that the law of multiple proportions should be accepted as the ultimate fact of chemistry.

While the vaguest ideas were held as to the possible structure of atoms, there was a general belief among the more philosophically minded that the atoms of the elements could not be regarded as simple unconnected units. The periodic variations of the properties of the elements brought out by Mendeleeff were only explicable if atoms were similar structures in some way constructed of similar material. We shall see that the problem of the constitution of atoms is intimately connected with our conception of the nature of electricity. The wonderful success of the electromagnetic theory had concentrated attention on the medium or ether surrounding the conductor of electricity, and little attention had been paid to the actual carriers of the electric current itself. At the same time the idea was generally gaining ground that an explanation of the results of Faraday's experiments on electrolysis was only possible on the assumption that electricity, like matter, was atomic in nature. The name 'electron' had even been given to this fundamental unit by Johnstone Stoney, and its magnitude roughly estimated, but the full recognition of the significance and importance of this conception belongs to the new epoch.

For the clarifying of these somewhat vague ideas, the proof in 1897 of the independent existence of the electron as a mobile electrified unit, of mass minute compared with that of the lightest atom, was of extraordinary importance. It was soon seen that the electron must be a constituent of all the atoms of matter, and that optical spectra had their origin in their vibrations. The discovery of the electron and the proof of its liberation by a variety of methods from all the atoms of matter was of the utmost significance, for it strengthened the view that the electron was probably the common unit in the structure of atoms which the periodic variation of the chemical properties had indicated. It gave for the first time some hope of the success of an attack on that most fundamental of all problems—the detailed structure of the atom. In the early development of the subject science owes much to the work of Sir J. J. Thomson,[1] both for the boldness of his ideas and for his ingenuity in developing methods for estimating the number of electrons in the atom, and in probing its structure. He early took the view that the atom must be an electrical structure, held together by electrical forces, and showed in a general way lines of possible explanation of the variation of physical and chemical properties of the elements, exemplified in the periodic law.

In the meantime our whole conception of the atom and of the magnitude of the forces which held it together were revolutionized by the study of radioactivity. The discovery of radium was a great step in advance, for it provided the experimenter with powerful sources of radiation specially suitable for examining the nature of the characteristic radiations which are emitted by the

1. Sir Joseph John Thomson (1856–1940), English physicist, discoverer of the electron. His model of the atom was superseded by that of Rutherford. (Ed.)

radioactive bodies in general. It was soon shown that the atoms of radioactive matter were undergoing spontaneous transformation, and that the characteristic radiations emitted, namely the α-, β-, and γ-rays, were an accompaniment and consequence of these atomic explosions. The wonderful succession of changes that occur in uranium and thorium, more than thirty in number, was soon disclosed and simply interpreted on the transformation theory. The radioactive elements provide us for the first time with a glimpse into Nature's laboratory, and allow us to watch and study, but not to control, the changes that have their origin in the heart of the radioactive atoms. These atomic explosions involve energies which are gigantic compared with those involved in any ordinary physical or chemical process. In the majority of cases an α-particle is expelled at high speed, but in others a swift electron is ejected often accompanied by a γ-ray, which is a very penetrating X-ray of high frequency. The proof that the α-particle is a charged helium atom for the first time disclosed the importance of helium as one of the units in the structure of the radioactive atoms, and probably also in that of the atoms of most of the ordinary elements. Not only then have the radioactive elements had the greatest direct influence on natural philosophy, but in subsidiary ways they have provided us with experimental methods of almost equal importance. The use of α-particles as projectiles with which to explore the interior of the atom has definitely exhibited its nuclear structure, has led to artificial disintegration of certain light atoms, and promises to yield more information yet as to the actual structure of the nucleus itself. [. . .]

We shall see that there is the strongest evidence that the atoms of matter are built up of these two electrical units, namely the electron and the hydrogen nucleus or proton, as it is usually called when it forms part of the structure of any atom. It is probable that these two are the fundamental and indivisible units which build up our universe, but we may reserve in our mind the possibility that further inquiry may some day show that these units are complex, and divisible into even more fundamental entities. On the views we have outlined, the mass of the atom is the sum of the electrical masses of the individual charged units composing its structure, and there is no need to assume that any other kind of mass exists. At the same time, it is to be borne in mind that the actual mass of an atom may be somewhat less than the sum of the masses of component positive and negative electrons when in the free state. On account of the very close proximity of the charged units in the nucleus of an atom, and the consequent disturbance of the electric and magnetic fields surrounding them, such a decrease of mass is to be anticipated on general theoretical grounds.

We must now look back again to the earlier stages of the present epoch in order to trace the development of our ideas on the detailed structure of the atom. That electrons as such were important constituents was clear by 1900, but little real progress followed until the part played by the positive charges was made clear. New light was thrown on the subject by examining the deviation of α-particles when they passed through the atoms of matter. It was found that occasionally a swift α-particle was deflected from its rectilinear path through more than a right angle by an encounter with a single atom. In such a collision the laws of dynamics ordinarily apply, and the relation between the velocities of the colliding atoms before and after collision are exactly the same as if the two colliding particles are regarded as perfectly elastic spheres of minute dimensions. It must, however, be borne in mind that in these atomic collisions there is no question of mechanical impacts such as we observe with ordinary matter. The reaction between the two particles occurs through the intermediary of the powerful electric fields that surround them. Beautiful photographs illustrating the accuracy of these laws of collision between an α-particle and an atom have been obtained by Messrs Wilson,[1] Blackett, and others, while Mr Wilson has recently obtained many striking illustrations of collisions between two electrons. Remembering the great kinetic energy of the α-particle, its deflection through a large angle in a single atomic encounter shows clearly that very intense deflecting forces exist inside the atom. It seemed clear that electric fields of the required magnitude could be obtained only if the main charge of the atom were concentrated in a minute nucleus. From this arose the conception of the nuclear atom, now so well known, in which the heart of the atom is supposed to consist of a minute but massive nucleus, carrying a positive charge of electricity, and surrounded at a distance by the requisite number of electrons to form a neutral atom.

A detailed study of the scattering of α-particles at different angles, by Geiger[2] and Marsden, showed that the results were in close accord with this theory, and that the intense electric forces near the nucleus varied according to the ordinary inverse square law. In addition, the experiments allowed us to fix an upper limit for the dimensions of the nucleus. For a heavy atom like that of gold the radius of the nucleus, if supposed to be spherical, was less than one-thousandth of the radius of the complete atom surrounded by its electrons, and certainly less than 4×10^{-12} cm. All the atoms were found to show this nuclear structure, and an approximate estimate was made of the nuclear charge of different atoms. This type of nuclear atom, based on direct experimental evidence, possesses some very simple properties. It is obvious that the number of units of resultant positive charge in the nucleus fixes the number of the outer planetary electrons in the neutral atom. In addition, since these outer electrons are in some way held in equilibrium by the attractive forces from the nucleus, and, since we are confident from general physical and chemical evidence that all atoms of any one element are identical in their external structure, it is clear that their arrangement and motion must be governed entirely by the magnitude of the nuclear charge. Since the ordinary chemical and physical properties are to be ascribed mainly to the configuration and motion of the outer electrons, it follows that the

1. C. T. R. Wilson (1869–1959), Scottish physicist, developed the Wilson 'cloud chamber' by which the paths of electrically charged particles become visible. (Ed.)
2. Hans Geiger (1882–1945), German physicist, known for his work on radioactivity and his construction of the 'Geiger-counter'. (Ed.)

properties of an atom are defined by a whole number representing its nuclear charge. [. . .]

Before leaving this subject it is desirable to say a few words on the important question of the energy relations involved in the formation and disintegration of atomic nuclei, first opened up by the study of radioactivity. For example, it is well known that the total evolution of energy during the complete disintegration of one gram of radium is many millions of times greater than in the complete combustion of an equal weight of coal. It is known that this energy is initially mostly emitted in the kinetic form of swift α- and β-particles, and the energy of motion of these bodies is ultimately converted into heat when they are stopped by matter. Since it is believed that the radioactive elements are analogous in structure to the ordinary inactive elements, the idea naturally arose that the atoms of all the elements contained a similar concentration of energy, which would be available for use if only some simple method could be discovered of promoting and controlling their disintegration. This possibility of obtaining new and cheap sources of energy for practical purposes was naturally an alluring prospect to the lay and scientific man alike. It is quite true that, if we were able to hasten the radioactive processes in uranium and thorium so that the whole cycle of their disintegration could be confined to a few days instead of being spread over thousands of millions of years, these elements would provide very convenient sources of energy on a sufficient scale to be of considerable practical importance. Unfortunately, although many experiments have been tried, there is no evidence that the rate of disintegration of these elements can be altered in the slightest degree by the most powerful laboratory agencies. With increase in our knowledge of atomic structure there has been a gradual change of our point of view on this important question, and there is by no means the same certainty today as a decade ago that the atoms of an element contain hidden stores of energy. It may be worth while to spend a few minutes in discussing the reason for this change in outlook. This can best be illustrated by considering an interesting analogy between the transformation of a radioactive nucleus and the changes in the electron arrangement of an ordinary atom. It is now well known that it is possible by means of electron bombardment or by appropriate radiation to excite an atom in such a way that one of its superficial electrons is displaced from its ordinary stable position to another temporarily stable position further removed from the nucleus. This electron in course of time falls back into its old position, and its potential energy is converted into radiation in the process. There is some reason for believing that the electron has a definite average life in the displaced position, and that the chance of its return to its original position is governed by the laws of probability. In some respects an 'excited' atom of this kind is thus analogous to a radioactive atom, but of course the energy released in the disintegration of a nucleus is of an entirely different order of magnitude from the energy released by return of the electron in the excited atom. It may be that the elements, uranium and thorium, represent the sole survivals in the earth today of types of elements that were common in the long-distant ages, when the atoms now composing the earth were in course of formation. A fraction of the atoms of uranium and thorium formed at that time has survived over the long interval on account of their very slow rate of transformation. It is thus possible to regard these atoms as having not yet completed the cycle of changes which the ordinary atoms have long since passed through, and that the atoms are still in the 'excited' state where the nuclear units have not yet arranged themselves in positions of ultimate equilibrium, but still have a surplus of energy which can only be released in the form of the characteristic radiation from active matter. On such a view, the presence of a store of energy ready for release is not a property of all atoms, but only a special class of atoms like the radioactive atoms which have not yet reached the final state for equilibrium.

It may be urged that the artificial disintegration of certain elements by bombardment with swift α-particles gives definite evidence of a store of energy in some of the ordinary elements, for it is known that a few of the hydrogen nuclei, released from aluminium for example, are expelled with such swiftness that the particle has a greater individual energy than the α-particle which causes their liberation. Unfortunately, it is very difficult to give a definite answer on this point until we know more of the details of this disintegration.

On the other hand, another method of attack on this question has become important during the last few years, based on the comparison of the relative masses of the elements. This new point of view can best be illustrated by a comparison of the atomic masses of hydrogen and helium. As we have seen, it seems very probable that helium is not an ultimate unit in the structure of nuclei, but is a very close combination of four hydrogen nuclei and two electrons. The mass of the helium nucleus, 4·00 in terms of O = 16, is considerably less than the mass, 4·03, of four hydrogen nuclei. On modern views there is believed to be a very close connection between mass and energy, and this loss in mass in the synthesis of the helium nucleus from hydrogen nuclei indicates that a large amount of energy in the form of radiation has been released in the building of the helium nucleus from its components. It is easy to calculate from this loss of mass that the energy set free in forming one gram of helium is large even compared with that liberated in the total disintegration of one gram of radium. For example, calculation shows that the energy released in the formation of one pound of helium gas is equivalent to the energy emitted in the complete combustion of about eight thousand tons of pure carbon. It has been suggested by Eddington and Perrin that it is mainly to this source of energy that we must look to maintain the heat emission of the sun and hot stars over long periods of time. Calculations of the loss of heat from the sun show that this synthesis of helium need only take place slowly in order to maintain the present rate of radiation for periods of the order of one thousand million years. It must be acknowledged that these arguments are somewhat speculative in character, for no certain experimental evidence has yet been obtained that helium can be formed from hydrogen.

The evidence of the slow rate of stellar evolution, however, certainly indicates that the synthesis of helium, and perhaps other elements of higher atomic weight, may take place slowly in the interior of hot stars. While in the electric discharge through hydrogen at low pressure we can easily reproduce the conditions of the interior of the hottest star so far as regards the energy of motion of the electrons and hydrogen nuclei, we cannot hope to reproduce that enormous density of radiation which must exist in the interior of a giant star. For this and other reasons it may be very difficult, or even impossible, to produce helium from hydrogen under laboratory conditions.

If this view of the great heat emission in the formation of helium be correct, it is clear that the helium nucleus is the most stable of all nuclei, for an amount of energy corresponding to three or four α-particles would be required to disrupt it into its components. In addition, since the mass of the proton in nuclei is nearly 1·000 instead of its mass 1·0072 in the free state, it follows that much more energy must be put into the atom than will be liberated by its disintegration into its ultimate units. At the same time, if we consider an atom of oxygen, which may be supposed to be built up of four helium nuclei as secondary units, the change of mass, if any, in its synthesis from already formed helium nuclei is so small that we cannot yet be certain whether there will be a gain or loss of energy by its distintegration into helium nuclei, but in any case we are certain that the magnitude of the energy will be much less than for the synthesis of helium from hydrogen. Our information on this subject of energy changes in the formation or disintegration of atoms in general is as yet too uncertain and speculative to give any decided opinion on future possibilities in this direction, but I have endeavoured to outline some of the main arguments which should be taken into account. [. . .]

In watching the rapidity of this tide of advance in physics I have become more and more impressed by the power of the scientific method of extending our knowledge of Nature. Experiment, directed by the disciplined imagination either of an individual, or still better, of a group of individuals of varied mental outlook, is able to achieve results which far transcend the imagination alone of the greatest natural philosopher. Experiment without imagination, or imagination without recourse to experiment, can accomplish little, but, for effective progress, a happy blend of these two powers is necessary. The unknown appears as a dense mist before the eyes of men. In penetrating this obscurity we cannot invoke the aid of supermen but must depend on the combined efforts of a number of adequately trained ordinary men of scientific imagination. Each in his own special field of inquiry is enabled by the scientific method to penetrate a short distance, and his work reacts upon and influences the whole body of other workers. From time to time there arises an illuminating conception, based on accumulated knowledge, which lights up a large region and shows the connection between these individual efforts so that a general advance follows. The attack begins anew on a wider front, and often with improved technical weapons. The conception which led to this advance often appears simple and obvious when once it has been put forward. This is a common experience, and the scientific man often feels a sense of disappointment that he himself had not foreseen a development which ultimately seems so clear and inevitable.

The intellectual interest due to the rapid growth of science today cannot fail to act as a stimulus to young men to join in scientific investigation. In every branch of science there are numerous problems of fundamental interest and importance which await solution. We may confidently predict an accelerated rate of progress of scientific discovery, beneficial to mankind certainly in a material, but possibly even more so in an intellectual sense.

NIELS BOHR

302

On the constitution of atoms and molecules

In order to explain the results of experiments on scattering of α-rays by matter Professor Rutherford has given a theory of the structure of atoms. According to this theory, the atoms consist of a positively charged nucleus surrounded by a system of electrons kept together by attractive forces from the nucleus; the total negative charge of the electrons is equal to the positive charge of the nucleus. Further, the nucleus is assumed to be the seat of the essential part of the mass of the atom, and to have linear dimensions exceedingly small compared with the linear dimensions of the whole atom. The number of electrons in an atom is deduced to be approximately equal to half the atomic weight. Great interest is to be attributed to this atom-model; for, as Rutherford has shown, the assumption of the existence of nuclei, as those in question, seems to be necessary in order to account for the results of the experiments on large angle scattering of the α-rays.

In an attempt to explain some of the properties of matter on the basis of this atom-model we meet, however, with difficulties of a serious nature arising from the apparent instability of the system of electrons: difficulties purposely avoided in atom-models previously considered, for instance, in the one proposed by Sir J. J. Thomson. According to the theory of the latter the atom consists of a sphere of uniform positive electrification, inside which the electrons move in circular orbits.

The principal difference between the atom-models proposed by Thomson and Rutherford consists in the circumstance that the forces acting on the electrons in the atom-model of Thomson allow of certain configurations and motions of the electrons for which the system is in a stable equilibrium; such configurations, however, apparently do not exist for the second atom-model. The nature of the difference in question will perhaps be most clearly seen by noticing that among the quantities characterizing the first atom a quantity appears—the radius of the positive sphere—of dimensions of a length and of the same order of magnitude as the linear extension of the atom, while such a length does not appear among the quantities characterizing the second atom, viz. the charges and masses of the electrons and the positive nucleus; nor can it be determined solely by help of the latter quantities.

The way of considering a problem of this kind has, however, undergone essential alterations in recent years owing to the development of the theory of the energy radiation, and the direct affirmation of the new assumptions introduced in this theory, found by experiments on very different phenomena such as specific heats, photoelectric effect, Röntgen rays, etc. The result of the discussion of these questions seems to be a general acknowledgement of the inadequacy of the classical electrodynamics in describing the behaviour of systems of atomic size. Whatever the alteration in the laws of motion of the electrons may be, it seems necessary to introduce in the laws in question a quantity foreign to the classical electrodynamics, i.e. Planck's constant, or as it often is called the elementary quantum of action. By the introduction of this quantity the question of the stable configuration of the electrons in the atoms is essentially changed, as this constant is of such dimensions and magnitude that it, together with the mass and charge of the particles, can determine a length of the order of magnitude required.

This paper is an attempt to show that the application of the above ideas to Rutherford's atom-model affords a basis for a theory of the constitution of atoms. It will further be shown that from this theory we are led to a theory of the constitution of molecules.

In the present first part of the paper the mechanism of the binding of electrons by a positive nucleus is discussed in relation to Planck's theory. It will be shown that it is possible from the point of view taken to account in a simple way for the law of the line spectrum of hydrogen. Further, reasons are given for a principal hypothesis on which the considerations contained in the following parts are based. [. . .]

PART I: BINDING OF ELECTRONS BY POSITIVE NUCLEI
The inadequacy of the classical electrodynamics in accounting for the properties of atoms from an atom-model such as Rutherford's, will appear very clearly if we consider a simple system consisting of a positively charged nucleus of very small dimensions and an electron describing closed orbits around it. For simplicity, let us assume that the mass of the electron is negligibly small in comparison with that of the nucleus, and further, that the velocity of the electron is small compared with that of light.

Let us at first assume that there is no energy radiation. In this case the electron will describe stationary elliptical orbits. The frequency of revolution ω and the major-axis of the orbit $2a$ will depend on the amount of energy W which must be transferred to the system in order to remove the electron to an infinitely great distance apart from the nucleus. [. . .]

Let us now, however, take the effect of the energy radiation into account, calculated in the ordinary way from the acceleration of the electron. In this case the electron will no longer describe stationary orbits. W will continuously increase, and the electron will approach the nucleus describing orbits of smaller and smaller dimensions, and with greater and greater frequency; the electron on the average gaining in kinetic energy at the same time as the whole system loses energy. This process will go on until the dimensions of the orbit are of the same order of magnitude as the dimensions of the electron or those of the nucleus. A simple calculation shows that the energy radiated out during the process considered will be enormously great compared with that radiated out by ordinary molecular processes.

It is obvious that the behaviour of such a system will be very different from that of an atomic system occurring in nature. In the first place, the actual atoms in their permanent state seem to have absolutely fixed dimensions and frequencies. Further, if we consider any molecular process, the result seems always to be that after a certain amount of energy characteristic for the systems in question is radiated out, the systems will again settle down in a stable state of equilibrium, in which the distances apart of the particles are of the same order of magnitude as before the process.

Now the essential point in Planck's theory of radiation is that the energy radiation from an atomic system does not take place in the continuous way assumed in the ordinary electrodynamics, but that it, on the contrary, takes place in distinctly separated emissions, the amount of energy radiated out from an atomic vibrator of frequency ν in a single emission being equal to $\tau h\nu$, where τ is an entire number, and h is a universal constant. [. . .]

Planck's theory deals with the emission and absorption of radiation from an atomic vibrator of a constant frequency, independent of the amount of energy possessed by the system in the moment considered. The assumption of such vibrators, however, involves the assumption of quasi-elastic forces and is inconsistent with Rutherford's theory, according to which all the forces between the particles of an atomic system vary inversely as the square of the distance apart. In order to apply the main results obtained by Planck it is therefore necessary to introduce new assumptions as to the emission and absorption of radiation by an atomic system.

The main assumptions used in the present paper are:

1. That energy radiation is not emitted (or absorbed) in the continuous way assumed in the ordinary electrodynamics, but only during the passing of the systems between different 'stationary' states.

2. That the dynamical equilibrium of the systems in the stationary states is

governed by the ordinary laws of mechanics, while these laws do not hold for the passing of the systems between the different stationary states.

3. That the radiation emitted during the transition of a system between two stationary states is homogeneous, and that the relation between the frequency ν and the total amount of energy emitted E is given by $E = h\nu$, where h is Planck's constant.

4. That the different stationary states of a simple system consisting of an electron rotating round a positive nucleus are determined by the condition that the ratio between the total energy, emitted during the formation of the configuration, and the frequency of revolution of the electron is an entire multiple of $\frac{h}{2}$. Assuming that the orbit of the electron is circular, this assumption is equivalent with the assumption that the angular momentum of the electron round the nucleus is equal to an entire multiple of $\frac{h}{2\pi}$.

5. That the 'permanent' state of any atomic system—i.e. the state in which the energy emitted is maximum—is determined by the condition that the angular momentum of every electron round the centre of its orbit is equal to $\frac{h}{2\pi}$.

It is shown that, applying these assumptions to Rutherford's atom model, it is possible to account for the laws of Balmer and Rydberg connecting the frequency of the different lines in the line-spectrum of an element. Further, outlines are given of a theory of the constitution of the atoms of the elements and of the formation of molecules of chemical combinations, which on several points is shown to be in approximate agreement with experiments.

LOUIS DE BROGLIE

303

Quantum relation and relativity

One of the most important new concepts introduced by the Theory of Relativity is the concept of inertia of energy. According to Einstein, energy has to be considered as having mass and all mass represents energy. Mass and energy are always tied to each other by the general relation:

$$\text{energy} = \text{mass} \times c^2$$

c being the constant called 'velocity of light', but we, for reasons to be expounded below, prefer to call it 'the limiting speed of energy'. As there is always a proportionality between mass and energy one has to consider matter and energy as two synonymous terms designating the same physical reality.

First the atomic theory and then the theory of the electron have taught us to consider matter as essentially discontinuous and this leads us to acknowledge, contrary to old ideas about light, that all forms of energy are, if not concentrated in small portions of space, at least condensed around certain points of singularity.

The principle of inertia of energy attributes to a body, the proper mass of which is m_0 (as measured by an observer moving with it), a proper energy of $m_0 c^2$. [. . .]

Having recalled all this to our mind, let us investigate in what form can we introduce the quanta into the dynamics of Relativity Theory. It seems to us that the fundamental idea of the theory of quanta is the impossibility to envisage an isolated quantity of energy without associating with it a certain frequency. This connection is expressed in a relation which I shall call the quantum relation:

$$\text{energy} = h \times \text{frequency}$$

h being Planck's constant. [. . .]

No doubt, the quantum relation would not make sense if energy were distributed in space in a continuous manner, but as we have just seen this is certainly not the case. Thus, one could conceive that, by a great law of Nature, with each parcel of energy of proper mass m_0 there is connected a periodic phenomenon of a frequency v_0 given by

$$h v_0 = m_0 c^2$$

where v_0 is of course measured in the system moving with that parcel of energy. This hypothesis forms the basis of our theory; as is the case with all hypotheses, its worth is exactly that of the consequences deducible from it. Are we to suppose that the periodic phenomenon is localized in the interior of the parcel of energy? This is not necessarily the case, and we shall see [. . .] that without doubt it is spread out over an extended part of space. But if so, what should we understand by the interior part of a parcel of energy? The electron is for us that type of an isolated parcel of energy which, perhaps mistakenly, we believe to understand best; but according to commonly accepted notions the energy of the electron is spread out over the whole of space with a very strong condensation in a region of very small dimensions about whose properties we know very little. What characterizes the electron as an atom of energy is not the small place it occupies in space—I repeat that it occupies all of it—but the fact that it is indivisible. That it forms a unity. [. . .]

Briefly, I have developed new ideas that could perhaps contribute to speeding up the necessary synthesis which would reunify the physics of radiations, nowadays so strangely split into two domains where two respectively opposing conceptions reign: the corpuscular conception and that of waves.

The wave nature of the electron

304

For a long time physicists had been wondering whether light was composed of small, rapidly moving corpuscles. This idea was put forward by the philosophers of antiquity and upheld by Newton in the eighteenth century. After Thomas Young's discovery of interference phenomena and following the admirable work of Augustin Fresnel, the hypothesis of a granular structure of light was entirely abandoned and the wave theory unanimously adopted. Thus the physicists of last century spurned absolutely the idea of an atomic structure of light. Although rejected by optics, the atomic theories began making great headway not only in chemistry, where they provided a simple interpretation of the laws of definite proportions, but also in the physics of matter where they made possible an interpretation of a large number of properties of solids, liquids, and gases. In particular they were instrumental in the elaboration of that admirable kinetic theory of gases which, generalized under the name of statistical mechanics, enables a clear meaning to be given to the abstract concepts of thermodynamics. Experiment also yielded decisive proof in favour of an atomic constitution of electricity; the concept of the electricity corpuscle owes its appearance to Sir J. J. Thomson and you will all be familiar with H. A. Lorentz's use of it in his theory of electrons.

Some thirty years ago, physics was hence divided into two: firstly the physics of matter based on the concept of corpuscles and atoms which were supposed to obey Newton's classical laws of mechanics, and secondly radiation physics based on the concept of wave propagation in a hypothetical continuous medium, i.e. the light ether or electromagnetic ether. But these two physics could not remain alien one to the other; they had to be fused together by devising a theory to explain the energy exchanges between matter and radiation —and that is where the difficulties arose. While seeking to link these two physics together, imprecise and even inadmissible conclusions were in fact arrived at in respect of the energy equilibrium between matter and radiation in a thermally insulated medium: matter, it came to be said, must yield all its energy to the radiation and so tend of its own accord to absolute zero temperature! This absurd conclusion had at all costs to be avoided. By an intuition of his genius Planck realized the way of avoiding it: instead of assuming, in common with the classical wave theory, that a light source emits its radiation continuously, it had to be assumed on the contrary that it emits equal and finite quantities, *quanta*. The energy of each quantum has, moreover, a value proportional the frequency v of the radiation. It is equal to hv, h being a universal constant since referred to as Planck's constant.

The success of Planck's ideas entailed serious consequences. If light is emitted as quanta, ought it not, once emitted, to have a granular structure? The existence of radiation quanta thus implies the corpuscular concept of light. On the other hand, as shown by Jeans and H. Poincaré, it is demonstrable that if the motion of the material particles in light sources obeyed the laws of classical mechanics it would be impossible to derive the exact law of black-body radiation, Planck's law. It must therefore be assumed that traditional dynamics, even as modified by Einstein's theory of relativity, is incapable of accounting for motion on a very small scale.

The existence of a granular structure of light and of other radiations was confirmed by the discovery of the photoelectric effect. If a beam of light or of X-rays falls on a piece of matter, the latter will emit rapidly moving electrons. The kinetic energy of these electrons increases linearly with the frequency of the incident radiation and is independent of its intensity. This phenomenon can be explained simply by assuming that the radiation is composed of quanta $h\nu$ capable of yielding all their energy to an electron of the irradiated body: one is thus led to the theory of light quanta proposed by Einstein in 1905 and which is, after all, a reversion to Newton's corpuscular theory, completed by the relation for the proportionality between the energy of the corpuscles and the frequency. A number of arguments were put forward by Einstein in support of his viewpoint, and in 1922 the discovery by A. H. Compton[1] of the X-ray scattering phenomenon which bears his name confirmed it. Nevertheless, it was still necessary to adopt the wave theory to account for interference and diffraction phenomena, and no way whatsoever of reconciling the wave theory with the existence of light corpuscles could be visualized.

As stated, Planck's investigations cast doubts on the validity of very small-scale mechanics. Let us consider a material point which describes a small trajectory which is closed or else turning back on itself. According to classical dynamics there are numberless motions of this type which are possible complying with the initial conditions, and the possible values for the energy of the moving body form a continuous sequence. On the other hand Planck was led to assume that only certain preferred motions, *quantized* motions, are possible or at least stable, since energy can only assume values forming a discontinuous sequence. This concept seemed rather strange at first but its value had to be recognized because it was this concept which brought Planck to the correct law of black-body radiation and because it then proved its fruitfulness in many other fields. Lastly, it was on the concept of atomic motion quantization that Bohr based his famous theory of the atom; it is so familiar to scientists that I shall not summarize it here.

The necessity of assuming for light two contradictory theories—that of waves and that of corpuscles—and the inability to understand why, among the infinity of motions which an electron ought to be able to have in the atom according to classical concepts, only certain ones were possible: such were the

1. Arthur H. Compton (1892–1962), American physicist, discoverer of the change in the wavelength of X-rays when scattered by electrons (the 'Compton-effect'). (Ed.)

enigmas confronting physicists at the time I resumed my studies of theoretical physics.

When I started to ponder these difficulties two things struck me in the main. Firstly the light-quantum theory cannot be regarded as satisfactory since it defines the energy of a light corpuscle by the relation $W = h\nu$ which contains a frequency ν. Now a purely corpuscular theory does not contain any element permitting the definition of a frequency. This reason alone renders it necessary in the case of light to introduce simultaneously the corpuscle concept and the concept of periodicity.

On the other hand the determination of the stable motions of the electrons in the atom involves whole numbers, and so far the only phenomena in which whole numbers were involved in physics were those of interference and of eigenvibrations. That suggested the idea to me that electrons themselves could not be represented as simple corpuscles either, but that a periodicity had also to be assigned to them too.

I thus arrived at the following overall concept which guided my studies: for both matter and radiations, light in particular, it is neccessary to introduce the corpuscle concept and the wave concept at the same time. In other words the existence of corpuscles accompanied by waves has to be assumed in all cases. However, since corpuscles and waves cannot be independent because, according to Bohr's expression, they constitute two complementary forces of reality, it must be possible to establish a certain parallelism between the motion of a corpuscle and the propagation of the associated wave. [. . .]

Such are the main ideas which I developed in my initial studies. They showed clearly that it was possible to establish a correspondence between waves and corpuscles such that the laws of mechanics correspond to the laws of geometrical optics. In the wave theory, however, as you will know, geometrical optics is only an approximation: this approximation has its limits of validity and particularly when interference and diffraction phenomena are involved, it is quite inadequate. This prompted the thought that classical mechanics is also only an approximation relative to a vaster wave mechanics. I stated as much almost at the outset of my studies, i.e. 'A new mechanics must be developed, which is to classical mechanics what wave optics is to geometrical optics'. This new mechanics has since been developed, thanks mainly to the fine work done by Schrödinger.[1] It is based on wave propagation equations and strictly defines the evolution in time of the wave associated with a corpuscle. It has in particular succeeded in giving a new and more satisfactory form to the quantization conditions of intra-atomic motion since the classical quantization conditions are justified, as we have seen, by the application of geometrical optics to the

1. Erwin Schrödinger (1887–1961), Austrian physicist, founder of wave mechanics which placed quantum theory on a new basis. (Ed.)

waves associated with the intra-atomic corpuscles, and this application is not strictly justified.

I cannot attempt even briefly to sum up here the development of the new mechanics. I merely wish to say that on examination it proved to be identical with a mechanics independently developed, first by Heisenberg, then by Born, Jordan, Pauli, Dirac, etc.: quantum mechanics. The two mechanics, wave and quantum, are equivalent from the mathematical point of view.

We shall content ourselves here by considering the general significance of the results obtained. To sum up the meaning of wave mechanics it can be stated that: 'A wave must be associated with each corpuscle and only the study of the wave's propagation will yield information to us on the successive positions of the corpuscle in space'. In conventional large-scale mechanical phenomena the anticipated positions lie along a curve which is the trajectory in the conventional meaning of the word. But what happens if the wave does not propagate according to the laws of optical geometry, if, say, there are interferences and diffraction? Then it is no longer possible to assign to the corpuscle a motion complying with classical dynamics, that much is certain. Is it even still possible to assume that at each moment the corpuscle occupies a well-defined position in the wave and that the wave in its propagation carries the corpuscle along in the same way as a wave would carry along a cork? These are difficult questions and to discuss them would take us too far and even to the confines of philosophy. All that I shall say about them here is that nowadays the tendency in general is to assume that it is not constantly possible to assign to the corpuscle a well-defined position in the wave. I must restrict myself to the assertion that when an observation is carried out enabling the localization of the corpuscle, the observer is invariably induced to assign to the corpuscle a position in the interior of the wave and the probability of it being at a particular point M of the wave is proportional to the square of the amplitude, that is to say the intensity at M.

WERNER HEISENBERG

305

On the intuitive content of quantum kinematics and mechanics

In order to be able to follow the quantum-mechanical behaviour of an object, one has to know the mass of the object and the forces of interaction with any sort of fields and with other objects. Only then can we set the Hamiltonian function[1] of the quantum-mechanical system. (The following remarks refer in general only to the non-relativistic quantum mechanics, as the laws of quantum-theoretical electrodynamics are still very little known.) It is unnecessary to say anything more about the 'Gestalt' of the object; it would best serve our

1. See footnote 1 on page 528. (Ed.)

purpose if we designate by this word the sum-total of all those forces of inter-action. If we wish to make clear what is understood when we speak of the 'position of an object', for example of the electron (relative to a given frame of reference), we have to indicate specific experiments by the help of which we intend to measure the 'position of the electron'; otherwise this word has no meaning. There are plenty of experiments which can, in principle, determine the 'position of the electron' with any desired accuracy; one can for instance illuminate the electron and observe it under a microscope. The highest possible accuracy attainable here for the position-measurement is essentially given by the wave-length of the light which is used. But in principle a γ-ray microscope can be built and with its help the position-measurement can be made as accurate as desired. Still, there is one secondary circumstance in this measurement which is essential: it is the Compton-effect. Every observation of the light scattered by the electron presupposes a photoelectric effect (in the eye, on the photographic plate, in the photocell), which can also be interpreted thus: the light-quantum hits an electron, is reflected from it or diffracted by it and then again deflected through the lenses of the microscope, triggering off the photo-effect. In the very moment of the determination of the position, that is at the instance when the light-quantum is diffracted by the electron, the latter changes its momentum abruptly. This change is the greater the smaller the wavelength of the light used, i.e. the more accurately the position is determined. At the moment when the position of the electron is known, its momentum can be known only up to the accuracy of an order of magnitude corresponding to this abrupt change; thus the more accurately the position is determined the less we know about the momentum, and vice versa [. . .] Let q_1 be the accuracy with which the value of q^1 is given (q_1 is about the average error in q), in our case the wavelength of the light; let p_1 be the accuracy with which the value of p is determinable, here the abrupt change of p in the Compton-effect; then according to the elementary formulas of the Compton-effect the relation between p_1 and q_1 is given by

$$p_1 q_1 \sim h \qquad\qquad (1) \qquad [. . .]$$

The word 'velocity' of an object can be defined easily through measurements, when treating motions not under the influence of forces. One could, for example, illuminate the object by red light and then ascertain the velocity of the particle through the Doppler-effect of the scattered light. The determination of the velocity becomes the more accurate the longer the wavelength of the light used, since then, as a result of the Compton-effect, the change in velocity of the particle for each light-quantum becomes smaller. The determination of the position becomes correspondingly less accurate, as required by equation (1) [. . .]

The results of the previous section could be summarized and generalized in the following assertion: All concepts, which are used in classical theory for the description of a mechanical system, can be exactly defined for atomic processes

1. q is the position of the electron, p its momentum (Ed.)

too, in accordance with classical concepts. However, we know from experience that the experiments underlying such a definition involve an uncertainty, if we wish to obtain through them the simultaneous determination of two canonically conjugate quantities.[1] The degree of this uncertainty is given by equation (1) (applicable to any two canonically conjugate quantities). In connection with this it springs to mind to compare quantum theory with the special theory of relativity. According to the theory of relativity the word 'simultaneous' cannot be defined otherwise than through experiments in which the velocity of the propagation of light plays an important part. If there existed a 'sharper' definition of simultaneity, e.g. signals which are propagated with infinite velocity, the theory of relativity would be impossible. Since, however, such signals do not exist, and because the velocity of light appears already in the definition of simultaneity, the postulate of the constancy of the velocity of light has become possible, and this is why a meaningful use of the words 'position, velocity, time' is not in contradiction with this postulate. A similar situation exists with regard to the definition of the concepts of 'position and velocity of the electron' in quantum theory. All the experiments which we can use for the definition of these words necessarily carry the uncertainty given by equation (1), although they allow us to define exactly each of the concepts of p or q by itself. It there existed experiments which could give a 'sharper' simultaneous determination of p and q than permitted by equation (1), quantum theory would be impossible.

306

The physical principles of the quantum theory

1. THEORY AND EXPERIMENT

The experiments of physics and their results can be described in the language of daily life. Thus if the physicist did not demand a theory to explain his results and could be content, say, with a description of the lines appearing on photographic plates, everything would be simple and there would be no need of an epistemological discussion. Difficulties arise only in the attempt to classify and synthesize the results, to establish the relation of cause and effect between them —in short, to construct a theory. This synthetic process has been applied not only to the results of scientific experiment, but, in the course of ages, also to the simplest experiences of daily life, and in this way all concepts have been formed. In the process, the solid ground of experimental proof has often been forsaken, and generalizations have been accepted uncritically, until finally contradictions between theory and experiment have become apparent. In order to avoid these contradictions, it seems necessary to demand that no concept enter a theory which has not been experimentally verified at least to the same degree of accuracy as the experiments to be explained by the theory. Unfortunately it is

1. The physical quantities in Hamilton's 'canonical equations' of the motion of a mechanical system (Ed.)

quite impossible to fulfil this requirement, since the commonest ideas and words would often be excluded. To avoid these insurmountable difficulties it is found advisable to introduce a great wealth of concepts into a physical theory, without attempting to justify them rigorously, and then to allow experiment to decide at what points a revision is necessary.

Thus it was characteristic of the special theory of relativity that the concepts 'measuring rod' and 'clock' were subject to searching criticism in the light of experiment; it appeared that these ordinary concepts involved the tacit assumption that there exist (in principle, at least) signals that are propagated with an infinite velocity. When it became evident that such signals were not to be found in nature, the task of eliminating this tacit assumption from all logical deductions was undertaken, with the result that a consistent interpretation was found for facts which had seemed irreconcilable. A much more radical departure from the classical conception of the world was brought about by the general theory of relativity, in which only the concept of coincidence in space-time was accepted uncritically. According to this theory, ordinary language (i.e. classical concepts) is applicable only to the description of experiments in which both the gravitational constant and the reciprocal of the velocity of light may be regarded as negligibly small.

Although the theory of relativity makes the greatest of demands on the ability for abstract thought, still it fulfils the traditional requirements of science insofar as it permits a division of the world into subject and object (observer and observed) and hence a clear formulation of the law of causality. This is the very point at which the difficulties of the quantum theory begin. In atomic physics, the concepts 'clock' and 'measuring rod' need no immediate consideration, for there is a large field of phenomena in which $\frac{1}{c}$ is negligible. The concepts 'space-time coincidence' and 'observation', on the other hand do require a thorough revision. Particularly characteristic of the discussions to follow is the interaction between observer and object; in classical physical theories it has always been assumed either that this interaction is negligibly small, or else that its effect can be eliminated from the result by calculations based on 'control' experiments. This assumption is not permissible in atomic physics; the interaction between observer and object causes uncontrollable and large changes in the system being observed, because of the discontinuous changes characteristic of atomic processes. The immediate consequence of this circumstance is that in general every experiment performed to determine some numerical quantity renders the knowledge of others illusory, since the uncontrollable perturbation of the observed system alters the values of previously determined quantities. If this perturbation be followed in its quantitative details, it appears that in many cases it is impossible to obtain an exact determination of the simultaneous values of two variables, but rather that there is a lower limit to the accuracy with which they can be known.

The starting-point of the critique of the relativity theory was the postulate that there is no signal velocity greater than that of light. In a similar manner, this lower limit to the accuracy with which certain variables can be known simultaneously may be postulated as a law of nature (in the form of the so-called uncertainty relations) and made the starting-point of the critique which forms the subject matter of the following pages. These uncertainty relations give us that measure of freedom from the limitations of classical concepts which is necessary for a consistent description of atomic processes. [...]

2. ILLUSTRATIONS OF THE UNCERTAINTY RELATIONS

The uncertainty principle refers to the degree of indeterminateness in the possible present knowledge of the simultaneous values of various quantities with which the quantum theory deals; it does not restrict, for example, the exactness of a position measurement alone or a velocity measurement alone. Thus suppose that the velocity of a free electron is precisely known, while the position is completely unknown. Then the principle states that every subsequent observation of the position will alter the momentum by an unknown and undeterminable amount such that after carrying out the experiment our knowledge of the electronic motion is restricted by the uncertainty relation. This may be expressed in concise and general terms by saying that every experiment destroys some of the knowledge of the system which was obtained by previous experiments. This formulation makes it clear that the uncertainty relation does not refer to the past; if the velocity of the electron is at first known and the position then exactly measured, the position for times previous to the measurement may be calculated. Then for these past times $\Delta p \Delta q^{1}$ is smaller than the usual limiting value, but this knowledge of the past is of a purely speculative character, since it can never (because of the unknown change in momentum caused by the position measurement) be used as an initial condition in any calculation of the future progress of the electron and thus cannot be subjected to experimental verification. It is a matter of personal belief whether such a calculation concerning the past history of the electron can be ascribed any physical reality or not.

307

The Copenhagen interpretation of quantum theory

The Copenhagen interpretation of quantum theory starts from a paradox. Any experiment in physics, whether it refers to the phenomena of daily life or to atomic events, is to be described in the terms of classical physics. The concepts of classical physics form the language by which we describe the arrangements of our experiments and state the results. We cannot and should not replace these concepts by any others. Still the application of these concepts is limited by the

1. Δq and Δp are respectively the uncertainties of the position and the momentum of the electron. (Ed.)

relations of uncertainty. We must keep in mind this limited range of applicability of the classical concepts while using them, but we cannot and should not try to improve them.

For a better understanding of this paradox it is useful to compare the procedure for the theoretical interpretation of an experiment in classical physics and in quantum theory. In Newton's mechanics, for instance, we may start by measuring the position and the velocity of the planet whose motion we are going to study. The result of the observation is translated into mathematics by deriving numbers for the coordinates and the momenta of the planet from the observation. Then the equations of motion are used to derive from these values of the coordinates and momenta at a given time the values of these coordinates or any other properties of the system at a later time, and in this way the astronomer can predict the properties of the system at a later time. He can, for instance, predict the exact time for an eclipse of the moon.

In quantum theory the procedure is slightly different. We could for instance be interested in the motion of an electron through a cloud chamber and could determine by some kind of observation the initial position and velocity of the electron. But this determination will not be accurate; it will at least contain the inaccuracies following from the uncertainty relations and will probably contain still larger errors due to the difficulty of the experiment. It is the first of these inaccuracies which allows us to translate the result of the observation into the mathematical scheme of quantum theory. A probability function is written down which represents the experimental situation at the time of the measurement, including even the possible errors of the measurement.

This probability function represents a mixture of two things, partly a fact and partly our knowledge of a fact. It represents a fact insofar as it assigns at the initial time the probability unity (i.e. complete certainty) to the initial situation: the electron moving with the observed velocity at the observed position; 'observed' means observed within the accuracy of the experiment. It represents our knowledge insofar as another observer could perhaps know the position of the electron more accurately. The error in the experiment does—at least to some extent—not represent a property of the electron but a deficiency in our knowledge of the electron. Also this deficiency of knowledge is expressed in the probability function.

In classical physics one should in a careful investigation also consider the error of the observation. As a result one would get a probability distribution for the initial values of the coordinates and velocities and therefore something very similar to the probability function in quantum mechanics. Only the necessary uncertainty due to the uncertainty relations is lacking in classical physics.

When the probability function in quantum theory has been determined at the initial time from the observation, one can from the laws of quantum theory calculate the probability function at any later time and can thereby determine the probability for a measurement giving a specified value of the measured quantity. We can, for instance, predict the probability for finding the electron at a later time at a given point in the cloud chamber. It should be emphasized, however, that the probability function does not in itself represent a course of events in the course of time. It represents a tendency for events and our know-ledge of events. The probability function can be connected with reality only if one essential condition is fulfilled: if a new measurement is made to determine a certain property of the system. Only then does the probability function allow us to calculate the probable result of the new measurement. The result of the measurement again will be stated in terms of classical physics.

Therefore, the theoretical interpretation of an experiment requires three distinct steps: (1) the translation of the initial experimental situation into a probability function; (2) the following up of this function in the course of time; (3) the statement of a new measurement to be made of the system, the result of which can then be calculated from the probability function. For the first step the fulfilment of the uncertainty relations is a necessary condition. The second step cannot be described in terms of the classical concepts; there is no descrip-tion of what happens to the system between the initial observation and the next measurement. It is only in the third step that we change over again from the 'possible' to the 'actual'.

Let us illustrate these three steps in a simple ideal experiment. It has been said that the atom consists of a nucleus and electrons moving around the nucleus; it has also been stated that the concept of an electronic orbit is doubtful. One could argue that it should at least in principle be possible to observe the electron in its orbit. One should simply look at the atom through a microscope of a very high resolving power, then one would see the electron moving in its orbit. Such a high resolving power could to be sure not be obtained by a microscope using ordinary light, since the inaccuracy of the measurement of the position can never be smaller than the wavelength of the light. But a microscope using γ-rays with a wavelength smaller than the size of the atom would do. Such a microscope has not yet been constructed but that should not prevent us from discussing the ideal experiment.

Is the first step, the translation of the result of the observation into a prob-ability function, possible? It is possible only if the uncertainty relation is ful-filled after the observation. The position of the electron will be known with an accuracy given by the wavelength of the γ-ray. The electron may have been practically at rest before the observation. But in the act of observation at least one light quantum of the γ-ray must have passed the microscope and must first have been deflected by the electron. Therefore, the electron has been pushed by the light quantum, it has changed its momentum and its velocity, and one can show that the uncertainty of this change is just big enough to guarantee the validity of the uncertainty relations. Therefore, there is no difficulty with the first step.

At the same time one can easily see that there is no way of observing the orbit of the electron around the nucleus. The second step shows a wave packet moving not around the nucleus but away from the atom, because the first light quantum will have knocked the electron out from the atom. The momentum of light quantum of the γ-ray is much bigger than the original momentum of the electron if the wavelength of the γ-ray is much smaller than the size of the atom. Therefore, the first light quantum is sufficient to knock the electron out of the atom and one can never observe more than one point in the orbit of the electron; therefore, there is no orbit in the ordinary sense. The next observation—the third step—will show the electron on its path from the atom. Quite generally there is no way of describing what happens between two consecutive observations. It is of course tempting to say that the electron must have been somewhere between the two observations and that therefore the electron must have described some kind of path or orbit even if it may be impossible to know which path. This would be a reasonable argument in classical physics. But in quantum theory it would be a misuse of the language which, as we will see later, cannot be justified. We can leave it open for the moment, whether this warning is a statement about the way in which we should talk about atomic events or a statement about the events themselves, whether it refers to epistemology or to ontology. In any case we have to be very cautious about the wording of any statement concerning the behaviour of atomic particles.

Actually we need not speak of particles at all. For many experiments it is more convenient to speak of matter waves; for instance, of stationary matter waves around the atomic nucleus. Such a description would directly contradict the other description if one does not pay attention to the limitations given by the uncertainty relations. Through the limitations the contradiction is avoided. The use of 'matter waves' is convenient, for example, when dealing with the radiation emitted by the atom. By means of its frequencies and intensities the radiation gives information about the oscillating charge distribution in the atom, and there the wave picture comes much nearer to the truth than the particle picture. Therefore, Bohr advocated the use of both pictures, which he called 'complementary' to each other. The two pictures are of course mutually exclusive, because a certain thing cannot at the same time be a particle (i.e. substance confined to a very small volume) and a wave (i.e. a field spread out over a large space), but the two complement each other. By playing with both pictures, by going from the one picture to the other and back again, we finally get the right impression of the strange kind of reality behind our atomic experiments. Bohr uses the concept of 'complementarity' at several places in the interpretation of quantum theory. The knowledge of the position of a particle is complementary to the knowledge of its velocity or momentum. If we know the one with high accuracy we cannot know the other with high accuracy; still we must know both for determining the behaviour of the system. The space-time description of the atomic events is complementary to their deterministic description. The probability function obeys an equation of motion as the coordinates did in Newtonian mechanics; its change in the course of time is completely determined by the quantum mechanical equation, but it does not allow a description in space and time. The observation, on the other hand, enforces the description in space and time but breaks the determined continuity of the probability function by changing our knowledge of the system.

Generally the dualism between two different descriptions of the same reality is no longer a difficulty since we know from the mathematical formulation of the theory that contradictions cannot arise. The dualism between the two complementary pictures—waves and particles—is also clearly brought out in the flexibility of the mathematical scheme. The formalism is normally written to resemble Newtonian mechanics, with equations of motion for the coordinates and the momenta of the particles. But by a simple transformation it can be rewritten to resemble wave equation for an ordinary three-dimensional matter wave. Therefore, this possibility of playing with different complementary pictures has its analogy in the different transformations of the mathematical scheme; it does not lead to any difficulties in the Copenhagen interpretation of quantum theory.

A real difficulty in the understanding of this interpretation arises, however, when one asks the famous question: But what happens 'really' in an atomic event? It has been said before that the mechanism and the results of an observation can always be stated in terms of the classical concepts. But what one deduces from an observation is a probability function, a mathematical expression that combines statements about possibilities or tendencies with statements about our knowledge of facts. So we cannot completely objectify the result of an observation; we cannot describe what 'happens' between this observation and the next. This looks as if we had introduced an element of subjectivism into the theory, as if we meant to say: what happens depends on our way of observing it or on the fact that we observe it. Before discussing this problem of subjectivism it is necessary to explain quite clearly why one would get into hopeless difficulties if one tried to describe what happens between two consecutive observations.

For this purpose it is convenient to discuss the following ideal experiment: We assume that a small source of monochromatic light radiates towards a black screen with two small holes in it. The diameter of the holes may be not much bigger than the wavelength of the light, but their distance will be very much bigger. At some distance behind the screen a photographic plate registers the incident light. If one describes this experiment in terms of the wave picture, one says that the primary wave penetrates through the two holes; there will be secondary spherical waves starting from the holes that interfere with one another, and the interference will produce a pattern of varying intensity on the photographic plate.

The blackening of the photographic plate is a quantum process, a chemical reaction produced by single light quanta. Therefore, it must also be possible to describe the experiment in terms of light quanta. If it would be permissible to say what happens to the single light quantum between its emission from the light source and its absorption in the photographic plate, one could argue as follows: The single light quantum can come through the first hole or through the second one. If it goes through the first hole and is scattered there, its probability for being absorbed at a certain point of the photographic plate cannot depend upon whether the second hole is closed or open. The probability distribution on the plate will be the same as if only the first hole was open. If the experiment is repeated many times and one takes together all cases in which the light quantum has gone through the first hole, the blackening of the plate due to these cases will correspond to this probability distribution. If one considers only those light quanta that go through the second hole, the blackening should correspond to a probability distribution derived from the assumption that only the second hole is open. The total blackening, therefore, should just be the sum of the blackenings in the two cases; in other words, there should be no interference pattern. But we know this is not correct, and the experiment will show the interference pattern. Therefore, the statement that any light quantum must have gone *either* through the first *or* through the second hole is problematic and leads to contradictions. This example shows clearly that the concept of the probability function does not allow a description of what happens between two observations. Any attempt to find such a description would lead to contradictions; this must mean that the term 'happens' is restricted to the observation.

Now, this is a very strange result, since it seems to indicate that the observation plays a decisive role in the event and that the reality varies, depending upon whether we observe it or not. To make this point clearer we have to analyse the process of observation more closely.

To begin with, it is important to remember that in natural science we are not interested in the universe as a whole, including ourselves, but we direct our attention to some part of the universe and make that the object of our studies. In atomic physics this part is usually a very small object, an atomic particle or a group of such particles, sometimes much larger—the size does not matter; but it is important that a large part of the universe, including ourselves, does *not* belong to the object.

Now, the theoretical interpretation of an experiment starts with the two steps that have been discussed. In the first step we have to describe the arrangement of the experiment, eventually combined with a first observation, in terms of classical physics and translate this description into a probability function. This probability function follows the laws of quantum theory, and its change in the course of time, which is continuous, can be calculated from the initial conditions; this is the second step. The probability function combines objective and subjective elements. It contains statements about possibilities or better tendencies ('potentia' in Aristotelian philosophy), and these statements are completely objective, they do not depend on any observer; and it contains statements about our knowledge of the system, which of course are subjective insofar as they may be different for different observers. In ideal cases the subjective element in the probability function may be practically negligible as compared with the objective one. The physicists then speak of a 'pure case'.

When we now come to the next observation, the result of which should be predicted from the theory, it is very important to realize that our object has to be in contact with the other part of the world, namely, the experimental arrangement, the measuring rod, etc., before or at least at the moment of observation. This means that the equation of motion for the probability function does now contain the influence of the interaction with the measuring device. This influence introduces a new element of uncertainty, since the measuring device is necessarily described in the terms of classical physics; such a description contains all the uncertainties concerning the microscopic structure of the device which we know from thermodynamics, and since the device is connected with the rest of the world, it contains in fact the uncertainties of the microscopic structure of the whole world. These uncertainties may be called objective insofar as they are simply a consequence of the description in the terms of classical physics and do not depend on any observer. They may be called subjective insofar as they refer to our incomplete knowledge of the world.

After this interaction has taken place, the probability function contains the objective element of tendency and the subjective element of incomplete knowledge, even if it has been a 'pure case' before. It is for this reason that the result of the observation cannot generally be predicted with certainty; what can be predicted is the probability of a certain result of the observation, and this statement about the probability can be checked by repeating the experiment many times. The probability function does—unlike the common procedure in Newtonian mechanics—not describe a certain event but, at least during the process of observation, a whole ensemble of possible events.

The observation itself changes the probability function discontinuously; it selects of all possible events the actual one that has taken place. Since through the observation our knowledge of the system has changed discontinuously, its mathematical representation also has undergone the discontinuous change and we speak of a 'quantum jump'. When the old adage 'Natura non facit saltus' is used as a basis for criticism of quantum theory, we can reply that certainly our knowledge can change suddenly and that this fact justifies the use of the term 'quantum jump'.

Therefore, the transition from the 'possible' to the 'actual' takes place during the act of observation. If we want to describe what happens in an atomic event, we have to realize that the word 'happens' can apply only to the observation, not to the state of affairs between two observations. It applies to the physical, not the psychical act of observation, and we may say that the transition from the 'possible' to the 'actual' takes place as soon as the interaction of the object with the measuring device, and thereby with the rest of the world, has come into play; it is not connected with the act of registration of the result by the mind of the observer. The discontinuous change in the probability function, however, takes place with the act of registration, because it is the discontinuous change of our knowledge in the instant of registration that has its image in the discontinuous change of the probability function.

To what extent, then, have we finally come to an objective description of the world, especially of the atomic world? In classical physics science started from the belief—or should one say from the illusion?—that we could describe the world or at least parts of the world without any reference to ourselves. This is actually possible to a large extent. We know that the city of London exists whether we see it or not. It may be said that classical physics is just that idealization in which we can speak about parts of the world without any reference to ourselves. Its success has led to the general ideal of an objective description of the world. Objectivity has become the first criterion for the value of any scientific result. Does the Copenhagen interpretation of quantum theory still comply with this ideal? One may perhaps say that quantum theory corresponds to this ideal as far as possible. Certainly quantum theory does not contain genuine subjective features, it does not introduce the mind of the physicist as a part of the atomic event. But it starts from the division of the world into the 'object' and the rest of the world, and from the fact that at least for the rest of the world we use the classical concepts in our description. This division is arbitrary and historically a direct consequence of our scientific method; the use of the classical concepts is finally a consequence of the general human way of thinking. But this is already a reference to ourselves and insofar our description is not completely objective.

It has been stated in the beginning that the Copenhagen interpretation of quantum theory starts with a paradox. It starts from the fact that we describe our experiments in the terms of classical physics and at the same time from the knowledge that these concepts do not fit nature accurately. The tension between these two starting points is the root of the statistical character of quantum theory. Therefore, it has sometimes been suggested that one should depart from the classical concepts altogether and that a radical change in the concepts used for describing the experiments might possibly lead back to a nonstatistical, completely objective description of nature.

This suggestion, however, rests upon a misunderstanding. The concepts of classical physics are just a refinement of the concepts of daily life and are an essential part of the language which forms the basis of all natural science. Our actual situation in science is such that we *do* use the classical concepts for the description of the experiments, and it was the problem of quantum theory to find theoretical interpretation of the experiments on this basis. There is no use in discussing what could be done if we were other beings than we are. At this point we have to realize, as von Weizsäcker has put it, that 'Nature is earlier than man, but man is earlier than natural science'. The first part of the sentence justifies classical physics, with its ideal of complete objectivity. The second part tells us why we cannot escape the paradox of quantum theory, namely, the necessity of using the classical concepts.

We have to add some comments on the actual procedure in the quantum-theoretical interpretation of atomic events. It has been said that we always start with a division of the world into an object, which we are going to study, and the rest of the world, and that this division is to some extent arbitrary. It should indeed not make any difference in the final result if we, e.g. add some part of the measuring device or the whole device to the object and apply the laws of quantum theory to this more complicated object. It can be shown that such an alteration of the theoretical treatment would not alter the predictions concerning a given experiment. This follows mathematically from the fact that the laws of quantum theory are for the phenomena in which Planck's constant can be considered as a very small quantity, approximately identical with the classical laws. But it would be a mistake to believe that this application of the quantum-theoretical laws to the measuring device could help to avoid the fundamental paradox of quantum theory.

The measuring device deserves this name only if it is in close contact with the rest of the world, if there is an interaction between the device and the observer. Therefore, the uncertainty with respect to the microscopic behaviour of the world will enter into the quantum-theoretical system here just as well as in the first interpretation. If the measuring device would be isolated from the rest of the world, it would be neither a measuring device nor could it be described in the terms of classical physics at all.

With regard to this situation Bohr has emphasized that it is more realistic to state that the division into the object and the rest of the world is not arbitrary. Our actual situation in research work in atomic physics is usually this: we wish to understand a certain phenomenon, we wish to recognize how this phenomenon follows from the general laws of nature. Therefore, that part of matter or radiation which takes part in the phenomenon is the natural 'object' in the theoretical treatment and should be separated in this respect from the tools used to study the phenomenon. This again emphasizes a subjective element in the description of atomic events, since the measuring device has been constructed by the observer, and we have to remember that what we observe is not nature in itself but nature exposed to our method of questioning. Our scientific work in physics consists in asking questions about nature in the language that we possess and trying to get an answer from experiment by the means that are at our disposal. In this way quantum theory reminds us, as Bohr has put it, of the old wisdom that when searching for harmony in life one must never forget that in the drama of existence we are ourselves both players and spectators. It is understandable that in our scientific relation to nature our own activity becomes very important when we have to deal with parts of nature into which we can penetrate only by using the most elaborate tools.

NIELS BOHR

308

The atomic theory and mechanics

I. THE CLASSICAL THEORIES

The analysis of the equilibrium and the motion of bodies not only forms the foundation of physics, but for mathematical reasoning has also furnished a rich field, which has been exceedingly fertile for the development of the methods of pure mathematics. This connection between mechanics and mathematics showed itself at an early date in the works of Archimedes, Galilei and Newton. In their hands the formation of concepts suitable for the analysis of mechanical phenomena was provisionally completed. Since the time of Newton, the development of the methods for treating mechanical problems has gone hand in hand with the evolution of mathematical analysis; we need only recall such names as Euler, Lagrange and Laplace. The later development of mechanics too, based on the work of Hamilton,[1] proceeded in very close association with the evolution of mathematical methods, the calculus of variations, and the theory of invariants, as appears clearly in recent times also in the papers of Poincaré.

Perhaps the greatest successes of mechanics lie in the domain of astronomy, but in the mechanical theory of heat an interesting application was also found in the course of the last century. The kinetic theory of gases, founded by Clausius and Maxwell, interprets the properties of gases to a large extent as results of the mechanical interactions of atoms and molecules flying about at random. We wish to recall especially the explanation of the two principles of thermodynamics given by this theory. The first principle is a direct result of the mechanical law of conservation of energy, while the second principle, the entropy law, can, following Boltzmann, be derived from the statistical behaviour of a large number of mechanical systems. It is of interest here that statistical considerations have permitted the description not only of the average behaviour of atoms, but also of the fluctuation phenomena, which have led by the investigation of the Brownian motion[2] to the unexpected possibility of counting atoms. The proper tools for the systematic development of statistical mechanics, to which especially Gibbs contributed, were furnished by the mathematical theory of canonical systems of differential equations.

The development of the electromagnetic theories in the second half of

---

1. Sir William Rowan Hamilton (1805–1865), Irish mathematican, who developed the differential equations of the motion of a mechanical system in a particularly symmetrical form. They are usually referred to as the 'canonical equations' of mechanics. (Ed.)

2. The irregular movement of microscopic particles arising from the thermal motion of the molecules of a liquid or gas in which they are suspended. It was first observed by the British botanist Robert Brown (1773–1858). (Ed.)

last century, following the discoveries of Oersted and Faraday, brought about a profound generalization of mechanical concepts. Although, to begin with, mechanical models played an essential part in Maxwell's electrodynamics, the advantages were soon realized of conversely deriving the mechanical concepts from the theory of the electromagnetic field. In this theory the conservation laws are explained by considering energy and momentum to be localized in the space surrounding the bodies. In particular, a natural explanation of radiation phenomena can be obtained in this way. The theory of the electromagnetic field was the direct cause of the discovery of electromagnetic waves, which today play so important a part in electrical engineering. Further, the electromagnetic theory of light founded by Maxwell provided a rational basis for the wave theory of light, which goes back to Huygens. With the aid of the atomic theory, it afforded a general description of the origin of light and of the phenomena taking place during the passage of light through matter. For this purpose, the atoms are supposed to be built up of electrical particles which can execute vibrations about positions of equilibrium. The free oscillations of the particles are the cause of the radiation, the composition of which we observe in the atomic spectra of the elements. Further, the particles will execute forced vibrations under the forces in the light waves and thus become centres of secondary wavelets which will interfere with the primary waves and produce the well-known phenomena of reflection and refraction of light. When the frequency of vibration of the incident waves approaches the frequency of one of the free oscillations of the atom, there results a resonance effect, by which the particles are thrown into specially strong forced vibrations. In this way a natural account was obtained of the phenomena of resonance radiation and the anomalous dispersion of a substance for light near one of its spectral lines.

Just as in the kinetic theory of gases, it is not merely the average effect of a large number of atoms that comes into consideration in the electromagnetic interpretation of optical phenomena. Thus, in the scattering of light the random distribution of the atoms makes the effects of the individual atoms appear in such a way that a direct counting of the atoms is possible. In fact, Rayleigh estimated from the intensity of the scattered blue light of the sky the number of atoms in the atmosphere, obtaining results in satisfactory agreement with the counting of atoms obtained by Perrin from a study of the Brownian motion. The rational mathematical representation of the electromagnetic theory is based on the application of vector analysis, or more generally tensor analysis of higher dimensional manifolds. This analysis founded by Riemann offered the proper means for the formulation of Einstein's fundamental theory of relativity which introduces concepts that go beyond Galilei's kinematics and may perhaps be considered as the natural completion of the classical theories.

2. THE QUANTUM THEORY OF ATOMIC CONSTITUTION

In spite of all the successful applications of mechanical and electrodynamical ideas to atomic theory, further development revealed profound difficulties. If these theories really provide a general description of thermal agitation and of the radiation connected with motion, then the general laws of heat radiation must be capable of a direct explanation. Contrary to all expectations, however, a calculation on this basis could not explain the empirical laws. Going beyond this, Planck demonstrated, retaining Boltzmann's account of the second law of thermodynamics, that the laws of heat radiation demand an element of discontinuity in the description of atomic processes quite foreign to the classical theories. Planck discovered that in the statistical behaviour of particles which execute simple harmonic oscillations about positions of equilibrium only such states of vibrations must be taken into account the energy of which is an integral multiple of a 'quantum', ωh, where ω is the frequency of the particle and h a universal constant, the so-called Planck's quantum of action.

The more precise formulation of the content of the quantum theory appears, however, to be extremely difficult when it is remembered that all the concepts of previous theories rest on pictures which demand the possibility of a continuous variation. This difficulty was especially emphasized by the fundamental researches of Einstein, according to which essential features of the interaction between light and matter suggest that the propagation of light does not take place by spreading waves but by 'light-quanta', which, concentrated in a small region of space, contain the energy $h\nu$, where ν is the frequency of light. The formal nature of this statement is evident because the definition and measurement of this frequency rests exclusively on the ideas of the wave theory.

The inadequacy of the classical theories was brought into prominence by the development of our knowledge of atomic structure. One formerly hoped that this knowledge might be gradually enlarged by an analysis of the properties of the elements based on the classical theories which had been fruitful in so many respects. This hope was supported shortly before the birth of the quantum theory by Zeeman's[1] discovery of the effect of magnetic fields on spectral lines. As Lorentz showed, this effect corresponds in many cases to just that action of magnetic fields on the motion of oscillating particles which is to be expected from classical electrodynamics. Besides, this account allowed conclusions to be drawn about the nature of the oscillating particles which agreed beautifully with the experimental discoveries of Lenard and Thomson in the field of electric discharges in gases. As a result, small negatively charged particles, the electrons, were recognized as units common to all atoms. It is true that the so-called 'anomalous' Zeeman effect of many spectral lines caused profound difficulties for the classical theory. These were similar to those which showed themselves in the attempts with the aid of electrodynamic models to explain the simple empirical regularities among the spectral frequencies which were brought to light through the work of Balmer, Rydberg, and Ritz. In particular, such an

---

1. Peter Zeeman (1865–1943), Dutch physicist, discovered the splitting up of the spectral lines under the influence of a magnetic field on the light source. (Ed.)

account of the spectral laws could scarcely be reconciled with the estimate of the number of electrons in the atom which Thomson obtained from observations on the scattering of X-rays by a direct application of the classical theory.

These difficulties could for a time be attributed to our imperfect knowledge of the origin of the forces by which the electrons are bound in the atom. The situation was, however, entirely changed by the experimental discoveries in the field of radioactivity, which furnished new means for the investigation of atomic structure. Thus, Rutherford obtained convincing support for the idea of the nuclear atom from experiments on the passage through matter of the particles ejected by radioactive substances. According to this idea, the greatest part of the atomic mass is localized in a positively charged nucleus, exceedingly small compared with the dimensions of the atom as a whole. Around the nucleus there move a number of light negative electrons. In this way, the problem of atomic structure took on a great similarity to the problems of celestial mechanics. A closer consideration, however, soon showed that, nevertheless, there exists a fundamental difference between an atom and a planetary system. The atom must have a stability which presents features quite foreign to mechanical theory. Thus, the mechanical laws permit a continuous variation of the possible motions, which is entirely at variance with the definiteness of the properties of the elements. The difference between an atom and an electrodynamic model appears also when one considers the composition of the emitted radiation. For, in models of the sort considered, where the natural frequencies of motion vary continuously with the energy, the frequency of the radiation will change continuously during emission according to classical theory and will therefore show no similarity to the line spectra of the elements.

The search for a more precise formulation of the concepts of the quantum theory which might be capable of overcoming these difficulties led to the enunciation of the following postulates:

1. An atomic system possesses a certain manifold of states, the 'stationary states', to which corresponds in general a discrete sequence of energy values and which have a peculiar stability. This latter shows itself in that every change in the energy of the atom must be due to a 'transition' of the atom from one stationary state to another.

2. The possibility of emission and absorption of radiation by the atom is conditioned by the possibility of energy changes of the atom, in such a way that the frequency of the radiation is connected with the energy differences between the initial and final states by the formal relation

$$h\nu = E_1 - E_2.$$

These postulates, which cannot be explained on classical ideas, seem to offer a suitable basis for the general account of the observed physical and chemical properties of the elements. In particular, an immediate explanation is given of a fundamental feature of the empirical spectral laws. This feature, the Ritz principle of combination of spectral lines, states that the frequency of every line in a spectrum can be represented as the difference between two terms of a manifold of spectral terms characteristic of the element; in fact, we see that these terms can be identified with the energy values of the stationary states of the atom, divided by h.

In addition, this account of the origin of spectra gives an immediate explanation of the fundamental difference between absorption and emission spectra. For, according to the postulates, the condition for selective absorption of a frequency which corresponds to the combination of two terms is that the atom is in the state of smaller energy, while for emission of such radiation it must be in the state of greater energy. In short, the picture described is in very close agreement with the experimental results on the excitation of spectra. This is shown especially in the discovery of Franck[1] and Hertz as to impacts between free electrons and atoms. They found that an energy transfer from the electron to the atom can take place only in amounts which are just the energy differences of the stationary states as computed from the spectral terms. In general, the atom is simultaneously excited to emit. Similarly, the excited atom can, according to Klein and Rosseland, lose its emissive power through an impact, and the colliding electron experiences a corresponding increase of its energy.

As Einstein has shown, the postulates also furnish a suitable basis for a rational treatment of statistical problems, especially for a very lucid derivation of Planck's law of radiation. This theory assumes that an atom, which can undergo a transition between two stationary states and is in the higher state, has a certain 'probability', depending only on the atom, of jumping spontaneously to the lower state in a given interval of time. Further, it assumes that external illumination with radiation of the frequency corresponding to the transition gives the atom a probability, proportional to the intensity of the radiation, of going from the lower to the higher state. It is also an essential feature of the theory that illumination with this frequency gives the atom in the higher state, besides its spontaneous probability, an induced probability of jumping down to the lower state. [. . .]

The constantly growing contrast between the wave theory of light, apparently required for the explanation of optical phenomena, and the light-quantum theory, which represents naturally so many features of the interaction between light and matter, suggested that the failure of classical theories may even affect the validity of the laws of conservation of energy and momentum. These laws, which hold so central a position in the classical theory, would, then, in the description of atomic processes, be only statistically valid. However, this suggestion does not offer a satisfactory escape from the dilemma, as is shown by the experiments on the scattering of X-rays which have been undertaken

---

1. James Franck (1882–1964), German Jewish physicist, whose experiments, together with Gustav Hertz, confirmed Bohr's theory of the existence of discrete energy levels in atoms. (Ed.)

recently with the beautiful methods permitting a direct observation of individual processes. For Geiger and Bothe have been able to show that the recoil electrons and photo-electrons which accompany the production and absorption of the scattered radiation are coupled in pairs just as one would expect from the picture of the light-quantum theory. With the method of the Wilson cloud chamber, Compton and Simon have even succeeded in demonstrating, besides this pairing, the connection demanded by the light-quantum theory between the direction in which the effect of the scattered radiation is observed and the direction of the velocity of the recoil electrons accompanying the scattering.

From these results it seems to follow that, in the general problem of the quantum theory, one is faced not with a modification of the mechanical and electrodynamical theories describable in terms of the usual physical concepts, but with an essential failure of the pictures in space and time on which the description of natural phenomena has hitherto been based. This failure appears also in a closer consideration of impact phenomena. In particular, for impacts in which the time of collision is short compared to the natural periods of the atom and for which very simple results are to be expected according to the usual mechanical ideas, the postulate of stationary states would seem to be irreconcilable with any description of the collision in space and time based on the accepted ideas of atomic structure.

309

*The atomic theory and the fundamental principles
underlying the description of nature*

As with classical mechanics, so quantum mechanics, too, claims to give an exhaustive account of all phenomena which come within its scope. Indeed, the inevitability of using, for atomic phenomena, a mode of description which is fundamentally statistical arises from a closer investigation of the information which we are able to obtain by direct measurement of these phenomena and of the meaning which we may ascribe, in this connection, to the application of the fundamental physical concepts. On one hand, we must bear in mind that the meaning of these concepts is wholly tied up with customary physical ideas. Thus, any reference to space-time relationships presupposes the permanence of the elementary particles, just as the laws of the conservation of energy and momentum form the basis of any application of the concepts of energy and momentum. On the other hand, the postulate of the indivisibility of the quantum of action represents an element which is completely foreign to the classical conceptions; an element which, in the case of measurements, demands not only a finite interaction between the object and the measuring instrument but even a definite latitude in our account of this mutual action. Because of this state of affairs, any measurement which aims at an ordering of the elementary particles in time and space requires us to forgo a strict account of the exchange of energy and momentum between the particles and the measuring rods and clocks used as a reference system. Similarly, any determination of the energy and the momentum of the particles demands that we renounce their exact coordination in time and space. In both cases, the invocation of classical ideas, necessitated by the very nature of measurement, is, beforehand, tantamount to a renunciation of a strictly causal description. Such considerations lead immediately to the reciprocal uncertainty relations set up by Heisenberg and applied by him as the basis of a thorough investigation of the logical consistency of quantum mechanics. The fundamental indeterminacy which we meet here may, as the writer has shown, be considered as a direct expression of the absolute limitation of the applicability of visualizable conceptions in the description of atomic phenomena, a limitation that appears in the apparent dilemma which presents itself in the question of the nature of light and of matter.

The resignation as regards visualization and causality, to which we are thus forced in our description of atomic phenomena, might well be regarded as a frustration of the hopes which formed the starting-point of the atomic conceptions. Nevertheless, from the present standpoint of the atomic theory, we must consider this very renunciation as an essential advance in our understanding. Indeed, there is no question of a failure of the general fundamental principles of science within the domain where we could justly expect them to apply. The discovery of the quantum of action shows us, in fact, not only the natural limitation of classical physics, but, by throwing a new light upon the old philosophical problem of the objective existence of phenomena independently of our observations, confronts us with a situation hitherto unknown in natural science. As we have seen, any observation necessitates an interference with the course of the phenomena, which is of such a nature that it deprives us of the foundation underlying the causal mode of description. The limit, which nature herself has thus imposed upon us, of the possibility of speaking about phenomena as existing objectively finds its expression, as far as we can judge, just in the formulation of quantum mechanics. However, this should not be regarded as a hindrance to further advance; we must only be prepared for the necessity of an ever extending abstraction from our customary demands for a directly visualizable description of nature. Above all, we may expect new surprises in the domain where the quantum theory meets with the theory of relativity and where unsolved difficulties still stand as a hindrance to a complete fusion of the extension of our knowledge and of the expedients to account for natural phenomena which these theories have given us.

Even though it be at the end of the lecture, yet I am glad to have the opportunity of emphasizing the great significance of Einstein's theory of relativity in the recent development of physics with respect to our emancipation from the demand for visualization. We have learned from the theory of relativity that the expediency of the sharp separation of space and time, required by our senses, depends merely upon the fact that the velocities commonly occurring are small compared with the velocity of light. Similarly, we may say that Planck's discovery has led us to recognize that the adequacy of our whole customary attitude, which is characterized by the demand for causality, depends solely upon the smallness of the quantum of action in comparison with the actions with which we are concerned in ordinary phenomena. While the theory of relativity reminds us of the subjective character of all physical phenomena, a character which depends essentially upon the state of motion of the observer, so does the linkage of the atomic phenomena and their observation, elucidated by the quantum theory, compel us to exercise a caution in the use of our means of expression similar to that necessary in psychological problems where we continually come upon the difficulty of demarcating the objective content. Hoping that I do not expose myself to the misunderstanding that it is my intention to introduce a mysticism which is incompatible with the spirit of natural science, I may perhaps in this connection remind you of the peculiar parallelism between the renewed discussion of the validity of the principle of causality and the discussion of a free will which has persisted from earliest times. Just as the freedom of the will is an experiential category of our psychic life, causality may be considered as a mode of perception by which we reduce our sense impressions to order. At the same time, however, we are concerned in both cases with idealizations whose natural limitations are open to investigation and which depend upon one another in the sense that the feeling of volition and the demand for causality are equally indispensable elements in the relation between subject and object which forms the core of the problem of knowledge. [. . .]

Besides, the fact that consciousness, as we know it, is inseparably connected with life ought to prepare us for finding that the very problem of the distinction between the living and the dead escapes comprehension in the ordinary sense of the word. That a physicist touches upon such questions may perhaps be excused on the ground that the new situation in physics has so forcibly reminded us of the old truth that we are both onlookers and actors in the great drama of existence.

310

Atoms and human knowledge

By means of the methods of quantum mechanics it was possible to account for a very large amount of experimental evidence on the physical and chemical properties of substances. Not only was the binding of electrons in atoms and molecules clarified in detail, but a deep insight was also obtained into the constitution and reactions of atomic nuclei. In this connection, we may mention that the probability laws for spontaneous radioactive transmutations have been harmoniously incorporated into the statistical quantum-mechanical description. Also the understanding of the properties of the new elementary particles, which have been observed in recent years in the study of transmutations of atomic nuclei at high energies, has been subject to continual progress resulting from the adaption of the formalism to the invariance requirements of relativity theory. Still, we are here confronted with new problems whose solution obviously demands further abstractions suited to combine the quantum of action with the elementary electric charge.

In spite of the fruitfulness of quantum mechanics within such a wide domain of experience, the renunciation of accustomed demands on physical explanation has caused many physicists and philosophers to doubt that we are here dealing with an exhaustive description of atomic phenomena. In particular, the view has been expressed that the statistical mode of description must be regarded as a temporary expedient which, in principle, ought to be replaceable by a deterministic description. The thorough discussion of this question has, however, led to that clarification of our position as observers in atomic physics which has given us the epistemological lesson referred to in the beginning of this lecture.

As the goal of science is to augment and order our experience, every analysis of the conditions of human knowledge must rest on considerations of the character and scope of our means of communication. Our basis is, of course, the language developed for orientation in our surroundings and for the organization of human communities. However, the increase of experience has repeatedly raised questions as to the sufficiency of the concepts and ideas incorporated in daily language. Because of the relative simplicity of physical problems, they are especially suited to investigate the use of our means of communication. Indeed, the development of atomic physics has taught us how, without leaving common language, it is possible to create a framework sufficiently wide for an exhaustive description of new experience.

In this connection it is imperative to realize that in every account of physical experience one must describe both experimental conditions and observations by the same means of communication as one used in classical physics. In the analysis of single atomic particles, this is made possible by irreversible amplification effects—such as a spot on a photographic plate left by the impact of an electron, or an electric discharge created in a counter device—and the observations concern only where and when the particle is registered on the plate or its energy on arrival at the counter. Of course, this information presupposes knowledge of the position of the photographic plate relative to the other parts of the experimental arrangement, such as regulating diaphragms and shutters defining space-time coordination or electrified and magnetized bodies which determine the external force fields acting on the particle and permit energy measurements. The experimental conditions can be varied in many ways, but the point is that in each case we must be able to communicate to others what we have done and what we have learned, and that therefore the functioning of the measuring instruments must be described within the framework of classical physical ideas.

As all measurements thus concern bodies sufficiently heavy to permit the quantum to be neglected in their description, there is, strictly speaking, no new observational problem in atomic physics. The amplification of atomic effects, which makes it possible to base the account on measurable quantities and which gives the phenomena a peculiar closed character, only emphasizes the irreversibility characteristic of the very concept of observation. While, within the frame of classical physics, there is no difference in principle between the description of the measuring instruments and the objects under investigation, the situation is essentially different when we study quantum phenomena, since the quantum of action imposes restrictions on the description of the state of the systems by means of space-time coordinates and momentum-energy quantities. Since the deterministic description of classical physics rests on the assumption of an unrestricted compatibility of space-time coordination and the dynamical conservation laws, we are obviously confronted here with the problem of whether, as regards atomic objects, such a description can be fully retained.

The role of the interaction between objects and measuring instruments in the description of quantum phenomena was found to be especially important for the clarification of this main point. Thus, as stressed by Heisenberg, the locating of an object in a limited space-time domain involves, according to quantum mechanics, an exchange of momentum and energy between instrument and object which is the greater the smaller the domain chosen. It was therefore of the utmost importance to investigate the extent to which the interaction entailed in observation can be taken into account separately in the description of phenomena. This question has been the focus of much discussion, and there have appeared many proposals which aim at the complete control of all interactions. In such considerations, however, due regard is not taken to the fact that the very account of the functioning of measuring instruments involves that any interaction, implied by the quantum, between these and the atomic objects, be inseparably entailed in the phenomena.

Indeed, every experimental arrangement permitting the registration of an atomic particle in a limited space-time domain demands fixed measuring rods and synchronized clocks which, from their very definition, exclude the control of momentum and energy transmitted to them. Conversely, any unambiguous application of the dynamical conservation laws in quantum physics requires that the description of the phenomena involve a renunciation in principle of detailed space-time coordination. This mutual exclusiveness of the experimental conditions implies that the whole experimental arrangement must be taken into account in a well-defined description of the phenomena. The indivisibility of quantum phenomena finds its consequent expression in the circumstance that every definable subdivision would require a change of the experimental arrangement with the appearance of new individual phenomena. Thus, the very foundation of a deterministic description has disappeared and the statistical character of the predictions is evidenced by the fact that in one and the same experimental arrangement there will in general appear observations corresponding to different individual processes.

Such considerations not only have clarified the above-mentioned dilemma with respect to the propagation of light, but have also completely solved the corresponding paradoxes confronting pictorial representation of the behaviour of material particles. Here, of course, we cannot seek a physical explanation in the customary sense, but all we can demand in a new field of experience is the removal of any apparent contradiction. However great the contrasts exhibited by atomic phenomena under different experimental conditions, such phenomena must be termed complementary in the sense that each is well defined and that together they exhaust all definable knowledge about the objects concerned. The quantum-mechanical formalism, the sole aim of which is the comprehension of observations obtained under experimental conditions described by simple physical concepts, gives just such an exhaustive complementary account of a very large domain of experience. The renunciation of pictorial representation involves only the state of atomic objects, while the foundation of the description of the experimental conditions, as well as our freedom to choose between them, is fully retained. The whole formalism which can be applied only to closed phenomena must in all such respects be considered a rational generalization of classical physics.

In view of the influence of the mechanical conception of nature on philosophical thinking, it is understandable that one has sometimes seen in the notion of complementarity a reference to the subjective observer, incompatible with the objectivity of scientific description. Of course, in every field of experience we must retain a sharp distinction between the observer and the content of the observations, but we must realize that the discovery of the quantum of action has thrown new light on the very foundation of the description of nature and revealed hitherto unnoticed presuppositions to the rational use of the concepts on which the communication of experience rests. In quantum physics, as we have seen, an account of the functioning of the measuring instruments is indispensable to the definition of phenomena and we must, so to speak, distinguish between subject and object in such a way that each single case secures the unambiguous application of the elementary physical concepts used in the description. Far from containing any mysticism foreign to the spirit of science, the notion of complementarity points to the logical conditions for description and comprehension of experience in atomic physics.

WOLFGANG PAULI

311

Phenomena and physical reality

In the following I will give a few indications as to which problems connected with the notions of 'phenomenon' and 'reality' play an important role in today's physics, without claiming to have even nearly mastered this inexhaustible subject. Still, I will touch on controversial issues also, for these are the ones which arouse the greatest interest. For the orientation of the philosophers I would like to point out at the outset, that I do not belong to any of those philosophical schools the names of which end with some sort of 'ism'. Moreover, I am very much against attributing 'isms' to any special physical theory. like the relativity theory, or the quantum or wave mechanics, even though this has been done from time to time by the physicists themselves. My general tendency is to keep to the middle between the two extremes. In this sense, the best course for us will be to realize first in what way do 'phenomenon' and 'reality' occur in the everyday life of the physicist.

I. PHENOMENON AND REALITY IN THE EVERYDAY LIFE OF THE PHYSICIST

A phenomenon can be elementary or rather complex. Data of consciousness belong to the direct phenomena. To describe these as perceptions is very one-sided, considering that thoughts and ideas are also generated spontaneously. One does talk about sudden insights (in German *Einfälle*) which means literally that something 'falls' (in German *fällt*) into our consciousness.

I would suggest therefore that we call ideas and thoughts also phenomena exactly as we do with sounds, colours and tactile impressions. Our ideas do not appear at random but follow each other in a sequence. It is the interdependence of the data of consciousness which enables us to separate dreaming from being awake and also to experience instinctively the existence of external objects and of the consciousness of other human beings. What happens to us, what is not subordinated to our will, and what we have to take into account, all this we designate as real. European languages have two words for this, derived from different roots; the one of Latin origin: reality from *res*, thing, the other from the German *Wirklichkeit*, from *wirken*, to act. The two are represented in English as 'reality' and 'actuality'. The more abstract one, the one derived from '*wirken*' is that which is nearer to scientific usage.

If we now try to formulate what a physical phenomenon is, and what physical actuality is, we will run into diverging opinions. Personally, I do not see how it can be possible in physics to give such a definition of phenomenon, which would seek to isolate data of perception from rational and ordering principles. Such a separation seems to me to be already the result of a highly critical thought process, which has removed all the unconscious-instinctive ingredients always present in thought. Life, as well as science, would become impossible if we limited ourselves solely to ascertained or ascertainable data of consciousness. At first involuntarily, and then on purpose, man posits the non-ascertainable, one could even say the transcendent—for example the consciousness of others, the reverse side of the moon, a history of the earth which in part has seen no life on it at all—only to derive from it the ascertainable.

From this position in the middle, it is the same distance to the elimination of the concept of reality on the one hand, and the acceptance of metaphysical, unconditional and eternal judgements on existence of entities, on the other. I believe that none of these is necessary for the natural sciences. Man will again and again have spontaneous experiences of reality and he will try to describe them in such words as will seem to him adequate. However he is capable of realizing that his judgements of what is existent are conditioned by aspirations, hopes, desires, in short by mental attitudes of the individual or the group which describe those experiences. To these attitudes belong, and especially for the scientist, the extent of his information, the measure of the general knowledge of his times. In such a way, a tension results between phenomena and reality which constitutes the charm of life and research alike.

The natural scientist has to deal with special sorts of phenomena and with a special sort of actuality. He must confine himself to *reproducible* events; here I include also those events, the reproduction of which has been provided by nature. I do not maintain that the reproducible as such is more important than the unique occurrence, but I do maintain that that which is intrinsically unique cannot be treated at all by the methods of natural science. The purpose and goal of these methods is to discover laws of nature and to examine them; it is towards this that the scientist's attention is, and must be, directed.

The coherent formulation of systems of thought, consisting of mathematical equations and rules to show the connection between these equations and data of experience we call a physical theory, which, within the limits of its range of applicability, one can designate as 'a model of reality'. As I have already said elsewhere, I find it idle to speculate which came first: the idea or the experiment. I hope that nobody is of the opinion any more that theories are derived as compelling logical conclusions from 'protocol-books', an opinion which was very much fashionable in my student days. Theories emerge through an *understanding* which is inspired by empirical material; this could be best explained, if, with Plato, we describe it as a state of congruence between inner images and external objects and their behaviours. That understanding is possible at all, indicates again that typical regulating patterns are present to which the internal as well as the external aspect of man are both subjected.

LIST OF SOURCES

Translations marked with an asterisk have been made especially for this anthology.

SECTION I ANTIQUITY

1 H. Diels, *Die Fragmente der Vorsokratiker*, 6th ed. Berlin 1951, 12 A 9 [In what follows from this collection of fragments, translations have been taken where available from: K. Freeman, *Ancilla to the Presocratic Philosophers*, B. Blackwell, Oxford 1948. Translations also by Rosemary Arundel* and R. Clapcott*.]
2 *ibid* 13 A5
3 *ibid* 13 B 2
4 *ibid* 22 B 49a
5 *ibid* 22 B 51
6 *ibid* 22 B 54
7 *ibid* 22 B 90
8 *ibid* 22 B 123
9 *ibid* 58 B 15
10 *ibid* 58 B 5
11 *ibid* 58 B 34
12 *ibid* 58 B 35
13 *ibid* 58 B 37
14 *ibid* 24 B 4
15 *ibid* 44 B 4
16 *ibid* 44 B 6
17 *ibid* 44 B 10
18 *ibid* 44 B 12
19 *ibid* 47 B 1
20 *ibid* 29 B 4
21 *ibid* 29 A 25
22 *ibid* 29 A 26
23 *ibid* 29 A 27
24 *ibid* 29 A 28
25 *ibid* 31 B 17
26 *ibid* 31 B 23
27 *ibid* 31 B 26
28 *ibid* 31 A 67
29 *ibid* 31 A 87
30 *ibid* 31 A 89
31 *ibid* 31 A 57
32 *ibid* 59 A 1
33 *ibid* 59 B 3
34 *ibid* 59 B 4
35 *ibid* 59 B 6

36   *ibid* 59 B 12
37   *ibid* 59 B 13
38   *ibid* 59 B 16
39   *ibid* 59 A 12
40   *ibid* 59 A 42
41   *ibid* 59 A 1
42   *ibid* 59 B 21
43   *ibid* 59 B 21a
44   *ibid* 64 B 3
45   *ibid* 67 A 6
46   *ibid* 67 B 2
47   *ibid* 68 A 37
48   *ibid* 68 A 135
49   *ibid* 68 B 7
50   *ibid* 68 B 8
51   *ibid* 68 B 11
52   *ibid* 68 B 125
53   Plato, *Timaeus* 48 A [Trs F. M. Cornford, *Plato's Cosmology*, Kegan Paul, London 1937]
54   Plato, *Philebus* 55 D–E [Trs R. Hackforth, *Plato's Examination of pleasure*, University Press, Cambridge 1945]
55   Plato, *Timaeus* 53 C–57 D [*Op cit*]
56   Plato, *Republic* 528 E–530 C, 530 E–531 C [Trs F. M. Cornford, *The Republic of Plato*, Clarendon Press, Oxford 1941]
57   Aristotle, *Physica* 184a 10–26 [Trs R. P. Hardie and R. K. Gaye, *The Works of Aristotle*, Ed W. D. Ross, Clarendon Press, Oxford 1908–1952]
58   Aristotle, *De caelo* 306a 9–11 [Trs J. L. Stocks, *The Works of Aristotle, op cit*]
59   Aristotle, *Physica* 192b 8–25 [*Op cit*]
60   ——*ibid* 194b 16–195a 3 [*Op cit*]
61   ——, *De interpretatione* 19a 24–36 [Trs E. M. Edghill, *The Works of Aristotle, op cit*]
62   ——, *De partibus animalium* 641b 10–27 [Trs W. Ogle, *The Works of Aristotle, op cit*]
63   ——, *Physica* 198b 17–199a 20 [*Op cit*]
64   ——, *ibid* 203b 16–35 [*Op cit*]
65   ——, *ibid* 205b 25–206a 7 [*Op cit*]
66   ——, *ibid* 208a 32–b 25 [*Op cit*]
67   ——, *ibid* 210b 32–211a 7 [*Op cit*]
68   ——, *ibid* 211b 6–21, 211b 30–212a 21 [*Op cit*]
69   ——, *ibid* 213a 12–19, 214b 28–215b 23, 216a 12–21 [*Op cit*]
70   ——, *ibid* 217b 29–218a 8 [*Op cit*]
71   ——, *ibid* 218b 9–20, 219a 1–b 5 [*Op cit*]
72   ——, *ibid* 221a 27–b 1, 222b 22–26 [*Op cit*]
73   ——, *ibid* 242a 16–20, 243a 3–17 [*Op cit*]
74   ——, *ibid* 249b 27–250a 20 [*Op cit*]
75   ——, *ibid* 266b 27–267a 18 [*Op cit*]
76   ——, *ibid* 267b 18–26 [*Op cit*]

77 ——, *ibid* 265a 13–34 [*Op cit*]

78 ——, *De caelo* 268b 27–269a 32, 269b 18–270b 31 [*Op cit*]

79 ——, *De generatione et corruptione* 323b 1–324a 9 [Trs H. Joachim, *The Works of Aristotle, op cit*]

80 ——, *ibid* 329b 17–33, 330a 30–b8 [*Op cit*]

81 ——, *ibid* 331a7–b 2 [*Op cit*]

82 ——, *Metaphysica* 1073b 17–1074a 12 [Trs W. O. Ross, *The Works of Aristotle, op cit*]

83 Epicurus, *Letter to Herodotus*, from Diogenes Laertius X 39–45 [Trs R. O. Hicks, *Lives of Eminent Philosophers*, Loeb Classical Library, Heinemann, London 1925]

84 ——, *ibid* 45 [*Op cit*]

85 ——, *ibid* 46 [*Op cit*]

86 ——, *ibid* 48–50 [*Op cit*]

87 ——, *ibid* 54–57 [*Op cit*]

88 ——, *ibid* 73–74 [*Op cit*]

89 ——, *ibid* 76–77 [*Op cit*]

90 ——, *ibid* 85–87 [*Op cit*]

91 Lucretius, *De rerum natura*, I 265–328 [Trs W. H. D. Rouse, Loeb Classical Library, Heinemann, London 1924]

92 ——, *ibid* II 62–141 [*Op cit*]

93 ——, *ibid* II 216–262 [*Op cit*]

94 ——, *ibid* III 177–230 [*Op cit*]

95 ——, *ibid* III 323–349 [*Op cit*]

96 Chrysippus, from Plutarch in *De Stoicorum repugnantiis* 1053F [Trs S. Sambursky, *Physics of the Stoics*, Routledge, London 1959]

97 ——, from Plautarch in *De communibus notitiis* 1085D [Trs S. Sambursky, *The Physical World of the Greeks*, Routledge, London 1956]

98 Sextus Empiricus, *Adversus mathematicos* IX 78 [Trs R. G. Bury, Loeb Classical Library, Heinemann, London 1933].

99 Philo, *Quod deus sit immutabilis* 35 [Trs S. Sambursky, *The Physical World of the Greeks*, Routledge, London 1956]

100 Alexander Aphrodisiensis, *Scripta minora* [Trs S. Sambursky, *Physics of the Stoics, op cit*]

101 Cleomedes, *De motu circulari corporum caelestium* [*Op cit*]

102 ——, *ibid* [*Op cit*]

103 Plutarch, *De communibus notitiis* 1079A [Trs S. Sambursky, *Physics of the Stoics op cit*]

104 ——, *ibid* 1078E [*Op cit*]

105 ——, *ibid* 1081C [*Op cit*]

106 Alexander Aphrodisiensis, *Scripta minora* [*Op cit*]

107 ——, *ibid* [*Op cit*]

108 Aulus Gellius, *Noctes Atticae*, VII, ii, II [Trs J. C. Rolfe, Loeb Classical Library, Heinemann, London 1927]

109 Alexander Aphrodisiensis, *Scripta minora* [Trs S. Sambursky, *Physics of the Stoics, op cit*]

110 Plutarch, *De Stoicorum repugnantiis* 1045C [*Op cit*]

111 Archimedes, *Arenarius* [Trs T. L. Heath, *The Works of Archimedes*, University Press, Cambridge 1897]

112 ——, [*The Method*. A supplement to *The Works of Archimedes, op cit*]

113 Seneca, *Naturales quaestiones* VII 25 [Trs J. Clarke, Macmillan, London 1910]

114 Plutarch, *De facie in orbe lunae* 923A–924F [Trs H. Cherniss, Loeb Classical Library, Heinemann, London 1957]

115 Ptolemy, *Almagest* I [Trs A. Wasserstein*]

116 ——, *ibid*, XIII [Trs A. Wasserstein*)

117 Proclus, *Hypotyposis astronomicarum positionum*, Ch 1 [Trs A. Wasserstein*]

118 ——, *ibid*, Ch 7 [Trs A. Wasserstein*]

119 Philoponus, *Comm. in Physica* [Trs S. Sambursky, *The Physical World of Late Antiquity*, Routledge, London 1962]

120 ——, *Comm. in De anima* [Op cit]

121 ——, *Comm. in Physica* [Op cit]

122 Simplicius, *Comm. in De caelo* [Op cit]

123 Philoponus, *De opificio mundi* [Op cit]

124 Simplicius, *Comm. in De caelo* [Op cit]

125 ——, *Comm. in Physica* [Op cit]

126 ——, *ibid* [Op cit]

127 ——, *ibid* [Op cit]

128 ——, *Comm. in De caelo* [Op cit]

SECTION II THE MIDDLE AGES

129 Rhazes, (Abu Bakr al-Razi) *Opera philosophica* [Ed. P. Kraus, Cairo 1939. Trs M. Schwarz*]

130, 131 ——, *ibid* [Op cit, Trs M. Schwarz*]

132 ——, *Doubts concerning Galen* [Trs S. Pines, *Actes VII Congrès internationale d'histoire des sciences*, Jerusalem 1953. Trs from French by Rita Instone*]

133 Al-Biruni, *Al-Quanun al Mas'udi* [Hyderabad 1954, Vol I. Trs M. Schwarz*]

134, 135 Alhazen (Ibn al-Haytham) [Abhandlung über das Licht, Ed. J. Baarmann, *Zeitschr. d. deutschen morgenländischen Gesellschaft*, Vol 36, 1882. Trs M. Schwarz*]

136 ——, *Critique of Ptolemy* [Trs into English by S. Pines, *Actes X Congrès internationale d'histoire des sciences*, Vol I, Ithaca 1962]

137 Avicenna (Ibn Sina), *Kitab al-Shifa* [Lithographed edition, Teheran. Trs M. Schwarz*]

138 Abu'l-Barakat al-Baghdadi, *Kitab al-Mu'tabar* [Hyderabad 1954, Vol II, p 101. Trs M. Schwarz*]

139 Averroes (Ibn Rushd), *Commentary on Aristotle's Metaphysics* [Ed. M. Bouyges, Beirut 1938–52, Vol III. Trs M. Schwarz*]

140 Moses Maimonides, *The Guide of the Perplexed* [Trs S. Pines, Chicago University Press 1963] part 2, chapter 24

141 ——, *ibid* [Op cit, part 1, chapter 73]

142 Robert Grosseteste, *Commentaria in libros posteriorum* [Trs A. C. Crombie, *Robert Grosseteste and the Origins of Experimental Science, 1100–1700*, Clarendon Press, Oxford 1953]

143, 144   Roger Bacon, *Opus maius* [Ed. J. H. Bridges, Oxford 1900, Vol I. Trs
Susan P. Hall*]

145   ——, *Epistola de secretis operibus* [Trs Lynn Thorndike, *History of Magic and
Experimental Science*, Macmillan, New York 1923–1958, Vol II]

146   Nicole Oresme, *Le livre du ciel et du monde* [*Medieval Studies* 1942, University of
Toronto, Trs B. Tolley*]

147   Nicolas Cusanus, *De docta ignorantia* (1440) [Trs G. Heron, Routledge, London
1954] Book II, chapter 4

148   *ibid* [*Op cit*] 11, 8

149   ——, *ibid* [*Op cit, 12*]

SECTION III   COPERNICUS TO PASCAL

150   Nicolaus Copernicus, *De revolutionibus orbium coelestium* (1543) Osiander's Introduction [Trs C. G. Wallis, *Great Books of the Western World*, Ed. R. M. Hutchins,
Encyclopaedia Britannica, Vol. 16, Chicago 1939]

151   ——, *ibid* Copernicus introduction [*Op cit*]

152   ——, *ibid* Ch 6 [*Op cit*]

153   ——, *ibid* Ch 7 [*Op cit*]

154   ——, *ibid* Ch 8 [*Op cit*]

155   ——, *ibid* Ch 9 [*Op cit*]

156   ——, *ibid* Ch 10 [*Op cit*]

157   Giordano Bruno, *De l'infinito universo et mundi* (1584) [Trs D. Singer, *Bruno,
his Life and Thought*, H. Shuman, New York 1950]

158–160   ——, *ibid* [*Op cit*]

161   William Gilbert, *De magnete* I, Ch 3 (1600) [Trs Sylvanus P. Thompson, issued by
the Gilbert Club, London 1900. Basic Books, New York 1958]

162   ——, *ibid* Ch 5 [*Op cit*]

163   ——, *ibid* Ch 17 [*Op cit*]

164   ——, *ibid* II Ch 14 [*Op cit*]

165   ——, *ibid* VI Ch 1 [Trs P. F. Mottelay, *Great Books of the Western World*, Ed.
R. M. Hutchins, *Encyclopaedia Britannica*, Vol. 28, Chicago 1952]

166   Francis Bacon, *Novum organum*, *Aphorisms* I (1620) Trs G. W. Kitchin,
University Press, Oxford 1855]

167–169   ——, *ibid* [*Op cit*]

170   ——, *ibid*, *Aphorisms* II [*Op cit*]

171   Johann Kepler, *Mysterium cosmographicum* (1596) [Trs James Shiel*]

172   ——, *Epitome astronomiae Copernicanae* (1618) IV, 3 [Trs C. G. Wallis, *Great Books
of the Western World*, Vol 16, Ed. R. M. Hutchins, *Encyclopaedia Britannica*,
Chicago 1939)

173–176   ——, *Harmonice mundi* (1619) V, Ch 3 [*Ibid*]

174   ——, *ibid* [*Op cit*]

175   ——, *ibid* V, Ch 9 [*Op cit*]

176   ——, *ibid* [*Op cit*]

177   Galileo Galilei, *Lettere a Madama Cristina di Lorena* (1614), *Opere*, Edizione nationale
Firenze 1933 [Trs S. Sambursky, *Three aspects of the historical significance of Galileo*,
Israel Academy of Sciences and Humanities, Jerusalem 1964]

178 ——, *Dialogo* (1632) *Op cit* [Trs Stillman Drake, University of California Press, 1953

179–182 ——, *ibid* [*Op cit*]

183 ——, *Discorsi* (1638), *Opere op cit* [Trs H. Crew and A. de Salvio, *Two New Sciences*, Macmillan, New York 1914]

184–189 ——, *ibid* [*Op cit*]

190 René Descartes, *Regulae ad directionem ingenii* (1701), Regula II. *Œuvres*, ed. C. Adams and P. Tannery, Paris 1901 [Trs Laurence Lafleur, Bobbs-Merrill Co, New York 1961]

191 ——, *ibid* Regula VII [*Op cit*]

192 ——, *ibid* Regula IX [*Op cit*]

193 ——, *Principia philosophiae* (1644) *Œuvres, op cit* [Trs E. S. Haldane and G. R. T. Ross, University Press, Cambridge 1931]

194–198 ——, *ibid* [*Op cit*]

199 ——, *ibid* Part IV [*Op cit*]

200 ——, *ibid* Part II [Trs E. Ascombe and P. T. Geach, Nelson, London 1954]

201–204 ——, *ibid* Part III

205–208 ——, *ibid* Part IV [Haldane and Ross, *op cit*]

209 Pierre Gassendi, *Opera omnia* (1658), Vol. 1, *Physica* [Trs M. R. G. Spiller*]

210 ——, *ibid*, Vol. III, *De motu impresso a motore translato* [Trs Susan Hall*]

211 Evangelista Torricelli, Letter to Ricci [*Opere*, Ed. G. Loria and G. Vassura, Faenza 1919, Trs Rachel Meeli*]

212 Blaise Pascal, *Traité de la pesanteur de la masse d'air* (1648) Ch 1

213 ——, *ibid*, Conclusion

SECTION IV THE ROYAL SOCIETY TO LAPLACE

214 Thomas Sprat, *The History of the Royal Society of London* (1667) 2nd edition 1702

215 ——, *ibid*

216 Robert Hooke, A Method for making the History of the Weather [in Thomas Sprat, *op cit*]

217 Robert Boyle, New Experiments Physico-Mechanical touching the spring of air, *Works* (1744) 2nd edition 1772, Vol I

218 Olaf Rømer, A Demonstration Concerning the Motion of Light, *Phil. Trans. Roy. Soc.*, Vol 12 (1677)

219 Christian Huygens, *Traité de la lumière* (1678) [Trs S. P. Thompson, *Treatise on Light*, Macmillan, London 1912]

220 Isaac Newton, *Philosophiae naturalis principia mathematica* (1687) Preface to the first edition [Trs F. Cajori, University of California Press 1947]

221 ——, *ibid* Cotes' preface to the second edition [*Op cit*]

222–227 ——, *ibid* [*Op cit*]

228–239 ——, *Opticks* (1704) [Fourth edition, 1730]

240 Gottfried Wilhelm Leibniz, *Philosophical Papers and Letters* [Ed. Leroy E. Loemker, University of Chicago Press 1956, Vol 11]

241–243 ——, *ibid* [*Op cit*]

244 Pierre Louis Moreau de Maupertuis, *Essai de cosmologie* (1751) [Trs R. R. Instone*]

245 Jakob Bernoulli, *Ars conjectandi* (1713) Part IV, chapter IV [Trs R. R. Instone*]

246 Daniel Bernoulli, *Hydrodynamica* (1738) Part X [Trs R. R. Instone*]

247 Johann Bernoulli, *Opera omnia* (1742) Vol III [Trs R. R. Instone*]

248 Jean d'Alembert, *Traité de dynamique* (1742) Discours préliminaire [Trs Y. Elkana*]

249 Immanuel Kant, *Allgemeine Naturgeschichte und Theorie des Himmels* (1755) [Trs W. Hastie, *Kant's Cosmogony*, Maclehose and Sons, Glasgow 1900]

250 ——, *ibid*

251 Roger Boscovich, *Theoria philosophiae naturalis* (1763) [Trs J. M. Child, MIT Press, Cambridge, Mass. 1966]

252 Leonhard Euler, *Lettres à une princesse d'Allemagne* (1768–1772) Lettre 134 [Trs H. Hunter, *Letters of Euler to a German Princess*, John Murray, London 1795, Vol II. Letter 19]

253 ——, *ibid* Lettre 136 [*Op cit* letter 21]

254 ——, *ibid* Lettre 139 [*Op cit* letter 24]

255 ——, *ibid* Lettre 140 [*Op cit* letter 25]

256 ——, *ibid* Lettre 142 [*Op cit* letter 27]

257 Pierre Simon Laplace, *Œuvres complètes* (1884) Vol VI [Trs Y. Elkana*]

258 ——, *ibid* [Y. Elkana*]

259 ——, *Essai philosophique sur les probabilitées* (1814) [Trs F. W. Truscott and F. L. Emory, Dover Publications, New York 1951]

SECTION V DALTON TO MACH

260 John Dalton, *A New System of Chemical Philosophy* (1808)

261 William Prout, *On the Relation between the Specific Gravities of Bodies in their Gaseous State and Weights of their Atoms* (1815)

262 Augustin Fresnel, *Œuvres completes* (1866) [Trs Y. Elkana*]

263 ——, *ibid* [Trs Y. Elkana*]

264 ——, *ibid* [Trs Y. Elkana*]

265 Christian Ørsted, Experiments on the effect of the current of electricity on the magnetic needle (*Annals of Philosophy*, 1820)

266 André Marie Ampère, *Recueil d'observations électrodynamiques* (1822) [Trs Y. Elkana*]

267 ——, *ibid* [Trs Y. Elkana*]

268 ——, *Mémoire sur la théorie mathématique des phénomènes électrodynamiques* (1827) [Trs Y. Elkana*]

269 Sadi Carnot, *Réflexions sur la puissance morrice du feu* (1824) [Trs R. H. Thurston, Dover Publications, New York 1960]

270 James Prescott Joule, *Scientific Papers* (1884) Vol I

271 Hermann Helmholtz, Über die Erhaltung der Kraft (1847) *Wissenschaftliche Abhandlungen* (1882) Vol I [Trs Y. Elkana*]

272 ——, On the application of the law of the conservation of force to organic nature (1861) *op cit* (1895) Vol III

273 Rudolf Clausius, *Die mechanische Wärmetheorie* (1867) [Trs T. Archer-Hirst, *The Mechanical Theory of Heat*, 1867]

274 Michael Faraday, *Experimental Researches in Electricity* (1839–1855) Vol I

275–279 ——, *ibid* Vol III

280 James Clerk Maxwell, *Scientific Papers*, Cambridge, University Press, 1890, Vol I

281–282 ——, *ibid*

283-285 ——, *ibid* Vol II

286 ——, *Theory of Heat* (1871) [8th edition, 1885]

287 Gustav Kirchhoff and Robert Bunsen, Chemische Analyse durch Spektral-beobachtungen (1860), *Ostwalds Klassiker der exakten Wissenschaften* No 72 [Trs Y. Elkana*]

288 Dmitri Mendeleev, Die periodische Gesetzmässigkeit der chemischen Elemente (1869), *Ostwalds Klassiker der exakten Wissenschaften*, No 68 [Trs Y. Elkana*]

289 Ludwig Boltzmann, *Vorlesungen über Gastheorie* (1912) Vol II [Trs Stephen G. Bush, *Lectures in Gas Theory*, University of California Press, 1964]

290 ——, Statistische Mechanik, *Populäre Schriften* (1905) [Trs A. van Helden*]

291 Albert Michelson, *Light Waves and their Uses*, University of Chicago Press, 1907

292 Heinrich Hertz, *Gesammelte Werke* (1895) Vol I [Trs Y. Elkana*]

293 Ernst Mach, *Die Mechanik in ihrer Entwicklung* (1883) 7th edition 1912 [Trs Philip E. B. Jourdaine, Open Court Publishing Co., Chicago and London 1915]

294 ——, *ibid* [*Op cit*]

SECTION VI PLANCK TO PAULI

295 Max Planck, Neue Bahnen der physikalischen Erkenntnis (1913) [Trs F. d'Albe, *Phil. Mag.* Vol 28, 1914]

296 Albert Einstein, Zur Elektrodynamik bewegter Körper (1905) [Trs W. Perret and G. B. Jeffery, *Side Lights on Relativity*, Methuen, London 1922]

297 ——, Die Grundlage der allgemeinen Relativitätstheorie (1916) [*Op cit*]

298 ——, Geometrie und Erfahrung (1921) [*Op cit*]

299 ——, Aether und Relativitätstheorie (1920) [*Op cit*]

300 ——, Über einen die Erzeugung and Verwandlung des Lichts betreffenden heuristischen Gesichtspunkt (1905) [Trs D. ter Haar, *The Old Quantum Theory*, The Pergamon Press, London 1967]

301 Ernest Rutherford, The electrical structure of matter (1923) *Nature*, Vol 112

302 Niels Bohr, On the constitution of atoms and molecules (1913) *Phil. Mag.* Vol 26

303 Louis de Broglie, Recherches sur la théorie des quanta (1925) *Annales de Physique* Vol 3 [Trs Y. Elkana*]

304 ——, The wave nature of the electron (1929) [*Nobel Lectures (Physics) 1922–1941*, Elsevier, Amsterdam 1965]

305 Werner Heisenberg, Über den anschaulichen Inhalt der quantentheoretischen Kinematik und Mechanik (1927) *Zeitschr. f. Physik*, Vol 43 [Trs Y. Elkana*]

306 ——, Die physikalischen Prinzipien der Quantentheorie (1930) [Trs Heath, *The Physical Principles of Quantum Theory*, University of Chicago Press, 1930]

307 ——, The Copenhagen interpretation of the quantum theory [*Physics and Philosophy*, Allen and Unwin, 1958]

308 Niels Bohr, Atomic theory and mechanics (1925) [*Atomic Theory and the Description of Nature*, Cambridge University Press, 1934]

309 ——, The atomic theory and the fundamental principles underlying the description of nature (1929) [*Op cit*]

310 ——, Atoms and Human Knowledge (1955) [*Atomic Physics and Human Knowledge*, Wiley, New York, 1958]

311 Wolfgang Pauli, Phänomen und physikalische Realität (1954) [Trs Y. Elkana*]

INDEX OF AUTHORS
Bio-bibliography

Abū Bakr al-Rāzī: *see* Rhazes

Abū 'L-Barakāt al-Baghdādī (*c.* 1077–*c.* 1165) Jewish philosopher, argued against Aristotle's concept of space, time and movement. He worked out a theory of the impetus differing in some respects from that of Avicenna.

Alcmaeon (flourished *c.* 500 B.C.) Pythagorean, physician, and one of the founders of ancient medicine.

d'Alembert, Jean le Rond (1717–1783) French mathematician and philosopher. In his *Traité de Dynamique* (1743) he formulated the mechanical principles (named after him) by which the laws of motion are reduced to those of equilibrium. With Diderot he edited the *Encyclopédie* (1751).

Alhazen (*c.* 965–1039) Islamic mathematician, physicist and astronomer, one of the founders of optics.

Anaxagoras (*c.* 499–*c.* 428 B.C.) His doctrine of matter was based on the conception of an infinite multitude of primary particles the mixture of which constitutes all material phenomena. These are dominated by the moving and ordering principle of Mind.

Anaximander (*c.* 610–*c.* 546 B.C.) One of the three first philosophers of nature, a younger contemporary of Thales of Miletus, the founder of the monistic school of nature. He regarded the Unlimited as the first principle of matter.

Anaximenes (died *c.* 525 B.C.) One of the three Milesian philosophers of nature; he regarded air as the first principle of matter.

Archimedes (287–212 B.C.) One of the greatest mathematicians of Greek antiquity. In his treatise *The Sandreckoner* he expounds the heliocentric doctrine of his contemporary Aristarchus.

Archytas (first half of the fourth century B.C.) A friend of Plato and a Pythagorean mathematician. He also dealt with problems of acoustics and mechanics.

Aristotle (384–322 B.C.) Greek philosopher and scientist, began as a pupil of Plato, but founded his own school (335 B.C.), later known as the Peripatetic School. His teachings influenced scientific thought until the beginning of modern times.

Averroes (1126–1198) Islamic philosopher, famous for his commentaries on Aristotle.

Avicenna (980–1037) Islamic philosopher, physician and physicist, elaborated a theory of the impetus similar to that of Philoponus.

Bacon, Francis (1561–1626) Philosopher and Lord Chancellor of England. His forceful rejection of scholasticism, his emphasis on the need for an empirical approach to nature and his 'new philosophy' basing scientific research on induction exerted an immense influence, in particular, on English scientists of the seventeenth century.

Bacon, Roger (*c.* 1214–1294) English philosopher who was also a Franciscan friar. His main scientific interest was in mathematics and its application to the sciences, in particular to optics and alchemy. He stressed the importance of the experimental method.

Bernoulli, Daniel (1700–1782) Swiss physicist and mathematician. His *Hydrodynamica* (1738) contains the beginnings of the kinetic theory of gases.

Bernoulli, Jakob (1654–1705) Swiss mathematician. His book on probability (*Ars conjectandi*), in which he stated the law of large numbers, appeared posthumously in 1712.

Bernoulli, Johann (1667–1748) Swiss mathematician, contributed to the clarification and formulation of the principles of mechanics.

Al-Biruni (*c.* 973–*c.* 1050) Islamic mathematician and scientist. Among his numerous treatises are works on the calendar, mathematics, astronomy and astrology as well as on geography and mineralogy.

Bohr, Niels (1885–1962) Danish physicist, one of the foremost theoreticians in the field of atomic and nuclear physics. His work on the structure of the atom (1913) is a fundamental contribution to the unification of atomic physics and quantum theory. He published important essays on the philosophical significance of quantum mechanics and introduced the concept of complementarity. Nobel Laureate of physics (1922).

Boltzmann, Ludwig (1844–1906) Austrian physicist and philosopher of science. He made important contributions to statistical mechanics and established the connection between the entropy and probability of a thermodynamical system.

Boscovich, Roger (1711–1787) Serbo-Croat Jesuit. He was noted for his work in many fields of science. In his book *Theoria philosophiae naturalis* (1758) he developed his hypothesis of atoms as singularities in a continuum, acting as centres of forces of repulsion and attraction.

Boyle, Robert (1627–1691) British physicist and chemist, one of the founders of the Royal Society. His experiments on the physical properties of air were published in 1660. He was one of the champions of the corpuscular theory of matter. His famous law of the relation between the pressure and the volume of a gas was formulated in 1662.

de Broglie, Louis (b. 1892) French physicist, originator of wave mechanics whose point of departure was the association of a frequency with every elementary particle (1924). His basic ideas were later developed by Schrödinger and led to a new interpretation of the wave–particle dualism. Nobel Laureate of physics (1929).

Bruno, Giordano (1548–1600) Dominican friar and Italian philosopher, an enthusiastic supporter of the Copernican theory which he publicized by lecturing throughout Europe. He was tried by the Inquisition and burnt at the stake. In his book *On the infinite universe and the infinity of worlds* (1584) he expounded the view that the fixed stars are suns.

Bunsen, Robert (1811–1899) German chemist. Together with Gustav Kirchhoff he developed (from 1859 onwards) spectroscopy as a discipline of physics and discovered by means of spectral analysis the elements caesium and rubidium.

Carnot, Sadi (1796–1832) French physicist. His *Réflexions sur la puissance motrice du feu* (1824) which analyses the efficiency of an ideal thermic engine, was the first essential contribution towards the second law of thermodynamics.

Clausius, Rudolf (1822–1888) German physicist who was the first to express the second law of thermodynamics (1850) stating that heat does not flow by itself from a colder to a warmer body. In the course of his investigations he introduced (1864) the term 'entropy'.

Copernicus, Nicolaus (1473–1543) Polish clergyman and astronomer, the creator of the heliocentric theory which paved the way for the new era of science. His *De revolutionibus*, on which he worked for over 30 years, was published in the year of his death.

Cusanus, Nicolas (1401–1464) Bishop of Brixen, one of the great mystics of the fifteenth century, expounded the idea of the infinity of the universe and of the axial rotation of the earth.

Dalton, John (1766–1844) English chemist and physicist, one of the pioneers of modern atomism. His principal contribution to this theory is included in his *A New System of Chemical Philosophy* (1808).

Democritus (flourished *c*. 420 B.C.) Developed the main concepts of the atomic doctrine into a system. He was also of importance as a philosopher.

Descartes, René (1596–1650) French philosopher and mathematician and one of the main founders of modern scientific thought. In his *Principia philosophiae* (1644), he expounded his mechanistic approach to nature, and in particular to physical phenomena (aether vortices, mechanical explanation of magnetism, propagation of light).

Diogenes of Apollonia (flourished *c*. 450 B.C.) One of the first philosophers to apply the teleological principle to natural phenomena.

Einstein, Albert (1879–1955) German-Jewish physicist, one of the greatest scientists of all times, originator of the special theory of relativity (1905) and the general theory of relativity (1916). His paper on the theory of light quanta (1905) was a decisive contribution to quantum theory. His mathematical derivation of the Brownian motion (1905) furnishes striking proof for the molecular constitution of matter. Nobel Laureate of physics (1921).

Empedocles (*c.* 482–422 B.C.) His doctrine of the four elements influenced science till the end of the Middle Ages. He introduced the concept of the opposite forces of Love and Hate as active principles of movement.

Epicurus (341–271 B.C.) Founder of the Epicurean School of philosophy which took over the Democritean doctrine of atomism and developed and amplified it. Among our main sources for his physical doctrine are his letters to Herodotus and Pythocles.

Euler, Leonhard (1707–1783) Swiss mathematician among whose numerous important publications are many dealing with problems of physics and astronomy. He adhered to the wave theory of light. His *Lettres à une Princesse d'Allemagne* (1768–1772) is a classic of popular science.

Faraday, Michael (1791–1867) English physicist and chemist, one of the greatest experimental scientists, originator of the concept of the physical field of force which put an end to the concept of action at a distance. His most important discoveries are electromagnetic induction (1831), the basic laws of electro-chemistry (1833–1834), and the rotation of the plane of polarization of light by a magnetic field (1845). In 1851, he published an account summarizing his ideas on the reality of lines of force.

Fresnel, Augustin (1788–1827) French physicist. After Thomas Young had discovered the interference of light waves (1801), Fresnel supplied a firm basis for the wave theory of light by his work on diffraction and polarization—founded on the assumption that light consists of transversal waves of the elastic aether.

Galilei, Galileo (1564–1642) Italian scientist and astronomer, a protagonist of the Copernican theory. His discoveries through the telescope (constructed by him in 1609) contributed essentially to the confirmation of the heliocentric theory. His *Dialogue Concerning the Two Chief World Systems* (1632) was published without the permission of the Church and led to his famous trial before the Inquisition (1633). Galilei's investigations of the laws of falling bodies and other mechanical problems (in his *Discourses and Demonstrations Concerning Two New Sciences*, 1638) together with his refutation of Aristotle, laid the foundation of modern mechanics.

Gassendi, Pierre (1592–1655) French scientist who revived the atomic hypothesis of Greek antiquity. His researches on motion and thrust contributed to the clarification of the concept of inertia.

Gilbert, William (1544–1603) Physician to Queen Elizabeth I. In his book, *De magnete* (1600), one of the first works of modern science, Gilbert expounds, on the basis of systematic experiments, the properties of the magnet and his theory of terrestrial magnetism. The book influenced the thinking of many scientists, among them Kepler.

Grosseteste, Robert (1175–1253) English philosopher, in particular a philosopher of nature. Developed a cosmology deriving from his conception of light and its properties.

Heisenberg, Werner (b. 1901) German physicist, originator (together with Born and Jordan) of quantum mechanics. He formulated the uncertainty relations which are also of fundamental significance for the epistemology of science. Nobel Laureate of physics (1932).

Helmholtz, Hermann (1821–1894) German physicist and physiologist. His famous publication *On the Conservation of Force* (1847) established the law of the conservation of energy the validity of which he also stressed for organic processes. His book *Sensations of Tone* (1862) began modern physiological acoustics.

Heracleitus (flourished *c.* 500 B.C.) His ideas on the role of fire in nature and on the continuous state of change of matter influenced the physical concepts of the Stoics.

Hertz, Heinrich (1857–1894) German physicist. His investigations of the properties of electric waves (1887) confirmed Maxwell's electromagnetic theory of light and formed the starting point for the development of wireless telegraphy.

Hooke, Robert (1635–1703) English physicist, one of the first Fellows of the Royal Society, discovered the law of elasticity named after him; worked on the application of mechanical principles to the planetary motions; suggested and discussed the development of meteorology as a scientific discipline. His most famous work is the *Micrographia* (1665).

Huygens, Christiaan (1629–1695) Dutch physicist, mathematician and astronomer. Among his important physical investigations are his wave theory of light (1678), and his work on the pendulum and on the centrifugal force.

Ibn al-Haytham: *see* Alhazen
Ibn Rushd: *see* Averroes
Ibn Sina: *see* Avicenna

Joule, James Prescott (1818–1889) English physicist. His work on the mechanical equivalent of heat (published 1843) was a decisive contribution to the establishment of the law of conservation of energy.

Kant, Immanuel (1724–1804) German philosopher whose *General History and Theory of the Heaven* (1755) was the first modern cosmology.

Kepler, Johann (1571–1630) German mathematician and astronomer. His three laws concerning the orbits and motions of the planets (published in 1609 and 1619) complemented and extended the Copernican system a protagonist of which Kepler had been all his life. These laws also laid the foundation for Newton's theory of gravitation.

Kirchhoff, Gustav (1824–1887) German physicist, noted for his laws concerning the electric current and heat radiation. Founder of spectroscopy (1859) with Bunsen.

Laplace, Pierre-Simon de (1749–1826) French astronomer, mathematician and physicist. One of his important contributions to astronomy was the theory of perturbations. His celebrated cosmogony was published in his *Système du monde* (1796). His *Théorie analytique des probabilitées* (1812) is one of the classical contributions to this field.

Leibniz, Gottfried Wilhelm (1646–1716) German philosopher and mathematician, invented (independently of Newton) the calculus. Unlike Newton, he conceived of space and time as relative, or relational, entities. He emphasized the importance of the concept of *vis viva* for mechanics, and the necessity for a teleological principle for the ultimate explanation of physical phenomena.

Leucippus (flourished *c.* 450 B.C.) Founder of the atomist school whose first principles are atoms and the void. He was the teacher of Democritus and probably the disciple of Zeno of Elea.

Titus Lucretius Carus (*c.* 96–55 B.C.) Roman poet whose didactic poem *De rerum natura* is a major source of information on the Epicurean doctrine.

Mach, Ernst (1838–1916) Austrian physicist and philosopher of science, one of the protagonists of positivism in science. His book *The Development of Mechanics* (1883) had a notable impact on scientific thought at the turn of the century.

Maimonides, Moses (1135–1204) Jewish philosopher and physician. His chief speculative work is the *Guide of the Perplexed*, which was translated into Latin in the 13th century and had a considerable influence on Christian scholasticism.

Maupertuis, Pierre Louis Moreau de (1698–1759) French scientist, the first to formulate the principle of least action (1744), which he attempted to support by metaphysical arguments.

Maxwell, James Clerk (1831–1879) British physicist, established the mathematical theory of the electromagnetic field (Maxwell's equations) which he developed in a series of papers (1861–1865). Through this theory, which showed that electromagnetic fields are propagated with the velocity of light, optics became a branch of electromagnetism. He also made important contributions to the kinetic theory of gases.

Mendeleev, Dmitri Ivanovich (1834–1907) Russian chemist, discoverer of the Periodic Law of the Elements (1869), which made possible the prediction and identification of new chemical elements.

Michelson, Albert Abraham (1852–1931) American physicist, constructed (1881) the interferometer named after him, in order to detect the influence of the movement of the earth on the velocity of light. His celebrated experiment together with Morley in 1887 showed that such an influence does not exist. Nobel Laureate for physics (1907).

Newton, Isaac (1642–1727) English physicist and mathematician, one of the greatest scientists of all times. His most famous achievements are the calculus, the universal law of gravitation and its applications, and the spectral decomposition of light. His *Philosophiae naturalis principia mathematica* (1687), the first systematic work on theoretical physics, opened the era of classical mechanics. In his *Opticks* (1704), he presupposes the corpuscular nature of light, combined with the hypothesis of periodic 'fits'.

Oresme, Hans Nicole (*c.* 1325–1382) French philosopher, economist and scientist, Bishop of Lisieux, noteworthy for his opposition to Aristotle's cosmology and for his anticipation of Copernican ideas; one of the leading theoreticians of the impetus.

Ørsted, Christian (1777–1851) Danish physicist, discovered (1820) the deflection of the magnetic needle by an electric current. On the basis of this discovery, Ampère developed his electrodynamic theory.

Pascal, Blaise (1623–1662) French scientist and philosopher. In the wake of Torricelli's discovery, he carried out a series of experiments (1647) which confirmed the laws of atmospheric pressure and in particular its dependence on altitude. This put an end to the conception of the '*horror vacui*' of nature.

Pauli, Wolfgang (1900–1958) Austrian physicist, discoverer (1925) of the exclusion principle (named after him) which explains the structure of the shell model of the atom and the periodic system of the elements. He was a penetrating interpreter and critic of modern physics. Nobel Laureate of physics (1945).

Philolaus (flourished *c.* 440 B.C.) One of the founders of the Pythagorean philosophy of numbers.

Philoponus, Johannes (first half of the sixth century) Christian Neoplatonist, professor of philosophy in Alexandria, famous Aristotelian commentator and known for his anti-Aristotelian views on motion. Developed the concept of impetus.

Planck, Max (1858–1947) German physicist, originator of quantum theory (1900), arrived at the concept of a quantum of energy (the product of Planck's quantum of action and the frequency of radiation) through his investigations of the black-body radiation. Published important essays on scientific epistemology. Nobel Laureate of physics (1918).

Plato (428–384 B.C.) Greek philosopher and founder (387 B.C.) of the Academy in Athens, which remained the spiritual centre of his philosophical doctrines until it was closed in A.D. 529. His main cosmological treatise is the *Timaeus*, but many of his important statements on science and astronomy are to be found in the *Republic*.

Plutarchus of Chaeronea (*c.* 45–*c.* 125) Prolific Greek writer and Platonic philosopher. His writings against the Epicureans and the Stoics are important sources for their philosophical systems. His book *On the Face in the Moon* reflects the astrophysical conceptions of his time.

Proclus (412–485) Neoplatonic philosopher and head of the Athenian Academy. His text book on the Ptolemaic astronomy is the last great scientific book in Greek antiquity.

Prout, William (1785–1850) English physician and chemist. His hypothesis that hydrogen atoms are the basic building-stones of all the chemical elements was first published anonymously (1815).

Ptolemy, Claudius (*c.* 90–168) Greek astronomer, author, among other important treatises on geography and astrology, of the famous *Syntaxis mathematica*, known as the *Almagest*, the comprehensive exposition of the geocentric system which was superseded by Copernicus (1543).

Rhazes (*c.* 864–925) Islamic physician, scientist and philosopher. His medical and scientific writings show an outspoken empirical approach. His physical conception was anti-Aristotelian.

Rømer, Olaf (1644–1710) Danish astronomer, determined (1676) the velocity of light by his observations of the eclipses of two satellites of Jupiter.

Rutherford, Ernest (1871–1937) British physicist, established the theory of radioactivity (in collaboration with Soddy). His investigations of the scattering of α-particles led him to his nuclear theory (1911) which describes the atom as composed of a small positive nucleus surrounded by an envelope of negative electrons. Nobel Laureate of chemistry (1908).

Seneca, Lucius Annaeus (4 B.C.–A.D. 65) Roman writer and philosopher, one of the prominent later Stoics. His *Naturales Questiones* are an important source for the study of his natural philosophy.

Sprat, Thomas (1635–1713) English bishop, one of the first Fellows of the Royal Society. His book *The History of the Royal Society* appeared in 1667, five years after the Royal Society's foundation. The book also contains summaries of researches by some of the first Fellows.

Stoic School, founded by Zeno of Citium in Cyprus (336–264 B.C.). The most important contributions to its physical and epistemological doctrines were made by Chrysippus (281–208 B.C.) who expounded, in particular, the concepts of the *pneuma*, of the dynamical continuum, and of determinism.

INDEX OF SUBJECTS

EDITOR'S NOTE

Acknowledgements

The continuity of science could not have been properly reflected in this work without the cooperation of my friend and colleague, Professor Shlomo Pines, of the Hebrew University of Jerusalem. With his profound knowledge of medieval thought and literature, he undertook the task of selecting the texts for the second section (The Middle Ages) and wrote the special introduction. My sincere thanks are owed to Professor Pines for this essential contribution. I also wish to thank Dr Z. Bechler of the Department of History and Philosophy of Science of the Hebrew University. Dr Bechler undertook the ardous work of checking the texts; he assisted me in composing the headings to many of them and in preparing the footnotes; in addition he provided many other helpful suggestions. Finally I thank my wife for her invaluable help in overcoming the innumerable difficulties encountered in the course of work. I also wish to express my gratitude to Hutchinson, in particular to Elaine Hopton, Simon Scott and Georgina Bannister for their patient cooperation during the many years of the *status nascendi* of this anthology and to various translators of passages which were not available in standard translations.

The editor and publisher are grateful to all the publishers who granted permission for the reprinting of copyright material here and to all those people who translated texts especially for this anthology. From time to time and for the sake of clarity, the editor has amended the text of the standard translations and of the translations prepared specially for this edition.

Technical Note

The reader will note that selected texts are numbered sequentially. These numbers are cross-referenced in the List of Sources on page 541. Headings, which have been provided for guidance by the editor and which are *not* part of the original author's work precede the text's serial number. Headings which form an integral part of the author's original text fall beneath the serial number. Omission of passages from an original text is usually marked by ellipses, thus (. . .).

GENERAL INDEX

Absolute Maximum, Cusanus on 163
Abū 'L-Barakāt, on impetus 142
Action, distant, Maxwell's summary of
  theories on 437–441
Actualization, Philoponus on 116
Aether,
  Descartes on vortices of 248
  Einstein on 496–498
  Newton on properties of 314–318
Air,
  Boyle on spring of 281
  cohesive properties of 46
  Pascal on compression and expansion of
    260
  Pascal on weight of 258–261
  Philoponus on properties of 117
Air pressure, Pascal on 259
Al-Biruni, on axial rotation of earth 133
Alcmaeon 49
  on preservation of health 39, 49
d'Alembert, Jean le Rond 330–336
  on equilibrium 333, 334
  on inertia 331
  on law of dynamics 330–336
Alhazen,
  on criticism of science 139
  on optics and light propagation 135
Al-Rāzī, Abū Bakr, see Rhazes, 3-dimen-
    sional space theories of 125
Ampère, André Marie 381–389
  on electrodynamical phenomena 386–
    389
  on electromotive action 381–384
Anaxagoras 52–55
  on astrology 38
  on doctrine of like parts 52
  on nature of celestial bodies 54
  on nature of Mind 38, 53
  on theory of equal parts 38
Anaximander 46
  first principle of the 'unlimited' 37, 46
Anaximenes 46

on composition of soul 46
on eternal motion 46
on properties of air 46
Archimedes 98–100
  on Aristarchus' heliocentric universe 99
  on expression of large numbers 43
  on idea of scientific progress 43
  on use of mechanical considerations 43
Archytas 50
  on importance of mathematics 50
Aristarchus, on a heliocentric universe 99
Aristotle 63–83
  on aether 42
  on biological basis of thought 40
  on causes of existence 64
  on cognition of nature 63
  on concentric spheres 82
  on definition of 'Now' 41
  on dichotomy 40
  on earth as centre of universe 41
  on the five elements 77
  on forced motion 74–76
  on the four elemental qualities 77–82
  on fulfilment of purpose 66
  on the infinite 67
  on law of force 74–76
  on mixing of components 41
  on motion 42, 70–72
  on necessity 65
  on place 68–70
  on principles 64
  on theory of spheres 42
  on time 72–74
  on the void 70–72
  physical doctrine of 14
Arithmetic, Descartes on purity of 239
Astronomy,
  applied, Copernicus on 170
  foundation of 4
  role in development of physical sciences
    5, 6
Atmospheric pressure, Torricelli on 257

575